Soil Ecosystems Management

Soil Ecosystems Management

Editor: Anthony Sherman

Callisto Reference
www.callistoreference.com

Callisto Reference,
118-35 Queens Blvd., Suite 400,
Forest Hills, NY 11375, USA

Visit us on the World Wide Web at:
www.callistoreference.com

ISBN: 978-1-63239-902-1 (Hardback)

The publisher's policy is to use permanent paper from mills that operate a sustainable forestry policy. Furthermore, the publisher ensures that the text paper and cover boards used have met acceptable environmental accreditation standards.

Printed in China.

Cataloging-in-Publication Data

Soil ecosystems management / edited by Anthony Sherman.
p. cm.
Includes bibliographical references and index.
ISBN 978-1-63239-902-1
1. Soil management. 2. Soil ecology. 3. Ecosystem management.
4. Soil science. I. Sherman, Anthony.
S591 .S65 2018
631.4--dc23

Table of Contents

Preface

Soil ecosystems are rich in biodiverse elements that contribute greatly to the sustenance of life on Earth. This book on soil ecosystems management discusses techniques and methods related to the restoration of soil health and vitality and the appearance of essential nutrients such as oxygen and hydrogen in soil. Soil also plays a crucial role in the decomposition of waste materials that could otherwise enter natural water sources. Maintaining soil ecological balance for the growth of plants and crops is crucial in this field. This book on soil ecosystems management includes contributions of experts and scientists that will provide innovative insights into this field. Students, researchers, and experts associated with this field will benefit alike from this book.

The information contained in this book is the result of intensive hard work done by researchers in this field. All due efforts have been made to make this book serve as a complete guiding source for students and researchers. The topics in this book have been comprehensively explained to help readers understand the growing trends in the field.

I would like to thank the entire group of writers who made sincere efforts in this book and my family who supported me in my efforts of working on this book. I take this opportunity to thank all those who have been a guiding force throughout my life.

Editor

Constitutive Model of Single Root System's Resistance to Tensile Stress - Taking *Pinus tabulaeformis, Betula platyphylla, Quercus mongolica* and *Larix gmelinii* as Experimental Objects

Lihua Chen*, Pinghua Wang, Yuanjun Yang, Jia He

Key Laboratory of Soil and Water Conservation & Desertification Combat of Ministry of Education, School of Water and Soil Conservation, Beijing Forest University, Beijing, China

Abstract

A constitutive model for the stress-strain relationship of single forest root system was developed in order to provide theoretical foundations for the mechanisms of soil-reinforcement by root system and offer a reliable basis for the analysis of root tensile strength character. This study started a general form of linear and non-linear stress-strain relation that was mathematically defined by four boundary conditions observed in typical tensile tests of single roots. The parameters of the model were determined by experiment data and had definite physical meaning. The model was verified by experiment data, which showed that the calculated values were in good agreement with the experimental single root tensile test results. The constitutive model was validated and found to be feasible for modeling single root tensile stress.

Editor: Derek Abbott, University of Adelaide, Australia

Funding: This study was funded by Project of Study and Demonstration of Construction Technology on Water Conservation Forest System in Three-North Area (No. 2011BAD38B05); National Natural Science Foundation of China (No. 30872067; No. 30901162) and Doctoral Fund of Ministry of Education of China (No. 20090014110002). The funders had no role in study design, data collection and analysis, decision to publish, or preparation of the manuscript.

Competing Interests: The authors have declared that no competing interests exist.

* E-mail: c_lihua@bjfu.edu.cn

Introduction

Roots are key plant organs that perform vital functions such as absorbing and storing soil moisture, nutrients, and energy, as well as anchoring the plant to the soil. One of the main ecological functions of forest roots is soil stabilization, which can enhance soil conservation, stabilize slopes, prevent gravity erosion, and prevent river and reservoir bank scouring [1–4]. Since 1990s, the mechanism of soil-reinforcement by root systems has received widespread attention from research community and led to many interdisciplinary studies [5–7]. Since single root of plants is the basic unit of soil-reinforcement, a series of studies has been carried out to study the mechanism of soil-reinforcement by single root system [8].

Currently indoor root tensile tests and *in situ* root-soil composite horizontal and the vertical pullout tests are used to determine the tensile strength, shear strength, and factors contributing to the root system's resistance to shear of different species [9–10]. Some models have been developed to describe the constitutive relationship(the stress-strain relationship) of single root system. Most models use different mathematical functions, such as linear, hyperbolic, polynomial, or power function [11–13] to describe the relationship between stress and strain. Those functional constitutive relationships are empirical in nature, since mathematical functions are selected based on the best curve fitting to experimental data. The experiment data determine the constants (or parameters) of the mathematical functions, which do not have clear physical meaning. The applicability and accuracy of those models are greatly limited [13–14].

In this study, boundary conditions that a constitutive model for tensile stress of a single root system should conform to are investigated by analyzing typical stress-strain curve of single root systems. A new constitutive model for the tensile stress resistance of single root systems is proposed and mathematically solved using these boundary conditions. Furthermore, the term of critical point (the definition will be presented in Section 2.3) is introduced and derived from the model. Finally, the model is verified by comparing the model calculated values to experimental data, which show good agreement between two. This study may provide theoretical foundations for the mechanisms of soil-reinforcement by root system and give a reliable basis for the analyses of the root tensile strength.

Materials and Methods

2.1 Ethics Statement

The sampling site is managed by Mulan Weichang, a state-owned forest Authority of He-bei province, China. Our study complies with the current laws of China and international rules. All necessary permits have been obtained before the actual sampling. The field study does not involve any endangered or protected species. Data will be made available upon request.

2.2 Sample Collection and Testing

The field samples were collected in He-bei province, which surrounds the Capital City of Beijing, at 36° 01′ to 42° 37′ N latitude and 113° 31′ to 119° 53′ E longitude. He-bei province covers a total area of 187,700 square kilometers and has a continental monsoon climate, with cold and dry winters and hot and humid summers. The annual precipitation is 400 to 800 mm, occurring mostly in summer. Roots of four common species of tall trees were chosen to test the proposed model. The trees were *Pinus tabulaeformis*, *Betula platyphylla*, *Quercus mongolica* and *Larix gmelinii*. Tensile tests were carried out for various root diameters and root lengths. To preserve the root and ensure a true reflection of the root mechanical properties, all root samples collected from the field were placed in sealed bags and stored in a refrigerator at 4°C before the test was taking place. All root samples were tested within one month and no pretreatment before testing. 22888 root samples in four gauges (50 mm, 100 mm, 150 mm, and 200 mm) were tested in total; among them, 6462 samples were in gauge 50 mm.

Tensile test was performed on WDW-100E electro-universal tester (Time Shijin, China; see Figure 1) equipped with microcomputer control. The test force range was 400–100 kN, and the speed range was 0.001–500 mm/min. The machine has functions of full automatic shift and stepless speed regulation. Its accuracy in measuring test load and displacement is ±0.5%.

This study uses and reports experiment data of single roots with different diameters and fixed gauge (50 mm) only. Root samples with intact bark and relatively uniformed diameter were selected. Each root was divided into four pieces; each piece was roughly 60 mm in length. The diameter at three dividing points was measured and the average diameter was recorded as the root diameter. The universal tester was adjusted to a unified standard distance of 50 mm. Root sample was inserted (50 mm at each end was fixed) onto the instrument and pulled (10 mm/min) until the root is completely pulled apart. Necking was observed during tests. Necking, in engineering and materials science, is a phenomenon that the tensile deformation (strain) has not reached to its extreme when the stress reaches the maximum. The sample statistics of four tree species are summarized in Table 1.

The test force (stress) and displacement (strain) data were collected by measuring the force and distance required to pull

Figure 1. Photo of the universal tester used for tensile test.

Table 1. Summary of sample's physical features and statistics.

Species	Diameter scope (mm)	Average diameter (mm)	Total root samples	Effective samples	Effective percentage %	Total samples of gauge 50 mm	Effective samples of gauge 50 mm	Effective percentage of gauge 50 mm %
Pinus	0.5–1.5	1.11	1539	985	64.00	471	362	76.86
tabulaeformis	1.5–2.5	2	2091	1203	57.53	647	341	52.70
	2.5–3.5	3.06	2101	1453	69.16	664	413	62.20
Larix gmelinii	3.5–4.5	3.9	2280	1400	61.40	664	278	41.87
	1.5–2.5	2.15	1659	1217	73.36	407	321	78.87
	2.5–3.5	2.99	2291	1782	77.78	741	498	67.21
Betula	3.5–4.5	3.98	2164	1225	56.61	702	266	37.89
platyphylla	0.5–1.5	1.06	1611	1026	63.69	490	295	60.20
	1.5–2.5	2.41	2500	1178	47.12	526	344	65.40
Quercus	2.5–3.5	3.03	1442	781	54.16	463	278	60.04
mongolica	1.5–2.5	2.1	1516	872	57.52	687	299	43.52
	3.5–4.5	4.39	1694	855	50.47	594	246	41.41
Total (Average)			22888	13977	61.07	6462	3941	57.35

apart a single root. The data were then plotted as stress-strain curves.

2.3 Boundary Condition

Before the discussion of the boundary condition, the dimensionless coordinates are defined for convenience:

$$x=\varepsilon/\varepsilon_p,\ y=\sigma/\sigma_p.$$

Where ε is Strain; σ is Stress; ε_p is Peak strain and σ_p is Peak stress.

The following terms are also used in this study:

$$E=E_0/E_p,\ E_P=\sigma_p/\varepsilon_p,\ E_0=d\sigma/d\varepsilon|_{x=0}.$$

Where E_0 is Linear elastic modulus; E_P is Peak secant modulus and E is Dimensionless elastic modulus. Since E_0 is larger than E_p, E is larger than 1.

A typical experiment stress-strain curve of single root is presented in Figure 2. Figure 2 uses dimensionless strain and stress for x and y axis, respectively. It shows that a stress-strain curve starts from zero stress and strain. As the load (stress) increases, the root deformation (strain) increases linearly in the initial stage. As the load continuously increases, the root deformation remain increasing, but the slope of the curve decreases monotonically. The curve is concave down and gradually plateaus to the maximum stress at which the root breaks. A critical point is marked on the curve in Figure 2. Before the critical point, the root linear deformation is dominated and after the critical point, the non-linear deformation becomes significant.

Above observations can be mathematically expressed as:

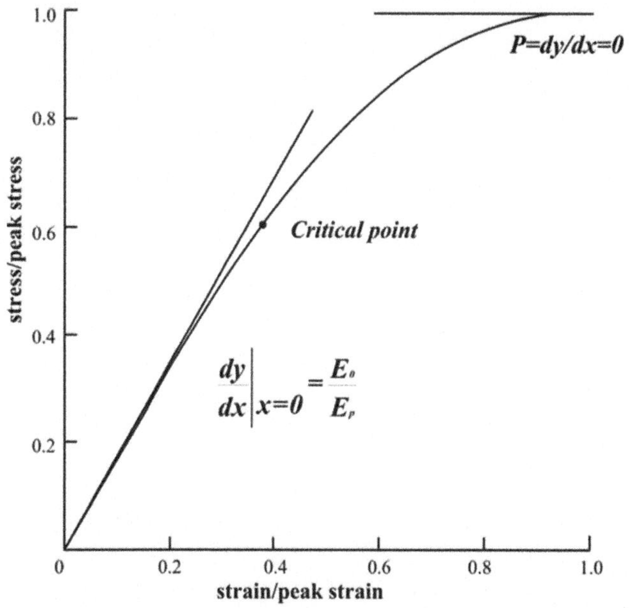

Figure 2. Typical stress-strain curve. (1) $y=0$, when $x=0$. The stress-strain curve passes the origin. (2) $dy/dx=\frac{d\sigma}{d\varepsilon}/\frac{\sigma_p}{\varepsilon_p}=\frac{E_0}{E_P}$, when $x=0$. The slope of the curve at the origin is equal to the dimensionless linear elastic modulus of the root. (3) $y=1$, when $x=1$. The peak stress happens $x=1$. (4) $\frac{dy}{dx}=0$, when $x=1$. The slope of the curve at the peak stress is zero.

1) $y=0$, when $x=0$. The stress-strain curve passes the origin.

2) $dy/dx=\frac{d\sigma}{d\varepsilon}/\frac{\sigma_p}{\varepsilon_p}=\frac{E_0}{E_P}$, when $x=0$. The slope of the curve at the origin is equal to the dimensionless linear elastic modulus of the root.

3) $y=1$, when $x=1$. The peak stress happens $x=1$.

4) $\frac{dy}{dx}=0$, when $x=1$. The slope of the curve at the peak stress is zero.

These four mathematical descriptions are the very boundary conditions that the proposed constitutive model should conform to.

2.4 Constitutive Model

As Figure 2 indicates, the root deformation of a single root is composed of both linear and non-linear deformations. Equation (1) is a general form of linear and non-linear deformation of single root.

$$y=ax^b+cx+d. \tag{1}$$

The four parameters of Equation (1) can be determined by applying the boundary conditions of (1) to (4) described in 2.3. Equation (1) is rewritten as:

$$y=Ex-(E-1)x^{\frac{E}{E-1}}. \tag{2}$$

Equation (2) is our proposed constitutive model for the stress-strain relationship of single root. The first term on the right hand side in Equation (2) describes the linear portion of root deformation, while the second term describes the non-linear portion of root deformation.

The second derivative of Equation (2) is:

$$\frac{d^2y}{dx^2}=-\frac{E}{E-1}x^{\frac{2-E}{E-1}}. \tag{3}$$

Since E is larger than 1, Equation (3) is less than zero, which indicates the curve is concave down, as observed in Figure 2.

As discussed in Section 2.3, the root deformation of single root system is mainly linear in the early stage and non-linear in the late stage, the critical point is the point that separates two stages. The proposed constitutive model of Equation (2) represents both linear and non-linear deformation. Strictly speaking, there is no absolute linear or non-linear deformation. In the early stage, the non-linear deformation, second term of the right hand side of Equation (2), is very small when compared to the linear deformation of the first term. As the curve passes the critical point, the non-linear deformation becomes more and more significant.

2.5 Results and Discussion

Data of tensile tests are used to determine physical parameters of single roots, E_0 and E_p, used in the proposed constitutive model of Equation (2) for four different root species. The linear elastic modulus (E_0) is the ratio of the material tensile stress to the corresponding tensile strain in the linear range and used to characterize the material resistance to linear deformation capacity. For instance, the linear elastic modulus of *Pinus tabulaeformis* root system is 50% to 70% of the tensile strength limit [15]. It is a safe assumption that the root deformation is mainly linear at 40% of the tensile strength limit for *Pinus tabulaeformis* root system.

Therefore, the secant modulus (the ratio of the stress and strain) at 40% of the tensile stress limit was taken as the elastic modulus ($E_0 = \sigma_{0.4}/\varepsilon_{0.4}$) for *Pinus tabulaeformis*. The same method is also applied to determine E_0 for other three root species. The peak scant modulus, $E_P = \sigma_p/\varepsilon_p$, is estimated by the ratio of the measured peak stress and peak strain data.

Traditionally, the root deformation has two stages, linear and non-linear or elastic and plastic deformation. The point that separates the two deformations is called the yield limit. It was commonly accepted that before the yield limit, the root deformation is linear or elastic and after the yield limit, the deformation is non-linear or plastic. However, data of our tensile tests show there is no clear point that separates linear or non-linear deformation. Instead, the dominated linear deformation is observed in early stage of tests, when the stress and strain are small. The weight of non-linear deformation gradually increases. At certain point, the non-linear deformation become significant and cannot be neglected. In this study, the term of the critical point is used to describe that point. Before the critical point, the non-linear deformation is very small when compared to the linear deformation (say less 5% of linear deformation). As the curve passes the critical point, the non-linear deformation becomes more and more significant. The critical point, x_e, is determined using Equation (2) and corresponding to the 40% of the tensile stress limit.

Once above parameters are determined using experiment data, the proposed model is used to calculate the stress-strain relation. The error between the model calculated values and experiment data is estimated using the residual mean and variance and summarized in Table 2. The most average residual values are less than 0.1 and the most residual variance values are less than 0.01 with only a few exceptions. Figure 3 presents eight figures of model calculated and experiment stress strain data. Seven of eight figures have a good match between model calculated values and experiment data, except the figure for *Latrix gmelinii* (diameter 3.98 mm). Table 2 also shows that an increase in root diameter corresponds to a decrease in elastic modulus (E_0), while the same trend between the root diameter and the peak secant modulus (E_p) is not obvious. As results, the changing trend of E versus the root diameter is not clear. Experiment data with different diameters and species may be needed to further verify these observations.

In summary, a constitutive model for the stress-strain relationship of single roots is proposed in this study. The model starts with a general form of linear and non-linear deformation and is defined with four boundary conditions based on typical stress-strain curves observed from the tensile tests of single roots. The definition of the model requires two physical characteristics of single root: the linear elastic modulus E_0 and the peak secant modulus E_p. These two parameters can be obtained from the laboratory tensile strength tests. The major difference between the proposed model and most of previous stress-strain models [13;16–17] is that the proposed model is based on mathematical expression of physical behavior of single root deformation and boundary conditions of a typical tensile test, instead of on experiment data only. The proposed model has a great potential to help the study of the mechanism of soil-reinforcement by root system.

Limitations

The mechanism of soil-reinforcement by root systems is a complex issue and factors that may affect it include 1) environmental factors: such as species, direction and position of slope, soil properties, and root moisture content; 2) experimental factors: such as loading speed, loading time, length and diameter

Table 2. Model parameters and error results.

Species	Diameter (mm)	Elastic Modulus E_0(MPa)	Peak Secant Modulus E_P(MPa)	E	CriticalPoint x_e	Average Residuals	Residual Variance
Pinus tabulaeformis	1.11	150.00	100.50	1.49241	0.39733	0.02445	0.00095
	2.00	107.12	61.46	1.74292	0.40057	0.06534	0.001366
	3.06	84.06	52.09	1.61375	0.41726	0.03231	0.002625
	3.90	81.06	56.90	1.4246	0.55640	0.081422	0.017675
Larix gmelinii	2.15	159.85	80.06	1.99673	0.40986	0.11016	0.006868
	2.99	75.40	39.09	1.92888	0.40330	0.099363	0.013765
	3.98	74.94	51.34	1.45980	0.62916	0.138209	0.01151
Betula platyphylla	1.06	183.26	116.93	1.56715	0.40629	0.004367	0.003816
	2.41	180.67	93.98	1.92247	0.40248	0.083052	0.012329
	3.03	180.56	135.70	1.33044	0.59350	0.051374	0.001545
Quercus mongolica	2.10	213.27	151.08	1.41163	0.39885	0.016976	0.002841
	4.39	209.96	122.61	1.71242	0.43298	0.057633	0.003057

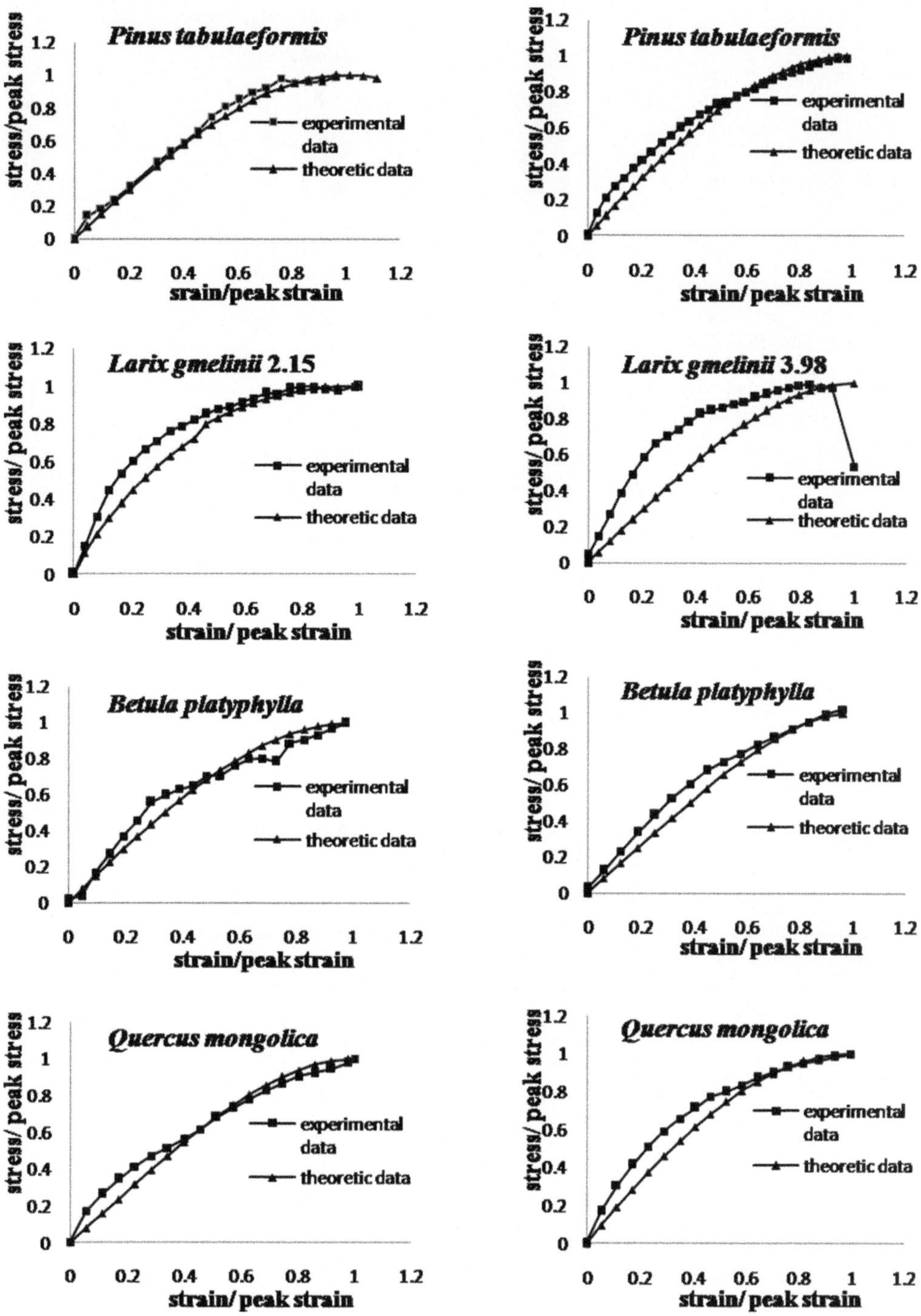

Figure 3. Comparison between model calculated values and experiment data of stress of selected testing. 1.11, 3.06, 2.15, 3.98, 1.06, 3.03 and 2.1 are root diameters. It can be seen that the theoretical stress-strain curves basically fit the experimental curves well.

of root, storage time of roots and root with or without bark; 3) micro-structural and chemical-composition factors: such as lignin and cellulose content, area ratio of wood fiber and phloem, and length-width ratio of wood fiber and tracheid. Our model developed in this study is based on laboratory root tensile tests. Although the proposed model has clear physical meanings and well defined boundary conditions, it is limited by test data and conditions. Following is discussion about some limitations.

3.1 Root Diameters and Tree Species

As discussed above, our tensile tests are carried out in a range of root diameters (1.1 to 4.39 mm) of four tree species (four kinds of tall trees commonly found in Northern China) in one gauge (50 mm) used. The results presented in this paper are limited by those root diameters and tree species. The proposed model needs to be further verified with test data before it is applied to different tree diameters and species and by different methods.

3.2 Loading Speed

The loading speed used in our experiments was a constant speed at 10 mm/min. Generally speaking, roots deform elastically before plastically. The root elastic deformation may not be fully developed if the loading speed is too high. Therefore, different loading speeds could affect the stress-strain relationship or the shape of curve. At extreme high loading speed, roots may suffer brittle fracture or necking in early stage of the test.

3.3 Storage Time and Moisture Loss

In our experiments, the maximum storage time was 30 days in order to avoid the moisture loss. Our data show that the moisture loss is insignificant within 30 days. The maximum moisture loss was 3% in weight of roots for all roots and 1.4% for roots with diameters larger than 3 mm at the end of 30 days. The moisture loss could be significant after 7 weeks for *Betula platyphylla*, 11 weeks for *Quercus mongolica*, and 8 weeks for *Larix gmelinii* and *Pinus tabulaeformis*.

3.4 With or Without Bark

Another reason for the maximum storage time of 30 days is to prevent the deterioration of roots. It is assumed that the deterioration of roots will happen from the bark. Our test samples were with bark and no deterioration was observed in our test samples at the time of testing. The model developed in this paper is applicable for roots with bark. However, root samples with and without bark could make some differences in the stress-strain relationship, which subjects to further study.

3.5 Boundary Conditions

The four boundary conditions discussed in Section 2.3 describe a typical stress-strain relation of single roots, such as observed in our tests. The third boundary condition indicates the peak stress and strain ($y = 1$, when $x = 1$) happen at the same time. The root breaks at the peak stress. This is true for almost all of test samples with a few exceptions. Some of test samples, although the number is small, do show the root deformation continued when the stress reached the peak value or necking. Figure 4 presents two stress-strain curves of such exceptions. It shows the peak stress does not correspond to the peak strain. The value of $x = 1$ does not represent the peak strain, instead it only represents the strain corresponding to the peak stress. The validity of the proposed model for this type of stress-strain relation is to be verified.

Additionally, the elastic modulus, E_0, is determined by the secant modulus at 40% of the tensile stress limit for all four tree species in this study. The selection of the secant modulus is based on previous study in which the elastic modulus of *Pinus tabulaeformis* root system is 50% to 70% of the tensile strength limit. This study applies the same method to determine the elastic modulus of other three tree species. Whether the selection of 40% can be used for determination of the elastic modulus for different tree species used is out of the scope of this study and deserves further study.

Even though with some limitations, the proposed model is promising, since it's deduced from mathematical procedures and verified by a large amount of test data. The application of the model is simple. It is one of first constitutive models of single root resistance to tensile stress with both physical and mathematical meaning. Systematic and detailed studies on above issues are subjects of further research.

Conclusions

Although the ultimate aim of our research is to study the role of tree roots in the slope stabilization, the specific aim of this paper is to build a constitutive model of single root system with a clear physical meaning through the rigorous mathematical derivation. The development of the constitutive model and the study of single root physical response to tensile stress are fundamentally significant to the understanding of soil-reinforcement by root systems. Since vegetation is one of major factors that affect slope stability, a common question is how exactly vegetation (especially the mechanical property of roots) affects the soil stabilization. The proposed model of single root resistance to tensile stress and the study reported in this paper provide a theoretical basis for the study of the slope stabilization by vegetation.

The mechanism of soil-reinforcement by root system is investigated by studying single roots of four common tall tree species from Northern China. The data of laboratory tensile tests is used to construct the root tensile constitutive model. The following is a summary of this study:

5) A constitutive model of single root system resistance to tensile stress is proposed.

6) The proposed model is mathematically defined by four boundary conditions observed in a typical tensile test of single root.

7) The term of the critical point is introduced. Before the critical point, the linear deformation is dominated and after the critical point, the non-linear deformation becomes significant.

8) The physical parameters of single root systems, linear elastic modulus (E_0) and peak secant modulus (E_p), can be determined by laboratory tensile tests.

9) Large amount of experiment data from the tensile tests of four common tall trees' root systems are used for the model verification.

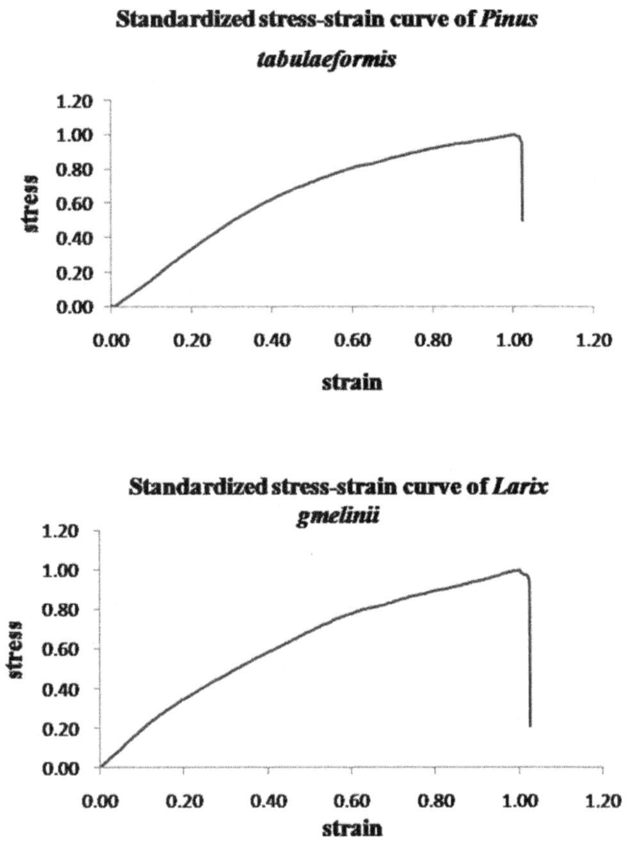

Figure 4. Examples of roots that broke after the stress reached the peak value.

The residual mean and variance between the model calculated values and experiment data indicates the proposed model can be used for the study of stress-strain relationships of single roots with good accuracy.

10) The proposed constitutive model of single roots provides theoretical basis for the research of the mechanisms of soil-reinforcement by root system and makes the quantitative analysis of the root tensile strength feasible.

Acknowledgments

We would like to express our gratitude to the Project of Study and Demonstration of Construction Technology on Water Conservation Forest System in Three-North Area (No. 2011BAD38B05) for the financial support. We sincerely thank Professor Yaping Liu who was a knowledgeable and patient discussion partner for her kind support.

Author Contributions

Conceived and designed the experiments: LHC JH. Performed the experiments: PHW YJY. Analyzed the data: PHW YJY. Contributed reagents/materials/analysis tools: PHW. Wrote the paper: LHC PHW.

References

1. Stokes A, Norris JE, van Beek LPH, Bogaard T, Cammeraat E, et al (2008) How vegetation reinforces the soil on slopes. In: Norris J, Stokes A, Mickovski S B, Cammeraat E, Van Beek R, Nicoll B C and Achim A (eds) Slope stability and erosion control: ecotechnological solutions. Springer, New York, 65–118. doi:10.1007/978-1-4020-6676-4_4.
2. Hubble TCT, Docker BB, Rutherfurd ID (2010) The role of riparian trees in maintaining riverbank stability: A review of Australian experience and practice. Ecol Eng 36, 292–304. doi:10.1016/j.ecoleng.2009.04.006.
3. Wu TH, Sidle RC (1995) A distributed slope stability model for steep forested basins. Water Resour. Res. (31): 2097–2110. doi:10.1029/95WR01136.
4. Tosi M (2007) Root tensile strength relationships and their slope stability implications of three shrub species in the Northern Apennines (Italy) Geomorphology 87(4): 268–283.
5. Song WF, XQ Wang (2007) Review of Research on Wood Root System. Journal of Southwest Forestry College, 27(5): 8–13. doi:10.3969/j.issn.2095-1914.2007.05.003.
6. Thorne CR (1990) Effects of vegetation on riverbank erosion and stability. Vegetation and erosion: 125–144.
7. Cofie P, Koolen AJ (2001) Test speed and other factors affecting the measurements of tree root properties used in soil reinforcement models. Soil and Tillage Research 63(1): 51–56.
8. Chen LH, XX Yu, WF Song, XP Liu (2008) Mechanics of Root-Soil [M]. Beijing: Science Press. ISBN: 9787030203915.
9. De Baets S, Poesen J, Reubens B, Wemans K, De Baerdemaeker J, et al (2008) Root tensile strength and root distribution of typical Mediterranean plant species and their contribution to soil shear strength. Plant Soil, 305: 207–226. doi:10.1007/s11104-008-9553-0.
10. Ji JN, N Kokutsec, M Genet, T Fourcaud, ZQ Zhang (2012) Effect of spatial variation of tree root characteristics on slope stability. A case study on Black Locust (Robinia pseudoacacia) and Arborvitae (Platycladus orientalis) stands on the Loess Plateau, China. CATENA, 92: 139–154. doi:10.1016/j.catena.2011.12.008.
11. Cheng H, T Xie, C Tang, F Li (2006) Overview of Mechanism of Plant Roots Improving Soil Reinforcement and Slope Stabilization. Bulletin of Soil and Water Conservation, 26(1): 97–102. doi:10.3969/j.issn.1000-288X.2006.01.023.
12. Chen LH, XX Yu, XP Liu, WF Song (2007), Study on the Constitutive Relation of Forest Root System, Journal of Mountain Science, 25(2): 224–228. doi: 10.3969/j.issn.1008-2786.2007.02.016.
13. Pollen N, A Simon (2005) Estimating the mechanical effects of riparian vegetation stream bank stability using a fiber bundle model. Water Resour. Res. 41(7): W07025. doi:10.1029/2004WR003801.
14. Schwarz M, Preti F, Giadrossich F, Lehmann P, Or D (2010) Quantifying the role of vegetation in slope stability: A case study in Tuscany (Italy) Ecological Engineering, 36: 285–291. doi:10.1016/j.ecoleng.2009.06.014.
15. XP Liu, LH Chen, WF Song, LY Rao, YC Hu (2006) Experimental on Resistance to Tensile Stress and Complete Stress-Strain Curves of Root System of *Pinus Tabulaeformis*. Science of Soil and Water Conservation, 4(2): 66–70. doi:10.3969/j.issn.1672-3007.2006.02.012.
16. Mickovski SB, Hallett PD, Bransby MF, Davies MCR, Sonnenberg R, et al (2009) Mechanical reinforcement of soil by willow roots: impacts of root properties and root failure mechanism. Soil Sci Soc Am J73: 1276–1285. doi:10.2136/sssaj2008.0172.
17. Wu TH (2007) Root reinforcement: analyses and experiments. In: Stokes, et al (ed) Eco and ground bioengineering: the use of vegetation to improve slope stability. Springer, 103, 21–30. doi:10.1007/978-1-4020-5593-5_3.

2

Seasonal Dynamics of Trace Elements in Tidal Salt Marsh Soils as Affected by the Flow-Sediment Regulation Regime

Junhong Bai[1,2]*, Rong Xiao[1], Qingqing Zhao[1], Qiongqiong Lu[1], Junjing Wang[1], K. Ramesh Reddy[2]

1 State Key Laboratory of Water Environment Stimulation, School of Environment, Beijing Normal University, Beijing, P. R. China, **2** Wetland Biogeochemistry Laboratory, Soil and Water Science Department, University of Florida, Gainesville, Florida, United States of America

Abstract

Soil profiles were collected in three salt marshes with different plant species (i.e. *Phragmites australis*, *Tamarix chinensis* and *Suaeda salsa*) in the Yellow River Delta (YRD) of China during three seasons (summer and fall of 2007 and the following spring of 2008) after the flow-sediment regulation regime. Total elemental contents of As, Cd, Cu, Pb and Zn were determined using inductively coupled plasma atomic absorption spectrometry to investigate temporal variations in trace elements in soil profiles of the three salt marshes, assess the enrichment levels and ecological risks of these trace elements in three sampling seasons and identify their influencing factors. Trace elements did not change significantly along soil profiles at each site in each sampling season. The highest value for each sampling site was observed in summer and the lowest one in fall. Soils in both *P. australis* and *S. salsa* wetlands tended to have higher trace element levels than those in *T. chinensis* wetland. Compared to other elements, both Cd and As had higher enrichment factors exceeding moderate enrichment levels. However, the toxic unit (TU) values of these trace elements did not exceed probable effect levels. Correlation analysis showed that these trace elements were closely linked to soil properties such as moisture, sulfur, salinity, soil organic matter, soil texture and pH values. Principal component analysis showed that the sampling season affected by the flow-sediment regulation regime was the dominant factor influencing the distribution patterns of these trace elements in soils, and plant community type was another important factor. The findings of this study could contribute to wetland conservation and management in coastal regions affected by the hydrological engineering.

Editor: Bo Li, Fudan University, China

Funding: This work was financially supported by the National Basic Research Program (2013CB430406), the National Natural Science Foundation of China (51379012 and 51179006), the Program for New Century Excellent Talents in University (NECT-10-0235) and the Fok Ying Tung Foundation (132009). The funders had no role in study design, data collection and analysis, decision to publish, or preparation of the manuscript.

Competing Interests: The authors have declared that no competing interests exist.

* Email: junhongbai@163.com

Introduction

Sediment contamination in rivers, lakes, reservoirs, and wetlands has been widely reported in the developing countries as a result of the intense land use in agricultural and urban environment. Contamination of sediments with trace elements is a major concern [1,2]. Moreover, these trace elements can be transferred and carried downstream into wetland ecosystems, and accumulate in wetland soils [3,4]. Wetland soils serve as "source" and "sink" of these chemical pollutants [5]. For example, salt marshes and estuarine sediments can retain these metals as metal sulfides [5,6]. However, seasonal hydrological changes and water level fluctuation can affect Eh and pH of wetlands, thus resulting in mobilizing trace elements [5,7].

Several studies on large river delta have focused on source identification of trace elements and organic pollutants in surface sediments [6,8,9,10]. Some studies have shown that heavy metals can be accumulated and retained in wetland soils for a long time period [11], and they would not vary significantly over seasons in salt marshes under natural conditions [12]. The global regulation of rivers and steams by building reservoirs and dams has brought great effects on downstream ecosystems [13]. Recently, Bai et al. [4] have reported that the flow-sediment regulation contributed to trace elements (i.e. Arsenic, Cadmium, and others) accumulation in surface wetland soils of the Yellow River Delta as bed sands can carry and settle these trace elements to the downstream due to strong hydraulic flushing [12,14]. However,less information is available on the dynamics and fate of these trace elements in estuarine wetlands after the flow- sediment regulation.

The Yellow River Delta (YRD) is one of the most active regions of land-ocean interaction, and a national nature reserve was established to better protect this newly-formed wetland ecosystem and maintain biological diversity [15]. However, with the rapid development of agriculture, fisheries, and the extensive exploitation of the Shengli oilfield (It is the second largest oilfield in China and is mainly located in both sides of the Yellow River Estuary, which was originally built in 1962 and the working area has covered approximately 4.4×10^4 km^2), the Yellow River Delta was greatly impacted by the intense human activities, leading to serious wetland degradation [16]. Moreover, the flow-sediment regulation regime has shown a significant influence on trace element accumulation [4] and wetland plant distribution [16] in the Yellow River Delta.

The primary objectives of this study were: (1) to determine the dynamic changes of selected trace elements including As, Cd, Cu, Pb, and Zn in wetland soils covered by different dominant plant species (i.e., *Phragmites australis*, *Suaeda salsa* and *Tamarix chinensis*) of the Yellow River Delta; (2) to assess enrichment levels of these trace elements and determine their ecological risks using enrichment and toxic indicators; and (3) to identify the relationships among trace elements and other selected soil properties.

Materials and Methods

Ethics statement

Our study area is located in the Yellow River Delta wetland nature reserve, which is owned by the Chinese government. We obtained a specific permit from the administration bureau of the Yellow River Delta National Nature Reserve to conduct this study in the reserve. Moreover, our sampling sites were not located in any strictly protected areas and the field studies did not involve endangered or protected species.

Study area

The study area is located in the Yellow River Delta of China (37°37′48″ to 37°48′36″ N and 119°01′ to 119°21′ E) (Figure 1). It has a temperate, continental monsoon climate with distinct seasons. The annual mean air temperature is 12°C, with 196 frost-free days, and the annual mean precipitation is 552 mm [17]. Soils are characterized as loamy fine sand or coarse sand in all soil layers with clay content of 5–7% and can be classified as aquents, aquolls and fluvaquents [18]. The dominant vegetation in tidal salt marshes include herbaceous plants such as *Phragmites australis* and *Suaeda salsa* and woody plant namely *Tamarix chinensis* [17]. The vegetation succession in this delta follows the order *Suaeda salsa* → *Tamarix chinensis* → *Phragmites australis*, and plant distributions are dominantly controlled by soil salinity. The study area was consistently influenced by the flow-sediment regulation of the upstream Xiaolangdi Reservoir from June and July since 2002 [4]. Xiaolangdi Reservoir began storing water in 1999, and considerable siltation occurred in the reservoir after commissioning, with a total sediment trapping of 32.47×10^{8} t from 1997 to 2007 [19]. The flow-sediment regulation of the upstream Xiaolangdi Reservoir has been enforced from June to July since 2002 [17]. For this study, three sampling sites with different dominant plant species (Table 1) were selected. The three sites are approximately 100 m away from each other and they were consistently inundated by tidal seawater (twice every day).

Sample collection and analysis

Soil samples were collected using a soil auger (4.8-cm diam.) from three sampling sites in summer (August, 2007), fall (November, 2007), and spring (April, 2008). Well-developed root systems were observed to depths of 50 cm at *P. australis* wetland and *T. chinensis* wetland, whereas the root zone was confined to depths of 15–20 cm at *S. salsa* wetland. Therefore, in each sampling season, the top 50 cm soils (sectioned into 0–10, 10–20, 20–30, and 30–50 cm) were collected with three or five replicates in *P. australis* wetland and *T. chinensis* wetland, and the top 20 cm soils (sectioned into 0–10 and 10–20 cm) were sampled with four replicates in *S. salsa* wetland. A total 120 soil samples were collected, including 36 samples in *P. australis* wetland, 60 samples in *T. chinensis* wetland, and 24 samples in *S. salsa* wetland. All soil samples were placed in polyethylene bags and brought to the laboratory, where they were air dried at room temperature for three weeks. Some air-dried soils of selected two profiles at each site in each sampling season were used for soil particle size analysis. All the other air-dried soil samples were sieved through a 2-mm nylon sieve to remove coarse debris and then ground with a pestle and mortar until all particles passed a 0.149-mm nylon sieve for the determination of soil chemical properties. Soil bulk density cores were also correspondingly collected using a 184 cm^3 cylinder from each soil layer of each soil profile, oven dried at 105°C for 24 h, and weighed for the determination of bulk density (BD) and soil moisture.

Plant samples for the aboveground parts (including stems and leaves) were also collected at each soil sampling site in summer. They were placed in paper bags after clean washing and transported to the laboratory. All plant samples were oven dried at 65°C for 48 h, and ground into fine powder for the determination of trace metals. Soil and plant samples were digested with an $HClO_4$-HNO_3-HF mixture in Teflon tubes to analyze total contents of Al, As, Cd, Cu, Pb, and Zn. The solutions of the digested samples were analyzed by inductively coupled plasma atomic absorption spectrometry (ICP-AAS). Quality assurance and quality control were assessed using duplicates, method blanks and standard reference soil materials (GBW07401) and standard reference plant materials (GBW07602) from the Chinese Academy of Measurement Sciences with each batch of samples (1 blank and 1 standard for each 10 samples). The recovery rates of samples spiked with standards ranged from 99 to 106% for soils, and from 82–119% for plants. Soil organic matter (SOM) was measured using dichromate oxidation [20]. Soil pH and salinity were determined in the supernatants of 1:5 soil and water mixtures using a pH meter or salinity meter. Soil particle size was analyzed using a Laser Particle Size Analyzer (Microtrac Inc., USA).

Enrichment factor (EF)

Enrichment factor (EF) was selected to evaluate possible anthropogenic inputs of observed sample metals[21]. EF was defined as $EF=(M/Al)_{sample}/(M/Al)_{background}$, where M_{sample} and $M_{background}$ are the determined contents of targeted elements (e.g., As, Cd, Cu, Pb, and Zn) in soil samples and their background contents, respectively; Al_{sample} and $Al_{background}$ are the determined content s of Al in soil samples and its background content, respectively; Al is used the reference element because it is a conservative lithogenic element presumed not to be enriched due to local contamination [22]. Contamination categories were classified based on EF values: EF<2, deficiency to minimal enrichment; EF = 2–5, moderate enrichment; EF = 5–20, significant enrichment; EF = 20–40, very high enrichment; EF>40, extremely high enrichment [23]. Background values for $M_{background}$ and $Al_{background}$ in this study were referenced to background contents in the loess materials of the Yellow River [24]. Background contents of 62700 μg g^{-1} for Al, 10.7 μg g^{-1} for As, 0.095 μg g^{-1} for Cd, 21.1 μg g^{-1} for Cu, 21.6 μg g^{-1} for Pb and 64.5 μg g^{-1} for Zn were used.

Toxic units (TUs)

Toxic units (TUs) are used to normalize the toxicities of various trace elements to allow for the comparison of the relative effects. TUs were defined as the ratios of the observed contents to the probable effect level (PEL) values [25]. PEL values represent the thresholds of chemical contents above which adverse effects are likely to occur. The PEL values for marine and estuarine ecosystem was used in this study (Table 2)[26].

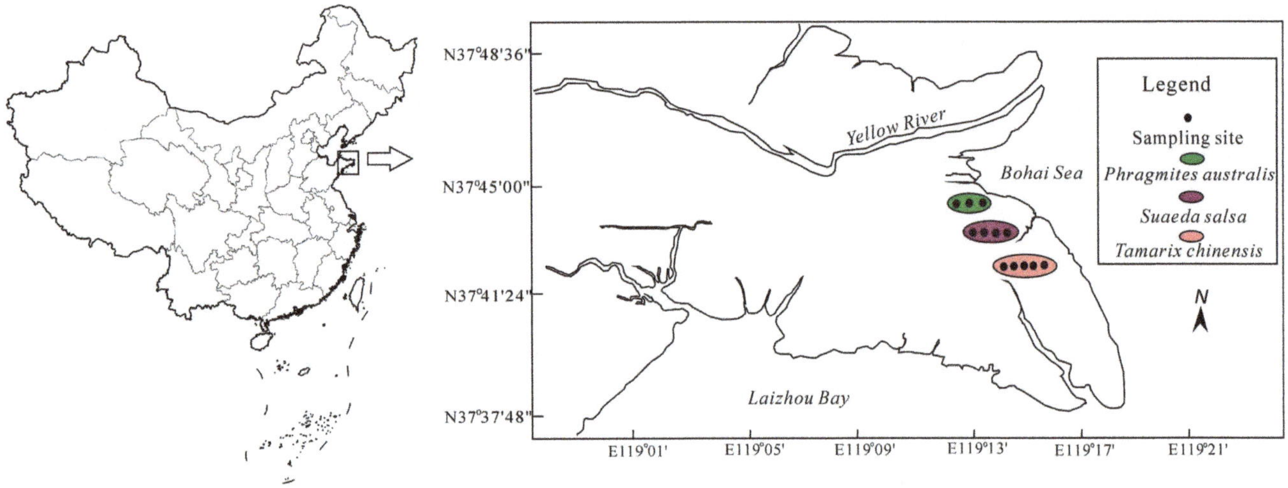

Figure 1. Location map of three sampling sites in the Yellow River Delta.

Biological Concentration Factors (BCFs)

Biological concentration factor is the ratio of metal content in the aboveground parts of plants to metal content in surrounding soils, which can more accurately describe the plant's uptake potential than plant metal content [27].

Statistical analysis

Pearson correlation analysis was performed to determine the relationships between trace elements and selected soil properties and the relationships between Al, trace elements, and sand and clay contents in selected soil profiles. Principal component analysis (PCA) was used to discriminate soil samples with similar or different contamination patterns and identify their influencing factors. Analysis of variance (ANOVA) analysis was implemented to analyze the differences in trace elements between sampling sites and seasons. Differences were considered to be significant at $p< 0.05$. Statistical analysis was performed using the SPSS 16.0 and Canoco 4.5 software packages for Windows.

Results and Discussion

Soil characteristics in three salt marshes

Selected physical-chemical properties in the top 20 cm soils of three salt marshes of the YRD are summarized in Table 2. No significant differences in soil moisture among three seasons in each sampling sites due to consistent tidal flooding condition. However, Soils in *P. australis* wetland contained higher moisture contents in spring and summer compared to *T. chinensis* wetland due to less sand contents in *P. australis* wetland ($p<0.05$; Table 2). Bulk density at each sampling site of the three was lower in spring than in summer and fall ($p<0.05$), but there were no significant differences among three sites in each season. With the exception of higher SOM level in *P. australis* wetland than those in *T. chinensis* and *S. salsa* wetlands in fall ($p<0.05$), no significant differences were observed among three sites in summer and spring. Compared to other seasons, higher SOM content appeared in *P. australis* wetland in fall, while it was higher in *T. chinensis* wetland in spring ($p<0.05$), which was likely caused by plant litter inputs and decomposition [5]. All soil samples of three sites exhibited higher salinities in spring than in summer or fall ($p<0.05$), which

Table 1. Ecological characteristics of three sampling sites in different sampling seasons.

Site	Location	Replicates	Sampling time	Plant height (m)	Plant density (m^{-2})
P. australis wetland	37°43′35.1″ to 37°43′59″N 119°12′42″ to 119°13′21″E	3	Spring	0.1–0.3	35–50
			Summer	1.8–2.2	230–280
			Fall	1.8–2.2	190–230
T. chinensis wetland	37°42′47″ to 37°42′52″N 119°13′49″ to 119°14′15″E	5	Spring	1–1.6	15–30
			Summer	1–2	15–30
			Fall	1–2	15–30
S. salsa wetland	37°43′16″ to 37°43′34″N 119°13′36″ to 119°13′53″E	4	Spring	0.05–0.15	100–220
			Summer	0.20–0.40	280–580
			Fall	0.30–0.45	136–520

Note: Spring = April; Summer = August; Fall = November.

Table 2. Summary of physical-chemical properties in the top 20 cm soils at three sampling sites during three sampling seasons.

	Moisture (%)	BD(g cm^{-3})	SOM(g kg^{-1})	Salinity(‰)	pH	Sand (%)	Clay (%)
P. australis sites							
Spring	28.8 ± 2.9^{a1}	1.4 ± 0.1^{a1}	6.9 ± 0.9^{a1}	1.5 ± 0.3^{a1}	8.4 ± 0.1^{a1}	52.1 ± 8.1^{a1}	5.2 ± 0.8^{a1}
Summer	30.7 ± 4.4^{a1}	1.8 ± 0.1^{b1}	6.7 ± 1.0^{a1}	0.6 ± 0.2^{b1}	8.4 ± 0.1^{a12}	23.8 ± 4.1^{b1}	13.9 ± 2.5^{b1}
Fall	28.0 ± 2.0^{a1}	1.8 ± 0.0^{b1}	8.2 ± 0.6^{b1}	0.7 ± 0.1^{b1}	8.6 ± 0.1^{a1}	55.7 ± 4.5^{a1}	5.0 ± 0.2^{a1}
T. chinensis sites							
Spring	23.6 ± 1.9^{a2}	1.5 ± 0.1^{a1}	7.0 ± 3.0^{a1}	3.4 ± 1.0^{a2}	8.3 ± 0.2^{a1}	81.1 ± 22.2^{a12}	0.7 ± 0.9^{a2}
Summer	23.0 ± 1.6^{a2}	1.9 ± 0.1^{b1}	5.2 ± 1.8^{b1}	2.8 ± 0.7^{ab2}	8.1 ± 0.2^{a1}	61.34 ± 2.8^{a2}	4.9 ± 1.6^{b2}
Fall	25.0 ± 2.2^{a12}	1.8 ± 0.0^{b1}	5.9 ± 2.2^{b2}	2.0 ± 0.5^{b2}	8.6 ± 0.1^{b1}	60.7 ± 6.5^{a1}	0.3 ± 0.2^{b1}
S. salsa sites							
Spring	26.2 ± 1.6^{ab12}	1.5 ± 0.1^{a1}	6.8 ± 1.4^{a1}	1.7 ± 0.4^{a1}	8.4 ± 0.1^{a1}	76.2 ± 9.1^{a2}	1.5 ± 1.2^{a2}
Summer	29.7 ± 4.0^{a1}	1.8 ± 0.1^{b1}	6.5 ± 0.5^{a1}	0.6 ± 0.4^{b1}	8.7 ± 0.3^{ab2}	23.12 ± 1.8^{b1}	14.2 ± 0.1^{b1}
Fall	23.1 ± 2.2^{b2}	1.9 ± 0.1^{b1}	5.5 ± 1.3^{b2}	2.1 ± 1.1^{a2}	8.7 ± 0.1^{b1}	81.1 ± 3.2^{a2}	0.6 ± 0.3^{a1}

Spring = April; Summer = August; Fall = November. BD = Bulk density; SOM = Soil organic matter.
^{a,b} different letters represent significant differences between three seasons at each site.
^{1,2} different numbers represent significant differences between three sites in each season.

was associated with the freshwater inputs from the flow-sediment regulation [4]. Moreover, *T. chinensis* wetland showed higher soil salinity ($p<0.05$) than *P. australis* wetland or *S. salsa* wetland in three seasons as *T. chinensis* wetland was further away from the river channel (Figure 1). However, no significant differences in pH values were observed among three wetlands in spring and fall except that *T. chinensis* wetland had lower pH values ($p<0.05$) than *S. salsa* wetland in summer. Both *T. chinensis* and *S. salsa* wetlands showed higher pH values ($p<0.05$) in fall compared to spring and summer, which might be associated with weaker soil respiration in fall as the dissolved CO_2 in overlaying water could decrease pH values [28]. All sampling sites showed higher clay contents in summer than in spring and fall and sand contents were also lower in both *P. australis* and *S. salsa* wetlands in summer ($p<0.05$; Table 2). This indicated that the flow and sediment regulation in summer brought more clay and silt contents to the tidal salt marshes [4]. However, no significant differences in sand or clay contents were observed between spring and fall, which was associated with tidal seawater erosion [4]. Moreover, soils in *P. australis* and *S. salsa* wetlands contained more clay contents and less sand contents in summer compared to *T. chinensis* wetland ($p<0.05$). This might be caused by the fact that *T. chinensis* wetland is located at the farther distances from the Yellow River Chanel (Figure 1). In both spring and fall, *P. australis* wetland had higher clay contents than *T. chinensis* and *S. salsa* wetlands.

Spatial and temporal variations in trace elements in marsh soils

Profile distributions of these trace elements in three seasons are shown in Figure 2. Generally, trace element contents exhibited a decreasing tendency with depth in each season except for As, Cd and Zn in *P. australis* wetland and Cd, Cu and Zn in *S. salsa* wetland in summer. Prusty et al. [29] also reported that trace elements such as As, Cu, Pb, and Zn decreased with increasing depths in wetland ecosystems. This was greatly associated with plant cycling which lead to trace elements upwards movement through plant litters and return to surface soils [29,30]. However, clear seasonal changes in the contents of these trace elements were observed along soil profiles (Figure 2). Among three seasons, most soil profiles of three sampling sites exhibited the highest levels of As, Cd and Zn in summer, followed by spring, whereas the lowest levels in fall. As for Cu and Pb, lower levels were also observed in fall compared to summer and spring. The average contents of As, Cd and Zn in the top 20 cm soils in *P. australis* and *T. chinensis* wetlands were significantly higher in summer than in fall ($p<0.05$; Table 3). Similarly, much higher average contents of these trace elements in the top 20 cm soils were also observed in *S. salsa* wetland in summer compared to fall ($p<0.05$; Table 3). This is inconsistent with the results by Roychoudhury [12], who presented that trace metal contents in surface soils were relatively lower in summer than in winter due to higher wetland plant uptake and standing stocks of metals in summer. Additionally, Ehlken and Kirchner [31] reported that higher water contents, lower salinities and bulk densities increased the potential mobility of metals in soil solutions. However, all trace elements generally reached their higher levels in summer in this study, although all soils contained lower salinities and higher water contents in three sampling sites after the flow-sediment regulation, and then fell to the lowest levels in fall. Therefore, the higher trace element contents observed in summer might be attributed to exogenous inputs of trace metals (e.g., water and sediment inputs from the flow-sediment regulation) [4,32]. The increased plant uptake of trace elements in the late fall could explain the lower trace element contents in these marsh soils in this season, as plants possess efficient root-to-shoot

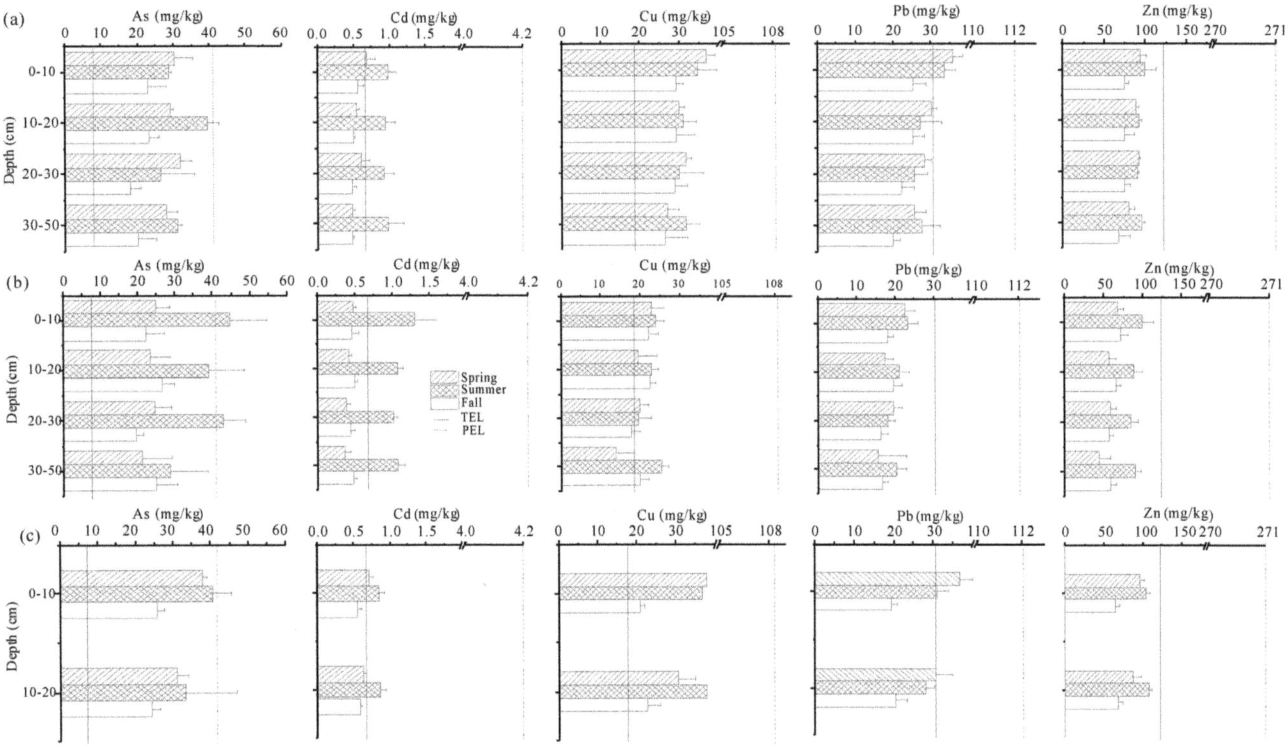

Figure 2. Profile distributions of As and heavy metals in marsh soils with *P. australis* (a), *T. chinensis* (b) and *S. salsa* (c) in three sampling seasons.

translocation systems that are activated at the end of the growing season [33].

Both Cu and Pb contents in soil profiles were generally lower in *T. chinensis* wetland than those in *P. australis* and *S. salsa* wetlands in spring and summer, whereas higher levels of As and Cd in soil profiles were observed in *T. chinensis* wetland in summer (Figure 2). Table 3 also showed that the average contents of trace elements in the top 20 cm soils were significantly higher in *P. australis* and *S. salsa* wetlands in spring compared to *T. chinensis* wetland ($p<0.05$). Meanwhile, higher levels of Cu and Pb in *P. australis* wetland and higher Cd contents in *S. salsa* wetland were observed in fall compared to other sites ($p<0.05$). In summer, *P. australis* and *S. salsa* wetlands exhibited greatly higher Cu and Pb contents than *T. chinensis* wetland ($p<0.05$, Table 3). The differences in Cu and Pb between sites or sampling seasons were most likely related to SOM, sand and clay contents (Tables 4 and 5). Additionally, lower Cu and Pb levels in the top 20 cm soils in *T. chinensis* wetland in summer compared to *S. salsa* wetland might be caused by higher plant uptake as the aboveground parts of *T. chinensis* had higher BCFs values for both Cu and Pb compared to *S. salsa* ($p<0.05$; Figure 3). However, it is difficult to explain much higher BCFs values of *T. chinensis* for Cd with higher soil Cd levels ($p<0.05$; Table 2) than *P. australis*. Therefore, the variations in trace element levels among the three sampling sites might be influenced by the multiple environmental factors including plant communities, soil properties, and water levels etc. [34,35,36]. As shown in Figure 4, all trace elements generally showed similar band-type distribution patterns, indicating higher levels of these trace elements in *P. australis* and *S. salsa* wetlands higher compared to *T. chinensis* wetland except for some soil samples with higher levels of As and Cd. These trace elements increased linearly with increasing Al content at each sampling site, which implies that these trace elements and Al might have common origin.

Assessment of trace element pollution by sediment quality guidelines (SQGs)

The extent of trace element pollution was assessed by comparing element contents in marsh soils to the sediment quality criteria developed by the Chinese Marine Sediment Quality Criteria (GB 18668-2002) [37] and the sediment quality guidelines (SQGs)[23] (Table 3). The Chinese Marine Sediment Quality Criteria indicated that approximately 85–95% of all collected soil samples in three sampling sites during three seasons exceeded the Class I criteria (suitable for fisheries and natural waters) but fell within the scope of Class II criteria (used for industrial and tourism sites) for As and Cd, and all samples were below the Class I criteria for Cu, Pb and Zn (Table 3). This supplies that most soil samples in this region have been moderately polluted by As and Cd. As shown in Figure 2, the SQGs also indicated that almost all soil samples along soil profiles at three sites fell within the range between the threshold effect levels (TELs) and probable effect levels (PELs) for As and Cu in each of the three seasons, even some samples of three sites exceed PELs in summer. Moreover, all soil profile samples at each site exceeded TELs in summer, whereas they were below TELs in both spring and fall. However, approximately 85% of soil profile samples in this region didn't exceed TELs for Pb in three seasons, and almost all soil profile samples at each site exhibited much lower Zn contents than TELs in each season. This implies higher potential risks of As, Cu and Cd in soil profiles of this region. Therefore, it is still required to control heavy metal pollution (i.e., As, Cu and Cd) in this region, as higher trace elements may make it more difficult to restore the degraded living habitat [36].

Table 3. Mean contents of As and heavy metals in the top 20 cm soils of three sampling sites during the whole sampling period (mg kg^{-1} dry weight).

	As	Cd	Cu	Pb	Zn
P. australis sites					
Summer	35.8±2.4[a1]	0.9±0.1[a1]	30.7±5.5[a1]	28.56±5.3[ab1]	95.1±11.0[a1]
Fall	23.0±4.8[b1]	0.5±0.1[b12]	29.2±1.5[a1]	25.27±1.3[b1]	74.7±5.7[b1]
Spring	29.8±3.5[c1]	0.6±0.1[b1]	33.5±4.3[a1]	32.9±2.4[ac1]	91.2±10.4[a1]
T. chinensis sites					
Summer	41.9±3.2[a2]	1.2±0.2[a2]	23.0±2.3[a2]	22.4±2.8[a2]	93.9±13.1[a2]
Fall	24.7±3.4[b1]	0.5±0.0[b2]	22.1±1.4[a2]	19.4±2.8[a2]	72.0±10.7[b1]
Spring	24.1±4.5[b2]	0.4±0.0[b2]	21.4±1.6[a2]	20.1±1.8[a2]	62.5±4.5[c2]
S. salsa sites					
Summer	40.5±6.4[a2]	0.9±0.1[a1]	37.7±1.7[a1]	29.9±1.9[a1]	106.3±3.8[a2]
Fall	25.2±1.8[b1]	0.6±0.0[b1]	22.0±2.1[b2]	19.9±2.4[b2]	66. 9±6.2[b1]
Spring	34.5±1.4[c1]	0.7±0.0[bc3]	34.5±3.7[a1]	33.7±4.5[a1]	92.1±9.5[c1]
Sediment Quality Criteria of China*					
Class I	20	0.5	35	60	150
Class II	65	1.5	100	130	350
Class III	93	5	200	250	600
SQGs†					
TEL	7.2	0.68	18.7	30.2	124.0
PEL	41.6	4.2	108.2	112.2	271.0

[a,b] different letters represent significant differences between three seasons at each site.
[1,2] different numbers represent significant differences between three sites in each season.
* National Standard of P.R. China [37].
† SQGs: Sediment Quality Guidelines for marine ecosystem; TEL: threshold effect level; PEL: probable effect level [23].

Enrichment factors of trace elements in marsh soils

Although the contents of trace elements were lower in the YRD than in most large rivers and estuaries [4], trace element pollution in the YRD is increasingly serious due to continuous disturbances caused by human activities and sediment movements [36,38]. Trace element contents in the upstream loess were used as indicators to determine the enrichment levels for trace elements in the soils. Figure 5 shows the proportions of enrichment levels in marsh soils at three sampling sites in each season. Cd levels in all soil samples were at the significant enrichment level in three seasons. As levels in almost 95% of soil samples were at a moderate enrichment level. Comparatively, more than 90% of soil samples showed minimal enrichment levels of Cu, Pb and Zn in three sampling sites in each season. Both *P. australis* and *T. chinensis* wetlands showed higher enrichment levels of Cd and lower enrichment of Cu, Pb, and Zn in summer than in both spring and fall, which was associated with higher BCFs of Cu, Pb and Zn and lower BCFs of As and Cd at both sites (Figure 3). Moreover, almost all EF values of As and Cd and more than 50% of EF values of Cu, Pb and Zn for all soils in three seasons exceeded 1.5, indicating that a significant portion of metal is delivered from non-crustal materials, or non-natural weathering processes, so anthropogenic sources might be an important contributor [39]. Bai et al. [4] also presented that the flow-sediment regulation could elevate the levels of trace metals.

Ecotoxicity assessments of trace elements

The potential acute toxicities of contaminants in soil samples can be estimated using the toxic unit (TU) index, which is the ratio of the determined content to the probable effect level (PEL) [26]. The TUs, the sum of toxic units ($\sum$TUs) and relative contributions of trace elements at each soil layer in the YRD are illustrated in Figure 6. The mean TUs value of each element at three sites decreased in the order of As > Zn > Cu > Pb > Cd in each of

Table 4. Relationships between soil texture and Al and heavy metals in typical profiles from three sampling sites in three sampling seasons.

	Al	As	Cd	Cu	Pb	Zn
Sand	−0.743**	−0.306*	−0.259*	−0.805**	−0.753**	−0.121
Clay	0.758**	0.305*	0.263*	0.850**	0.756**	0.276*

* represents significant correlation at the level of $p < 0.05$;
** represents significant correlation at the level of $p < 0.01$.
n = 60.

Table 5. Correlation matrix among trace elements and selected soil properties for three sampling sites.

	Moisture	Bd	SOM	Salinity	pH	S	Al	As	Cd	Cu	Pb	Zn
Moisture	1.000											
Bd	0.210*	1.000										
SOM	−0.006	−0.189*	1.000									
Salinity	0.143	−0.179	−0.151	1.000								
pH	0.238**	0.201*	−0.038	−0.376**	1.000							
S	0.034	−0.262**	0.478**	0.099	0.014	1.000						
Al	−0.028	0.212*	0.144	−0.265**	−0.141	0.140	1.000					
As	−0.097	0.020	0.037	0.009	−0.295**	0.137	0.507**	1.000				
Cd	−0.082	0.277**	0.131	−0.087	−0.346**	0.062	0.745**	0.618**	1.000			
Cu	−0.139	−0.193*	0.496**	−0.297**	0.110	0.455**	0.617**	0.343**	0.377**	1.000		
Pb	−0.147	−0.348**	0.440**	−0.203*	0.021	0.455**	0.529**	0.371**	0.347**	0.883**	1.000	
Zn	−0.155	0.021	0.278**	−0.237**	−0.101	0.262**	0.783**	0.571**	0.723**	0.772**	0.726**	1.000

* represents significant correlation at the level of $p < 0.05$;
** represents significant correlation at the level of $p < 0.01$.
n = 144.

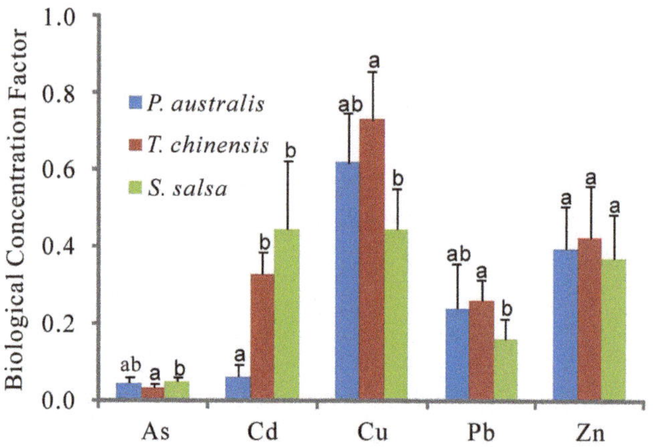

Figure 3. Biological concentration factors of the aboveground parts of plants in each site. [abc] Different letters on the error bars represent significant differences ($p<0.05$).

three seasons. As and Cd in all soil samples had much higher TUs in summer compared to spring and fall. Generally, the ∑TUs decreased with depth along soil profiles. The ∑TUs for surface soils (0–10 cm) in *S. salsa* wetland in spring or in *T. chinensis* wetland in summer were much higher than those for all the other soil samples of the three sites, but neither exceeded the moderate toxicity level (∑TUs = 4;) [25]. The order of the ∑TUs in the top 20 cm soils of three sampling sites in spring was *S. salsa* wetland > *P. australis* wetland >*T. chinensis* wetland. Additionally, the ∑TUs for all soil layers at the three sites reached higher levels in summer than in fall and spring. The ∑TUs for each layer were higher in *P. australis* and *S. salsa* wetlands than *T. chinensis* wetland in spring, whereas they exhibited similar ∑TUs levels at three sites in summer and fall.

The values of the ∑TUs increased from spring to summer and then decreased to their lowest levels in fall. This corresponded to the temporal variations in trace element contaminations during the three sampling seasons which was closely related to the input flow containing trace elements from upstream during the flow-sediment regulation period from June to July. Tixier et al. [40] presented that the storm water exhibited the seasonally varying toxicity of trace elements with high toxicity in spring and recovery in fall, which was caused by the Terraview–Willow field storm water management facility. The ∑TUs for all soil layers of the three sampling sites were less than 4, suggesting low ecotoxicity in the current YRD soils. This is consistent with the result by Gao and Chen [41], who presented that the surface sediments showed no or low ecotoxicity of trace elements in the coastal Bohai Bay. However, As manifested the highest contribution ratios (ranging from 24.4 to 53.7%) to the ∑TUs at three sites in each of the three seasons compared to other trace elements (Figure 6). Therefore, the soils were 'As toxic' (with TUs for As greater than 0.5), indicating that As might have greatly negative effects and ecological risks on some or all of the environmental components [42].

Relationships between trace elements and soil physico-chemical properties

The correlation analysis between trace elements in the marsh soils of each sampling site was performed to assess possible co-contamination from similar sources. As shown in Table 4, significantly positive correlations among all trace elements (i.e., As, Cd, Pb, Cu, and Zn) were observed, suggesting that these trace

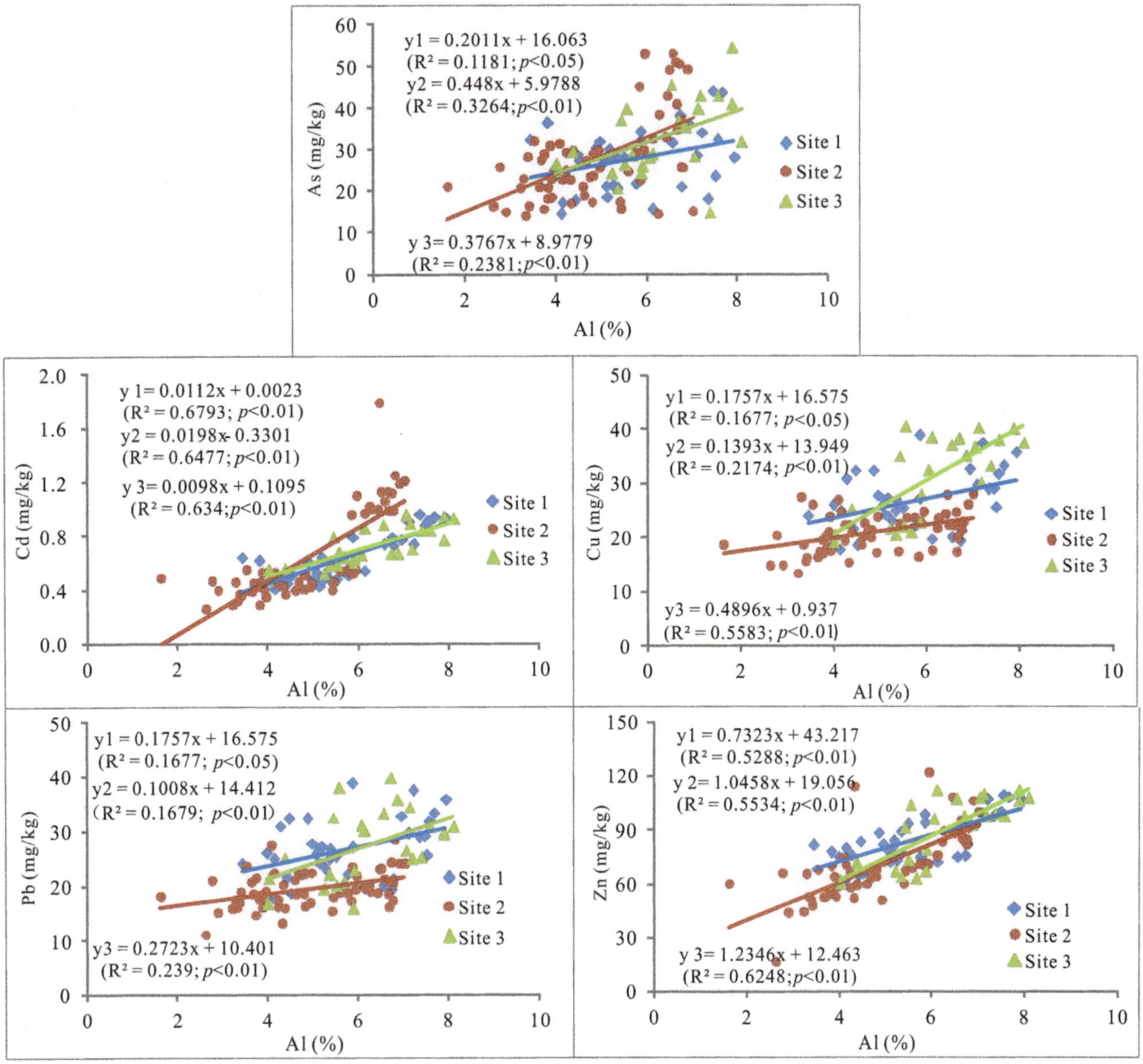

Figure 4. Relationships between the contents of aluminum (%) and trace elements (mg/kg) in all soil samples in the Yellow River Delta.

elements in marsh soils might originate from common source. Figure 4 also exhibited that the linear relationships between Al and these trace elements at each sampling site, which implies that these trace elements and Al were greatly associated with the upstream loess materials with higher Al contents of the Yellow River (62700 μg g^{-1}) [24]. Moreover, Al and these trace elements were significantly correlated with sand and clay contents (Table 4), indicating that higher clay contents in August (Table 2) would contribute to these trace element accumulation. Therefore, the upstream flow-sediment regulation in June could take main responsibility for the accumulation of these trace elements in the delta in the studied period due to no other pollution source inputs in the study area. This is also supported by the result of Bai et al. [4], who presented that the upstream flow-sediment regulation contributed to elevating trace element contents in surface soils in this delta.

Table 5 showed the negatively linear correlations between salinity and Al, Cu, Pb and Zn in the study area ($p<0.01$). This is in agreement with the results presented by Du Laing et al. [43], indicating increasing salinity might enhance the mobility of these metals in soils, as the chloro complex formation and the substitution of Ca^{2+} and Mg^{2+} in the exchange positions could promote the mobility of heavy metals [44]. Additionally, Manousaki et al. [45] found that salinity could favor metal accumulation in shoots of salt cedars, thus decreasing soil metal contents. Bulk densities exhibited negative correlations with Cu and Pb in all soils, implying that lower bulk densities in saturated substances with higher SOM favored the accumulation of both metals. Numerous studies reported that SOM and soil pH played important roles in determining the fates of heavy metals in soils [11,46]. SOM can act as a major sink for heavy metals due to its strong complexing capacity for metallic contaminants [47]. In this

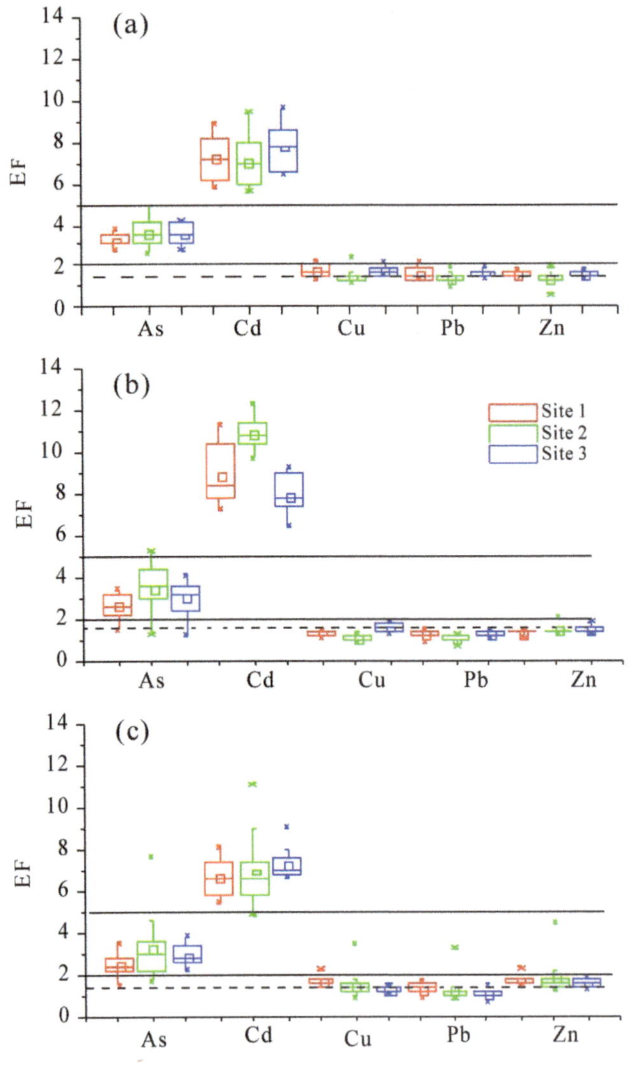

Figure 5. Box-plots of enrichment factors of trace elements in all soil samples of each site of the three in three seasons (a) Spring; (b) Summer; (c) Fall. Straight lines represent the EF = 2 or 5; dash line represents EF = 1.5.

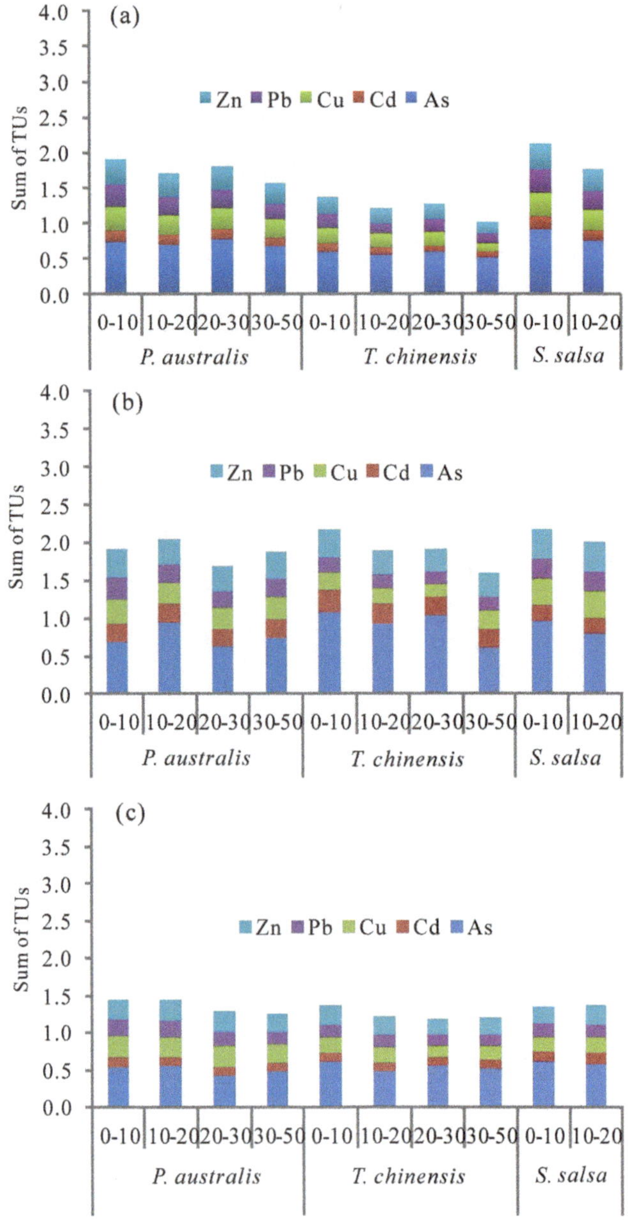

Figure 6. The sum of the toxic units (∑TUs) and toxic contribution of each metal at each soil layer of three sampling sites in three seasons (a) Spring; (b) Summer; (c) Fall.

study, SOM showed a positive linear relationship with Cu, Pb and Zn ($p<0.01$; Table 5). Most researchers presented that subalkaline environments could decrease the mobility of metals and increase the ability of soil to stabilize metals [11,36,48]. However, in this study, soil pH values were negatively correlated with As and Cd ($p<0.01$; Table 5). This was most likely due to plant uptake or leaching as alkalization will increase metal leaching at higher pH ranges (8–10) [46,49]. Sulfur exhibited significantly positive correlations with Cu, Pb and Zn ($p<0.01$) due to the deposition of metal sulphates (i.e., CuS, PbS and ZnS) under anaerobic conditions [4]. Moreover, Kumpiene et al. [46] reported that the presence of one contaminant (e.g., Cu or Pb) can decrease the stabilization efficiency of another contaminant (e.g., Zn) due to competition for sorption sites. Therefore, heavy metals levels in the marsh soils could be impacted by such factors as moisture, salinity, SOM, pH values, sulfur and the interactions between trace elements.

To assess the relationships between spatial distribution patterns of trace elements and several environmental factors (including 5 soil parameters), PCA was implemented to characterize the variations in trace element contents among sampling sites. The factor loading scores in the biplot (Figure 7) exhibited a clear separation amongst *P. australis*, *T. chinensis* and *S. salsa* wetlands and amongst spring, summer, and autumn along the first principal component (PC) accounting for 75.1% and the second PC accounting for 16.0% of the variance. As shown in Figure 7, three main groups can be clearly identified. The red plate contained the soils collected in spring, the yellow plate contained the soils collected in summer, and the blue plate included the soils collected in fall. It appears that the sampling seasons play an important role in shaping the spatial distribution patterns of As and heavy metals in marsh soils. However, the soils in *T. chinensis* wetland in spring did not fall into the red plate but entered or attached to the blue plate, suggesting that the distribution patterns

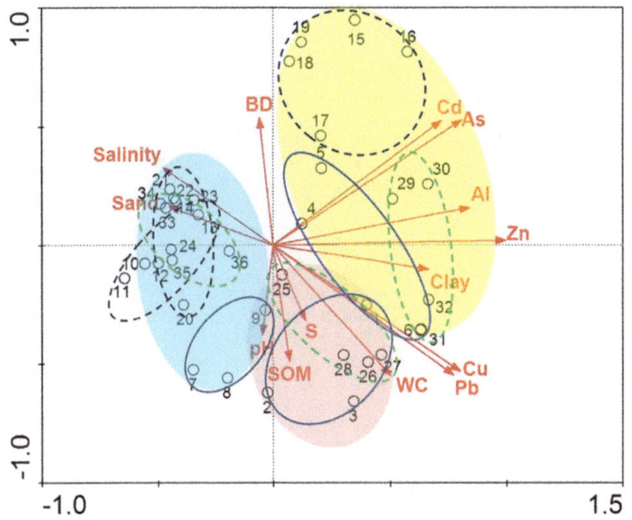

Figure 7. Ordination plots of the PCA results for surface soil samples. The direction of an arrow indicates the steepest increase in the variable, and the length indicates the strength relative to other variables. S, sulfur; BD, bulk density; SOM, soil organic matter; WC, moisture. No. 1, 2, 3 represent three replicates of *P. australis* wetland in spring; No. 4, 5, 6 represent three replicates of *P. australis* wetland in summer; No. 7, 8, 9 represent three replicates of *P. australis* wetland in fall; No. 10, 11, 12, 13, 14 represent five replicates of *T. chinensis* wetland in spring; No. 15, 16, 17, 18, 19 represent five replicates of *T. chinensis* wetland in summer; No. 20, 21, 22, 23, 24 represent five replicates of *T. chinensis* wetland in fall; No. 25, 26, 27, 28 represent four replicates of *S. salsa* wetland in spring; No. 29, 30, 31, 32 represent four replicates of *S. salsa* wetland in summer; No. 33, 34, 35, 36 represent four replicates of *S. salsa* wetland in fall.

of trace elements in those soils are more similar to those in *T. chinensis* and *S. salsa* wetlands in fall than those in *P. australis* and *S. salsa* wetlands in spring.

PCA exhibited that the strongest determinants of the distribution patterns of trace elements in spring were soil moisture, S, SOM and pH value, while the strongest influencing factors in fall were soil salinity and soil texture. Obviously, salinity, water content and soil texture controlled the distribution patterns of trace elements in *T. chinensis* and *S. salsa* wetlands, while SOM, S and soil pH values were the dominant factor in *P. australis* wetland. All trace elements with the longest arrows in the summer group testified the highest accumulations of trace elements in summer, especially in *P. australis* and *S. salsa* wetlands with herbaceous plants. The results obtained from the PCA were consistent with the results derived from the correlation analysis.

Although spatial distribution patterns of trace elements were dominantly controlled by sampling seasons due to the flow-sediment regulation, the effects of plant species on their distributions in soils could not be ignored due to different plant accumulation capabilities for different species (Figure 3). As shown in Figure 7, in the same season plate, sampling sites with the same plant community were clustered together. This was because plant roots can significantly affect plants' abilities to transfer metals [50,51], as they employed a variety of ways to alleviate the stresses from root zone pollution through increasing the rate of root exudation as a result of losses of membrane integrity or of breakdowns in normal cell metabolism [52,53]. The fixation and stabilization of trace elements by plants originating from physiological characteristics (e.g., root exudation in response to underground stress) were different between herbaceous plants (*P. australis* and *S. salsa*)) and woody plant (*T. chinensis*). Additionally, eclogical characteristics of plants (e.g., plant density) might also affect the responses of plants to trace elements, especially under the interactive impacts of water and salt in the tidal salt marshes. This implies that marsh soils with different plant communities might maintain different capabilities for immobilizing trace elements at the same pollution level (Figure 3). Further studies are needed to investigate the impacts of land cover changes on these trace elements.

Conclusions

Obvious spatial and seasonal changes in trace elements in wetland soils in the intertidal salt marshes were observed. The higher trace element contents (especially As and Cd) in summer compared to spring and fall suggested the contribution of the flow-sediment regulation (from June to July) to pollutants from upper reaches deposition and accumulation in the delta and higher self-purifying capabilities of these salt marshes. This implies that these tidal salt marshes can alleviate the effects of the upstream flow-sediment regulation. Higher trace element contents in these soils were closely correlated to soil properties such as soil moisture, salinity, sulfur, SOM, soil texture and pH value. It is noted that sampling seasons had dominant impacts on the spatial distribution patterns and ecotoxicity of trace elements in soils in this study. Therefore, seasonal changes in trace element contamination should be taken into account in tidal wetlands especially in tidal wetlands with great hydrological fluctuations. We also observed that marsh soils with herbaceous plants (*P. australis* and *S. salsa*) had higher trace element accumulation than woody plant (*T. chinensis*). Although all soil samples showed lower ecotoxicity, higher ecotoxicity of As in each of the three seasons should attract our concerns in this region. Moreover,it is important to continuously monitor these trace elements in coastal wetland soils to identify the ecological and environmental effects of hydrological engineering. The findings of this study could contribute to wetland conservation and management in coastal regions affected by the hydrological engineering. Further studies of the final fate and transport of these trace elements and plant uptake mechanism under water and salt stress in coastal wetlands will greatly contribute to trace element pollution control in coastal region.

Acknowledgments

The authors are grateful to the Editor and two anonymous reviewers for their valuable comments and suggestions. We also express our thanks to other members of our research group for their field work.

Author Contributions

Conceived and designed the experiments: JB RX. Performed the experiments: QZ QL JW. Analyzed the data: JB KR. Contributed reagents/materials/analysis tools: JB KR. Wrote the paper: JB RX KR.

References

1. Carpenter B, Haltmeier R, Wilde C (1997) Dredged sediments disposal. Journal of Water Environment Technology 9: 47–50.
2. Yeloff DE, Labadz JC, Hunt CO (2005) The suitability of a south Pennine (UK) reservoir as an archive of recent environmental change. Journal of Paleolimnology 34: 339–348.
3. Bai JH, Cui BS, Xu XF, Gao HF, Ding QY (2009) Heavy metal contamination in riverine soils upstream and downstream of a hydroelectric dam on the Lancang River, China. Environmental Engineering Science 26: 941–946.
4. Bai JH, Xiao R, Zhang K, Gao H (2012) Arsenic and heavy metal pollution in wetland soils from tidal freshwater and salt marshes before and after the

flow-sediment regulation regime in the Yellow River Delta, China. Journal of Hydrology 450–451: 244–253.
5. Reddy KR, DeLaune RD (2008) Biogeochemistry of Wetlands: Science and Applications, 1st Ed., CRC Press, Boca Raton, FL.
6. Bai JH, Xiao R, Cui BS, Zhang KJ, Wang QG, et al. (2011) Assessment of heavy metal pollution in wetland soils from the young and old reclaimed regions in the Pearl River Estuary, South China. Environmental Pollution 159: 817–824.
7. Gambrell RP (1994) Trace and toxic metals in wetlands—A review. Journal of Environmental Quality 23:883–891.
8. Grabowski LA, Houpis JLJ, Woods WI, Johnson KA (2001) Seasonal bioavailability of sediment-associated heavy metals along the Mississippi river floodplain. Chemosphere 45: 643–651.
9. Zhang WG, Feng H, Chang JN, Qu JG, Xie HX, et al. (2009) Heavy metalcontamination in surface sediments of Yangtze River intertidal zone: an assessment from different indexes. Environmental Pollution 157: 1533–1543.
10. Wang LL, Yang ZF, Niu JF, Wang JY (2009) Characterization, ecological risk assessment and source diagnostics of polycyclic aromatic hydrocarbons in water column of the Yellow River Delta, one of the most plenty biodiversity zones in the world. Journal of Hazard Materials 169: 460–465.
11. Bai JH, Yang ZF, Cui BS, Gao HF, Ding Q Y(2010) Some heavy metals distribution in wetland soils under different land use types along a typical plateau lake, China. Soil &Tillage Research 106: 344–348.
12. Roychoudhury AN (2007) Spatial and seasonal variations in depth profile of trace metals in saltmarsh sediments from Sapelo Island, Georgia, USA. Estuarry, Coast and Shelf Science 72: 675–689.
13. Brandt SA (2000) Classification of geomorphological effects downstream of dams. Catena 40: 375–401.
14. Hollert H, Keiter S, König N, Rudolf M, Ulrich M, et al. (2003) A new sediment contact assay to assess particulate-bound pollutants using Zebrafish (Daniorerio) embryos. Journal of Soils and Sediments 3: 197–207.
15. Zhao YM, Song C S (1995) Scientific Survey of the Yellow River Delta Natural Reserve. China Forestry Press, Beijing, pp. 26–30.
16. Li SN, Wang GX, Deng W, Hu YM, Hu WW (2009) Influence of hydrology process on wetland landscape pattern: A case study in the Yellow River Delta. Ecological Engineering 35: 1719–1726.
17. Cui BS, Tang N, Zhao XS, Bai JH (2009) A management-oriented valuation method to determine ecological water requirement for wetlands in the Yellow River Delta of China. Journal for Natural Conservation 17:129–141.
18. Soil Survey Staff (2010) Keys to Soil Taxonomy, 11th ed. USDA-Natural Resources Conservation Service, Washington, DC.
19. Peng J, Chen S, Dong P (2010) Temporal variation of sediment load in the Yellow River basin, China, and its impacts on the lower reaches and the river delta. Catena 83, 135–147.
20. Page AL, Miller RH, Keeney DR (1982) Methods of Soil Analysis Part 2: Chemical and Microbiological Properties, 2nd edn., American Society of Agronomy, and Soil Science Society of America, Madison. WI., pp. 539–579.
21. Selvaraj K, Ram Mohan V, Szefer P (2004) Evaluation of metal contamination in coastal sediments of the Bay of Bengal, India: geochemical and statistical approaches. Marine Pollution Bulletin 49: 174–185.
22. Chabukdhara M, Nema AK (2012) Assessment of heavy metal contamination in Hindon River sediments: A chemometric and geochemical approach. Chemosphere 87: 945–953.
23. Han Y, Du P, Cao J, Eric SP (2006) Multivariate analysis of heavy metal contamination in urban dusts of Xi'an City, China. Science of the Total Environment 355, 176–186.
24. China National Environmental Monitoring Center (CNEMC)(1990) Chinese Elemental Background Values for Soils. Chinese Environmental Science Press, Beijing.
25. Pedersen F, Sjbrnestad E, Andersen HV, Kjlholt J, Poll C (1998) Characterization of sediments from Copenhagen harbour by use of biotests. Water Science and Technology 37: 233-240.
26. Long ER, MacDonald DD (1998) Recommended uses of empirically derived, sediment quality guidelines for marine and estuarine ecosystems. Human and Ecological Risk Assessment 4: 1019–1039.
27. Madejón P, Murillo J, Maranon T, Cabrera F, Soriano M (2003) Trace element and nutrient accumulation in sunflower plants two years after the aznalcollar mine spill. Science of the Total Environment 307: 239–257.
28. Mitsch WJ, Gosselink JG (2007) Wetlands, 4th ed. John Wiley Inc, New York.
29. Prusty B, Kumar A, Azeez PA (2007) Vertical distribution of alkali and alkaline earth metals in the soil profile of a wetland-terrestrial ecosystem complex in India. Austrianl Journal of Soil Research 45: 533–542.
30. Gregorauskiene V, Kadunas V (2006) Vertical distribution patterns of trace and major elements within soil profile in Lithuania. Geological Quarterly 50: 229–237.
31. Ehlken S, Kirchner G (2002) Environmental processes affecting plant root uptake of radioactive trace elements and variability of transfer factor data: a review. Journal of Environmental Radioactivity 58: 97–112.
32. Lassabatere L, Winiarski T (2004) Retention of three heavy metals (Zn, Pb, and Cd) in a calcareous soil controlled by the modification of flow with geotextiles. Environmental Science & Technology 38: 4215–4221.
33. Bragato C, Schiavon M, Polese R, Ertani A, Pittarello M, et al. (2009) Seasonal variations of Cu, Zn, Ni and Cr concentration in Phragmites australis (Cav.) Trin ex steudel in a constructed wetland of North Italy. Desalination 246: 35–44.
34. Beth R, Ehrenfeld JG, Haggblom MM (2003) A comparison of sediment microbial communities associated with *Phragmites australis* and *Spartina alterniflora* in two brackish wetlands of New Jersey. Estuaries 26: 465–474.
35. Zehetner F, Rosenfellner U, Mentler A, Gerzabek MH (2009) Distribution of road salt residues, heavy metals and polycyclic aromatic hydrocarbons across a highway-forest interface. Water, Air and Soil Pollution 198: 125–132.
36. Nie M, Xian NX, Fu XH, Chen XF, Li B (2010) The interactive effects of petroleum-hydrocarbon spillage and plant rhizosphere on concentrations and distribution of heavy metals in sediments in the Yellow River Delta, China. Journal of Hazard Materials 174: 156–161.
37. National Standard of PR China (2002) Marine Sediment Quality (GB18668-2002).Standards Press of China, Beijing (in Chinese).
38. Zhang J, Huang WW, Martin JM (1988) Trace metals distribution in huanghe (Yellow River) estuarine sediments. Estuary Coast and Shelf Science 26: 499–516.
39. Feng H, Han X, Zhang W, Yu L (2004) A preliminary study of heavy metal contamination in Yangtze River intertidal zone due to urbanization. Marine Pollution Bulltin 49: 910–915.
40. Tixier G, Rochfort Q, Grapentine L, Marsalek J, Lafont M (2012) Spatial and seasonal toxicity in a stormwater management facility: Evidence obtained by adapting an integrated sediment quality assessment approach. Water Research 46(20): 6671–6682.
41. Gao XL, Chen CT (2012) Heavy metal pollution status in surface sediments of the coastal Bohai Bay. Water Research 46: 1901–1911.
42. Lacutusu R (2000) Appraising levels of soil contamination and pollution with heavy metals. Research Report No. 4, European Soil Bureau.
43. Du Laing G, De Vos R, Vandecasteele B, Lesage E, Tack FMG, et al. (2008) Effect of salinity on heavy metal mobility and availability in intertidal sediments of the Scheldt estuary. Estuary, Coast, Shelf Science 77: 589–602.
44. Doner HE (1978) Chloride as a factor in mobilities of Ni (II), Cu (II), and Cd (II) in soil. Soil Science Society of American Journal 42: 882–885.
45. Manousaki E, Kokkali F, Kalogerakis N (2009) Influence of salinity on lead and cadmium accumulation by the salt cedar (Tamarix smyrnensis Bunge). Journal of Chemical Technology & Biotechnology 84: 877–883.
46. Kumpiene J, Lagerkvist A, Maurice C (2008) Stabilization of As, Cr, Cu, Pb and Zn in soil using amendments-a review. Waste Management 28: 215–225.
47. Gonzalez ZI, Krachler M, Cheburkin AK, Shotyk W (2006) Spatial distribution of natural enrichments of arsenic, selenium, and uranium in a minerotrophic peatland, Gola di Lago, Canton Ticino, Switzerland. Environmental Science & Technology 40: 6568–6574.
48. Ullrich SM, Ramsey MH, Helios-Rybicka E (1999) Total and exchangeable concentrations of heavy metals in soils near Bytom, an area of Pb/Zn mining and smelting in Upper Silesia, Poland. Appllied Geochemistry 14: 187–196.
49. Hartley W, Edwards R, Lepp NW (2004) Arsenic and heavy metal mobility in iron oxide-amended contaminated soils as evaluated by short- and long-term leaching tests. Environmental Pollution 131: 495–504.
50. Audet P, Charest C (2008) Allocation plasticity and plant-metal partitioning: meta-analytical perspectives in phytoremediation. Environmental Pollution 156: 290–296.
51. Mench M, Morel JL, Guckert A, Guillet B (1988) Metal-binding with root exudates of low-molecular weight. Journal of Soil Science 39: 521–527.
52. Neumann G, Römheld V (2001) The Release of Root Exudates as Affected by the Plant's P hysiological Status. Marcel Dekker Inc, New York.
53. Chiang PN, Wang MK, Chiu CY, Chou SY (2006) Effects of cadmium amendments on low-molecular-weight organic acid exudates in rhizosphere soils of tobacco and sunflower. Environmental Toxicology 21: 479–488.

3

Interaction of Soil Heavy Metal Pollution with Industrialisation and the Landscape Pattern in Taiyuan City, China

Yong Liu[1]*, Chao Su[1], Hong Zhang[1], Xiaoting Li[1], Jingfei Pei[2]

1 Institute of Loess Plateau, Shanxi University, Taiyuan, Shanxi Province, China, 2 School of Mathematical Sciences, Shanxi University, Taiyuan, Shanxi Province, China

Abstract

Many studies indicated that industrialization and urbanization caused serious soil heavy metal pollution from industrialized age. However, fewer previous studies have conducted a combined analysis of the landscape pattern, urbanization, industrialization, and heavy metal pollution. This paper was aimed at exploring the relationships of heavy metals in the soil (Pb, Cu, Ni, As, Cd, Cr, Hg, and Zn) with landscape pattern, industrialisation, urbanisation in Taiyuan city using multivariate analysis. The multivariate analysis included correlation analysis, analysis of variance (ANOVA), independent-sample T test, and principal component analysis (PCA). Geographic information system (GIS) was also applied to determine the spatial distribution of the heavy metals. The spatial distribution maps showed that the heavy metal pollution of the soil was more serious in the centre of the study area. The results of the multivariate analysis indicated that the correlations among heavy metals were significant, and industrialisation could significantly affect the concentrations of some heavy metals. Landscape diversity showed a significant negative correlation with the heavy metal concentrations. The PCA showed that a two-factor model for heavy metal pollution, industrialisation, and the landscape pattern could effectively demonstrate the relationships between these variables. The model explained 86.71% of the total variance of the data. Moreover, the first factor was mainly loaded with the comprehensive pollution index (*P*), and the second factor was primarily loaded with landscape diversity and dominance (*H* and *D*). An ordination of 80 samples could show the pollution pattern of all the samples. The results revealed that local industrialisation caused heavy metal pollution of the soil, but such pollution could respond negatively to the landscape pattern. The results of the study could provide a basis for agricultural, suburban, and urban planning.

Editor: Shyamal D. Peddada, National Institute of Environmental and Health Sciences, United States of America

Funding: This work was financially supported by the National Natural Science Foundation of China under Grants 41271513, 41101558 and by the Important Specialized Science and Technology Item of Shanxi Province, China under Grant 20121101011. The funders had no role in study design, data collection and analysis, decision to publish, or preparation of the manuscript.

Competing Interests: The authors have declared that no competing interests exist.

* Email: liuyong@sxu.edu.cn

Introduction

From industrialized age, the unreasonable utilization of natural resources by humans caused many ecological and environmental issues, including land resources. Particularly, in the recent two decades, with the process of urbanization and industrialization, human activities, such as excessive use of fertilizers and pesticides, sewage irrigation, and discharge of waste influenced soil environment significantly. Soil in many regions has been polluted by various pollutants in different levels. Soil pollution could be harmful for human health. For example, it could cause toxicity, cancer, and gene mutation. Specifically, the soil heavy metal pollution is one of the most important issues because of the innate traits of heavy metals. Soil pollution by heavy metals may cause by waste water irrigation. The waste water is mainly from heavy industries discharging in urban areas. Pollution, products, and information are dispersed from urban area to suburban area. This process causes the pollution of the soil by heavy metals.

Landscape pattern is also the result of human activities. Landscape pattern is the mixture of natural and human-managed patches that vary in shape, size and arrangement. In other words, landscape pattern is the arrangement of landscape components with different sizes and shapes [1,2]. Composition and configuration of landscape components are basic properties of landscape pattern [1,2]. Composition describes the number and relative frequency of components, and configuration refers to the spatial arrangement of the landscape components [1,2]. Landscape patterns result from complex interactions among social, biological, and physical forces [3–6]. Non-natural landscapes, e.g., agricultural landscapes, reflect not only natural foundations but also social conditions and human activities [7,8]. Human activities strongly affect landscape, creating a mosaic of natural and human-managed patches [9]. Moreover, urban, suburban, and agricultural areas may interact with each other [6]. Suburban areas and agricultural landscapes are strongly influenced by the understanding and management of landscape patterns and structures [6].

Landscape indices, including patch shape, patch size, landscape diversity (*H*), relative richness (*R*), landscape aggregation (*C*), and dominance (*D*), have been widely used to describe landscape patterns. For example, landscape indices were used to analyse the spatial pattern of agricultural landscapes on the Loess Plateau [8]. In particular, the patch size index could represent the landscape

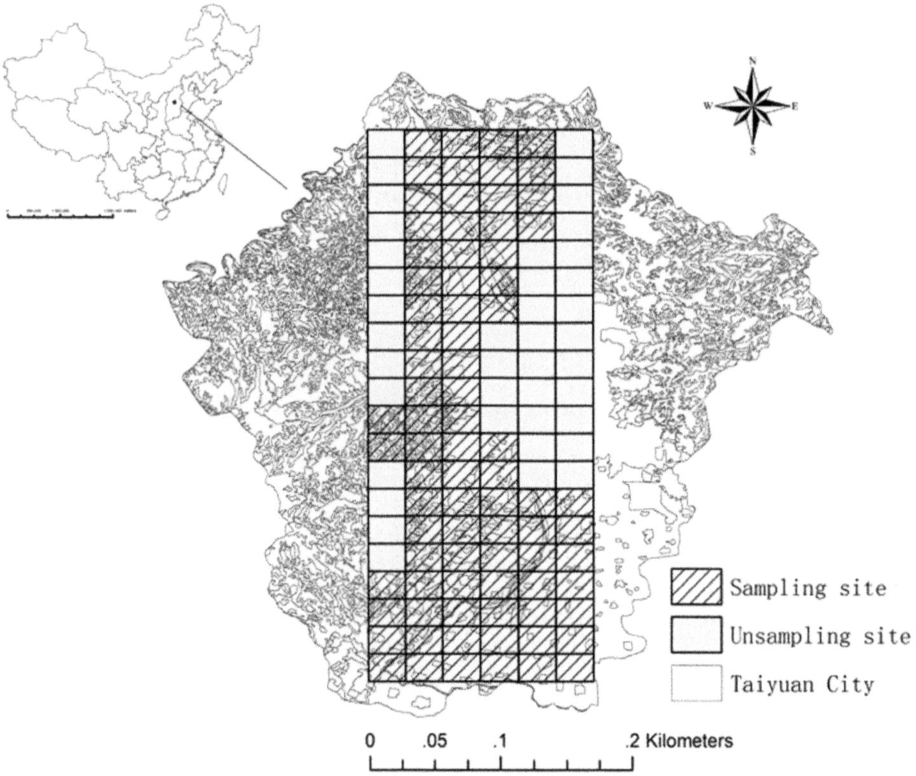

Figure 1. Study area and sampling sites.

pattern in a direct and clear way. Increases in the landscape diversity index reflect increases in the diversity of the landscape. The relative richness index, describing the richness of a landscape and dominance, can measure the extent to which one or a few patch types dominate a certain landscape [8,10]. However, few previous studies have conducted a combined analysis of the landscape pattern and heavy metal pollution. The comprehensive consideration of landscape patterns and pollution may offer a new perspective for research in soil pollution. It expected to provide important basis for agricultural, suburban, and urban planning.

Multivariate analysis provides tools for classifying relationships among measured variables. Principal component analysis (PCA) and factor analysis are the two most common multivariate analysis methods. Principal component analysis, a statistical method, linearly transforms an original set of variables into a substantially smaller set of uncorrelated new variables that represent most of the information in the original data set [11,12]. A smaller set of uncorrelated variables is easier to understand and apply in further analysis [11,12]. Factor analysis is based on the assumption that certain underlying factors influence the covariation of the observed variables. Numbers of underlying factors are fewer than observed variables [11,13]. Many ecological and environmental studies used PCA. For example, PCA was used to study soil pollution, was applied to surface water quality assessment, and was combined with the geoaccumulation index to assess heavy metal pollution of the soil in the vicinity of a copper mine tailings area, and so on [12,14–17].

During the past two decades, urbanisation and industrialisation in Shanxi, especially in Taiyuan city, have polluted many agricultural areas by discharging waste water through irrigation

Table 1. Industrial plants and heavy metals.

Metals	Description of sources
Pb	Batteries, glass, gasoline, cement, ammunition, paints, insecticides, pigments,
Cu	Brass, wire, alloys, dyes, plating, pipes, paints
Ni	Steel and alloys, cosmetics, electroplating, batteries, pigments
As	Pesticides, glass, pigments, wood preservatives, fireworks, textiles, printing, lubricating oil, alloys, oil cloth, semiconductors, photo-conductors
Cd	Electroplating, alloys, cement, pigments, batteries, plastics, rubber, textiles
Cr	Pigments, electroplating, chrome-plating, chrome, varnishes, dye fixers, tanning, photography emulsion
Hg	Paints, fungicides, plastics, paper products, cement, catalysts, pharmaceuticals, batteries
Zn	Alloys, paints, glass, textiles, rubber, metal coatings, cosmetics

Table 2. Background content of heavy metals in soils in Shanxi province (mg/kg).

Metals	Pb	Cu	Ni	As	Cd	Hg	Zn
Background content	14.7	22.9	29.9	9.1	0.102	0.023	63.5

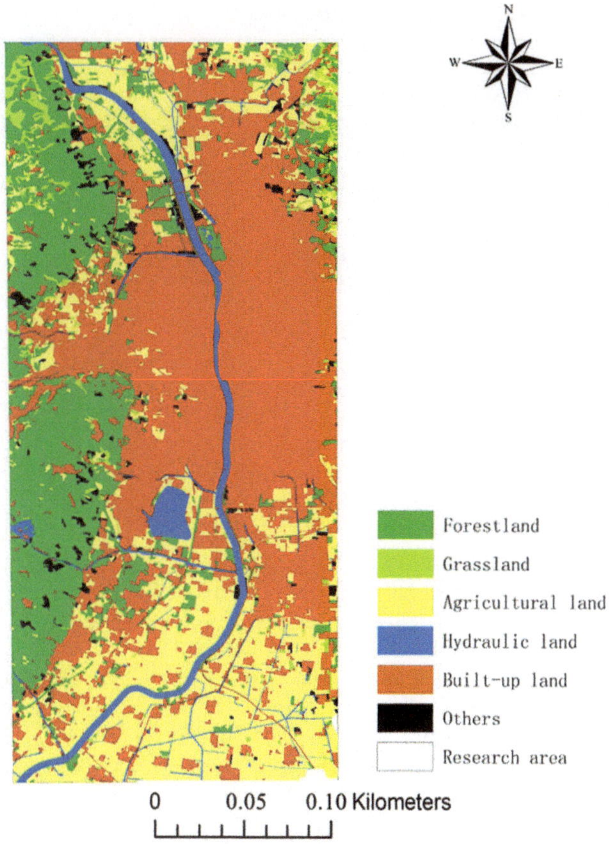

Figure 2. Land use in the study area.

systems. In this polluted area, the heavy metal pollution of the soil is one of the greatest concerns. This study used multivariate analysis to reveal the relationships of soil heavy metal pollution, industrialization, urbanization, and landscape pattern. Firstly, correlation analysis was used to explore the correlations among heavy metals. Secondly, ANOVA and independent-sample T test were applied to reveal the impact of industrialization on soil heavy metal pollution. Then, correlations between soil heavy metals and urbanization, landscape indices were analysed to describe the characteristics of heavy metal pollution of the soil and the patterns of urbanisation at the sampling sites. Finally, PCA was used to group the heavy metal pollution and urbanization, industrialization, landscape indices. These analyses aimed to identify the interactions among soil heavy metals, landscape pattern, and human activities.

The objectives of this paper are to: (1) illustrate the spatial distribution of the eight heavy metals in the soil by constructing maps of the suburban area of Taiyuan city; (2) explore the relationships between heavy metal pollution and landscape pattern, industrialisation; and (3) characterise the factor patterns of heavy metal pollution, landscape pattern, and industrial plants. This study was expected to provide a basis for preventing soil pollution and protecting the ecological environment in local area.

Materials and Methods

Ethics statement

The work does not require any permits or approvals of any authorities and does not involve endangered or protected species,

Table 3. The industrialisation indices.

The industrialisation index	Representation
NEP	the number of electroplating plants
NP1	the number of power plants
NCM	the number of coal and mineral plants
NP2	the number of paper plants
NC	the number of chemical industries
NM	the number of metal plants
NP	the total number of industrial plants
Rup	the urban planning area ratio

the samples were collected from ordinary farmland, and those lands are not privately or protected.

Sampling description

The study was conducted in the suburban area of Taiyuan city, Shanxi, China (Fig. 1), except the west and east of Taiyuan. The west, east, and parts of north of Taiyuan city are highly mountainous, so that these areas couldn't be sampled. Besides, Fen River is the main source of drinking water and irrigation system of Taiyuan city, and it runs through the north and south in Taiyuan. Thus, the spatial domain is irregularly shaped, being much narrower in the East-West direction and much longer in the North-South direction. The longitude of the study area is from 112.57° E to 112.74° E and the latitude is from 36.62° N to 37.03° N. Taiyuan city is the capital of Shanxi province and is a base of heavy industry in China. Certain local sites may be polluted by waste water discharging from industrial plants through irrigation systems.

Samples were collected in 2008 from geographically distributed sites arranged in a network pattern. Eighty sampling sites were sampled in the research area (Fig. 1). All samples were taken between 6th April, 2008 and 8th April, 2008. The topsoil was sampled at depths of 0–20 cm. The average pH value of the topsoil was 8.6.

Comprehensive pollution index

Table 1 lists the heavy metals that might be discharged with waste water from different types of plants [6]. Table 2 shows the

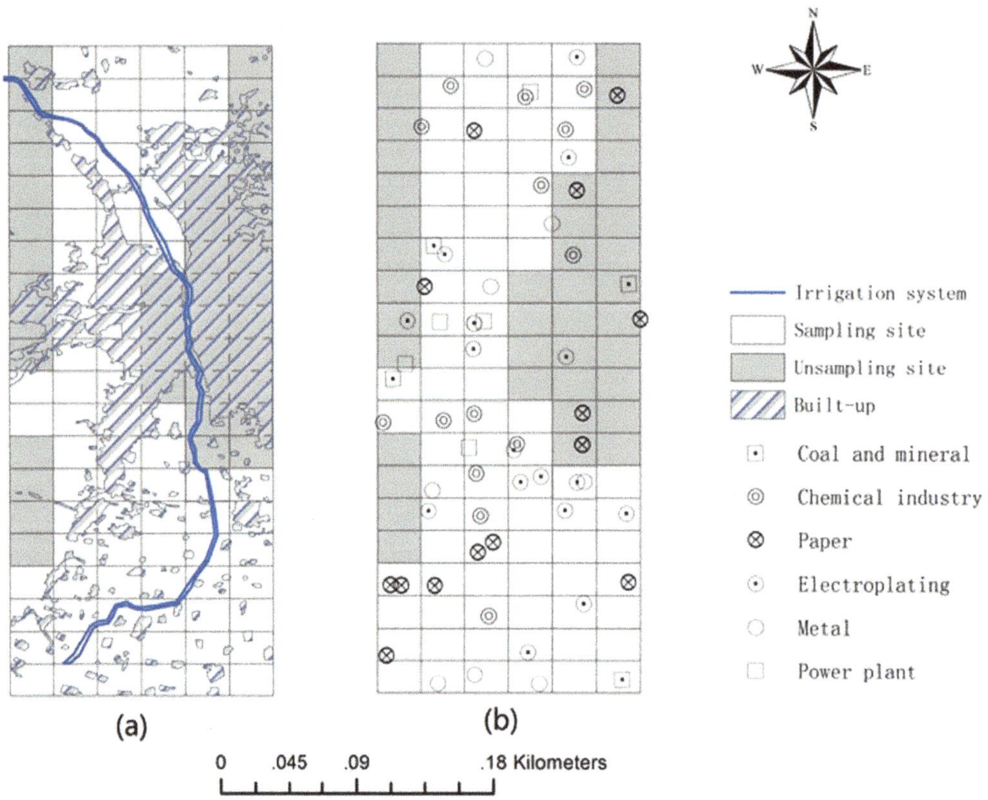

Figure 3. Spatial distribution of: (a) urban planning area and (b) industrial plants.

Table 4. Statistics of the concentrations of eight heavy metals (mg/kg).

Metal	N	Mean	Standard deviation	Median	Maximum	Minimum	CV (%)
Pb	80	26.24	10.70	23.95	86.00	17.40	0.409
Cu	80	28.87	5.28	27.90	47.60	17.20	0.184
Ni	80	29.76	5.10	29.90	50.2	20.5	0.172
As	80	10.96	2.81	11.15	17.80	4.50	0.258
Cd	80	0.21	0.05	0.20	0.40	0.10	0.248
Cr	80	73.69	9.22	72.55	109.00	54.70	0.126
Hg	80	0.12	0.08	0.10	0.39	0.02	0.628
Zn	80	86.08	11.68	83.89	121.00	58.60	0.137

background values for all heavy metals found in soils of Shanxi province [18].

The Comprehensive Index Evaluation method was applied to measure the comprehensive pollution index (P) of each sampling site.

C_i is the measured value for heavy metal i and C_{i0} is the background value of Shanxi province for heavy metal i in the soils [18]; i ranges from 1 to 8. Then,

$$P_i = {}^{C_i}/_{C_{i0}} \tag{1}$$

Comprehensive pollution index P is given by [19]

$$P = \sqrt{(P_{i\max}{}^2 + P_{iave}{}^2)/2} \tag{2}$$

where $P_{i\max}$ is the maximum value of P_i in all sites and P_{iave} is the average value of P_i in all sites.

Landscape indices

The following landscape indices were used to describe the landscape pattern for each sampling site: landscape diversity (H), relative richness (R), dominance (D). Fig. 2 shows the land use types of Taiyuan city in 2008. Six land use types–green space, hydraulic, built-up, industrial, agricultural, and others–were used to calculate the landscape indices for 80 sampling sites by ArcGIS 10.0 [20].

The landscape diversity index used was the Shannon-Weaver Diversity (H) [8],

$$H = -\sum_{i=1}^{m} Q_i \ln Q_i \tag{3}$$

where Q_i is the proportion of land use type i at a site and m is the number of observed land use types.

Relative richness (R) is given by [21]

$$R = ({}^{N}/_{N_{\max}}) * 100\% \tag{4}$$

where N is the number of different land use types present and $N_{\max}$ is the maximum number of land use types possible. The larger the value of R is, the richer the landscape will be.

Dominance (D) could measure the extent to which one or a few patch types dominate the landscape, and it is defined as [8]

$$D = H_{\max} + \sum_{i=1}^{m} Q_i \ln Q_i \tag{5}$$

where H_{max} is the maximum value of diversity among all the sites, m is the number of landscape types and Q_i is the proportion of area of type i.

Industrialisation indices

The following industrialisation indices were used to describe the landscape pattern for each sampling site (Table 3).

Rup is the proportion of each sampling site covered by urban planning area, it is calculated from:

$$Rup = \frac{Aup}{S} \tag{6}$$

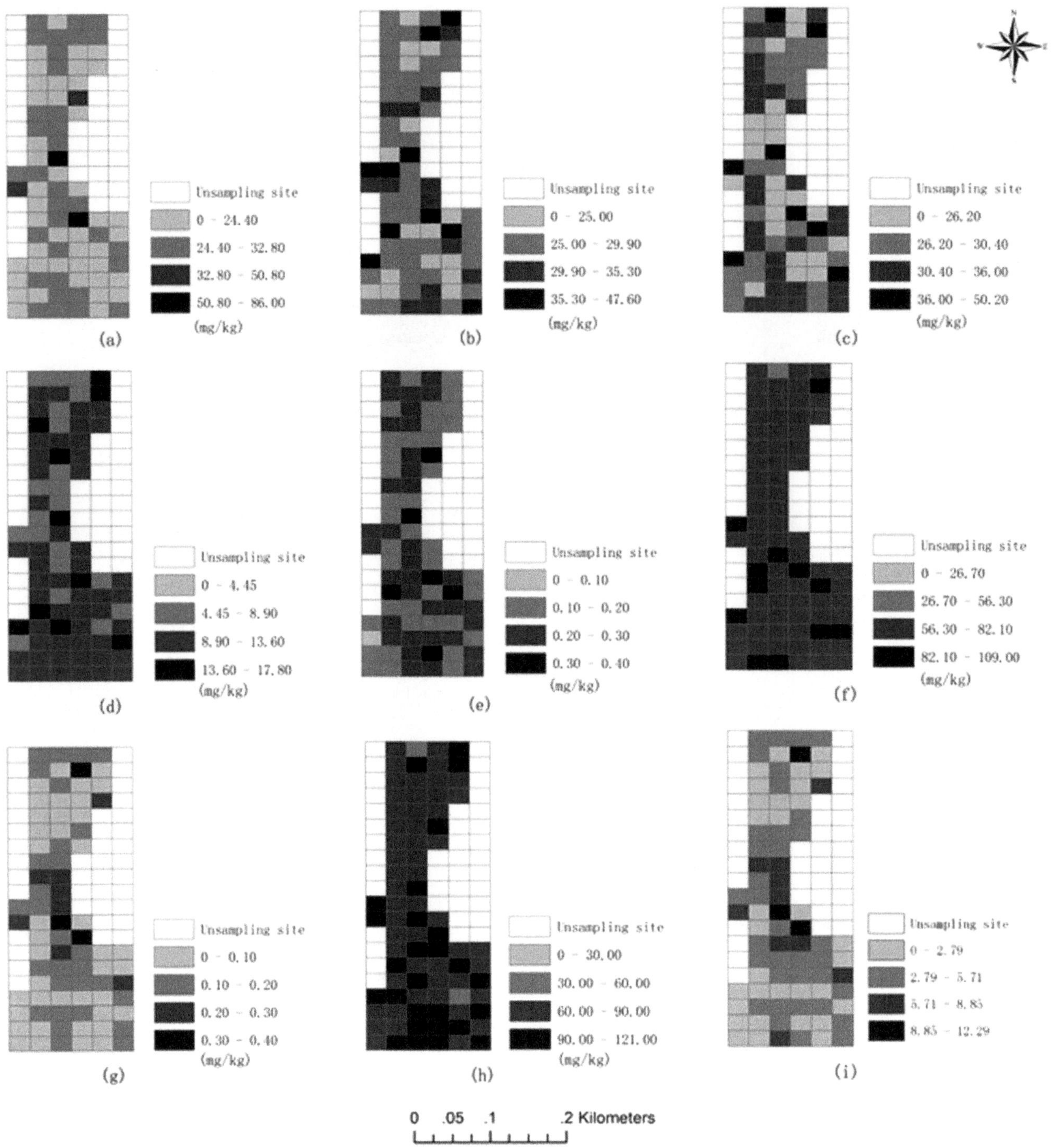

Figure 4. Spatial distributions of heavy metals in soil and comprehensive pollution index: (a) Pb, (b) Cu, (c) Ni, (d) As, (e) Cd, (f) Cr, (g) Hg, (h) Zn and (i) *P*.

where *Aup* is the urban planning area at the sampling site and *S* is the area of the sampling site.

The industrialisation indices for 80 sampling sites were calculated using ArcGIS 10.0 [20].

Fig. 3(a) shows the urban planning area in research area, corresponding to the built-up zone. The highly urbanised and industrialised area located in the centre of Taiyuan. The Fen River flows through Taiyuan. Fig. 3(b) shows the location of the major large-scale industries in the study area. The pollutant emissions from these industries were more than 80% of the total emissions in the study area. Industrial plants were evenly distributed among the sampling sites. Several paper plants, two chemical plants, and one electroplating plant were located at sites that were not sampled (Fig. 3 (b)). It is possible that these industrial plants could have affected the concentrations of heavy metals at the sampling sites.

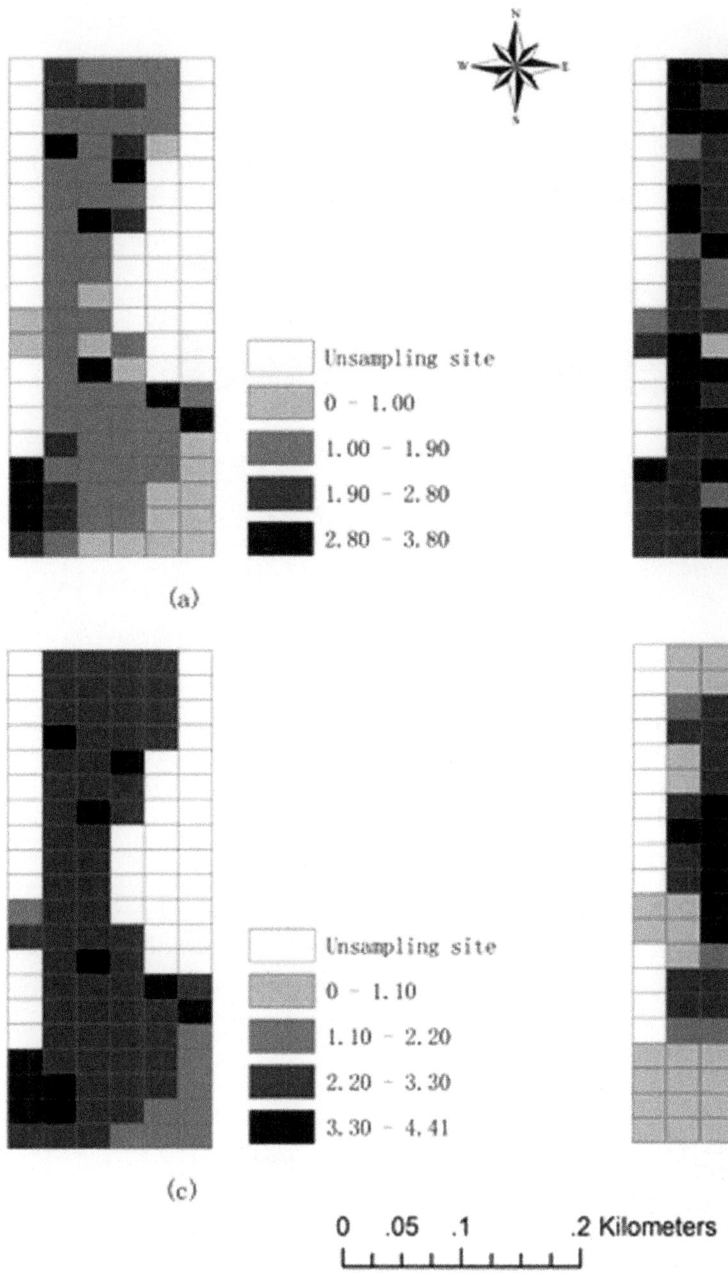

Figure 5. Spatial distribution of: (a) landscape diversity, (b) relative richness, (c) landscape dominance, and (d) urban planning ratio.

Statistical analysis

Descriptive statistics were used to identify the main features of the concentrations of heavy metals. Descriptive statistics are used to summarise the general trends in a data set. This analysis is fundamental in statistical studies. The indicators used in this study included mean, standard deviation, range for each heavy metal, CV% value, and correlation analysis.

Pearson correlation analysis was used to measure the correlations between two numeric variables. Analysis of variance (ANOVA) and independent-sample T test were applied to test if the industrialization had significant influences on soil heavy metal pollution. They were performed with SPSS 18.0 statistical software.

The principal component analysis was performed with **SIMCA-P software**. In this method, the first factor explains the greatest variation in the variables of interest, the second factor explains the next highest amount of variation and the other factors similarly explain the remaining variation [15,16,22]. Finally, the factor scores are calculated from the variables using a regression method with a matrix of factor-score coefficients [15].

In this study, PCA was applied to explore the factor pattern of soil heavy metal (Pb, Cu, Ni, As, Cd, Cr, Hg, and Zn) pollution at the 80 sampling sites. The comprehensive pollution index, industrialization indices, and landscape indices were then grouped using PCA to show the relationships among them.

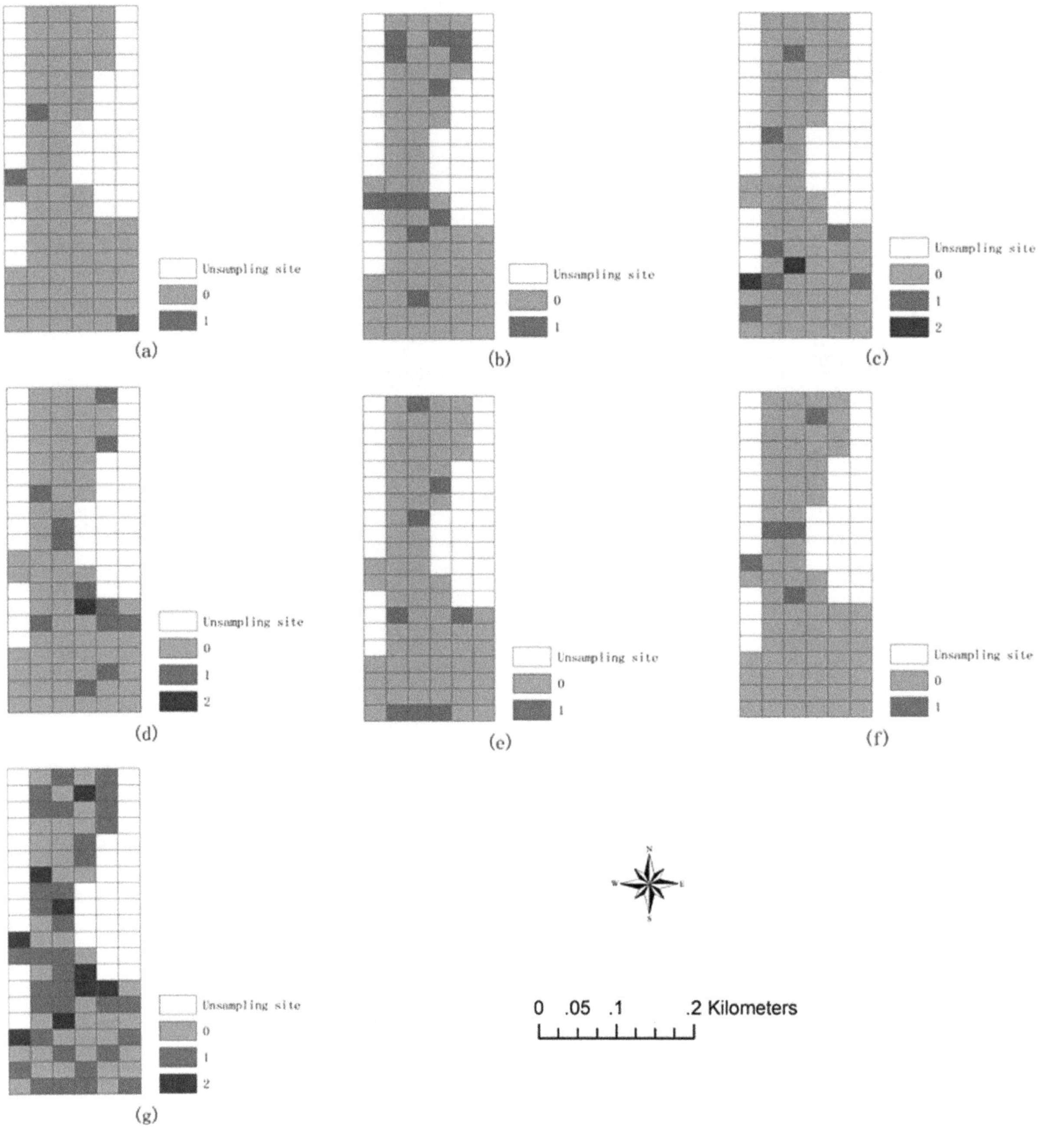

Figure 6. Number of: (a) coal and mineral industrial plants, (b) chemical industrial plants, (c) paper industrial plants, (d) electroplating industrial plants, (e) metal industrial plants and (f) power plants, and (g) total number of industrial plants at each sampling site.

Results

Descriptive statistical analysis of eight heavy metals

The mean, standard deviation, and range for each heavy metal are presented in Table 4. The CV% values reflect the mean variation of each sampling site in the population. The order of the CV%s for each element, from high to low, was Hg>Pb>As>Cd>Cu>Ni>Zn>Cr. This result showed that the variation of Hg and Pb in the soil was larger than other metals.

Fig. 4 shows the spatial distribution of the measured values for the eight heavy metals at each sampling site and the spatial distribution of the comprehensive pollution index (*P*). It shows that the concentrations of Pb, Cu, and Hg in the centre of the study area were higher than in other areas. The concentrations of Cr and Zn were uniformly distributed, and the concentrations of Ni, As, and Cd were randomly distributed throughout the domain.

Fig. 4 (i) shows the spatial distribution of the comprehensive pollution index values for the sampling sites. This pattern was similar to the spatial distribution pattern of the heavy metal Hg. The most severely polluted sites were those of highest Hg concentration. These results showed that the concentrations of Hg substantially exceeded the background value, and Hg pollution is relatively serious.

Table 5. Correlations among the eight heavy metals.

	Pb	Cu	Ni	As	Cd	Cr	Hg	Zn
Pb	1.00							
Cu	0.598**	1.00						
Ni	0.269**	0.425**	1.00					
As	0.290**	0.493**	0.509**	1.00				
Cd	0.225*	0.174	0.078	0.075	1.00			
Cr	0.101	0.377**	0.337**	0.477**	0.071	1.00		
Hg	0.392**	0.300**	−0.234*	−0.204	0.022	−0.228*	1.00	
Zn	0.671**	0.735**	0.308**	0.391**	0.245*	0.370**	0.349**	1.00

** *Sig.*<0.01 (two-tailed),
* *Sig.*<0.05 (two-tailed).

Relationships between landscape indices and industrialisation

The spatial distribution of landscape diversity, relative richness, landscape dominance and the urban planning ratio at the sampling sites were shown in Fig. 5. The sampling sites with greater landscape diversity and lower landscape dominance located in the central and south-western parts of the study area (Fig. 5 (a) and (c)). Sampling sites with greater landscape dominance were primarily in the south-eastern portions of the study area (Fig. 5 (c)). The sites with high values of relative richness were located in the northern and central parts of the study area (Fig. 5 (b)). The sites with high values of the urban planning ratio were located in the centre of the study area (Fig. 5 (d)). The comparison of the relative richness, landscape diversity, and landscape dominance maps (Fig. 5 (a)–(c)) illustrates that the landscape diversity showed a more significant pattern than the relative richness and dominance. The spatial distribution of the urban planning ratio appears to display a north-south orientation corresponding to the urban planning area (Fig. 3 (a)). Higher values of the urban planning ratio were found in the central sampling sites. These sites are located near the centre of Taiyuan.

Fig. 6 displays the number of each type of pollution source by sampling site. These values range from 0 to 2 because the study area was part of the urban district and was the site of relatively few large industries, which usually located primarily in suburban districts close to the study area. This distribution pattern was similar to the spatial distribution pattern of the eight heavy metals (Fig. 4).

Correlations among the eight heavy metals

The Pearson Correlation was conducted in SPSS 18.0 [23]. The Pearson correlations among the eight heavy metals were shown in Table 5. The correlations among the eight heavy metals were generally significant. The correlations among Pb, Cu, Ni, As, Cr, and Zn were highly significant (*Sig.*<0.01), whereas the correlations between Cd and the other heavy metals were not significant. Additionally, Hg was positively correlated with Pb and Cu (*Sig.*< 0.01), and negatively correlated with Ni and Cr (*Sig.*<0.05).

Effects of industrialization on soil heavy metal pollution

The variables, the number of industrial plants, were ordinal variables. We used analysis of variance (ANOVA) and independent-sample T test to analyze if the industrial plants had significant effects on soil heavy metal pollution [23]. The results were shown in Table 6.

The results of one-way ANOVA showed that the variable *NP* had highly significant effects on Cu, Hg, Zn, and *P* (*Sig.*<0.01), and significant effects on Pb, Cd (*Sig.*<0.05). Furthermore, Bonferroni multiple comparisons test was performed at 0.01 level after the ANOVA (Table 6). It revealed that the grouping means of Cu, Hg and the comprehensive value *P* in the sites with more industrial plants were significantly higher than those with less industrial plants.

For variables *NEP*, *NP1*, *NCM*, *NP2*, *NC* and *NM*, independent-sample T test was used to explore the relationships between them and soil heavy metal pollution. Table 6 indicated that, the variable *NEP* had significant effects on heavy metals Cu (*Sig.*<0.05) and Zn (*Sig.*<0.01). The variable *NCM* had a significant influence on heavy metal Cu (*Sig.*<0.01). And the variable *NP2* had significant influences on heavy metals As, Hg, and comprehensive pollution index *P* (*Sig.*<0.05). In other words, the number of electroplating plants had significant influences on concentrations of Cu and Zn. The number of coal and mineral

Table 6. Grouping mean concentrations of heavy metals and significance tests of differences among industrialisation indices.

Grouping variable		Pb	Cu	Ni	As	Cd	Cr	Hg	Zn	*P*
NP	0	22.85a	27.11b	29.24	10.77	0.19a	71.86	0.10b	81.82a	3.18b
	1	28.04a	29.48a	30.01	11.11	0.23a	74.88	0.13a	88.88a	4.37a
	2	33.37a	33.84a	30.92	11.14	0.22a	76.74	0.18a	93.00a	5.91a
Sig.(ANOVA)		0.012*	0.001**	0.634	0.860	0.017*	0.228	0.004**	0.006**	0.003**
NEP	0	24.66	28.13	29.35	10.69	0.20	73.13	0.11	84.20	3.79
	1	30.07	31.50	31.26	11.87	0.24	75.80	0.15	93.91	4.96
Sig. (T-test)		0.317	0.028*	0.232	0.173	0.116	0.359	0.277	0.005**	0.258
NP1	0	26.31	28.91	29.81	10.99	0.21	73.99	0.12	85.74	4.05
	1	25.94	27.46	27.87	9.70	0.23	69.90	0.11	87.36	3.70
Sig. (T-test)		0.933	0.495	0.341	0.244	0.190	0.271	0.729	0.733	0.718
NCM	0	26.09	28.47	29.56	10.93	0.21	73.12	0.12	85.78	3.98
	1	30.00	39.17	34.73	11.67	0.22	88.47	0.14	93.80	4.48
Sig. (T-test)		0.540	0.00**	0.087	0.661	0.727	0.279	0.734	0.249	0.725
NP2	0	26.11	28.74	29.94	11.15	0.21	73.11	0.12	86.01	3.82
	1	28.06	30.94	27.04	8.08	0.20	82.46	0.21	87.10	6.63
Sig. (T-test)		0.698	0.372	0.224	0.018*	0.706	0.395	0.010*	0.843	0.010*
NC	0	26.03	28.59	29.99	10.96	0.21	74.04	0.11	85.89	3.68
	1	27.43	30.48	28.44	10.95	0.21	71.73	0.18	87.18	5.78
Sig. (T-test)		0.681	0.257	0.339	0.992	0.819	0.431	0.112	0.735	0.111
NM	0	26.23	29.02	29.57	11.03	0.21	73.63	0.12	85.93	4.00
	1	26.31	27.56	31.40	10.35	0.22	74.21	0.12	87.46	3.89
Sig. (T-test)		0.983	0.465	0.343	0.526	0.555	0.912	0.903	0.728	0.891

** *Sig.*<0.01 (two-tailed),
* *Sig.*<0.05 (two-tailed).
Letters (a, b) indicated the results of Bonferroni test (*Sig.*<0.01), no significant differences (same letter) or significant differences (different letter).

Table 7. Correlations of the heavy metals with landscape indices.

	Pb	Cu	Ni	As	Cd	Cr	Hg	Zn	P
H	−0.178	−0.180	−0.077	−0.073	−0.200	−0.100	−0.294**	−0.330**	−0.296**
R	−0.172	−0.110	0.100	−0.011	−0.125	0.003	−0.141	−0.141	−0.108
D	0.178	0.180	0.077	0.073	0.200	0.100	0.294**	0.330**	0.296**
Rup	0.015	−0.093	−0.226*	−0.048	−0.029	−0.195	0.226**	−0.105	0.223*

** *Sig.*<0.01 (two-tailed),
* *Sig.*<0.05 (two-tailed).

plants had a significant influence on concentration of Cu, and the number of paper plants had significant influence on concentrations of As, Hg, and value of *P*. However, the variables *NP1*, *NC*, and *NM* had no significant effects on any heavy metal (*Sig.*>0.05).

In conclusion, the total number of industrial plants had highly significant effects on many heavy metals (Cu, Hg, and Zn) and the comprehensive pollution index *P*. Cu was mainly affected significantly by the number of electroplating plants and coal and mineral plants. And the number of electroplating plants also had a significant influence on concentration of Zn. The number of paper plants had significant influence on concentrations of As, Hg, and value of *P*.

Relationships of heavy metals and landscape pattern, urbanization

The Pearson correlation coefficients of the heavy metals with the landscape indices were shown in Table 7. These coefficients were used to identify the relationships between the pairs of variables analysed for the 80 sampling sites.

Table 7 shows that the variables *H*, *D*, and *Rup* had significant correlations with Hg, Zn, and *P*. According to the two-tailed test, the correlation coefficients of Hg with *H*, *D*, and *Rup* were −0.294, 0.294, and 0.226, respectively (*Sig.*<0.01). It reflected a significant relationship. The landscape diversity had a significant negative correlation with Hg, and the urban planning ratio had a significant positive correlation with Hg. Similarly, Zn showed significant correlations with the landscape indices. The correlation coefficients of Zn with *H* and *D* were −0.330 and 0.330, respectively (*Sig.*<0.01). Zn also had a significant negative correlation with landscape diversity. Besides, the comprehensive pollution index *P* had a significant correlation with *H*, *D*, (*Sig.*< 0.01), and *Rup* (*Sig.*<0.05). As with Hg, the variable *P* also had a negative correlation with landscape diversity and a positive correlation with urban planning ratio. The possible reason for the correlations between the heavy metals Zn, Hg and the indices *H*, *D* was that fewer industrial plants were located in areas with higher diversity and lower dominance.

In conclusion, the heavy metal Hg showed significant negative correlations with landscape diversity, as did Zn. These results showed that increases in landscape diversity could potentially mitigate the soil pollution caused by heavy metals. Hg was significantly correlated with the urban planning ratio. This result showed that urbanisation may be related to higher concentrations of Hg.

Relationships among landscape diversity, urbanization and industrialization

Fig. 7 shows the overlay map combining landscape diversity (*H*), the urban planning area, and industrial plants to explore their relationships. Because the variables *H* and *Rup* are numeric variables, the Pearson correlation analysis can be done between them. But the correlation between them was not significant (*Sig.*> 0.05) and the plot between them didn't have an obvious trend. However, in Fig. 7, it divided the landscape diversity index into four classes, and it could be seen that the four sites with the fourth landscape diversity class didn't have much higher urban planning ratio. In order to analyze their relationship further, we analyzed if the division has significant influences on *Rup* using ANOVA (Analysis of variance) [23]. The results showed there was no significant influence (*Sig.*>0.05), but the group mean values showed that the second and third classes had the much higher urban planning ratio. In addition, this map also shows that most industrial plants were located in non-urbanised areas. Several

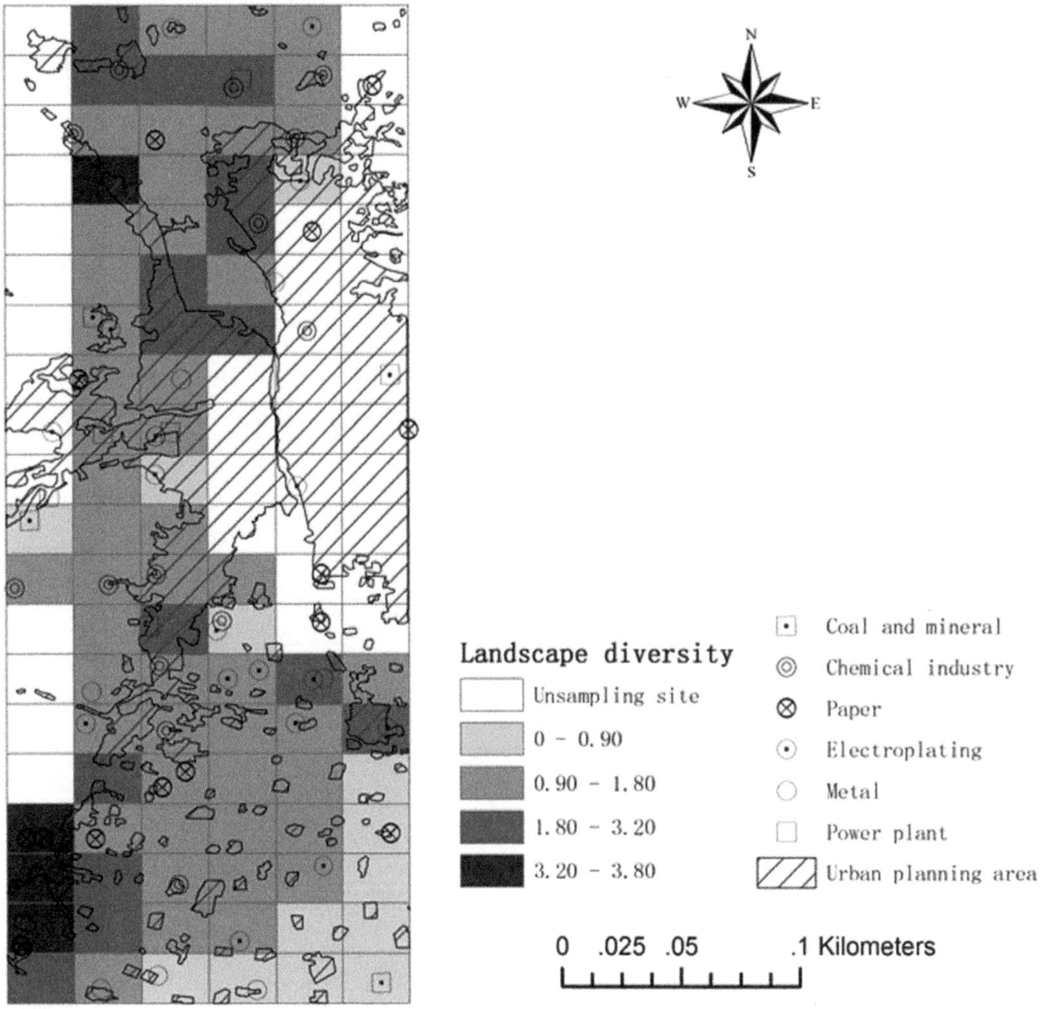

Figure 7. Overlay map of landscape diversity, industrial plants and the urban planning area.

plants were located on the boundaries of the urban planning area. These results showed that industrialisation was not always highly corresponding to urbanisation. It reflected the policy of sustainable development.

PCA of soil heavy metal pollution, landscape and industrialisation

A Principal Component Analysis (PCA) provides an overview of the lower-dimensional structure of multivariate data and can reduce the dimensionality of a complex data set to facilitate more effective analysis. In this paper, its process was conducted in SIMCA-P [24]. In this process, the raw data had been centralized (File S1). The centralization method is subtracting mean value from original data. The aim is to integrate different variables data scale.

PCA showed that a two-group model explained 86.71% of the total variance of all the variables (*P*, *H*, *R*, *D*, *Rup*, *NEP*, *NP1*, *NCM*, *NP2*, *NC*, *NM*, and *NP*). This result is highly satisfactory. The first component explained 71.23% of the total variance. Fig. 8 (a) shows a plot of the 12 variables against their values on principal axes *X* and *Y*. The first factor reflected the comprehensive pollution index (*P*), and the second factor primarily reflected landscape diversity and dominance (*H* and *D*). But the loading of landscape diversity is positive and that of the other is negative. Other variables plotted closer to the original point. This showed that the contributions of these variables were quite small.

Fig. 8 (b) shows the PCA scores for the 80 sampling sites. A relatively large number of sampling sites were negative on the first factor. And a relatively large number of sampling sites were positive on the second factor. Because the origin represents the average values of all samples, these results showed that the comprehensive pollution index *P* of all the samples was higher than the mean value. Additionally, the landscape diversity was higher than the mean value and the dominance of most samples was lower than the mean value. The highest score on the first factor was that of sample 41. The second highest score on the first factor was sample 38, in the centre of the study research area. The sampling sites scoring highest on the second factor was 57, followed by 63, in the central to lower portion of the study area. These results were consistent with the relatively high values of the comprehensive pollution index for samples 41 and 38, and the relatively high landscape diversity and low dominance value of sample 57. Except for samples 41, 38, 7, and 57, other samples could be clustered to form a single class.

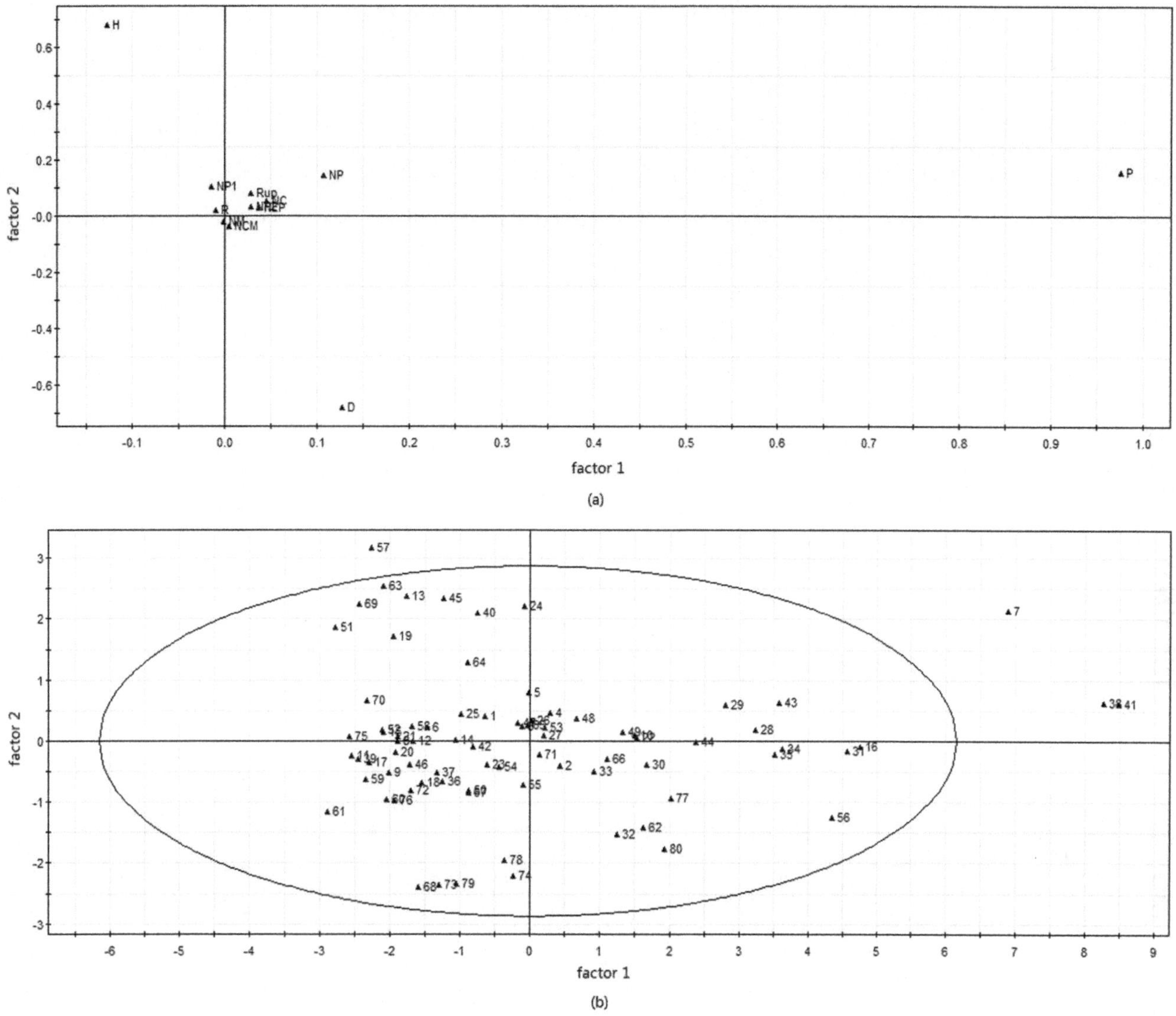

Figure 8. PCA for soil heavy metal pollution, industrialisation and landscape pattern.

Conclusions and Discussion

This paper used correlation analysis, principal component analysis, and geographic information system methodology to investigate and understand the impacts of landscape pattern, industrialisation, and urbanisation on the soil heavy metal pollution. Combining landscape and industrialisation indices and the urbanisation ratio, this study revealed the relationships. The landscape diversity, relative richness, and landscape dominance indices could effectively represent the landscape pattern of the study area. It is possible that other landscape indices, e.g., fractal dimension and landscape evenness, could also serve to describe this pattern. In this study, the data on the number of coal and mineral industries, chemical industries, paper plants, electroplating plants, metal industries, power plants, industrial plants, and the urban planning ratio played a important role in the spatial analysis of the pattern of industrialisation and urbanisation as well as in the multivariate analyses.

The analysis of the correlations among the heavy metals showed that the correlations among these variables were significant. Moreover, the correlations among Pb, Cu, Ni, As, Cr, and Zn were extremely significant. The exploration of the relationships between heavy metals and industrialisation, landscape indices showed that industrialisation could remarkably affect the concentrations of some heavy metals, such as Cu, Cd, Hg, Zn, and *P*. However, the landscape pattern (landscape diversity) was significantly negatively correlated with the concentrations of heavy metals, Hg, Zn, and *P* in the soil. This result is not consistent with the results of Lin et al., who investigated the relationship between heavy metals in the soil and the landscape pattern (landscape diversity, fractal dimension and landscape dominance) in Changhua County in Taiwan in 2002. A possible explanation of this discrepancy is that fewer plants were located at the sites with higher landscape diversity. Because there were fewer industrial plants in the area with a high urban planning ratio, the heavy metal pollution of the soil was not serious there. Additionally, the heavy metal pollution of the soil was affected by human activities in the study area.

Finally, the PCA showed that a two-factor model for heavy metals, industrialisation, and the landscape indices could effectively reveal the relationships of heavy metal pollution of the soil with industrialisation and the landscape pattern. The first factor was mainly loaded with the comprehensive pollution index (*P*), explaining 71.23% of the total variance, and the second factor was primarily loaded with landscape diversity and dominance (*H* and *D*). Other variables (*NEP*, *NP1*, *NCM*, *NP2*, *NC*, and *NM*) were closer to the original point. In future research, the number of industrial plants should be the subject of a more comprehensive investigation. The analysis of 80 sampling sites showed that (with the exception of samples 41, 38, 7, and 57) the samples could be clustered to form a single class. The comprehensive pollution index of samples 41 and 38 was relatively high, and the landscape diversity of sample 57 was the highest of all the samples and the dominance of 57 is the lowest.

Acknowledgments

We thank editors and reviewers for comments on earlier versions of this manuscript.

Author Contributions

Conceived and designed the experiments: YL CS. Performed the experiments: XTL YL. Analyzed the data: CS HZ JFP. Contributed reagents/materials/analysis tools: YL HZ. Wrote the paper: YL CS.

References

1. Han M, Sun YN, Xu SG, Tang XL (2005) Study on changes of marsh landscape pattern in Zhalong wetland assisted by RS and GIS. Progress in Geography, 6: 42–49.
2. Chen LD, Liu Y, Lu YH, Feng XM, Fu BJ (2008) Pattern analysis in landscape ecology: progress, challenges and outlook. Acta Ecologica Sinica, 11: 5521–5531.
3. Forman RTT, Godron M (1986) Landscape Ecology. Wiley: New York.
4. Turner MG (1990) Landscape changes in nine rural counties in Georgia. Photogram. Eng. Remote Sens, 3: 379–386.
5. Hulshoff RM (1995) Landscape indices describing a Dutch landscape. Landscape Ecol, 2: 101–111.
6. Lin YP, Teng TP, Chang TK (2002) Multivariate analysis of soil heavy metal pollution and landscape pattern in Changhua county in Taiwan. Landscape and Urban Planning, 2002: 19–35.
7. Urban DL, O'Neill RV, Shugart HH (1987) Landscape ecology. Bioscience, 37: 119–127.
8. Fu BJ, Chen LD (2000) Agricultural landscape spatial pattern analysis in the semiarid hill area of the loess plateau, China. J. Arid Environ, 3: 291–303.
9. Leduc A, Prairie Y, Bergeron Y (1994) Fractal dimension estimates of a fragmented landscape: source of variability. Landscape Ecol, 4: 279–286.
10. O'Neill RV, Krummel JR, Gardner RH, Sugihara G, Jackson B, et al. (1988) Indices of landscape pattern. Landscape Ecol, 2: 153–162.
11. Lewis-Beck MS (1994) Factor analysis and Related Techniques. SAGE Publications, Toppan Publishing, London.
12. Andrade JM, Kubista M, Carlosena A, Prada D (2007) 3-way characterization of soils by Procrustes rotation, matrix-augmented principle components analysis and parallel factor analysis. Analytica Chimica Acta, 2007: 20–29.
13. Liu PWG (2009) Simulation of the daily average PM10 concentrations at Ta-Liao with Box-Jenkins time series models and multivariate analysis. Atmospheric Environment, 2009: 2104–2113.
14. Briz-Kishore BH, Murali G (1992) Factor analysis for revealing hydrochemical characteristics of a watershed. Environ Geol, 19: 3–9.
15. Carlon C, Critto A, Marcomini A, Nathanail P (2001) Risk based characterization of contaminated industrial site using multivariate and geostatistical tools. Environ Pollut, 111: 417–427.
16. Roger LO, Rick WC, Jim CL (2012) Water quality sample collection, data treatment and results presentation for principle components analysis – literature review and Illinois River watershed case study. Water Research, 2012: 3110–3122.
17. Wei ZY, Wang DF, Zhou HP, Qi ZP (2011) Assessment of Soil Heavy Metal Pollution with Principle Component Analysis and Geoaccumulation Index. Procedia Environment Sciences, 2011: 1946–1952.
18. Shi CW, Zhao LZ, Guo XB, Gao S, Yang JP, et al. (1994) Background values of soil elements in Shanxi and there distribution feature. Jour Geol & Min Res North China, 2: 188–196.
19. Liu Y (2008) The research of land ecological risk assessment based on spatial information technology. Doctor's dissertation.
20. Song XD, Niu XY (2007) Geographic information system practice tutorial. Beijing: Science Press.
21. Turner MG (1989) Landscape ecology: the effect of pattern on process. Annual Review on Ecological Systems, 20: 171–197.
22. Zhang H, Lu YL, Dawson RW, Shi YJ, Wang TY (2005) Classification and ordination of DDT and HCH in soil samples from the Guanting Reservoir, China. Chemosphere, 2005: 762–769.
23. Zhang WT, Kuang CW (2011) SPSS statistical analysis foundation. Beijing: Higher Education Press. pp. 265–342.
24. Wang HW, Wu ZB, Meng J (2006) Linear and nonlinear approach of partial least square regression. Beijing: National Defence Industry Press.

An Integrated Study to Analyze Soil Microbial Community Structure and Metabolic Potential in Two Forest Types

Yuguang Zhang[1]*, Jing Cong[1,2], Hui Lu[1], Caiyun Yang[3], Yunfeng Yang[4], Jizhong Zhou[3,4,5], Diqiang Li[1]*

1 Institute of Forestry Ecology, Environment and Protection, and the Key Laboratory of Forest Ecology and Environment of State Forestry Administration, Chinese Academy of Forestry, Beijing, China, **2** School of Mineral Processing and Bioengineering, Central South University, Changsha, China, **3** Institute for Environmental Genomics and Department of Microbiology and Plant Biology, University of Oklahoma, Norman, Oklahoma, United States of America, **4** State Key Joint Laboratory of Environment Simulation and Pollution Control, School of Environment, Tsinghua University, Beijing, China, **5** Earth Sciences Division, Lawrence Berkeley National Laboratory, Berkeley, California, United States of America

Abstract

Soil microbial metabolic potential and ecosystem function have received little attention owing to difficulties in methodology. In this study, we selected natural mature forest and natural secondary forest and analyzed the soil microbial community and metabolic potential combing the high-throughput sequencing and GeoChip technologies. Phylogenetic analysis based on 16S rRNA sequencing showed that one known archaeal phylum and 15 known bacterial phyla as well as unclassified phylotypes were presented in these forest soils, and *Acidobacteria*, *Protecobacteria*, and *Actinobacteria* were three of most abundant phyla. The detected microbial functional gene groups were related to different biogeochemical processes, including carbon degradation, carbon fixation, methane metabolism, nitrogen cycling, phosphorus utilization, sulfur cycling, etc. The Shannon index for detected functional gene probes was significantly higher ($P<0.05$) at natural secondary forest site. The regression analysis showed that a strong positive ($P<0.05$) correlation was existed between the soil microbial functional gene diversity and phylogenetic diversity. Mantel test showed that soil oxidizable organic carbon, soil total nitrogen and cellulose, glucanase, and amylase activities were significantly linked ($P<0.05$) to the relative abundance of corresponded functional gene groups. Variance partitioning analysis showed that a total of 81.58% of the variation in community structure was explained by soil chemical factors, soil temperature, and plant diversity. Therefore, the positive link of soil microbial structure and composition to functional activity related to ecosystem functioning was existed, and the natural secondary forest soil may occur the high microbial metabolic potential. Although the results can't directly reflect the actual microbial populations and functional activities, this study provides insight into the potential activity of the microbial community and associated feedback responses of the terrestrial ecosystem to environmental changes.

Editor: A. Mark Ibekwe, U. S. Salinity Lab, United States of America

Funding: This research was supported by the public welfare project of the national scientific research institution (CAFRIFEEP201101), China, and National Biological Specimens and Resources Sharing Platform in Nature Reserve (2005DKA21404).However, the funders had no role in study design, data collection and analysis, decision to publish, or preparation of the manuscript.

Competing Interests: The authors have declared that no competing interests exist.

* E-mail: yuguang@sina.com.cn (YZ); lidq@caf.ac.cn (DL)

Introduction

Microorganisms are the most abundant organisms on earth and play key roles in natural ecosystem, including the biogeochemical cycling of carbon, nitrogen, sulfur, phosphorus, and metals, and biodegradation or stabilization of environmental contaminants [1–3]. Because of their important roles, changes in soil microbial community may directly affect soil ecosystem function, particularly carbon and nitrogen cycling [4]. A deeply analysis of microbial community structure and their roles in ecological processes would improve our understanding of the biogeochemical elemental cycles affected by microbial communities in natural or man-made environments [5]. However, our understanding of soil microbial communities in terms of structure, composition, and functional activity are still limited, especially for soil microbial metabolic activity and ecosystem function have received little attention to date.

Numerous studies have investigated the structure and diversity of microbial communities and their relationships with the surrounding environments [4–8]. Previous studies have shown inconsistent or contradictory results, which might be caused by 1) high microbial community complexity and plasticity in natural environments, 2) different methods used to assess microbial diversity, 3) frequency and magnitude of environmental disturbance, and 4) limited knowledge on the temporal and spatial scales of microbial community in ecosystems [4–7,9]. To date, most studies have only described microbial community complexity at the phylogenetic level using 16S rRNA or single functional gene analysis [8]. Further studies of the metabolic potential and physiological traits of microorganisms are necessary to detect specific metabolic processes and functions [4]. Functional gene microarrays, with quantitative and high resolution capabilities, may be a useful tool for this purpose [10,11].

With the increasing efforts made towards implementing sustainable forest management and conservation strategies since last century [12], natural secondary forest, which form naturally after disturbance of the primary forest by human activities or by extreme natural events, are evident in temperate regions around the world [13]. Shennongjia Mountain is located in the northwestern region of Hubei Province, Central China, and this region is considered to have one of the highest levels of biodiversity in China owing to its unique geographic location and complex terrain [14]. In this study, we selected natural mature forest and a natural secondary forest, which formed in the 1970s following the fall of the primary deciduous broadleaf forest on Shennongjia Mountain and analyzed the soil microbial community and metabolic potential using both GeoChip and high-throughput 16S rRNA sequencing approaches. The aims of this study were to determine (i) the composition and structure of soil microbial community in two deciduous broadleaf forest types, (ii) the soil microbial functional gene diversity and the metabolic potential involved in carbon and nitrogen cycling, and (iii) the major environmental factors governing the soil microbial community structure.

Materials and Methods

Site and sampling

The study site was located on Shennongjia Mountain, Hubei Province, China, where deciduous broadleaf forest is a typical and important forest type. Two forest types were selected in this study: natural mature forest (MAT), and natural secondary forest (SEC), which was formed in the 1970s following the felling of the primary deciduous broadleaf forest. The two study sites had similar geography in terms of slope (30°), aspect (NE30°), and elevation (1715–1813 m), and were within 10 km of each other. The soil type was mountain yellow brown soil [15]. The plant survey and soil collected were permitted by the administrative bureau of Shennongjia National Nature Reserve. The detail location for the study site is N 31°29′, E110°21′ for MAT and N 31°25′, E11°20′ for SEC, respectively. At each study site, four plots (20 m×20 m) were selected with about 20 meters between adjacent plots. Ten to fifteen soil cores, at a depth of 0–10 cm, were taken from each plot and combined to obtain about 400 g of soil. Samples were sieved with 2 mm mesh to remove roots and stones, then mixed thoroughly. Soil samples were preserved at −80°C until DNA extraction.

Plant diversity, soil geochemical properties and microbial biomass analyses

Plant diversity was surveyed at each plot, including the plant species, number, height and canopy of each tree or shrub, and diameter at breast (1.3 m) height of trees (DBH>5 cm) and shrubs (DBH>1 cm). Average soil temperature at each plot was measured by placing the Long-Stem Thermomter (SPECTRUM, US) at a depth of 10 cm in relatively open patches. Soil moisture, soil pH, total soil organic carbon and nitrogen, available phosphorous, available water-dissolved organic carbon, and oxidizable organic carbon were measured as previously described [16]. Microbial biomass carbon was determined by the chloroform fumigation-extraction method [17].

DNA extraction, purification, and quantification

Soil microbial genomic DNA was extracted directly from each soil sample (5 g) using a protocol that included liquid nitrogen grinding, freezing and thawing, and treatment with sodium dodecyl sulfate for cell lysis, as previously described [18]. The freshly extracted DNA was purified twice using 0.5% low melting point agarose gel. Final DNA concentrations were quantified by the PicoGreen method [19] using a FLUOstar Optima microplate reader (BMG Labtech, Jena, Germany).

DNA amplification, sequencing, and data analysis

The V4 hypervariable region of the 16S rRNA gene was amplified in each soil sample using PCR primers 515F (5′-GTGCCAGCMGCCGCGGTAA-3′) and 806R (5′-GGACTACH VGGGTWTCTAAT-3′), using a sample tagging approach [20,21]. The amplification mix contained 10 units of AccuPrime High Fidelity *Taq* polymerase (Invitrogen, Grand Island, NY), 1× AccuPrime PCR buffer II (Invitrogen), 200 µM dNTPs (Amersham, Piscataway, NJ), and 10 µM of each primer in a volume of 25 µl. Genomic DNA (10 ng) was added to the PCR mix. Each sample was amplified under the following conditions: 94°C for 1 min, 30 cycles of 94°C for 20 s, 53°C for 25 s, and 68°C for 45 s, then 10 min at 68°C. The purified PCR amplicons were analyzed using a Miseq Benchtop Sequencer for 2×150 bp paired-end sequencing (Illumina, San Diego, CA).

To minimize the effect of random sequencing errors, only the first 250 bp after the proximal PCR primer of each sequence was analyzed. Sequence that did not perfectly match the PCR primer, had non-assigned tags, or had reads <250 bp was removed. All sequences were aligned using the Ribosomal Database Project Infernal Aligner [22], and the complete linkage clustering method was used to define operational taxonomic units (OTUs) using 97% identity as a cutoff [21]. The number of detected OTUs and sequences at different levels of classification were counted. Details of amplicon preparations, sequencing, and data analysis (e.g. classification, OTU identification) were described in previous reports [20,21]. To standardize samples, a sub-sample of 20,000 sequences per soil sample was used to compare relative difference among samples. Singletons were removed for downstream analyses. All the 16S rRNA sequences were deposited in GenBank database and the accession number is SRP035449.

GeoChip hybridization and data analysis

GeoChip 4.0 was used for DNA hybridization. This microarray contains >83,000 oligonucleotide probes targeting >150,000 genes in 410 gene categories involved in biogeochemical cycling of carbon, nitrogen, phosphorus, and sulfur, and bioremediation of metal and organic contaminants [24]. To produce consistent hybridizations from all samples, the amplification was used to generate approximately 3.0 µg of DNA from 50 ng of purified DNA using a TempliPhi Kit (GE Healthcare, Piscataway, NJ) following the manufacturer's instructions. Amplified DNA was labeled with Cy5 fluorescent dye (GE Healthcare) using a random priming method [23]. All hybridizations were carried out at 45°C for 10 h with 50% formamide using a TECAN HS4800 (TECAN, US) microarray hybridization station, and arrays were scanned using a ScanArray 5000 analysis system (Perkin-Elmer, Wellesley, MA).

Signal intensities of each spot were measured using ImaGene 6.0 (Biodiscovery, El Segundo, CA), and only the spots automatically scored as positive in the raw data output were used for further data analysis [25]. Spots with a signal-to-noise ratio (SNR = (signal intensity - background intensity)/standard deviation of the background) greater than 2.0 were used for further analysis [23]. The GeoChip data were further analyzed using the following steps: (i) genes detected in only two out of four samples from the same forest type were removed; (ii) the signal intensity of each spot was normalized by dividing by the mean value of eight samples of total signal intensity; and (iii) the data were transformed to the natural logarithmic form.

Statistical analyses

Relative abundance of the 16S rRNA gene was based on the proportional frequencies of those DNA sequences that could be classified at different OTUs. Diversity of the microbial community based on sequencing analysis and diversity of functional gene from GeoChip analysis were calculated using Simpson's reciprocal index (1/D) and the Shannon index (H′) using online software (http://ieg.ou.edu/). Detrended correspondence analysis (DCA) was used to determine the differences in overall microbial community structure and functional gene diversity between the two forest types. To examine the differences in soil microbial community structure and functional diversity between the two study sites, we used the Bray-Curtis similarity index to calculate distance matrices from OTUs and GeoChip hybridization data with the multi-response permutation procedure (MRPP) [26] and Adonis [27]. Mantel tests, canonical correspondence analysis (CCA), and variation partitioning analysis were used to evaluate the linkages between microbial community structure, metabolic potential related to carbon and nitrogen cycles, and environmental factors. The variance inflation factor was used for step-wise removal of redundant variables in CCA modeling [23]. All the analyses were performed by functions in the Vegan package (v. 1.15-1) in R (v. 2.9.1) (http://www.r-project.org/).

Results

Plant diversity and soil geochemical properties

Plant diversity and soil properties in the MAT and SEC forests were analyzed (Table 1). The number of plant species (including trees and shrubs) was 63 and 91 in MAT and SEC, respectively. The dominant trees were *Fagus engleriana*, *Quercus aliena var. acuteserrata*, and *Carpinus viminea* at the MAT forest site, and *Toxicodendron verniciflua*, *Tilia paucicostata*, *Juglans cathayensis*, and *Q. aliena var. acuteserrata* at the SEC site, respectively. The Shannon index of plant was significantly lower for MAT (1.18) than for SEC (2.12).

Soil organic carbon, oxidizable organic carbon, and microbial biomass carbon were significantly higher ($P<0.01$) in the SEC samples than in the MAT samples. Soil geochemical properties, such as total nitrogen, available nitrogen, available phosphorus, pH, and soil moisture, were also significantly different ($P<0.01$) between the two forest types (Table 1).

Soil microbial community composition and structure

To determine the overall composition of the soil microbial community in the two forest types, soil microbial communities were analyzed by 16S rRNA high throughput sequencing. After preprocessing of all reads, 65,666 and 65,584 high-quality sequences were obtained from the MAT and SEC sites, respectively (Table S1). The number of sequences ranged from 15,521 to 16,919 in MAT samples and from 16,002 to 16,970 in SEC samples. 10,160 OTUs were detected using 97% identity as a cutoff. Phylogenetic analysis showed that one known archaeal phylum and 15 known bacterial phyla as well as unclassified phylotypes were presented in these forest types (Table 2), and that all of the phylotypes were detected in both MAT and SEC, except for BRC1, which was only found in SEC. At the phylum level, the majority of OTUs were derived from *Proteobacteria* (4,048, 39.8%), followed by *Acidobacteria* (1,787, 17.6%), *Actinobacteria* (1,019, 10.0%), and *Verrucomicrobia* (896, 8.82%) (Table 2).

The Shannon index of 16S rRNA sequences was 8.05 and 8.28 for the MAT and SEC sites, respectively (Table S1). The number of OTUs and the Simpson index were significantly ($P<0.05$) different between MAT and SEC samples (Table S1). DCA was performed using the relative abundance values of sequencing data, resulting in two distinct clusters (Figure 1A). The results of nonparametric multivariate statistical tests, Adonis and MRPP, showed significant differences ($R^2=0.688$, $P=0.001$, and $\delta=0.387$, $P=0.026$, respectively) based on the abundance of all OTUs detected in MAT and SEC samples (Table S2). Significant ($P<0.05$) differences were also observed at the phylum level for *Acidobacteria, Actinobacteria, Armatimonadetes, Bacteroidetes, Proteobacteria, Planctomycetes*, and BRC1, except for WS3 (Table 2). The differences in microbial diversity at the phylum level in the two forest types varied. Ten known microbial phyla and unclassified taxa were significantly ($P<0.05$) higher in SEC samples, including the dominant phyla *Proteobacteria* and *Actinobacteria*. However, *Acidobacteria*, *Chlamydiae*, and *Planctomycetes* were significantly ($P<0.05$) lower in SEC soils. These results indicated that the

Table 1. Plant diversity and soil geochemical properties in two forest types.

Forest type	MAT	SEC	Forest type	MAT	SEC
non index of tree and shrubs**	1.18±0.27	2.12±0.17	Soil organic carbon (g/kg)**	36.40±3.41	86.07±9.70
Species number of tree*	6.00±0.41	12.50±2.10	Dissolved organic carbon (g/kg)	0.17±0.01	0.26±0.08
Height of tree**	14.14±0.54	8.73±0.34	Liable organic carbon (g/kg)**	2.00±0.48	12.37±1.58
Diameter at breast height of tree**	19.85±1.02	12.06±0.75	Microbial biomass carbon (g/kg)**	0.81±0.05	1.31±0.07
Soil Moisture (%) **	50.12±1.23	27.32±2.45	Total nitrogen (g/kg)**	2.33±0.45	5.83±0.45
Temperature at 10 cm depth (°C)	15.92±0.15	15.32±0.09	Available nitrogen (g/kg)**	0.22±0.02	0.41±0.03
Soil pH**	4.81±0.05	5.91±0.28	Available phosphorus (g/kg) *	0.004±0.00	0.015±0.00

Data present the mean value and standard error.
*, $P<0.05$,
**, $P<0.01$.

Table 2. Numbers of soil microbial OTUs and composition in two forest types.

Domain and phylum	Total[a] (%)	Average[b]		P (unpaired t test)
		MAT	SEC	
Archaea				
Crenarchaeota	19 (0.19)	4.50±0.96	10.00±0.82	0.005
Bacteria				
Acidobacteria	1787 (17.59)	777.75±22.21	685.50±14.06	0.017
Actinobacteria	1019 (10.03)	362.25±19.69	443.75±17.09	0.021
Armatimonadetes	22 (0.22)	5.25±0.63	8.50±1.19	0.065
Bacteroidetes	621 (6.11)	91.50±5.87	379.50±10.81	0.000
BRC1	3 (0.03)	0.00	2.25±0.25	0.000
Chlamydiae	17 (0.17)	8.00±1.47	1.50±0.28	0.019
Chloroflexi	109 (1.07)	26.25±2.32	42.25±3.71	0.014
Cyanobacteria	3 (0.03)	1.00±0.00	2.50±0.29	0.014
Firmicutes	129 (1.27)	45.50±4.17	60.75±5.07	0.061
Gemmatimonadetes	97 (0.95)	23.00±1.08	53.50±5.74	0.011
Nitrospirae	7 (0.07)	1.00±0.41	4.50±0.29	0.001
Planctomycetes	657 (6.47)	288.50±9.91	173.75±9.29	0.000
Proteobacteria	4048 (39.84)	1409.00±31.39	1826.25±52.01	0.001
Verrucomicrobia	896 (8.82)	459.00±41.12	349.75±3.07	0.076
WS3	12 (0.12)	0.50±0.28	6.50±0.87	0.004
Unclassified	714 (7.03)	212.25±5.96	326.75±11.06	0.000
Total	**10160**	**3623.00±77.66**	**4469.75±53.03**	**0.000**

[a]Data represent total numbers of OTUs detected by sequencing across all 8 samples.
[b]Data represent the mean value and standard error of OTUs detected using 4 samples in different forest types.

overall soil microbial taxonomic composition and structure were significantly different between the SEC and MAT samples.

Soil microbial functional gene diversity

A total of 38,925 and 50,068 gene probes were detected using GeoChip 4.0 in the MAT and SEC samples, respectively (Table S3). The detected gene groups were related to different biogeochemical processes, such as carbon degradation, carbon fixation, methane metabolism, nitrogen cycling, phosphorus utilization, stress responses, sulfur cycling, contaminant remediation, metal remediation, and energy processing (Table S4). The average number of gene probes detected was 43,427.3 for SEC

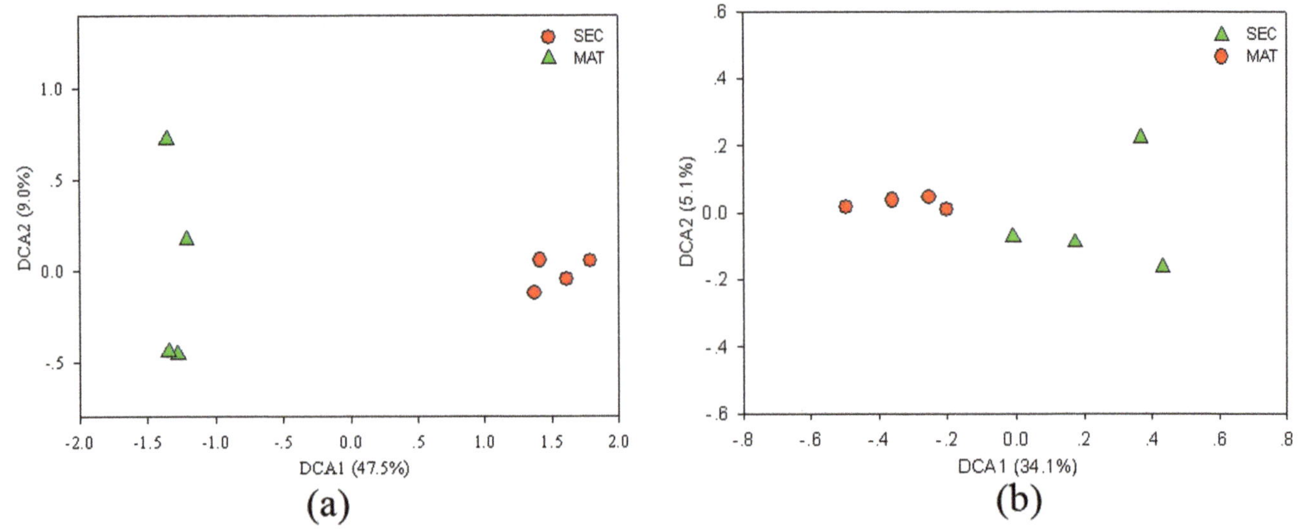

Figure 1. Detrended correspondence analysis (DCA) of soil microbial community structure. The DCA was analyzed based on the relative abundances of OTUs (A) and functional genes based on GeoChip 4.0 (B).

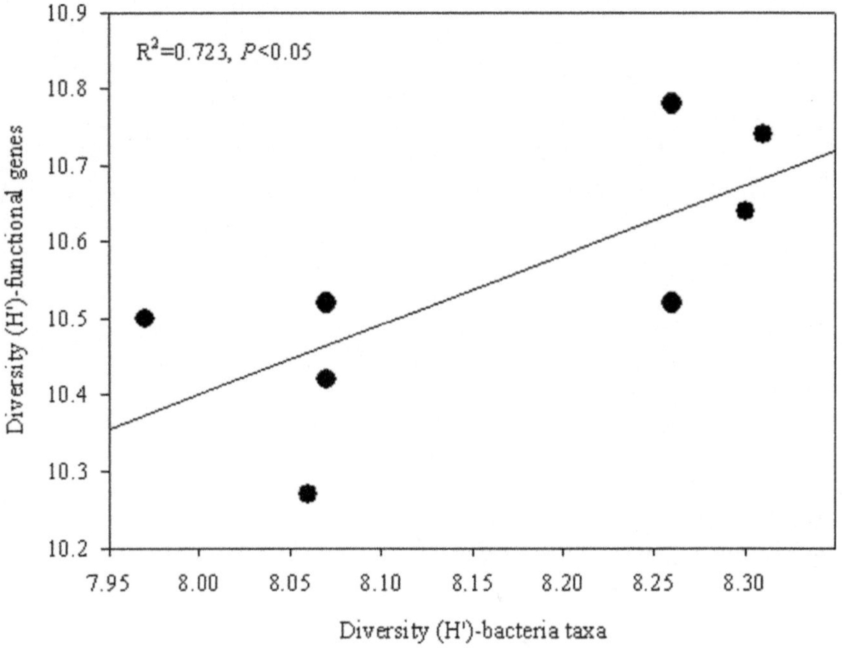

Figure 2. Relationship between soil microbial phylogenetic diversity index and functional gene diversity index.

samples and 33,982.8 for MAT samples. The Shannon index for detected functional microbial gene probes were significantly higher ($P<0.05$) at SEC sites (10.67) than at MAT sites (10.43) (Table S3). DCA showed that MAT samples were well separated from SEC samples (Figure 1B). The regression analysis showed that a strong positive ($P<0.05$) correlation was existed between the soil microbial functional gene diversity and phylogenetic diversity (Fig. 2).

Microbial functional genes and metabolic potential related to carbon cycling

To gain insight into the soil microbial metabolic potential related to carbon cycle, GeoChip 4.0 data for genes related to carbon cycling was analyzed further. Among the 5,792 gene probes related to carbon degradation, 4,276 (74%) were common between the two forest types, while 59 (1%) were presented only in MAT samples and 1,457 (25%) were presented only in SEC samples. The relative signal intensities of the different gene categories related to degradation of relatively active carbon (e.g. starch, hemicelluloses, pectin, and cellulose) and recalcitrant carbon (e.g. chitin and lignin) were significantly ($P<0.05$) higher in SEC than in MAT samples (Figure 1). This included genes encoding amylase and pullulanase (related to starch degradation), arabinofuranosidase and xylanase (related to hemicellulose degradation), cellobiase and exoglucanase (related to cellulose degradation), pectinase (related to pectin degradation), endochitinase and exochitinase (related to chitin degradation), and glyoxal oxidase and phenol oxidase (related to lignin degradation). These results indicated that the soil microbial metabolic capacity involved in carbon degradation were significantly different between the two different forest types.

Among the 1,085 genes involved in carbon fixation, the relative abundance of three important functional genes, ribulose-1,5-bisphosphate carboxylase/oxygenase (Rubisco), carbon monoxide dehydrogenase (CODH), and propionyl-coA (Pcc), was substantially different. A total of 306, 650 and 230 gene probes were detected for the Rubisco, Pcc, and CODH pathways, respectively. These genes had significantly higher signal intensity in SEC soils than in MAT soils (Figure S1). Geochip 4.0 also detected three enzymes involved in methane metabolism, namely the alpha-subunit of methyl coenzyme M reductase (mcrA), for methane production, and particulate methane monooxygenase (pmoA) and methane monooxygenase (mmoX) for methane consumption. Over 80% of the gene probes were associated with uncultured bacteria and archaea. The signal intensity of all the genes involved in methane metabolism was higher in SEC samples than in MAT samples (Figure 3). These results indicated that the microbial functional capacity related to carbon fixation and methane metabolism may be significantly affected by forest types. The relationships between soil microbial metabolic capacity related to carbon cycling and soil carbon components or enzyme activities were analyzed using the Mantel test (Table 3). The results showed that soil organic carbon and soil microbial biomass carbon were significantly linked ($P<0.05$) to the relative abundance of genes related to carbon cycling. Oxidizable organic carbon was significantly linked ($P<0.05$) to the relative abundance of genes involved in active organic carbon degradation (cellulase, hemicellulase, and starch). As expected, cellulase, glucanase, and amylase activities were significantly linked ($P<0.05$) to the relative abundance of genes encoding cellulase, endoglucanase and exoglucanase, and amylase, respectively.

Microbial functional genes and metabolic potential related to nitrogen cycling

Among the gene probes detected in the forest soil samples, 4,238 were involved in nitrogen cycling, including 767 related to nitrogen fixation, 513 related to ammonification (including 250 gene probes related to assimilatory N reduction), 1,734 related to denitrification, 351 related to dissimilatory N reduction, and 621 related to nitrification. These key functional gene categories were associated with different phylogenetic groups, for example, almost all gene probes related to the ammonification were derived from

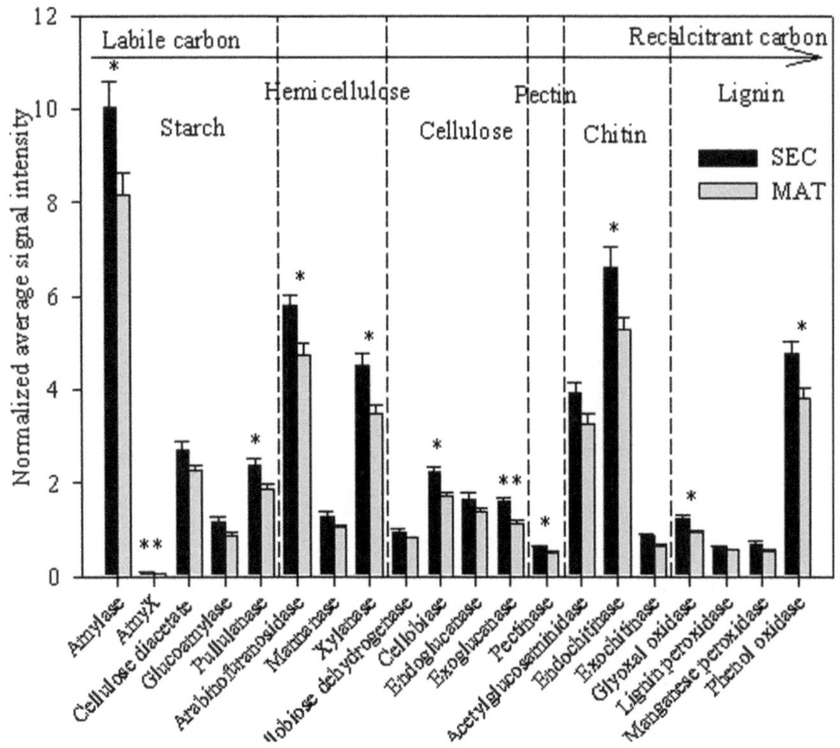

Figure 3. The normalized gene relative signal intensity of the detected key gene categories involved in carbon degradation. The signal intensities were the sum of detected individual gene for each functional gene, averaged among 4 soil samples. The complexity of carbon is presented in order from active to recalcitrant. All data are presented as mean ± standard error. Significant differences between two forest types are indicated above the bars. *, $P<0.05$, **, $P<0.01$.

cultured bacteria, including *Rhodococcus erythropolis*, *Pseudomonas entomophila*, and *Acinetobacter baumannii*. A total of 1,533 (88%) gene probes related to the denitrification were linked to unclassified bacteria, and 137 (22%) gene probes related to the nitrification were derived from archaea, which suggests that archaea maybe play an important role in nitrification process.

The abundance of several genes involved in nitrogen cycling were significantly ($P<0.05$) higher in SEC samples than in MAT samples (Figure 4), including *nifH* (related to nitrogen fixation), *amoA* (related to nitrification), *narG*, *nirS*, *nirK*, and *nosZ* (related to denitrification), *napA* and *nrfA* (related to dissimilatory N reduction to ammonium), *gdh* and *ureC* (related to ammonification), and *niR* and *nirB* (related to assimilatory N reduction). The relationships between microbial functional genes and soil nitrogen components were analyzed using the Mantel test (Table 4). The results showed that soil total nitrogen was significantly linked ($P<0.05$) to the relative signal intensity of nitrogen cycling genes. The NH_4^+-N and NO_3-N concentrations were linked ($P<0.1$) to the signal

Table 3. The relationships between relative abundances of microbial functional genes involved in carbon and nitrogen cycling and soil carbon or nitrogen components and enzyme activity by Mantel-test.

Soil carbon parameters	Microbial relative abundance	R value	P
Soil organic carbon	Carbon cycling genes	0.389	**0.039**
Oxidizable organic carbon	Oxidizable organic carbon degradation genes(Cellulose, Hemicellulose, Starch)	0.403	**0.046**
Water dissolved organic carbon	Carbon cycling genes	0.114	0.216
Soil microbial biomass carbon	Carbon cycling genes	0.500	**0.020**
Total soil nitrogen	Nitrogen cycling genes	0.382	**0.050**
NH_4^+-N	Ammonification genes	0.296	**0.089**
No_3^-N	Nitrification genes	0.302	**0.079**
Cellulase activity	Cellulase genes	0.342	**0.046**
Glucanase activity	Endo- and exoglucanase genes	0.351	**0.022**
Polyphenol oxidase activity	Polyphenol oxidase genes	0.046	0.626
Amylase activity	Amylase genes	0.358	**0.027**

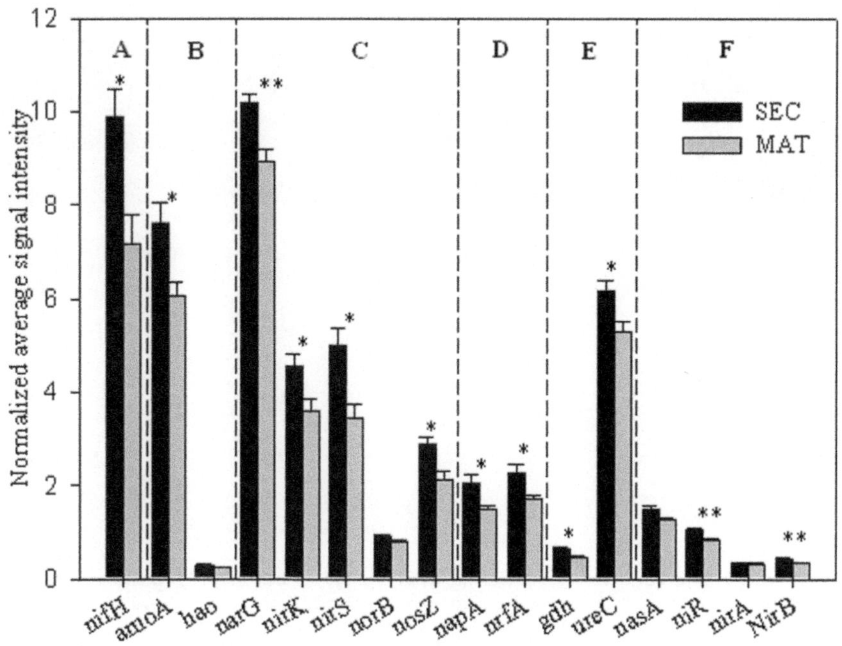

Figure 4. The normalized average signal intensity of detected key gene categories involved in the N cycling between SEC and MAT. The signal intensities were the sum of all detected individual gene for each gene category, and then averaged among 4 samples. (A). N_2 fixation, including nifH encoding nitrogenase; (B). Nitrification, including amoA encoding ammonia monooxygenase, hao for hydroxylamine oxidoreductase; (C). Denitrification, including narG for nitrate reductase, nirS for nitrite reductase, norB for nitric oxide reductase, and nosZ for nitrate reductase; (D). Dissimilatory N reduction to ammonium, including napA for nitrate reductase, and nrfA for c-type cytochrome nitrite reductase; (E). Ammonification, including gdh for glutamate dehydrogenase and ureC encoding urease; (F). Assimilatory N reduction, including nasA encoding nitrate reductase, niR, nirA and nirB encoding dissimilatory nitrite reductase. All data are presented as the mean±SE (error bars). **, $P<0.01$, *, $P<0.05$.

intensity of genes involved in the ammonification and nitrification processes, respectively.

Linking microbial community structure and environmental factors

Mantel tests and CCA were performed to analyze the major environmental factors responsible for shaping the microbial community structure and functional diversity. The Mantel test showed that microbial community structure and functional gene diversity were significantly ($P<0.05$) linked to soil and plant factors (Table 4). CCA was used to identify the major environmental variables controlling the soil microbial community structure, resulting in a model significant at a confidence level of $P=0.048$. The plant Shannon index and tree DBH appeared to be important environmental factors controlling the microbial community structure, as they had a long projection on Axis 1 ($P<0.01$), which represented the major variation among microbial communities (Figure 5A). Similarly, soil temperature and soil organic carbon were critical for shaping microbial communities.

Variance partitioning analysis was used to quantify the contributions of soil chemical factors (Soil), soil temperature (Temp), and plant diversity (Plant) to the variation in the microbial community. A total of 81.58% of the variation was significantly explained by these three environmental variables (Figure 5B). Soil, Temp, and Plant explained 26.02%, 8.58%, and 19.35% of variation, respectively, verifying that they were major factors in shaping the microbial community structure.

Table 4. Mantel test and partial Mantel test between 16S rRNA OTUs and Geochip data with environmental properties.

Environmental variable	16S OTUs		GeoChip 4.0	
	r value	P value	r value	P value
Whole variable	0.868	0.005	0.412	0.008
Soil organic carbon	0.770	0.003	0.392	0.013
Total nitrogen	0.832	0.003	0.402	0.019
pH	0.622	0.006	0.322	0.043
Temperature at the 10 cm depth	0.463	0.002	0.085	0.272
Shannon Index for trees and shrubs	0.779	0.006	0.337	0.028
Diameter at breast height of trees	0.532	0.008	0.125	0.158

The Mantel test was only performed for whole variable, which included all variables presented in this table.

Discussion

The development and increasing availably of high-throughput molecular techniques has helped improve our understanding of microbial composition, especially through the use of direct sequencing of metagenomic DNA. Fierer et al. indicated that all biomes were dominated by *Acidobacteria*, *Actinobacteria*, *Proteobacteria* and *Bacteroidetes*, and the bacterial community composition does not vary significantly across different biomes [1].Our results showed that microbial diversity at the phylum level differed between the two forest types, and *Acidobacteria*, *Protecobacteria*, and *Actinobacteria* were three of most abundant soil bacterial phyla. *Acidobacteria* is known to be dominant in other soil types, although they are often difficult to cultivate [8,28]. The relative abundance of *Acidobacteria* was also shown to be higher in forest, desert, and prairie soils compared with agricultural soils [28]. Recent studies

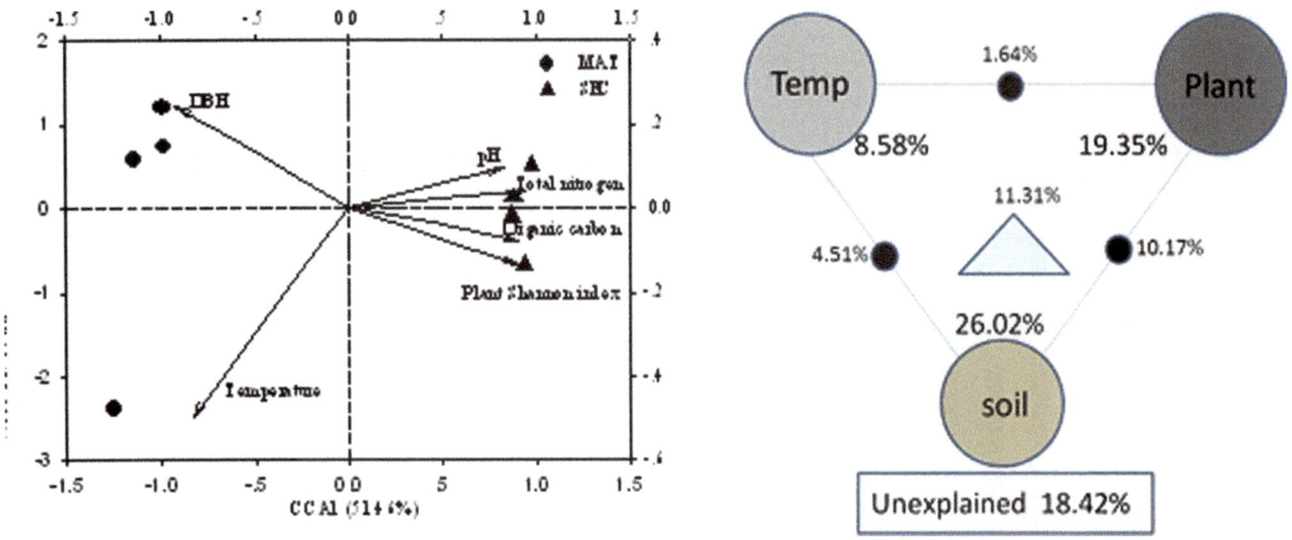

Figure 5. The Canonical correspondence analysis (CCA) and Variation partitioning analysis (VPA) between OTUs data and soil environmental variables. The left was CCA between microbial OTUs data and soil environmental factors, and the right was the VPA. DBH, diameter at breast height of tree. Soil, including soil organic carbon, total nitrogen and pH, soil temperature (Temp) and plant diversity (Plant, including the Shannon index of trees and shrubs, diameter at breast height of trees), and their relationships. Each diagram represents the biological variation partitioned into the relative effects of each factor or a combination of factors, in which geometric areas were proportional to the percentages of explained variation.

showed that *Acidobacteria* is in general oligotrophic, consistent with results showing that the proportion of *Acidobacteria* was significantly lower in nutrient-rich rhizosphere and agricultural soils than in bulk soil [4,29]. Some studies also found that the relative abundance of *Acidobacteria* was negatively linked to the level of nitrogen input [30]. In this study, the relative abundance of *Acidobacteria* was high in the natural forest, which had low soil organic carbon and nitrogen contents.

Due to microbial highly diversity and natural ecosystem complexity, it is a great challenging to set up the link of soil microbial structure and composition to functional activity related to ecosystem functioning in microbial ecology. Many studies have described microbial community structure in a range of different environments, but have not identified critical groups related to detailed functional processes [4–8,31]. Therefore, it remains unclear the relationship between microbial community structure and microbial functional gene diversity [32–34]. Functional redundancy is considerable debate in ecology as to how the relationship between the taxonomic diversity and observed functional diversity [35]. This study combines high-throughput sequencing to study the microbial community structure at the phylogenetic level and GeoChip to microbial potential metabolic activities at functional gene level. 16S rRNA gene-based high-throughput sequencing data provide information on microbial phylogenetic structure and composition, and GeoChip-based data provide quantitative information on various microbial functional groups [23]. These integrated technologies provide a reliable method for detection soil microbial structure, composition and potential functional activity [23]. In this study, a strong positive correlation were found between the soil microbial functional and phylogenetic diversity among sampling sites, which suggesting microbial phylogenetic and functional diversity aligned well and a low degree of functional redundancy were existed. Our result was consistent with a recent work on reconstructing the microbial diversity and function of pre-agricultural tallgrass prairie soils in the United States [36]. Furthermore, the higher phylogenetic diversity detected in SEC was consistent with the higher number of functional genes involved in the carbon and nitrogen cycles, unveiling the larger functional potential and capacity of the secondary forest soil.

Soil microbial metabolic potential and ecosystem function have received little attention owing to difficulties in methodology. As we all known, most of the biogeochemical processes were controlled and related to diverse and various functional genes, therefore, the conjunction between microbial community identified by using normal molecular methods (i.e. DGGE, T-RFLP) with soil enzyme activities or components is difficult. GeoChip-based data provides large-scale quantitative information on various biogeochemical functional groups, thus making it possible to link the functional structure of microbial communities with ecosystem processes [23]. Previous studies showed that change observed in the microbial community structure and functional groups has significant effects on soil enzyme activities or components related to carbon and nitrogen cycles. Yergeau et al. [37] and Reeve et al. [34] showed a significant correlation between cellulase enzyme activity and cellulose gene variants, detected using GeoChip, as well as significant correlations between dehydrogenase genes and enzyme activity, urease genes and soil ammonium, and sulfite reduction genes and soil sulfur level [34]. In this study, we also found the linked of soil microbial functional gene diversity to soil enzyme activities and components, such as the soil organic carbon and soil microbial biomass carbon were significantly linked ($P<0.05$) to the relative abundance of genes related to carbon cycling, oxidizable organic carbon was significantly linked ($P<0.05$) to the relative abundance of genes involved in active organic carbon degradation (cellulase, hemicellulase, and starch), respectively. Although the GeoChip data can't directly reflect the actual microbial populations and functional activities, it implied that the microbial populations carrying those genes could exist and have the functional capacity. Our GeoChip data showed that metabolic genes involved in the degradation of starch, cellulose, hemicellulose, chitin, lignin, and pectin, as well as those involved in nitrogen cycling, were significantly higher ($P<0.05$) in SEC than in MAT soils. Therefore, SEC soil had high microbial gene diversity and microbial metabolic potential. The use of GeoChip 4.0 may allow us to rapidly obtain community-wide information and observe differences functional capacity in different environments.

Only about 1% of the microorganisms in soil can be cultured, and researchers have increased our understanding of microbes in the natural environments by use of PCR amplification methods which reduce our dependence on the presence of culturable microbes. In our study, the majority of the genes related to carbon and nitrogen cycles were derived from the uncultured bacteria, suggesting that they are unique and may represent novel genes in the study area. For example, 88% gene probes related to the denitrification process were linked to unclassified bacteria. In previous studies, researchers received the same results, for example, Xie et al found 64.28% sequences were from uncultivated bacteria in the acid mine drainage [38], Zhang et al found the 320/372 genes were derived from the unidentified or uncultured organisms in the alpine meadows soils in the Qinghai-Tibetan plateau [39]. Therefore, a lot of the microbial functional gene resources were existed in natural environments.

Much recent work in soil microbial ecology has showed that soil microbial community structures are affected by specific environmental changes or disturbances [11,25]. Soil properties and plant diversity are key factors in shaping the microbial community composition and structure [40,41]. Organic carbon availability limited microbial communities in most soils [37], and addition of labile organic material rapidly alters microbial communities by selecting for populations that are most competitive in terms of growth rates and ability to absorb nutrients [41]. Thus, it was not considered to be the limiting factor for the microbial communities. Our data showed that plant diversity also significantly influenced the microbial community structure at the phylum level. Plant communities can influence the associated soil microbial communities through the types and amounts of organic carbon and nutrient inputs, and by altering the temperature and water content of the soil [42,43]. Therefore, changes in plant diversity, composition, and production affect the composition and diversity of soil microbial communities [9]. However, our understanding of differences in soil microbial diversity depending on various vegetation types is still poor.

In summary, this work studied the microbial community structure and metabolic potential by combining high-throughput sequencing and GeoChip technologies, and showed the soil microbial community structure and functional metabolic potential had a positive correlation. Although the microbial populations and metabolic activities could not be characterized comprehensively, the detected functional gene groups may represent the *in situ* microbial metabolic potential at the functional cluster level to some degree. Further analysis of the functional activity with different approaches such as mRNA-based analysis is needed to understand the biogeochemical processes related to microbial community.

Supporting Information

Figure S1 The normalized gene relative signal intensity of the detected key gene categories involved in carbon fixation and methane cycling. The signal intensities were the sum of detected individual gene for each functional gene, averaged among 4 soil samples. CODH: carbon monoxide dehydrogenase; Pcc: propionyl-coA; Rubisco: ribulose-1,5 -bisphosphate carboxylase/oxygenase; pmoA: methane monooxygenase; mcrA: methyl coenzyme M reductase. All data are presented as mean ± standard error. Significant differences between two forest types are indicated above the bars. *, $P<0.05$, **, $P<0.01$.

Table S1 16S rRNA diversity index. Summary of sequence numbers and OTUs, Shannon index and Simpson index for each sample at MAT and SEC based on sequencing data.

Table S2 The difference of soil microbial community composition and structure. Statistical analysis of differences in the microbial community composition and structure between SEC and MAT at the phylum level.

Table S3 The functional gene diversity index. Summary of numbers of detected gene probes, Shannon index and Simpson index for each sample at MAT and NAF based on GeoChip 4.0 data.

Table S4 The detected functional genes families. Summary of numbers of detected gene probes for different functional gene families at SEC and MAT based on GeoChip 4.0 data.

Acknowledgments

We gratefully acknowledge Dr. Ye Deng, Dr. Junjun Ding, Mr. Jingyuan Yang, Mr. Xiujiang Su, Dr. Guangliang Li and Dr. Yadong Xue for helping data analysis and field survey the plant plots.

Author Contributions

Conceived and designed the experiments: YZ JZ DL. Performed the experiments: YZ JC HL CY. Analyzed the data: YZ JC YY. Contributed reagents/materials/analysis tools: YZ DL. Wrote the paper: YZ.

References

1. Fierer N, Jackson RB (2006) The diversity and biogeography of soil bacterial communities. Proc Natl Acad Sci USA 103: 626–631.
2. Green JL, Bohannan BJM, Whitaker RJ (2008) Microbial biogeography: from taxonomy to traits, Science 320: 1039–1043.
3. He Z, Nostrnd JDV, Deng Y, Zhou J (2011) Development and applications of functional gene microarrays in the analysis of the functional diversity, composition, and structure of microbial communities. Front. Environ. Sci Engin China 5: 1–20.
4. López-Lozano NE, Heidelberg KB, Nelson WC, Garcia-Oliva F, Eguiarte LE, et al (2013) Microbial secondary succession in soil microcosms of a desert oasis in the Cuatro Cienegas Basin, Mexico. PeerJ 1: e47. Available: http://dx.doi.org/10.7717/peerj.47. Accessed 2013 March 5.
5. Fierer N, Nemergut D, Knight R, Craine JM (2010) Changes through time: integrating microorganisms into the study of succession. Res Microbiol 161: 635–642.
6. Kuramae EE, Gamper HA, Yergeau E, Piceno YM, Brodie EL, et al (2010) Microbial secondary succession in a chronosequence of chalk grasslands. ISME J 4: 711–715.
7. Banning NC, Gleeson DB, Grigg AH, Grant CD, Andersen GL, et al (2011) Soil microbial community successional patterns during forest ecosystem restoration. Appl Environ Microbiol 77: 6158–6164.
8. Ding G, Piceno YM, Heuer H, Weinert N (2013) Changes of soil bacterial diversity as a consequence of agricultural land use in a semi-arid ecosystem. PLOS one 8: e59497.
9. Zhu W, Cai X, Liu X, Wang J, Cheng S, et al (2010) Soil microbial population dynamics along a chronosequence of moist evergreen broad-leaved forest succession in southwestern China. J M Sci 7: 327–338.
10. He Z, Gentry TJ, Schadt CW, Wu L, Liebich J, et al (2007) Geochip: a comprehensive microarray for investigating biogeochemical, ecological and environmental processes. ISME J 1: 67–77.

11. Zhou J, Kang S, Schadt CD, Garten CTJ (2008) Spatial scaling of functional gene diversity across various microbial taxa. Proc Natl Acad Sci USA 105: 7768–7773.
12. Aerts R, Honnay O (2011) Forest restoration, biodiversity and ecosystem functioning. BMC Ecol 11: 29.
13. Wang CK, Yang JY (2007) Rhizospheric and heterotrophic components of soil respiration in six Chinese temperature forests. Global Change Biol 13: 123–131.
14. Ma C, Zhu C, Zheng C, Wu C, Guan Y, et al (2008) High-resolution geochemistry records of climate changes since late-glacial from Dajiuhu peat in Shennongjia Mountains, Central China. Chinese Sci Bull 53 (Supp.1): 28–41.
15. Zhao C, Chen W, Tian Z, Xie Z (2005) Altitudinal pattern of plant species diversity in Shennongjia Mountains, Central China. J Integr Plant Biol 47(12): 1431–1449.
16. Bao SD (1999) Soil and agricultural chemistry analysis. Beijing: China Agriculture Press, 25–150 p.
17. Chen CR, Xu ZH, Mathers NJ (2004) Soil carbon pools in adjacent natural and plantation forests of subtropical Australia. Soil Sci Soc Am J 68: 282–291.
18. Zhou JZ, Bruns MA, Tiedje JM (1996) DNA recovery from soils of diverse composition. Appl Environ Microbiol 62: 316–322.
19. Ahn S, Costa J, Emanuel J (1996) PicoGreen quantitation of DNA: effective evaluation of samples pre- or post-PCR. Nucleic Acids Res 24: 2623–2625.
20. Caporaso JG, Lauber CL, Walters WA, Berg-Lyons D, Lozupone CA, et al (2011) Global patterns of 16S rRNA diversity at a depth of millions of sequences per sample. Proc Natl Acad Sci USA 108: 4516–4522.
21. Caporaso JG, Lauber CL, Walters WA, Berg-Lyons D, Huntley J, et al (2012) Ultra-high-throughput microbial community analysis on the Illumina HiSeq and MiSeq platforms. ISME J 6: 1621–1624.
22. Cole JR, Wang Q, Cardenas E, Fish J, Chai B, et al (2009) The ribosomal database project: improved alignments and new tools for rRNA analysis. Nucleic Acids Res 37: 141–145.
23. He Z, Xu M, Deng Y, Kang S, Kellogg L, et al (2010) Metagenomic analysis reveals a marked divergence in the structure of belowground microbial communities at elevated CO_2. Ecol Lett 13: 564–575.
24. Lu Z, Deng Y, Van Nostrand JD, He Z, Voordeckers J, et al (2011) Microbial gene functions enriched in the deepwater horizon deep-sea oil plume. ISME J 5: 1–10.
25. Liang Y, Nostrand JD, Deng Y, He Z, Wu L, et al (2011) Functional gene diversity of soil microbial communities from five oil-contaminated fields in China. ISME J 5: 403–413.
26. McCune B, Grace JB (2002) Analysis of ecological communities. MJM Software Design, Gleneden Beach, OR.
27. Anderson MJ (2001) A new method for non-parametric multivariate analysis of variance. Aust J Ecol 26: 32–46.
28. Janssen PH (2006) Identifying the dominant soil bacterial taxa in libraries of 16S rRNA and 16S rRNA genes. Appl Environ Microbiol 72: 1719–1728.
29. Kielak A, Pijl AS, van Veen JA, Kowalchuk GA (2009) Phylogenetic diversity of Acidobacteria in a former agricultural soil. ISME J 3: 378–382.
30. Fierer N, Lauber CL, Ramirez KS, Zaneveld J, Bradford MA, et al (2012) Comparative metagenomic, phylogenetic and physiological analyses of soil microbial communities across nitrogen gradients. ISME J 6: 1007–1017.
31. Kirk JL, Beaudette LA, Hart M, Moutoglis P, Klironomos JN, et al (2004) Methods of studying soil microbial diversity. J Microbiol Meth 58: 169–188.
32. Kelly JJ (2003) Molecular techniques for the analysis of soil microbial processes: functional gene analysis and the utility of DNA microarrays. Soil Sci 168: 597–605.
33. Little AEF, Robinson CJ, Peterson SB, Raffa KE, Handelsman J (2008) Rules of engagement: inter-species interaction that regulate microbial communities. Annu Rev Microbiol 62: 375–401.
34. Reeve JR, Schadt CW, Carpenter-Boggs L, Kang S, Zhou J, et al (2010) Effects of soil type and farm management on soil ecological functional genes and microbial activities. ISME J 4: 1–9.
35. Hillebrand H, Matthiessen B (2009) Biodiversity in a complex world: consolidation and progress in functional biodiversity research. Ecol Lett 12: 1405–1419.
36. Fierer N, Ladau J, Clemente JC, Leff JW, Owens SM, et al (2013) Reconstructing the microbial diversity and function of pre-agricultural tallgrass prairie soils in the United States. Science 342: 621–624.
37. Yergeau E, Kang S, He Z, Zhou J, Kowalchuk GA (2007) Functional microarray analysis of nitrogen and carbon cycling genes across an Antarctic latitudinal transect. ISME J 1: 1–17.
38. Xie JP, He ZL, Liu XX, Liu XD, Van Nostrand JD, et al (2011) GeoChip-Based Analysis of the Functional Gene Diversity and Metabolic Potential of Microbial Communities in Acid Mine Drainage. Appl Environ Microbiol 77: 991–999.
39. Zhang Y, Lu Z, Liu S, Yang Y, He Z, et al (2013) Geochip-based analysis of microbial communities in alpine meadow soils in the Qinghai-Tibetan plateau. BMC Microbiol 13(1): 72.
40. Carney K, Maston P (2005) Plant communities, soil microorganism, and soil carbon cycling: does altering the world belowground matter to ecosystem functioning? Ecosystem 8: 928–940.
41. Drenovsky RE, Vo D, Graham KJ, Scow KM (2004) Soil water content and organic carbon availability are major determinants of soil microbial community composition. Microbial Ecol 48: 424–430.
42. Myers RT, Zak DR, White DC, Peacock A (2001) Landscape-level patterns of microbial community composition and structure in upland forest ecosystems. Soil Sci Soc Am J 65: 359–367.
43. Waldrop MP, Firestone MK (2004) Microbial community utilization of recalcitrant and simple carbon compounds: impact of oak-woodland plant communities. Oecologia 138: 275–284.

5

A Framework for Identifying Plant Species to Be Used as 'Ecological Engineers' for Fixing Soil on Unstable Slopes

Murielle Ghestem[1*¤a], Kunfang Cao[2¤b], Wenzhang Ma[3], Nick Rowe[4], Raphaëlle Leclerc[5], Clément Gadenne[6], Alexia Stokes[7]

1 AgroParis Tech, UMR AMAP, Montpellier, France, **2** Xishuangbanna Tropical Botanical Garden, Chinese Academy of Sciences, Menglun, Mengla, Yunnan, China, **3** Kunming Institute of Botany, Kunming, Yunnan, China, **4** CNRS, UMR AMAP, Montpellier, France, **5** CNRS, UMR CEFE, Montpellier, France, **6** AgroParisTech, Centre de Nancy, Nancy, France, **7** INRA, UMR AMAP, Montpellier, France

Abstract

Major reforestation programs have been initiated on hillsides prone to erosion and landslides in China, but no framework exists to guide managers in the choice of plant species. We developed such a framework based on the suitability of given plant traits for fixing soil on steep slopes in western Yunnan, China. We examined the utility of 55 native and exotic species with regard to the services they provided. We then chose nine species differing in life form. Plant root system architecture, root mechanical and physiological traits were then measured at two adjacent field sites. One site was highly unstable, with severe soil slippage and erosion. The second site had been replanted 8 years previously and appeared to be physically stable. How root traits differed between sites, season, depth in soil and distance from the plant stem were determined. Root system morphology was analysed by considering architectural traits (root angle, depth, diameter and volume) both up- and downslope. Significant differences between all factors were found, depending on species. We estimated the most useful architectural and mechanical traits for physically fixing soil in place. We then combined these results with those concerning root physiological traits, which were used as a proxy for root metabolic activity. Scores were assigned to each species based on traits. No one species possessed a suite of highly desirable traits, therefore mixtures of species should be used on vulnerable slopes. We also propose a conceptual model describing how to position plants on an unstable site, based on root system traits.

Editor: Vanesa Magar, Centro de Investigacion Cientifica y Educacion Superior de Ensenada, Mexico

Funding: Funding was provided by a PhD bursary (Murielle Ghestem) from the French Government and AgroParisTech, the CNRS funded project 'EcoPente,' an INRA 'Jeune Equipe' and the BMU International Climate Initiative funded project 'Ecosystems Protecting Infrastructure and Communities' (EPIC), coordinated by IUCN and ProAct Network. The funders had no role in study design, data collection and analysis, decision to publish, or preparation of the manuscript.

Competing Interests: The authors have declared that no competing interests exist.

* Email: yellownosed@gmail.com

¤a Current address: CETE SO, DALETT, Toulouse, France
¤b Current address: Guangxi University, State Key Laboratory of Conservation and Utilization of Subtropical Agro-bioresources, College of Forestry, Nanning, Guangxi, China

Introduction

Eco-engineering has been defined as the long-term, ecological strategy to manage a site with regard to natural or man-made hazards [1]. Vegetation has long been recognized as useful for increasing slope stability with regard to shallow landslides and erosion [2]. Plant root systems have predominant effects on slope stability, compared to aboveground attributes [3]. To improve slope stability on a large scale, managers could focus initially on relatively small areas, or 'hotspots' [4]. With regard to soil erosion, these areas are defined as sites with soil erosion rates well above soil loss tolerance levels [5]. Hotspots often only occupy a small fraction of a catchment's area, but may be held responsible for a very significant contribution to overall sediment production, thus leading to off-site problems [6]. Reducing erosion or soil slippage on these degradation hotspots via an appropriate species or mixture of species would be an economic and efficient method to protect against large-scale landslides. Therefore, the appropriate root characteristics for fixing soil in hotspots should be identified. The optimum spatial positions of species within a hotspot could also be determined, depending on local soil and climatic conditions.

If plants could be used as 'ecological engineers,' i.e. can be used to create a sustainable ecosystem that adds value to both the environment and society [7], the most preferred species for slope stabilization would be native with local economic value. Individuals should possess root traits known to improve slope stability, where a trait is a well-defined, measurable property of an organism, usually measured at the individual level and used comparatively across species [8]. Lists of desirable root system traits for fixing soil on slopes are available [9,10,11] (Table 1). Individuals should possess extensive and deep root systems with strong and fast-growing roots that are slow to decompose [10]. However, the most commonly used traits for quantifying the contribution of plant root systems to slope stability are root area ratio (RAR), which is the surface area of roots over a given area of soil, and root maximal tensile strength. These traits are used to calculate the root mechanical contribution as additional soil cohesion [12,13,14,15]. Nevertheless, it is recognised that these simple traits cannot describe the full root reinforcement mechanism on slopes, and that we should consider more complex suites of traits [11]. With regard to root mechanical traits, it has recently been shown that the mechanical behaviour of a root during a deformation, and not only its maximal tensile strength at breakage, should be used to calculate more realistic values of additional cohesion [16,17,18]. To improve slope stability, roots with a high tensile strength at breakage and a small tensile strain at breakage would be most efficient, because soil can bear only a very small displacement before it ruptures [11,19,20]. Strain in a root held in tension is composed of three phases: (i) an initial phase of stretching followed by (ii) a phase of reversible deformation called the elastic phase, which is sometimes, but not always, followed by (iii) a phase of non-linear and irreversible deformation called the plastic phase leading to breakage. To improve slope stability, the irreversible phase of deformation should be as small as possible because roots which can be re-mobilized will be more efficient in the long-term. Ghestem [21] showed that the maximal tensile strength required to cause root failure is correlated with the elastic characteristics (elastic strength, elastic strain and tensile modulus of elasticity) and thus represents the capacity of a root to deform reversibly. The ultimate tensile strain is correlated with the plastic characteristic of roots (plastic strain and plastic modulus) and thus represents the irreversible deformation of a root. Very few studies have examined such detailed root mechanical traits in the context of slope stability analyses [22,23]. In most previous studies, where the contribution of vegetation to slope stability was determined, only the tensile strength of thin and fine roots was determined, and found to be correlated with cellulose and lignin content [24,25]. Yet, thicker roots act as soil nails, pinning the root systems into the substrate. Therefore, [26] stipulated the importance of also measuring the bending resistance of thick roots, so that they could be incorporated into slope stability models [27].

Root architectural traits allow for the description of root system morphology and topology, each of which influence slope stability [11,28]. The individual soil volume (ISV) is the root system's overall envelope, given by its maximum radius (horizontal extension) and its maximum depth (vertical extension) [29,30,31], and thus quantifies root spread of an individual on a slope.

Root physiological traits provide useful information about root system functioning and uptake capacity. Root metabolic activity (respiration rate and nutrient uptake) is correlated with root nitrogen (N) concentration; root longevity is correlated with root cellulose content [24,32]. The ultimate tensile strain, correlated with the irreversible deformation of a root as described hereabove, also worth considering as root physiological trait: a short ultimate tensile strain means that the root is able to recover promptly after deformation without structural damage.

To determine the plant species useful for engineering slope stability, we analysed traits of plants growing along a degraded slope in southern China. The number of shallow landslides in China has increased enormously over the last 50 years, due mainly to deforestation, infrastructure and road construction [1,33,34]. The Chinese government has therefore launched a major landslide inventory named "monitoring and preventing of landslides by the masses" in 1990 [33] and two afforestation programs: the Natural Forest Protection Program (NFPP) in 1998 and the Sloping Land Conversion Programme in 1999 (SLCP, or "Grain for Green programme," which aims at planting trees on existing agricultural land, concentrating on zones where slopes are >25°) [1,35]. The results of these two programs are contrasted [35,36,37]. Case studies recorded high seedling mortality where tree species used for replanting were not suitable for the local environment [38,39], degradation of local population welfare [40], more superficial erosion [41] and increased slope instability [42]. Within such a major socio-economic context, information on how species can be used to engineer slope stability is vital, especially during the early years after plantation on a bare slope, when the window of landslide susceptibility is greatest [43].

We discuss therefore, how plants can be screened for use as 'ecological engineers', through a better understanding of mechanical, architectural and physiological traits of their root systems. Results are discussed with regard to the optimisation of species mixtures and planting patterns on degraded slopes. As developed by [44] and [45] studying soil and gully erosion in Ethiopia, we aimed at producing a generic framework which can be used by managers to determine suites of plant traits useful for engineering slope stability with regard to shallow landslides.

Material and Methods

Ethics statement

Access to the field site was obtained from the Daxingdi forest service and Town Hall. Field sites were publicly owned and not protected. No protected species were sampled during this study.

Table 1. Desirable traits to compare species efficiency for stabilizing slopes and abbreviations used in the text.

Plant properties	Desirable traits	Abbreviations
Root abundance in soil	High number of plant stems per m^2 of degraded slope	Nb stems.m^{-2}
	High individual soil volume standardized by the collar diameter of the plant	ISV/Dc
	High root area ratio	RAR
Root mechanical resistance	High maximal tensile strength x high proportion of fine roots	T_{max} x RAR_f
	Short ultimate tensile strain	ε_{ult}
	High bending rigidity x high proportion of coarse roots	EI x RAR_c
Root metabolic activity	Short ultimate tensile strain, proxy for high reversibility in tension	ε_{ult}
	High root nitrogen content, proxy for high metabolic activity	N
	High cellulose content, proxy for long lifespan	CELL

Ultimate tensile strain is a trait representing both a mechanical property and a proxy for a physiological property: the ability of a root to recover after a deformation (reversible deformation).

Study site

We studied root morphological, mechanical and physiological traits for species growing in the Yunnan province, southern China, where erosion and landslides are severe [46]. The study area (26°01′N, 98°50′E) was located near Daxingdi village, north of Liuku town in the Salween river valley. This part of China is under the influence of the Indian monsoon, and described as a "warm-dry climate", being a combination of subtropical and alpine climates. Annual mean temperature (from 1961 to 2002) is 15.2°C, and mean annual precipitation is 1200 mm, the majority of which falls between May and October [14,41]. Numerous landslides occur during the monsoon season (May-October) and soil erosion is severe, largely due to the cutting of roads through the steep slopes [1,34]. The use of *Agava americana* L. to fix soil on steep slopes after road building is common practice in the region. We carried out fieldwork in 2009 and 2010. In 2010, precipitation during the summer months was particularly high (Figure 1) and shallow landslides throughout the area were numerous. At our study sites, corn (*Zea mays* L.) was cultivated from 1980, after deforestation, until 1999 when the SLCP was initiated. Several species of trees and shrubs were then planted, including *Agava americana* L., *Jatropha curcas* L., *Pueraria stricta* Kurz., *Ricinus communis* L., and *Vernicia fordii* Helmsl.

We identified two study sites. One site was an active shallow landslide approximately 30 m wide and 50 m long. The origin of the landslide may have been due to severe erosion leading to gully formation and eventual soil slippage. As soil slippage at this site was active, we considered it as a degradation hotspot. The second site was located at 3 m from the first site. A shallow landslide had occurred at the second site in 2000, after an extreme precipitation event. The type of soil slippage observed was typical of western Yunnan [34]. The specific vegetation composition and distribution was similar to other sites along the Salween river valley (data not shown). Since 2000, species have colonized this second site naturally or have been planted within the SLCP, therefore we presumed that it was relatively stable. The two sites were located at an altitude of 1010 m. Slope angle was 35–45° at the degradation hotspot and 50–60° at the stable site. Both sites were oriented at 300° from due north.

Soil characteristics

Soil profiles. We determined soil profiles at the hotspot, stable site and also at a third site taken as reference. This reference site was situated 200 m from our sites at same altitude, covered with the same type of vegetation and was used for comparison of soil profiles (Figure 2). This reference site was considered highly stable because no evidence of previous landslides or erosion was found. Soil profiles to a depth of 1.0 m were examined at each site and described using colour charts [47]. Potential shear surfaces were identified as the limit between soil and bedrock horizons as is generally observed in the field and especially where percolating water stagnates [48,49]. The soil was represented by a ferrallitic red carbonated soil with many mineral coloured spots, e.g. iron and manganese. In the third reference site with no previous evidence of a landslide, humus thickness was <1 cm, soil thickness was 0.7–2.0 m and the source rock was limestone. Humus was classified as a mesomull [50]. Source rock (limestone) emergence occurred at 0.5 m at the hotspot and at 0.4 m on the relatively stable site where a landslide had occurred 8 years previously (Figure 2).

Soil texture and chemical characteristics. Soil texture and chemical characteristics were measured at the hotspot and stable site. Samples were taken randomly throughout each plot from representative typical A- and at B-horizons i.e. n = 8 samples at 0.05 m and n = 8 samples at 0.35 m. Soil analyses were carried out on soil fractions finer than 2 mm. Sand (2.00–0.05 mm), silt (0.050–0.002 mm), and clay (<0.002 mm) contents were determined using the sedimentation and sieving method (Table 2), NF P 94–056 and 94–057 [51]. Soil pH was measured using a potentiometry method (LY/T1239–1999) [52], cation exchange

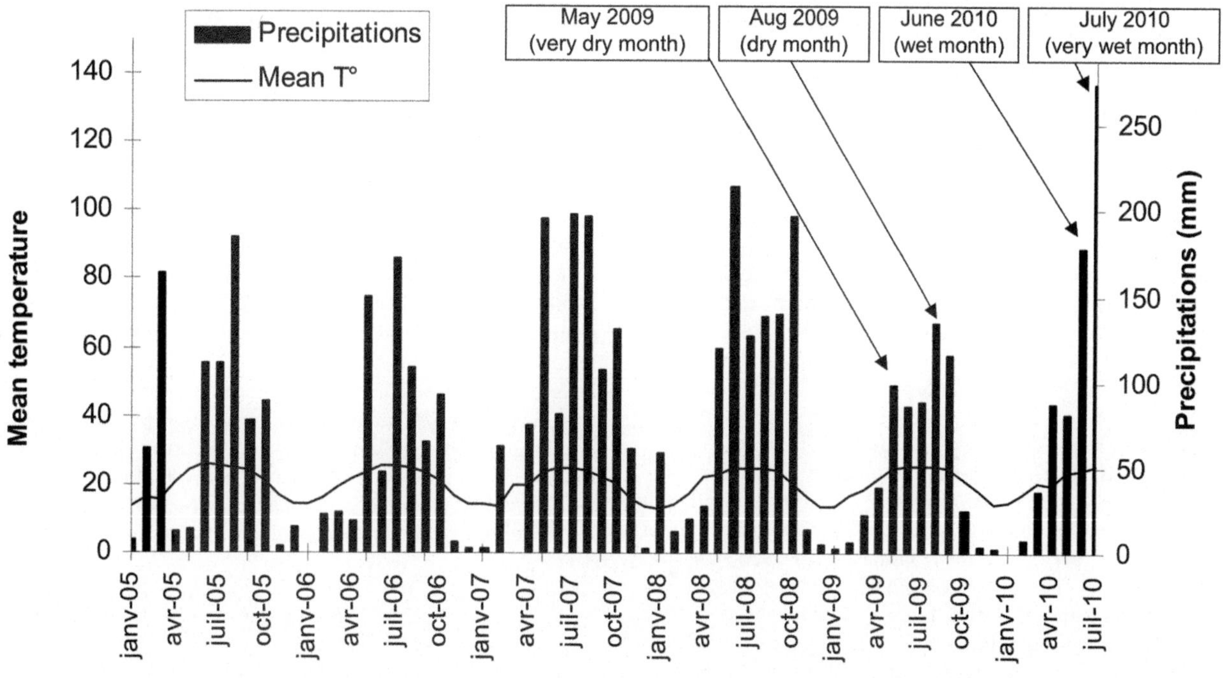

Figure 1. Mean temperature and precipitations in Liuku. Diagram from Liuku meteorological station situated 30 km south from the field site (source: Meteorological Bureau of Yunnan Province). Data from January 2005 to May 2010. Arrows show the months when roots were collected.

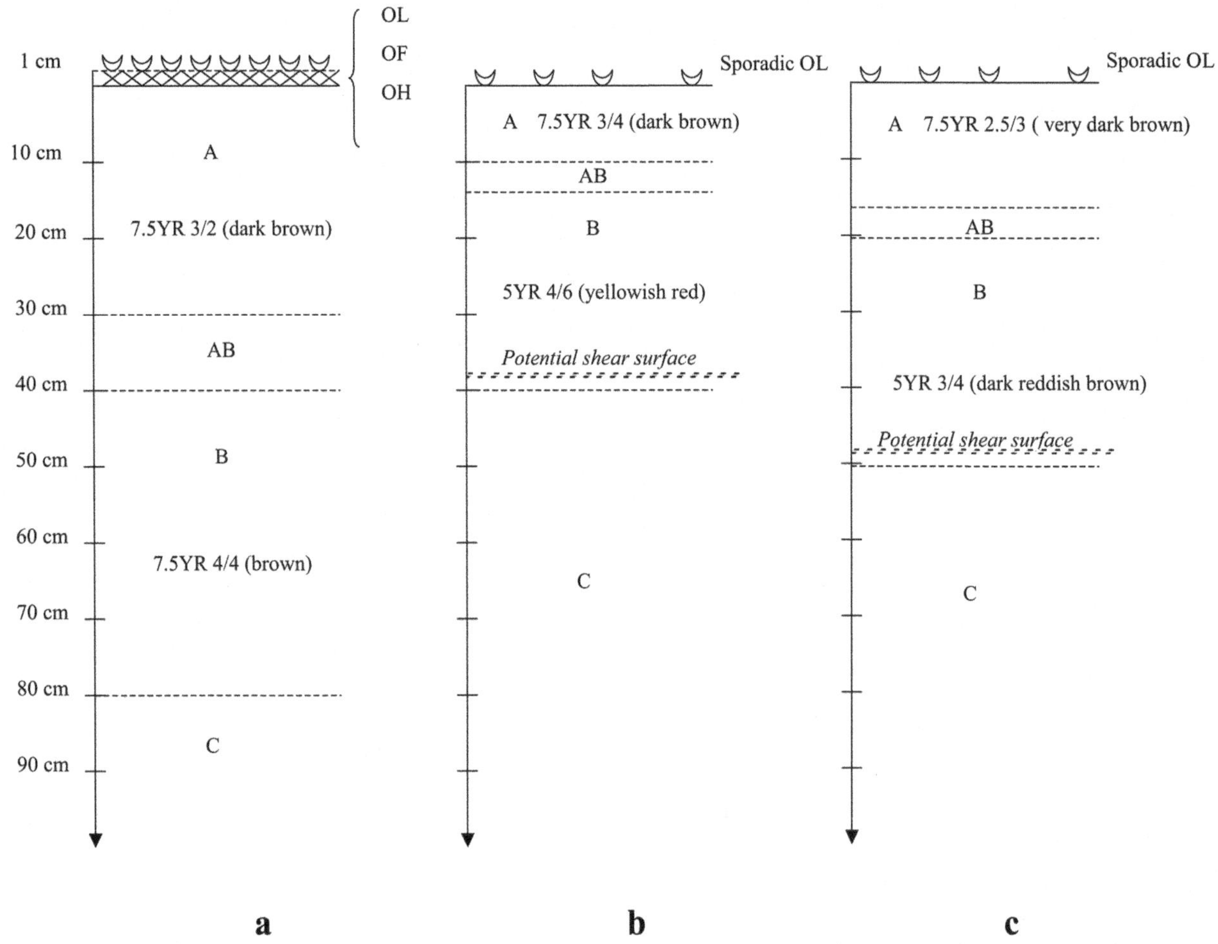

Figure 2. Soil profiles at the study site. Soil horizons at a) a reference site with no previous evidence of landslides or erosion; b) the stable site where a shallow landslide had occurred eight years previously and c) the hotspot. Colours were identified using a Munsell colour chart (Munsell 1947). OL: fresh litter, OF: fermenting litter, OH: litter with humic substances and well-transformed organic matter, A: organico-mineral layer, AB: mixture between A and B, B: layer of bedrock alteration, pieces of bedrock are visible, C: bedrock, mineral layer (Legros 2000; Baize and Girard 1995).

capacity (CEC) was measured using the distillation method (LY/T1243–1999) and soil organic carbon and total nitrogen (N) contents were measured using a C-N analyzer (Thermo-Finnigan CHN analyser). The two sites possessed similar basic pH and CEC values (Table 2). CEC was high compared to typical values of soils containing 50% clay [53], indicating a high potential fertility. Soil organic carbon and total N concentrations (Table 2) were similar to those found in forest or pasture soils [53]. Soil organic carbon, representing soil organic matter (Soil organic matter = Soil organic carbon*1.7) [53], was more abundant at the hotspot compared to the stable site. The proportion of organic carbon on total N was higher at the hotspot compared to the stable site, indicating that mineralisation of organic matter was slower at the hotspot.

Soil physical characteristics. A manual shear tore vane (IMG I01 025) was used to estimate soil shear strength in situ at a depth of 0.20 m at 20 randomly located positions within each plot (Table 2). A manual penetrometer (Elmeg PEN-3960) was also used to estimate penetration at a depth of 0.20 m at 20 locations within each plot. Strain-controlled direct shear tests were carried out on eight reconstituted, drained 60 mm×60 mm×20 mm soil samples from each site and each horizon. Roots were removed during the reconstitution process [41]. Samples were not saturated prior to testing, and as they were kept sealed at 4°C after removal from the field, it can be assumed that soil moisture content was similar to that in field conditions. Samples were placed in a shear testing device (VJTech 2760A) and normal loads of 200, 300 and 500 N were applied as weights on three separate samples taken from the same block of soil [54]. A lateral displacement was applied at a rate of 0.8 mm min^{-1} until failure occurred and the peak shear force recorded. Soil cohesion and the angle of internal friction were obtained by the Mohr-Coulomb theory [54]. To obtain soil initial moisture content (w_i), soil moisture content at saturation (w_s) and soil dry bulk density (ρ_d), a modified method from [53] was used on seven samples at the two sites (in 2009) and at A and B horizons (in 2010). First, samples were weighed while they were fresh (m_i, initial mass). Samples were then dipped in paraffin and the volume of water occupied by the sample measured in a graduated cylinder (v, soil volume). Samples were then saturated with water and weighed (m_s, saturated mass), before drying at 105°C until constant weight (m_d, dry mass).

$$w_i(\%) = \frac{(m_i - m_d) \times 100}{m_d} \quad \text{(Eq.1)}$$

$$w_s(\%) = \frac{(m_s - m_d) \times 100}{m_d} \quad \text{(Eq.2)}$$

Table 2. Soil textural, chemical and physical characteristics in horizons A and B at the hotspot and stable site.

Soil property	Soil horizon	Stable site	Unstable hotspot
		n = 8	n = 8
Clay (%)	A	41.70 ± 6.73	49.21 ± 2.66
Silt (%)		44.04 ± 4.46	41.12 ± 4.25
Sand (%)		14.27 ± 1.69	9.68 ± 3.61
Clay (%)	B	47.62 ± 8.52	46.71 ± 11.49
Silt (%)		42.20 ± 6.59	39.03 ± 8.37
Sand (%)		10.18 ± 2.41	14.26 ± 7.34
		n = 8	n = 8
pH		8.42	8.33
CEC cmol(+).kg^{-1}		51.51	43.81
Organic carbon (g.kg^{-1})	A	12.31	25.8 ± 1.9
	B		28.7 ± 1.4
Total nitrogen (g.kg^{-1})	A	0.91	1.5 ± 0.1
	B		0.7 ± 0.8
		n = 7	n = 7
Soil water content (%)in August 2009		27 ± 8.23	28 ± 6.62
Soil water content (%) in July 2010	A	18.28 ± 3.73	23.29 ± 8.28
	B	17.48 ± 3.26	15.3 ± 3.50
Soil water content (%)		60.0 ± 7.44	70.0 ± 3.00
Dry bulk density (g.cm^{-3})	A	0.88 ± 0.23	0.78 ± 0.03
	B	1.01 ± 0.18	0.99 ± 0.16
		n = 20	n = 20
Resistance with manual shear tore vane (kPa)		0.27 ± 0.02	0.17 ± 0.02***
Resistance with manual penetrometer (kPa)		3.25 ± 0.20	1.48 ± 0.13***
		n = 8	n = 8
Soil cohesion (kPa)		5.35 ± 1.97	0.49 ± 0.92
Angle of internal friction (°)		20. 21 ± 3.02	26.98 ± 1.41

The number (n) of samples is indicated for each test. When n<10, no statistical analyses were carried out. Data are means ± standard error.
***: Differences were significant at >0.001 level.

$$\rho_d = \frac{m_d}{v} \quad \text{(Eq.3)}$$

Choice of species

A species inventory was carried out in 2008 and 60 species identified (Species List S1). Before we selected species for the study, we obtained information on species' ecology and economic value from the literature, field observations and discussions with local people. We then screened for two necessary criteria: the species must be present on disturbed sites in the region and must be compatible with other species, i.e. not be invasive. Amongst species answering to those two criteria, ecological characteristics were considered e.g. plant lifespan (annual, biannual, perennial), rooting type (through visual observations) and growth form (herb, shrub, climber, creeping plant, tree). Any economic and ethnobotanical properties were noted e.g. food and fodder use, medicinal purposes, fertilizer, fuel or handcrafts. Species were then classified according to these properties (Species List S1). From this initial screening, we selected nine species for this study: *Agava americana*, *Arthraxon hispidus* (Thunb. ex Murray) Makino, *Artemisia codonocephala* DC., *Bauhinia championii* Benth., *Chloris anomala* B. S. Sun & Z. H. Hu, *Ficus tikoua* Bureau, *Jatropha curcas*, *Pueraria stricta*, *Rhus chinensis* Miller. All species were pioneers. They were present at both sites at the beginning of the rainy season, when slopes are more prone to shallow landslides and water erosion. Species were not invasive and were of economic value for the local population. Five different growth forms were also represented i.e. herb, shrub, creeper, climber, and tree forms.

Plant abundance

In August 2009 and July 2010, the density of each of the nine species was measured at both sites in two transects along the slope. Eight 1 m^2 quadrats were placed every 4 m along each transect. Within each quadrat, we counted the number of individuals of each of the nine species and the percentage of soil covered by the vertical projection of their canopy at a height of 30 cm above the soil surface.

Choice of individuals

For each of the nine species, 12 individuals were chosen i.e. six at each site. All individuals were chosen within the same size range. Plant height and stem basal diameter were measured

(Table 3), but are not indicative of age because of occasional cattle grazing. By comparing these individuals with reference plants that germinated during the 3 years we worked at the site, we estimated that the individuals we studied were 3–6 years old. Root systems were excavated by hand (Figure 3). Excavations were carried out with extreme caution, so as to not damage root systems (Table 3).

During excavation, the orientation of roots in a circular sector was considered qualitatively: all roots in the upper half of the sector, with regard to slope direction, were noted as upslope roots, and those in the lower sector noted as downslope roots. Roots were also classed into depth classes of 10 cm perpendicular to the soil surface. During excavation, roots were covered with wet towels to prevent desiccation. Root systems were then transported to the laboratory and stored at 4°C.

Architectural traits

Soil volume was calculated as the corresponding portion of individual soil volume (ISV) in each 10 cm layer of soil and in each upslope or downslope sector. Spatial x, y, z coordinates of each structural root allowed us to determine the maximum radius and depth of each root system. Using these coordinates for each plant, ISV was calculated as a quarter of ellipse upslope plus a quarter of ellipse downslope, with the size of the ellipse determined by the maximum radius and depth. However, in *A. americana* and *F. tikoua*, root systems were organised linearly along the main plant stem (Figure 3). Therefore, the ISV of these species was calculated as a quarter of a cylinder upslope plus a quarter of a cylinder downslope. The maximum radius and ISV volume were then deduced for each soil layer from geometrical equations. Standardizing ISV by the collar diameter allows plant individuals of different sizes to be compared [55].

Root area ratio (RAR) is the cumulated cross sectional area (CSA) of all roots crossing the potential shear surface per unit of soil surface. RAR can also be calculated as the cumulated volume of roots per unit of soil volume, if we consider that all roots within a layer cross the surface of this layer perpendicularly [56].

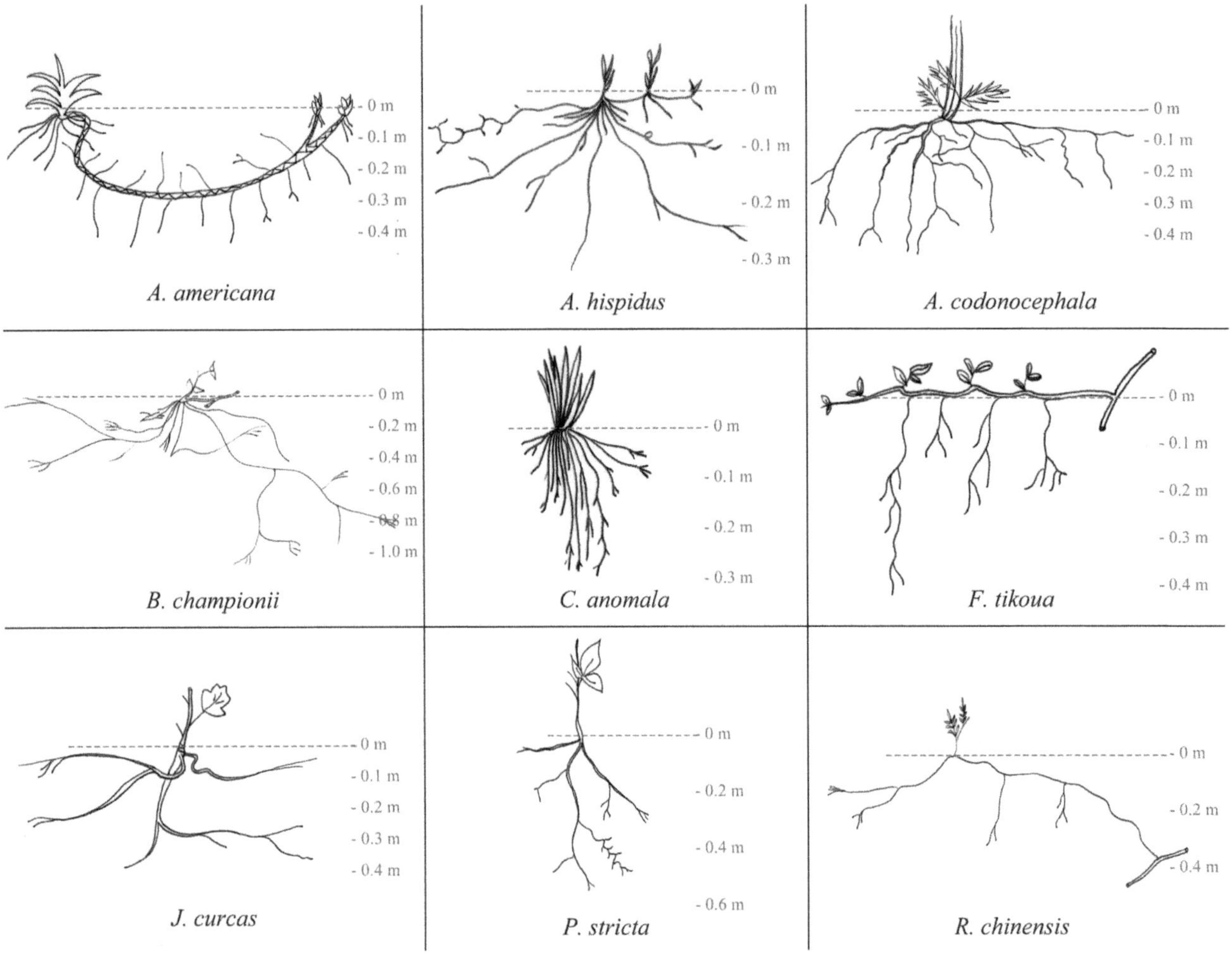

Figure 3. Description of the root systems of the nine studied species. Root systems of the nine species and their rooting depth (m). *A. americana*'s root system was composed of an underground stem from which emerged thin roots; *A. hispidus*'s root system comprised only a few thin roots emerging from the plant collar; *A. codonocephala* possessed a root system with long lateral roots turning downwards over time; *B. championii* had long and deep roots, which deviated on encountering an obstacle and which were densely branched; *C. anomala* possessed a tufted and shallow root system; vertical roots of *F. tikoua* emerged from creeping stems; *J. curcas* and *P. stricta* both possessed taproot systems, but roots of *P. stricta* were deeper and more densely branched; *R. chinensis* had a sprouting root system comprising long, deep and scarcely branched roots. Note that the scale for soil depth (y axis) differs between species for easier viewing.

Table 3. Shoot characteristics and time of harvest for individuals studied (means ± se).

Species	Family	n	Height of plant crown (cm)	Width of plant crown (cm)	Width of collar diameter Dc (cm)	Mode of reproduction	Months when roots were collected (total precipitations for the month)
Agava americana	Asparagaceae	14	16.81 ± 20.45	22.91 ± 39.77	0.43 ± 0.56	vegetative	July 2010 (275 mm)
Artemisia codonocephala	Asteraceae	12	12.74 ± 4.37	22.27 ± 8.45	0.75 ± 0.35	sexual	May 2009 (97 mm) and July 2010 (275 mm)
Arthraxon hispidus	Poaceae	12	17.25 ± 11.29	15.00 ± 6.15	0.43 ± 0.16	sexual, vegetative	June (181 mm) and July 2010 (275 mm)
Bauhinia championii	Leguminosae	12	20.82 ± 12.38	27.77 ± 28.20	0.61 ± 0.30	sexual, vegetative	June (181 mm) and July 2010 (275 mm)
Chloris anomala	Poaceae	12	11.78 ± 6.92	17.18 ± 6.18	5.41 ± 6.70 (tufts)	sexual, vegetative	May 2009 (97 mm) and June 2010 (181 mm)
Ficus tikoua	Moraceae	11	7.60 ± 3.03	36.38 ± 31.27	0.71 ± 0.33	sexual	August 2009 (137 mm) and July 2010 (275 mm)
Jatropha curcas	Euphorbiaceae	5	22.80 ± 9.39	15.30 ± 10.77	0.17 ± 0.06	sexual	July 2010 (275 mm)
Pueraria stricta	Leguminosae	12	33.75 ± 26.06	22.98 ± 11.32	0.41 ± 0.17	sexual	May 2009 (97 mm) and June 2010 (181 mm)
Rhus chinensis	Anacardiaceae	13	16.54 ± 8.73	22.54 ± 11.46	0.50 ± 0.23	sexual, vegetative	August 2009 (137 mm), June (181 mm) and July 2010 (275 mm)

Table 4. Different statistical tests were used depending on the variables tested.

Statistical test	Presentation of results and parameters	Analyzed data
Linear correlation between two continuous variables	r = correlation coefficient, $R^2 = r^2$ = coeff of determination	Log(T_{max}) and ε_{ult} depending on diameter
Anova: parametric test of analysis of variance	$F_{factor\ df,\ errors\ df}$ value; P value	N and CELL depending on species and diameter classes
Ancova: parametric test of analysis of covariance with root diameter as covariable	$Fcov_{factor\ df,\ errors\ df}$ value; P value	T_{max}, ε_{ult} and *EI* depending on species; T_{max} and ε_{ult} depending on depth, up/downslope, sites and season
Kruskal-Wallis non parametric test for independant variables	$H_{n = nb\ cases,\ N = nb\ of\ observations}$ value; P value	Number of stems m^{-2}, ISV/Dc and RAR depending on species; *EI*, N and CELL depending on depth
Friedman Anova non parametric test for dependant variables (>2)	$X^2_{N = nb\ of\ observations,\ factor\ df}$ value; P value	RAR depending on depth
Mann-Whitney U non parametric test for two independant variables	$Z_{nb\ of\ valid\ observations\ in\ one\ case,\ nb\ of\ valid\ observations\ in\ the\ other\ case}$ value; P value	*In situ* soil resistance, *in situ* soil penetration, Number of stems m^{-2}, ISV/Dc, RAR, EI, N and CELL depending on sites; Number of stems m^{-2}, ISV/Dc and RAR depending on seasons

Abbreviations and parameters used in the text are also indicated.

$$RAR = \frac{\sum_{i=0}^{j} CSA}{A} = \frac{\left(\sum_{i=0}^{j} CSA\right) \times h}{A \times h} = \frac{V_r}{V} \qquad \text{(Eq.4)}$$

where $\sum_{i=0}^{j} CSA$: sum of CSA in diameter classes [0; j] of all roots crossing a soil surface A, h: height of the considered layer, V_r: cumulated volume of roots within the corresponding volume of soil V.

RAR was calculated for roots separated into two diameter classes: fine [0; 2 mm], and coarse [2 mm; + ∞]. Within 24 hours after harvesting, roots were washed, paper-dried and scanned at a resolution of 700 dpi using an EPSON V700 Pro scanner. Root length, diameter and volume were measured using the image analysis software WinRHIZO (Pro version 3.0, Regent Instruments, Canada) [57], within each 10 cm depth class and within up- and down-slope sectors.

Mechanical traits

After scanning, and within 24 hours after harvesting, tensile testing of a sub-sample of individual roots was performed for each species. Testing was successfully carried out on 1116 roots, using a portable testing machine (In-Spec 2200 BT, Instron Corporation) equipped with three force transducers (maximum capacities of 250, 50 and 10 N and accuracy of 0.25%) chosen according to the size of the root. Span tests were carried out for each species to check whether the length of each sample had to be at least 30 times its central diameter [58]. Root diameter was measured at three points along each root using a binocular microscope and the mean diameter calculated. Crosshead speed was kept constant at 1.0 mm min^{-1} and both force and speed were measured constantly during each test (Instron Series IX software). Tests to failure were considered successful only when specimens failed in the middle third of the root and if the root did not slip inside the clamps, because we were measuring tensile stresses as well as tensile strains. In order to avoid slippage of roots out of the clamps, the clamps were chosen according to the diameter of the root and two pieces of sandpaper were placed on either side of the root within the jaws of the clamp. Each stress-strain curve was analysed to obtain the mechanical properties of the root [12,21], in particular its maximal tensile stress (T_{max}) and the ultimate strain at failure (ε_{ult}). T_{max} was calculated as the force required to cause breakage, divided by the root CSA at the point of breakage. ε_{ult} was calculated as the displacement between the clamps at breakage divided by the initial distance between the clamps. For measuring bending properties of thick roots, three-point bending tests were performed. Span tests were carried out for each species to check the necessary span to depth ratio in order to avoid including significant amounts of shear [59]. The depth diameter (d) and the width diameter (w) were measured at three points along the root using a binocular microscope. The axial second moment of inertia (I) was calculated using:

$$I = \left(\frac{d}{2}\right)^3 \times \left(\frac{w}{2}\right) \times \left(\frac{\Pi}{4}\right) \qquad \text{(Eq.5)}$$

The stiffness or bending modulus of elasticity E was calculated from the stress-strain curve of 137 bending tests as the linear slope shortly after the beginning of the test [60]. The bending modulus is not equal to the above-cited tensile modulus of elasticity [61]. The bending rigidity (EI) of each sample was calculated as the product of $E \times I$.

Physiological traits

After mechanical testing, roots were dried at 40 °C until constant weight. Nitrogen concentration on seven (*A. hispidus*) up to 77 (*B. championii*) roots per species was determined using an elemental analyser (CHN model EA 1108; Carlo Erba Instruments, Milan, Italy). The concentration of cellulose was measured on six (*A. hispidus*) up to 50 (*B. championii*) roots per species, using the Van Soest method [62], and with a Fibersac 24 fiber analyser (Ankom, Macedon, NJ, USA). The number of analysed roots per species depended on the quantity of available material for each species: the amount of collected roots was very light for some species (e.g. *A. hispidus*) whereas they were much heavier for other species (e.g. *B. championii*).

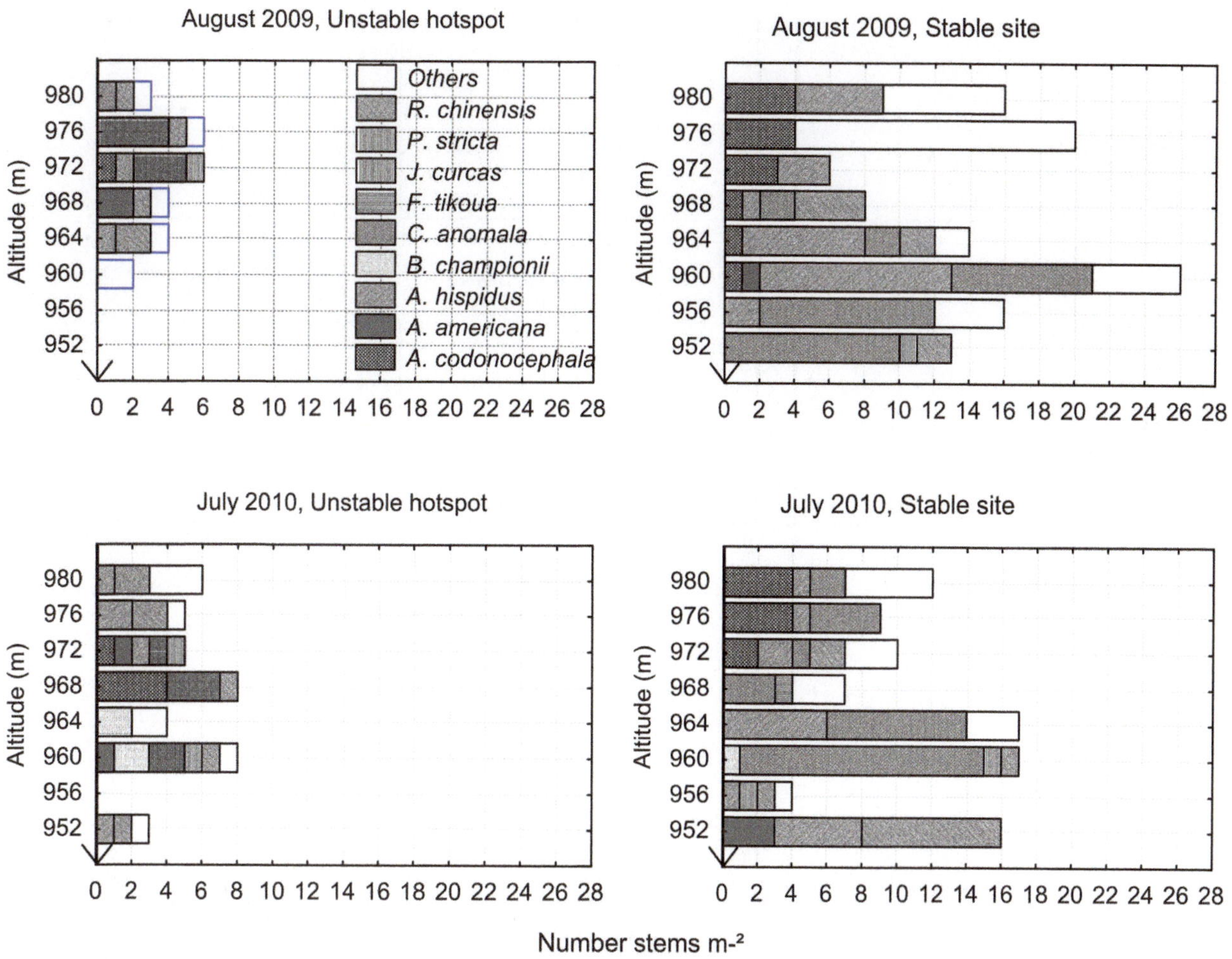

Figure 4. Plant abundance. Number of stems per square meter of each species present on the unstable hotspot and the stable site in 2009 and 2010. "Others" were composed of the following species: *Bidens pilosa* L., *Celosia argentea* L., *Elsholtzia winitiana* Craib, *Indigofera* sp., *Malvastrum coromandelianum* (L.) Garcke, *Convolvulus arvensis* L., *Solanum verbascifolium* L.

Statistical analysis

Different statistical tests were used depending on the number and type of factors (Table 4). In the majority of analyses, root diameter was considered a continuous variable. In some analyses, root diameters were separated into two classes: fine [0; 2 mm], and coarse [2 mm; + ∞] and are indicated as "diameter class".

Each time parametric tests are used, the following assumptions were checked: residuals are independent, they have homogeneous variance (homoscedasticity) and they are identically distributed following the normal law N $(0, \sigma^2)$. If these assumptions were not met, parametric tests were still performed on raw data if the amount of data was >30 in each treatment within each data set (Central Limit Theorem) [63]. Data were not log-transformed because of the implied data distortion [63]. Post-hoc tests used Tukey Honestly Significant Difference (HSD) tests to discriminate among treatments. When parametrical tests could not be used, non-parametrical tests were performed (Table 4).

Scores for species

We aimed at designating scores for each species with regard to their utility as 'ecological engineers.' The traits of each species were assigned a score depending on their suitability for fixing soil on slopes: for each trait, a score equal to 1 (poor performance), 2 (average), 3 (good performance) was attributed to each species. A global score was then attributed for each of the three studied properties: (i) ability to occupy soil with roots; (ii) mechanical resistance; and (iii) root physiological properties (Table 1). To obtain a more accurate estimation for root resistance in tension, the score obtained for the maximal tensile strength was multiplied with the score obtained for the proportion of fine roots (RAR_f). Similarly, to determine a more accurate estimation for root resistance in bending, the score obtained for *EI* was multiplied with the score obtained for the proportion of coarse roots (RAR_c).

The rules used for synthesizing scores were:

Poor performance (score 1) combined with poor performance (1) = poor global performance (1);

Good performance (3) combined with good performance (3) = good global performance (3);

Poor performance (1) combined with good performance (3) = average global performance (2);

Poor performance (1) combined with average performance (2) = poor global performance (1);

Good performance (3) combined with average performance (2) = good global performance (3).

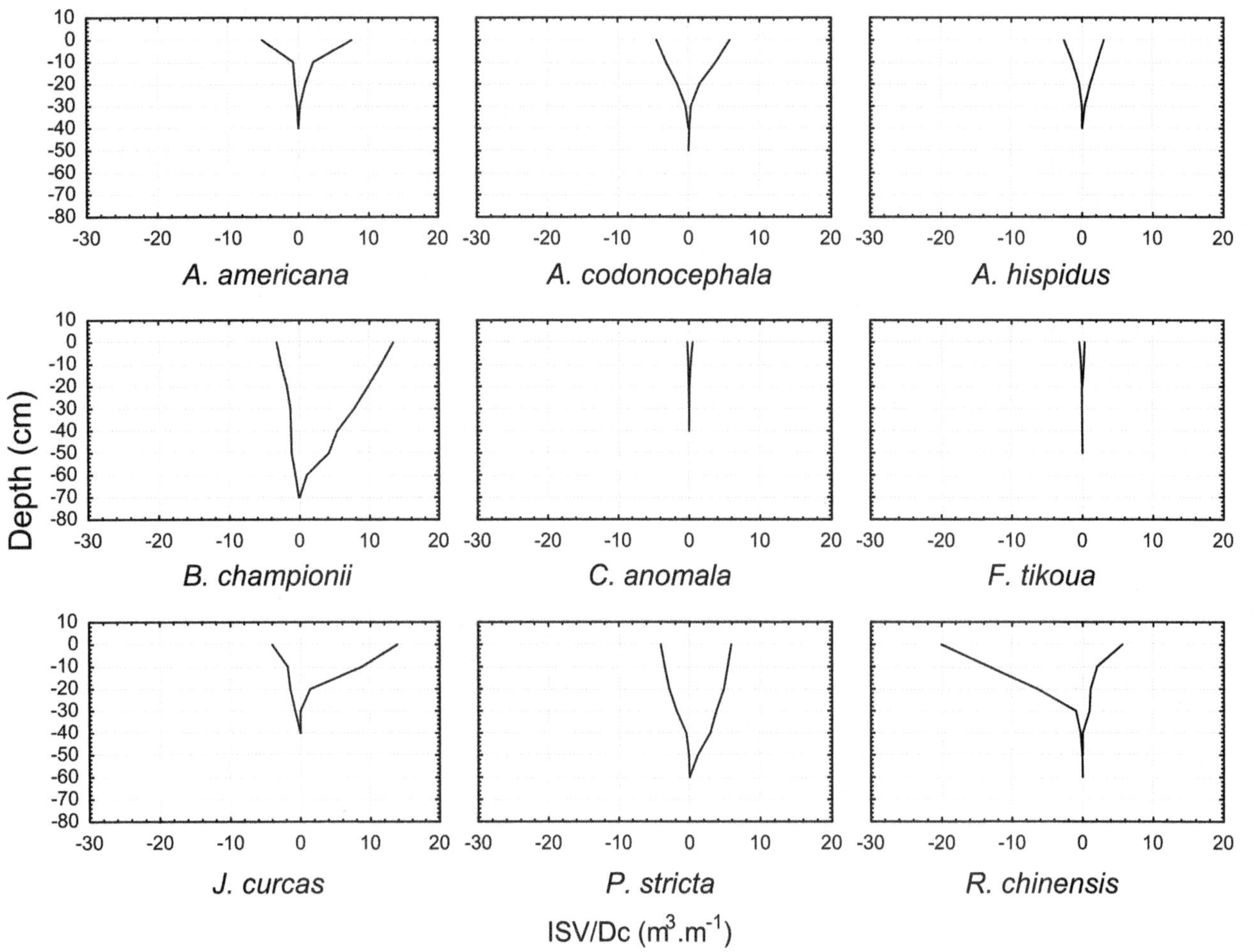

Figure 5. Mean individual soil volume standardized by collar diameter (ISV/Dc). Negative values of ISV/Dc represent downslope orientation.

Results

Soil characteristics

In situ shear resistance (measured using a tore vane) was significantly lower in the hotspot compared to the stable site ($Z_{20,20} = -3.28$, P<0.001; Table 2). In situ resistance to penetration was significantly lower at the hotspot ($Z_{20,20} = -5.25$, P< 0.001; Table 2). Soil cohesion (measured via direct shear tests in the laboratory) was extremely low at both sites, in particular the hotspot (Table 2), whereas the angle of internal friction was similar between sites (Table 2).

Species abundance

Regardless of site and year, the most abundant species were the two herbaceous species *C. anomala* and *A. hispidus*, followed by the creeping liana *F. tikoua* and then the tree *R. chinensis* ($H_{9,460} = 95.88$; P<0.001; Figure 4). All four species were observed to reproduce through clonal reproduction. *A. codonocephala* was relatively abundant. *A. americana*, *J. curcas* and *P. stricta* had been planted on the same slope where our fieldsites were located but at a lower altitude. Nevertheless, results showed that these species had begun to spread up the slope and colonize our sites. *P. stricta* spreads through sexual reproduction, producing numerous, light seeds [64], whereas *A. americana* reproduces largely through the production of underground stems [65]. *J. curcas* reproduces through large and heavy seeds when adult [66]. *B. championii* was not highly abundant (Figure 4).

A. codonocephala, A. hispidus and *C. anomala* produced a significantly higher number of stems m^{-2} at the stable site compared to the unstable hotspot ($H_{2,46} = 6.36$; P = 0.042; $H_{2,46} = 7.49$; P = 0.024 and $H_{2,46} = 12.13$; P = 0.002 respectively). On the contrary, *F. tikoua* possessed a significantly higher number of stems m^{-2} at the unstable hotspot compared to the stable site ($H_{2,46} = 6.54$; P = 0.038). *B. championii* produced a higher number of stems m^{-2} in July 2010 (very wet period) compared to August 2009 (dryer period; $Z_{22,24} = -2.46$; P = 0.013). No other differences were observed with regard to site or year.

Root spread (ISV/Dc)

When standardized by collar diameter (Dc), ISV depended on species ($H_{8,104} = 54.44$; P<0.001; Figure 5). In particular, *B. championii* possessed deeper and wider ISV/Dc compared to other species. *P. stricta*'s root system was also very deep (up to 60 cm). *F. tikoua* and *C. anomala* root systems occupied a very narrow ISV/Dc (Figure 5).

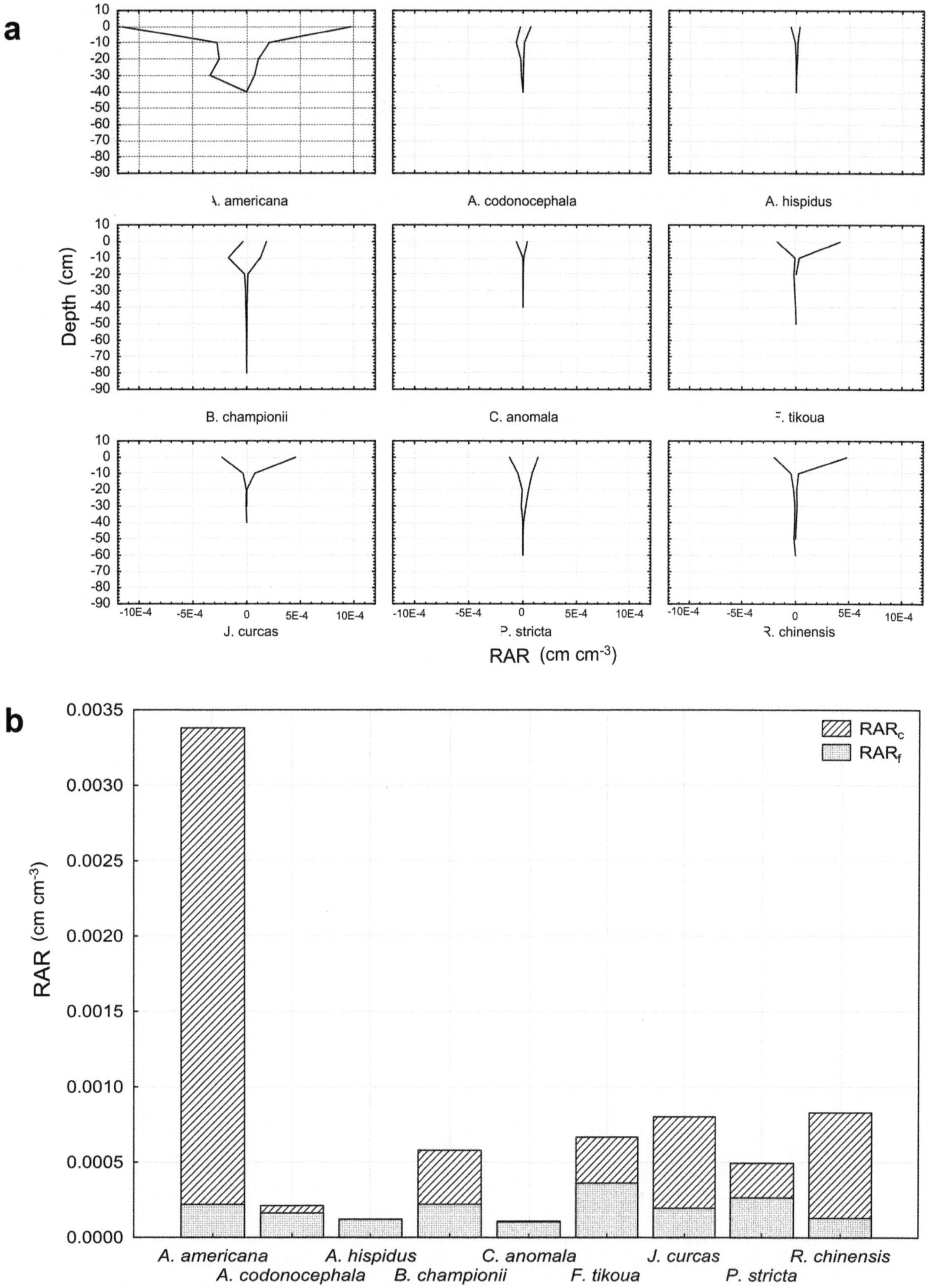

Figure 6. Root density of all species and proportion of coarse and fine roots. (a) Root area ratio (RAR) of all species. Negative values of RAR represent downslope orientation; (b) Coarse and fine roots area ratio (RAR).

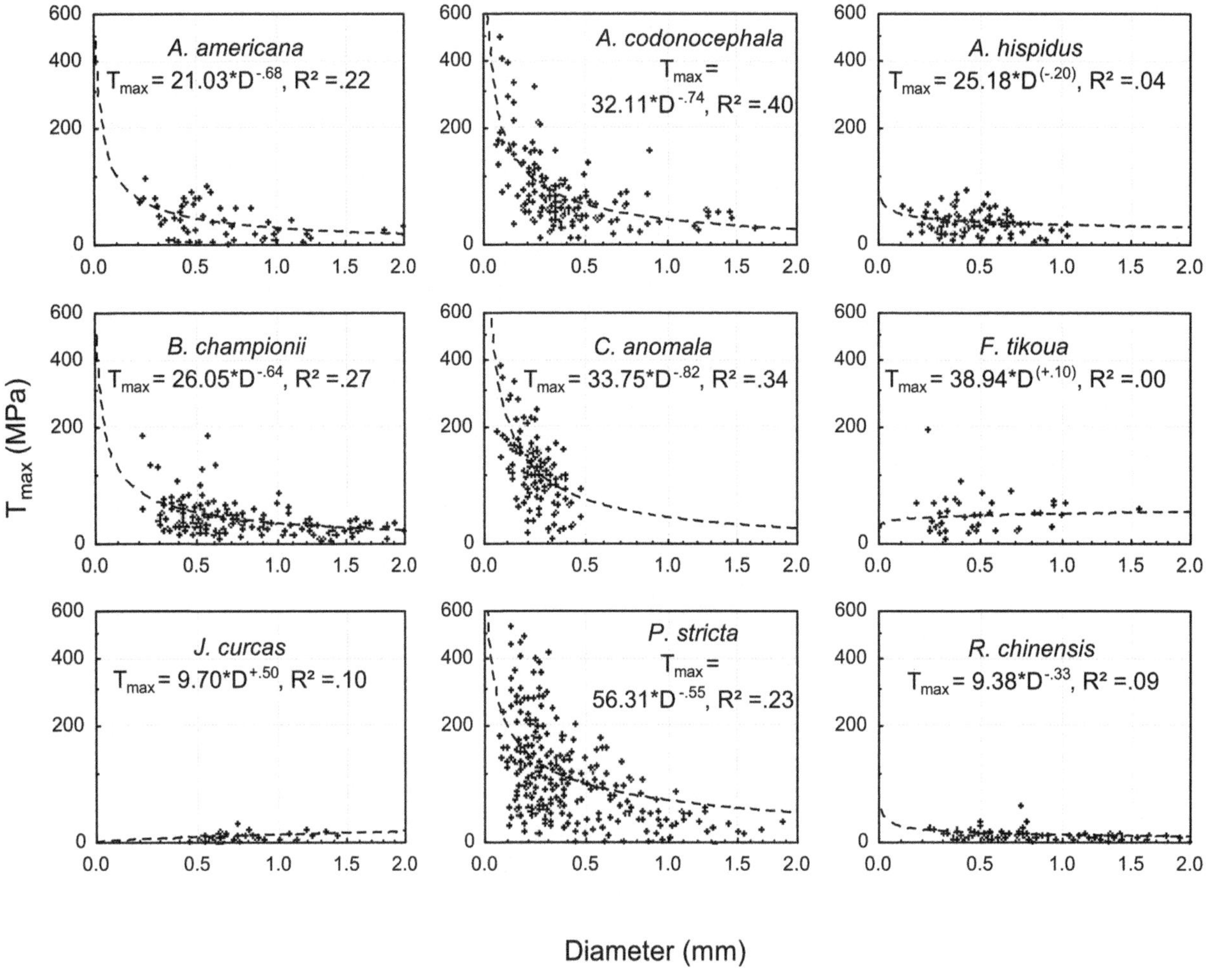

Figure 7. Tensile stress at failure. Tensile stress at failure (T_{max}) for root diameters from 0 to 2 mm. Logarithmic scales. Fitting curves: $T_{max} = \alpha * D^{-\beta}$, equations are presented on the graphs with the determination coefficient (R^2), parameters in brackets are not significant.

B. championii, *C. anomala* and *P. stricta* occupied a significantly higher ISV/Dc upslope compared to downslope from the stem ($Z_{13} = 2.20$; $P = 0.028$; $Z_{12} = 3.06$; $P = 0.002$ and $Z_{12} = 1.80$; $P = 0.007$ respectively, Figure 5). *R. chinensis* exploited a significantly larger ISV/Dc at the unstable hotspot compared to the stable site ($Z_{6,7} = -1.93$; $P = 0.005$, Figure 5). No other differences were observed with regard to site or slope sector.

Root area ratio (RAR)

RAR was species-dependant ($H_{8,100} = 50.25$; $P<0.001$; Figure 6a), with *A. americana* having significantly higher RAR than all other species. When *A. americana* was removed from the analysis, the species effect was still significant ($H_{7,86} = 38.78$; $P< 0.001$): *F. tikoua*, *J. curcas*, *B. championii*, *P. stricta* and *R. chinensis* all possessed significantly higher RAR than *A. codonocephala*, *A. hispidus* and *C. anomala* (Figure 6a). *B. championii*, *P. stricta* and *R. chinensis* all produced roots at depth of >50 cm.

For all species, RAR decreased significantly with increasing depth ($X^2_{14,3} = 14.36$; $P = 0.002$ for 0–10 cm compared to 10–20 cm and $X^2_{12,7} = 68.97$; $P<0.001$ for 10–20 cm compared to 20–30 cm, Figure 6a) except for *A. americana* and *B. championii*, for which RAR at 10–20 cm was not significantly lower than RAR at 0–10 cm. *P. stricta* possessed a significantly higher RAR in the upslope compared to downslope sector ($Z_{12} = 1.65$; $P = 0.009$, Figure 6a). *A. codonocephala* had a higher RAR at the unstable hotspot compared to the stable site ($Z_{4,6} = 2.02$; $P = 0.043$). No other significant differences between depth, slope sector or site were found.

With regard to the RAR of fine and coarse roots, the RAR of coarse roots was significantly greater in *A. americana* (Figure 6b) because we made no distinction between coarse roots and underground stems. The proportion of coarse roots was also higher than the proportion of fine roots for *J. curcas* and *R. chinensis*. *A. codonocephala*, *A. hispidus* and *C. anomala* root systems were composed largely of fine roots, therefore RAR of coarse roots was low (Figure 6b). For *B. championii*, *F. tikoua* and *P. stricta*, RAR of fine and coarse roots were similar (Figure 6b).

Root strength in tension (T_{max})

T_{max} increased with decreasing root diameter in all species except *F. tikoua* and *J. curcas* in which T_{max} increased with increasing root diameter (Figure 7). These trends were not significant for *A. hispidus* and *F. tikoua*. T_{max} differed significantly between species when root diameter was used as a covariate ($Fcov_{9,1105} = 29.85$; $P<0.001$; Figure 7). All species possessed

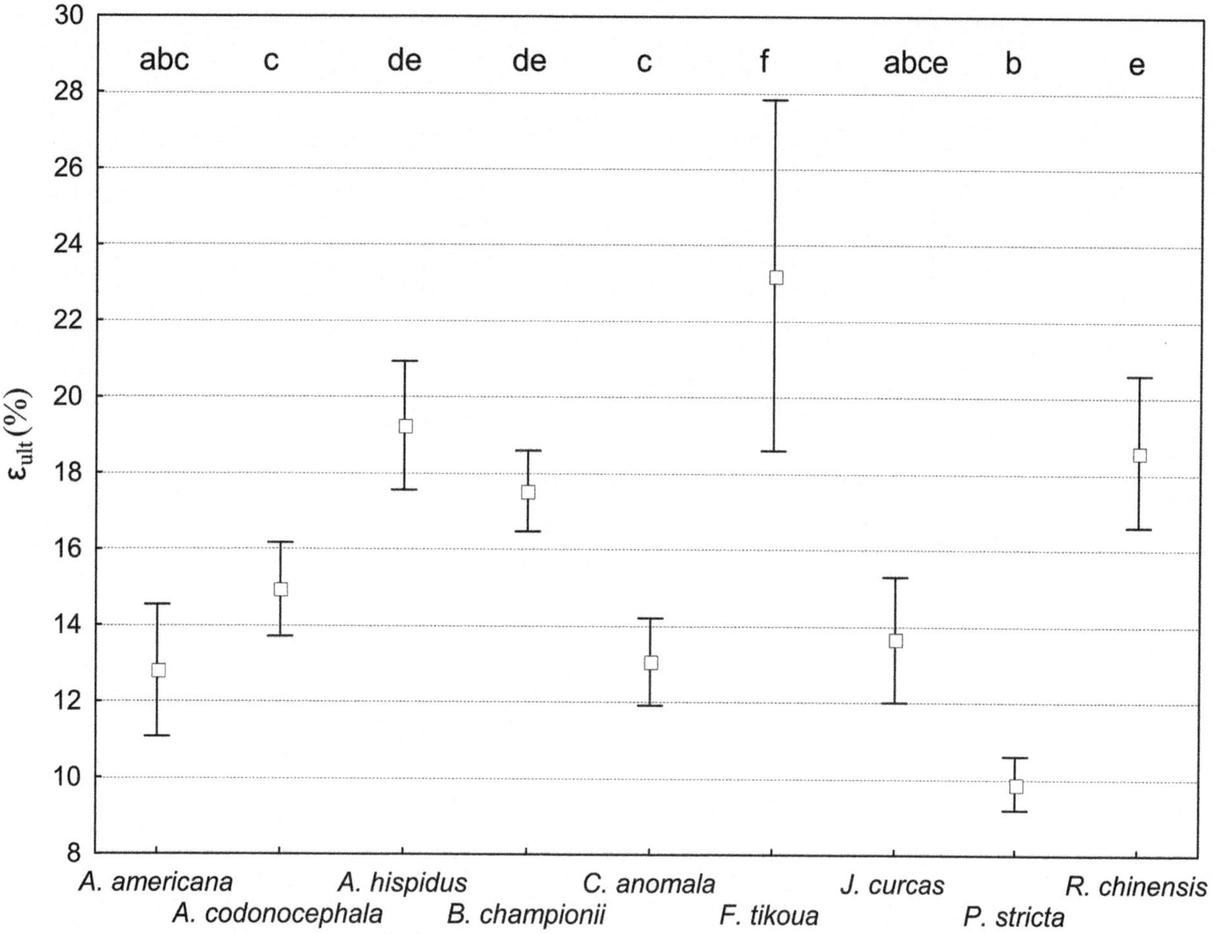

Figure 8. Ultimate strain at failure. Ultimate strain at failure (ε_{ult}) for each species. Vertical bars denote 0.95 confidence intervals, letters indicate significant differences between species (P<0.05).

similar values for the T_{max} of coarse roots (15–20 MPa) but T_{max} was greater in coarse roots of *F. tikoua* and *P. stricta* (approximately 40 MPa) and lower in coarse roots of *J. curcas* and *R. chinensis* (7–15 MPa; Figure 7). For fine roots, *P. stricta* had the strongest T_{max} (up to 200 MPa for roots 0.1 mm in diameter), followed by *A. codonocephala* and *C. anomala*. *J. curcas* and *R. chinensis* possessed very low T_{max} for fine roots (Figure 7).

With regard to depth in the soil, roots <1 mm in diameter of *B. championii* growing at a depth >30 cm possessed a significantly greater mean T_{max} compared to roots growing in shallower depth layers ($Fcov_{1,192}=3.91$; P<0.001). For a given diameter, roots of *A. americana* and *C. anomala* were significantly stronger at the stable site compared to the unstable hotspot ($Fcov_{1,50}=14.66$; P< 0.001 and $Fcov_{1,118}=4.18$; P = 0.043 respectively). For *P. stricta*, roots thicker than 0.5 mm diameter had a higher mean T_{max} at the unstable hotspot compared to the stable site but the tendency was inversed for roots thinner than 0.5 mm ($Fcov_{1,286}=5.21$; P = 0.023). *R. chinensis* possessed higher mean T_{max} for roots > 1.5 mm diameter at the unstable hotspot than at the stable site and again the tendency was inversed for thinner roots ($Fcov_{1,93}=6.80$; P = 0.011). With regard to the season when the roots were harvested for mechanical testing, roots of *A. codonocephala*, *C. anomala* and *P. stricta* were significantly stronger during the dry season (May 2009) compared to the rainy season (July 2010; $F_{1,46}=26.61$; P<0.001; $Fcov_{1,118}=14.74$; P<0.001 and $Fcov_{1,286}=12.31$; P<0.001 respectively). No other differences in mean T_{max} were found with regard to depth, slope sector, field sites and the season when roots were harvested.

Root strain in tension (ε_{ult})

The mechanical behaviour of roots differed between species with regard to strain. A positive relationship between ε_{ult} and root diameter was significant for *A. codonocephala* (r = 0.23, P<0.001) and *P. stricta* only (r = 0.48, P<0.001). For all other species, correlations between ε_{ult} and root diameter were not significant. Mean ε_{ult} was significantly different depending on species ($F_{8,1105}=32.34$; P<0.001; Figure 8). *F. tikoua* had the highest ε_{ult} (mean ε_{ult} = 23,24%), followed by *A. hispidus*, *B. championii* and *R. chinensis* (Figure 8). *A. americana*, *A. codonocephala*, *C. anomala*, *J. curcas* had relatively small mean ε_{ult} with *P. stricta* having the the smallest ε_{ult} (mean ε_{ult} = 9,81%).

Roots of *A. codonocephala* had significantly smaller mean ε_{ult} at depths >30 cm compared to roots growing at depths <30 cm ($Fcov_{4,146}=3.65$; P<0.001). On the contrary, in both *P. stricta* and *R. chinensis*, mean ε_{ult} was significantly greater in roots located at depths >30 cm than in shallow roots ($Fcov_{5,286}=5.64$; P<0.001 and $F_{5,91}=3.93$; P = 0.003, respectively). With regard to differences between upslope and downslope sectors, mean ε_{ult} in

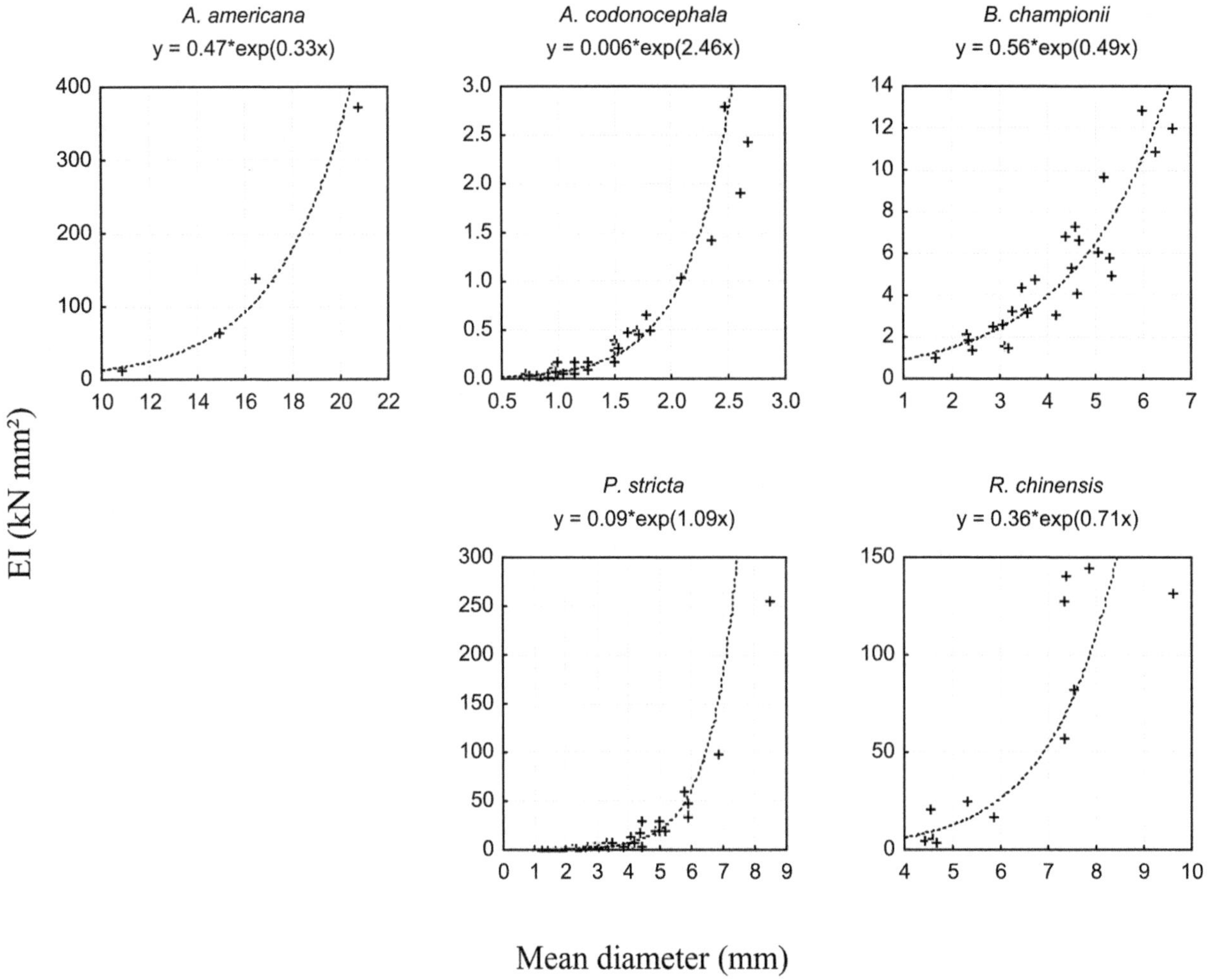

Figure 9. Bending rigidity. Bending rigidity (*EI*) as a function of root mean diameter for each species where data were sufficient for statistical analyses. Mean root diameter is the mean of depth and width diameters for each root. Note that scales differ between graphs for easier viewing.

roots of *P. stricta* was significantly shorter upslope compared to downslope ($Fcov_{1,286} = 5.88$; $P = 0.016$). Roots of *P. stricta* possessed a mean ε_{ult} that was significantly smaller at the unstable hotspot compared to the stable site ($Fcov_{1,286} = 6.76$; $P = 0.023$) and also significantly smaller during the dry season (May 2009) compared to the rainy season (July 2010; $Fcov_{1,286} = 9.75$; $P = 0.002$). On the contrary, roots of *A. hispidus* possessed significantly smaller mean ε_{ult} in July 2010, compared to the drier month of June 2010; $F_{1,89} = 295.29$; $P = 0.041$). No other significant differences in mean ε_{ult} were found with regard to depth, slope sector, site and the season when roots were harvested.

Resistance in bending (*EI*)

In species where enough large and stiff roots were found, i.e. all species except *A. americana*, *A. hispidus*, *C. anomala*, *F. tikoua* and *J. curcas*, *EI* increased with root diameter (Figure 9). *EI* differed significantly between species when root diameter was used as a covariate ($Fcov_{6,\ 128} = 240.41$; $P < 0.001$; Figure 9). Roots which were the most resistant in bending belonged to *P. stricta* (300 kN mm^2), *R. chinensis* (maximal values around 1.50 kN mm^2), followed by *B. championii* (15 kN mm^2) and *A. codonocephala* (3 kN mm^2).

The only significant difference with regard to site was found for roots of *P. stricta*, where mean *EI* was higher at the unstable hotspot compared to the stable site ($Z_{21,5} = -2.08$; $P = 0.037$). Data were too few to obtain significant comparisons between root depth in the soil and slope sector.

Chemical composition

The quantity of N present in roots depended on species ($F_{8,231} = 33.48$; $P < 0.001$, Figure 10a) and the interaction between species and root diameter ($F_{15,224} = 26.32$; $P < 0.001$, Fig; 10a). *A. codonocephala* had the highest quantity of N present in roots, regardless of root diameter (Figure 10a). The leguminous *P. stricta* also possessed high levels of N in fine roots (Figure 10a), as did the fine roots of the leguminous *B. championii*, but to a lesser exent (Figure 10a).

With regard to slope sector, all roots of *A. codonocephala* growing downslope contained more N than those growing upslope ($Z_{9,7} = -2.12$; $P = 0.034$). In *J. curcas*, N was significantly higher in roots growing upslope compared to those growing downslope ($Z_{5,5} = 2.30$; $P = 0.021$). With regard to site, root N was

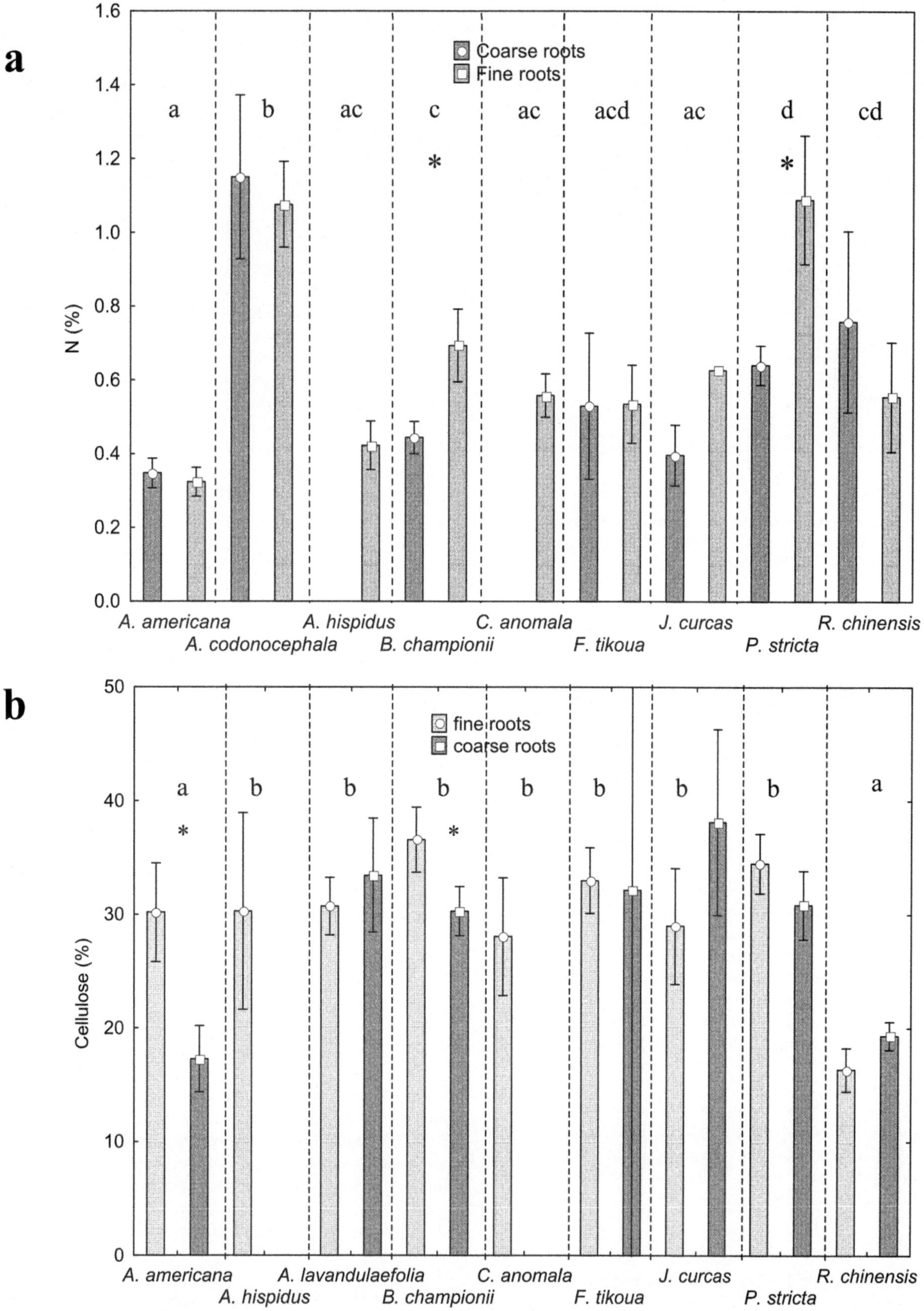

Figure 10. Nitrogen and cellulose contents. a) Nitrogen and b) cellulose concentrations in coarse and fine roots for each species. Letters indicate significant differences between species ($P<0.05$) when all (coarse and fine) roots are taken into account. Asterisks denote significant differences between coarse and fine roots within a species. Vertical bars denote 0.95 confidence interval.

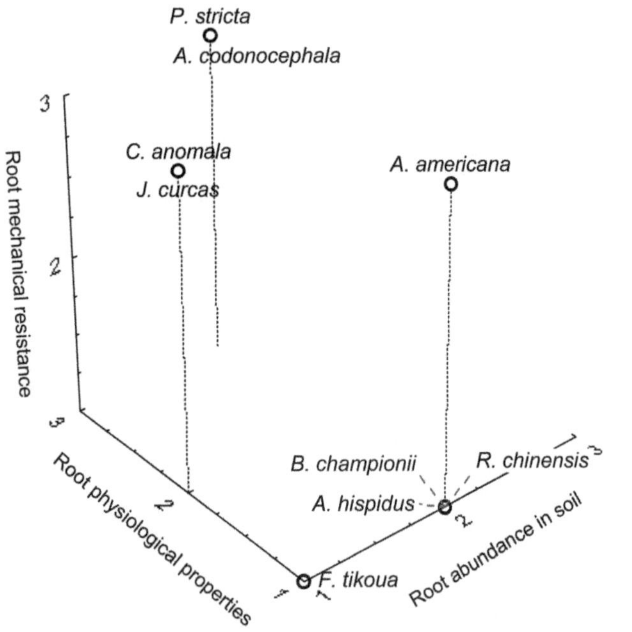

Figure 11. 3D diagram of species' scores. Summary of species' performance according to root abundance in soil, root mechanical resistance and root physiological properties.

significantly greater in all roots of *A. americana*, *B. championii*, *P. stricta* and *R. chinensis* growing on the unstable hotspot ($Z_{17,32}=4.19$; $P<0.001$; $Z_{24,44}=3.29$; $P<0.001$ and $Z_{29,21}=3.54$; $P<0.001$; $Z_{5,6}=2.46$; $P=0.014$ respectively). No other differences in root N were found with regard to slope sector or site and no differences were found with regard to depth.

The quantity of cellulose in roots depended on species ($F_{8,214}=20.90$; $P<0.001$, Figure 10b) and on the interaction between species and root size ($F_{6,333}=8.33$; $P<0.001$, *A. hispidus* and *C. anomala* having no coarse roots and root size not being defined for some samples, Figure 10b). For all species, cellulose content was significantly higher in fine roots compared to coarse roots ($F_{169.02}=-4.55$; $P<0.001$), *A. americana* and *B. championii* having significantly higher cellulose in fine roots than in coarse roots ($F_{8,214}=20.903$; $P<0.01$ and $F_{6,190}=8.33$; $P<0.01$ respectively). The highest cellulose content was found in coarse roots of *J. curcas*. The lowest cellulose content was found in fine and coarse roots of *R. chinensis* and in coarse roots (i.e. coarse roots and underground stems) of *A. americana* (Figure 10b). With regard to root depth in the soil, *P. stricta* possessed significantly more cellulose in roots of at a depth 0.0 to 0.10 m compared to those deeper than 0.10 m ($H_{3,42}=0.00$; $P=0.014$). The only differences between sites were in roots of *A. americana*, which had more cellulose at the stable site compared to the unstable hotspot ($Z_{28,13}=3.07$; $P=0.002$). No other significant differences in cellulose content were found with regard to depth, site and upslope/downslope sectors.

Discussion

Differences in traits between species and sites

The traits of each species were examined with regard to their desirability for fixing soil on slopes (Table 5 and Figure 11). Each species possessed one or several traits, which were desirable for improving slope stability, but no one species possessed a suite of traits that were ideal for fixing soil. *P. stricta* and *A. codonocephala* obtained the highest scores, as they possessed roots which were mechanically resistant and had suitable physiological traits, but soil occupation was average (Table 5 and Figure 11). In the case of *P. stricta*, the poor occupation of soil by roots was mainly due to a low number of stems per square meter. It can be assumed that increasing the number of individuals would increase root density in the soil over time. *J. curcas* and *A. americana* have been planted at many sites in the world to counteract slope instability and erosion processes [67,68,69]. However we found that although root mechanical properties were suitable, these species were not among the most useful for reinforcing soil (Table 5 and Figure 11), because of a poor capacity to occupy soil (*J. curcas*) and unsuitable physiological traits (*A. americana*). *A. hispidus* (herb), *B. championii* (leguminous liana) and *R. chinensis* (tree with vegetative multiplication by roots) had poor scores. They did not possess the same root system morphologies (Figure 3). All three species had poor mechanical and physiological root traits, and only an average soil occupation. The scoring system we developed allows us therefore to consider very different species and estimate their performance with regard to slope stability.

Seasonality and root traits

Several species possessed different properties depending on the season. Among them, *B. championii* did show an increase in stem abundance from August 2009 to July 2010, but this was probably due to progressive colonization of the slope rather than a seasonal effect. The mechanical resistance of roots of *A. codonocephala*, *C. anomala* and *P. stricta* (estimated by an increase of T_{max} in the three species and a decrease of ε_{ult} in *P. stricta*) decreased during the wet season. However, [24] showed that during winter and wet conditions, the tensile strength of *Salix* and *Populus* clones was higher. The only species with a slight advantage during the monsoon season was *A. hispidus*, whose roots deformed less in the wetter conditions. Mao et al., [70] showed that fine roots of temperate tree species produced during the summer were thinner and more short-lived than those produced during the winter months. The investment of resources in short-lived roots is usually less than that in longer-lived roots [32]. Therefore, the differences in mechanical properties that we observed are probably due to inherent differences between species and linked to seasonality in root production. The chemical composition of roots will also influence mechanical properties depending on water content.

Which species for slope short-term restoration?

All species showed significant differences in root traits whether they grew in stable or in unstable conditions (except *J. curcas* which was not present on the unstable hotspot). *F. tikoua* was more abundant on the unstable hostpot than on the stable site. *P. stricta* and *R. chinensis* both possessed stronger roots in unstable soil conditions. *A. codonocephala* had a higher RAR on the unstable site, but was not as abundant at the unstable hotspot, thus human action may be needed to enhance colonization. Roots of *B. championii*, *A. americana*, *P. stricta* and *R. chinensis* at the unstable hotspot had higher N content, suggesting that these roots had greater metabolic activity. These species may be more suitable for growing on a site undergoing restoration or a site requiring rapid protection.

Where is the most efficient location for species on hotspots?

To reinforce a slope against landslides, roots have to cross the potential shear plane. The potential shear plane of a slope can be circular or parallel to the soil surface (Figure 12). Therefore, as it is

Table 5. Multi-criteria table summarizing performances of each species: traits and related functions.

Functional traits & related functions	*A. americana*	*A. codonocephala*	*A. hispidus*	*B. championii*	*C. anomala*	*F. tikoua*	*J. curcas*	*P. stricta*	*R. chinensis*
High presence on hotspots (high number stems m^{-2})	1	2	3	1	3	2	1	1	2
High propagation of roots (high ISV/Dc)	2	2	2	3	1	1	2	3	2
High root density (high RAR)	3	2	1	2	1	2	2	2	2
Root abundance in soil	2	2	2	2	1	1	1	2	2
High resistance in tension (high fine roots RAR x high T_{max})	1*2 = 1	3*3 = 3	3*2 = 3	2*2 = 2	3*3 = 3	2*2 = 2	1*1 = 1	2*3 = 3	1*1 = 1
High resistance in bending (high coarse roots RAR x high *EI*)	3*3 = 3	1*2 = 1	1*1 = 1	2*2 = 2	1*1 = 1	2*1 = 1	3*3 = 3	2*3 = 3	3*3 = 3
Short deformation (short ε_{ult})	3	3	1	1	3	1	3	3	1
Root mechanical resistance	3	3	1	1	3	1	3	3	1
Short deformation (short ε_{ult})	2	2	2	2	2	1	2	3	2
High metabolism (high N)	1	3	1	2	1	1	1	3	2
High longevity (high CELL)	2	3	1	1	3	3	3	1	1
Root physiological properties	1	3	1	1	2	1	2	3	1

1/2/3: poor/average/good performance.

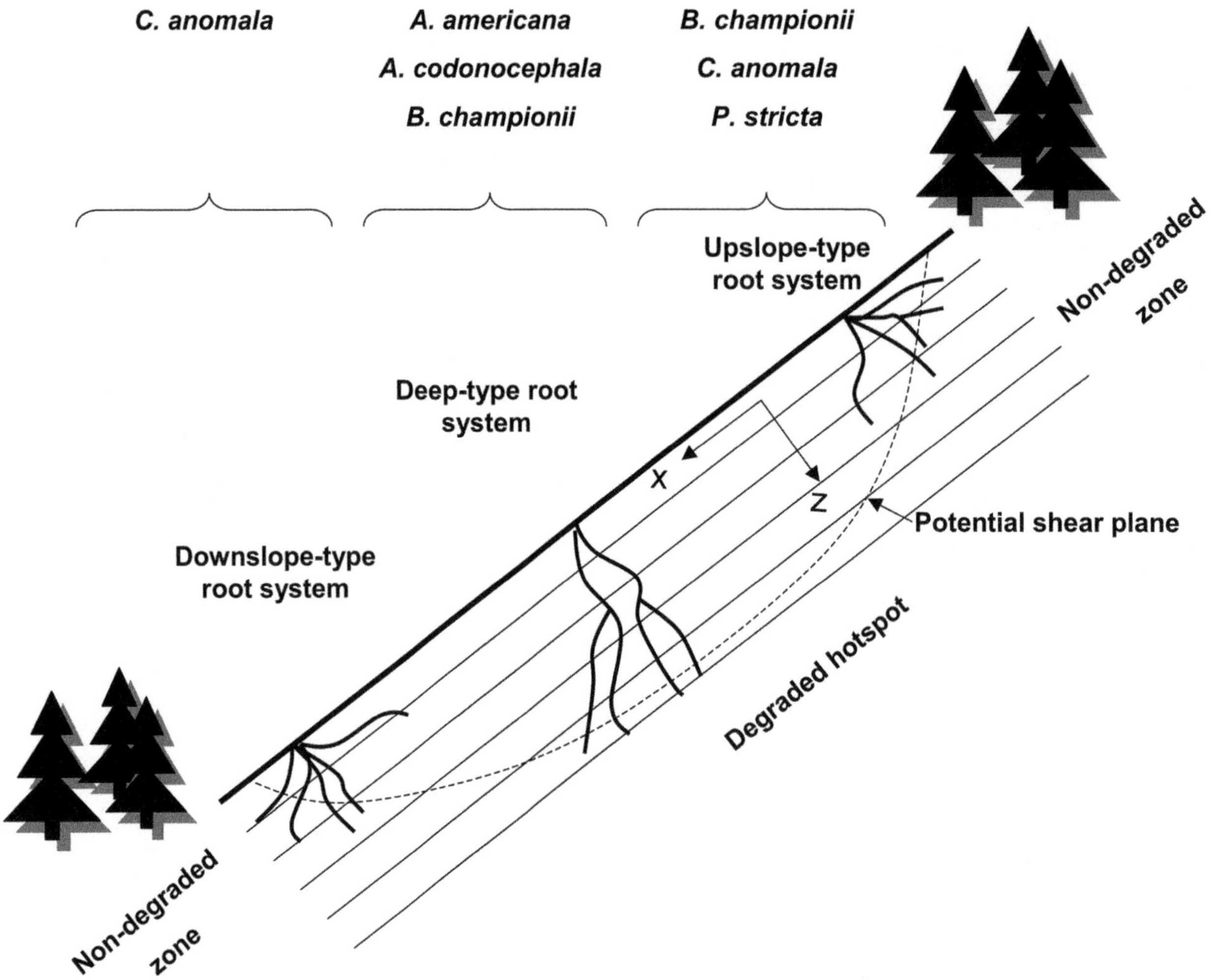

Figure 12. Best species depending on the location on the hotspot of instability. On degraded hotspots, the diversity of root systems will play an important role for slope stability. Roots growing upslope from the stem will have more chance to cross the potential shear plane if the plant grows at the top of the slope. Thus, root systems with desirable traits upslope of the stem will act more efficiently if they are located at the top of the slope, whereas the inverse is applicable for downslope roots. Root systems with desirable traits deeper in the soil will act more efficiently in the middle of the hotspot.

unlikely that any one species can possess an entire suite of traits that are optimal for increasing slope stability, different species can be planted at different positions along a slope, to optimise soil reinforcement. For example, species with vertical and strong roots will fix soil better in the middle of the slope, whereas plants with more and stronger roots upslope or downslope will better reinforce the top or toe of the slope, respectively (Figure 12), [28]. In our study on small areas of soil slippage, or hotspots, the precise location of species is all the more relevant because the depth of the potential shear zone may increase or decrease rapidly from the top to the bottom of the hotspot. Except for *A. hispidus*, *F. tikoua* and *J. curcas*, all species had roots which possessed different traits, depending on the depth of the root as well as its orientation with regard to slope direction.

A. americana and *B. championii* both had a high density of roots deeper in the soil and roots of the latter species were significantly stronger than shallower roots. Deeper roots of *A. codonocephala* had significantly smaller values of strain. Therefore these three species are able to increase the reinforcement of the potential shear zone at depth.

The root system of *C. anomala* occupied a larger soil volume near the soil surface. The cellulose content in roots of *P. stricta* was higher at surface. Therefore these two species would be more appropriate at the top and at the toe of the landslide rather than in the middle.

To select between top and toe, root systems of *B. championii* and *P. stricta* occupied a greater soil volume deeper in the soil upslope compared to downslope. *P. stricta* had more numerous and stronger roots upslope thus increasing lateral root reinforcement [18]. These two species will therefore be more efficient for improving slope stability when planted at the top of a hotspot (Figure 12).

Conclusions

In this study, we demonstrated that it is unlikely that any one species possesses an entire suite of root traits necessary to efficiently stabilise a slope with regard to shallow landslides. We suggest the use of mixtures of species, as well as a targeted spatial use of species in particularly fragile hotspots. Such mixtures have been shown to be more efficient at decreasing soil erodibility with regard to water erosion [44,71], but to our knowledge, this is the first study whereby an in-depth study of desirable root traits with regard to slope stability has been performed. As this was a short-term study, it is now necessary to project root traits' efficacy over time, especially as the woody species we examined were in the juvenile stage.

Acknowledgments

We are grateful to Xishuangbanna Tropical Botanical Garden (XTBG) in Kunming, and the Daxingdi village inhabitants who gave us a warm welcome. Thanks are due to J. Xie (Yunnan Agricultural University, China), K. Luo (Yunnan University, China), F.X. Mine (ISARA, France), for their help in the field, to Pr Peng Hua (Kunming Institute of Botany, China) for his help in species identification, to C. Hartmann (IRD, France) and D. Cluzeau (CNRS, France) for their help in describing soils and to E. Rutishauser (CIRAD, France) for his help in statistics.

Author Contributions

Conceived and designed the experiments: MG KC WM NR RL AS. Performed the experiments: MG WM RL CG AS. Analyzed the data: MG NR RL CG AS. Contributed reagents/materials/analysis tools: MG KC WM NR RL AS. Wrote the paper: MG KC NR AS.

References

1. Stokes A, Sotir R, Chen W, Ghestem M (2010) Soil bio- and eco-engineering in China: Past experience and future priorities. Ecological Engineering 36: 247–257.
2. Greenway DR (1987) Vegetation and slope stability. In John Wiley and Sons, New York, editors. Slope stability: geotechnical engineering and geomorphology. pp 187–230.
3. Sidle RC, Ochiai H (2006) Landslides: Processes, Prediction, and Land Use. American Geophysical Union.
4. Baigorria GA, Romero CC (2007) Assessment of erosion hotspots in a watershed: Integrating the WEPP model and GIS in a case study in the Peruvian Andes. Environmental Modelling & Software 22: 1175–1183.
5. Poesen J, Nachtergaele J, Verstraeten G, Valentin C (2003) Gully erosion and environmental change: importance and research needs. Catena 50: 91–133.
6. Smets T, Poesen J, Knapen A (2008) Spatial scale effects on the effectiveness of organic mulches in reducing soil erosion by water. Earth Science Reviews 89: 1–12.
7. Mitsch W (2012) What is ecological engineering? Ecological Engineering 45: 5–12.
8. MacGill BJ, Enquist BJ, Weiher E, Westoby M (2006) Rebuilding community ecology from functional traits. Trends in Ecology & Evolution 21: 178–185.
9. Burylo M (2010) Relations entre les traits fonctionnels des espèces végétales et leurs fonctions de protection contre l'érosion dans les milieux marneux restaurés de montagne. PhD Thesis. Grenoble, France: University Joseph Fourier. 262 p.
10. Stokes A, Atger C, Bengough A, Fourcaud T, Sidle RC (2009) Desirable plant root traits for protecting natural and engineered slopes against landslides. Plant and Soil 324: 1–30.
11. Ghestem M, Veylon G, Bernard A, Vanel Q, Stokes A (2014) Influence of plant root system morphology and architectural traits on soil shear resistance. Plant and Soil 377: 43–61.
12. Waldron LJ (1977) The shear resistance of root-permeated homogeneous and stratified soil. Soil Science Society of America Journal 41: 843–849.
13. Wu TH, McKinnell WP, Swanston DN (1979) Strength of tree roots and landslides on Prince of Wales Island, Alaska. Canada Geotechnical Journal 16: 19–33.
14. Ghestem M, Stokes A, Cao KF, Ma WZ, Xie JL (2009) Using vegetation to stabilize steep slopes in Southern China: Root biomechanics as a factor in the choice of species. In Thibaut B. editor. Proceedings of the 6th Plant Biomechanics Conference. Cayenne, French Guyana. pp 450–455.
15. Genet M, Stokes A, Fourcaud T, Norris JE (2010) The influence of plant diversity on slope stability in a moist evergreen deciduous forest. Ecological Engineering 36: 265–275.
16. Köhler L, Speck T, Spatz HC (2000) Micromechanics and anatomical changes during early ontogeny of two lianescent *Aristolochia* species. Planta 210: 691–700.
17. Pollen N, Simon A (2005) Estimating the mechanical effects of riparian vegetation on stream bank stability using a fibre bundle model. Water Resources Research 41: 1–11.
18. Schwarz M, Lehmann P, Or D (2010) Quantifying lateral root reinforcement in steep slopes - from a bundle of roots to tree stands. Earth Surface Processes and Landforms 35: 354–367.
19. Ekanayake JC, Phillips C (1999) A method for stability analysis of vegetated hillslopes: An energy approach. Canadian Geotechnical Journal 36: 1172–1184.
20. Docker BB, Hubble TCT (2008). Quantifying root-reinforcement of river bank soils by four Australian tree species. Geomorphology 100: 401–418.
21. Ghestem M (2012) Quelles propriétés racinaires et quelles espèces-outils pour la stabilisation des points chauds de dégradation en Chine du Sud? Which root properties and which tool-species can best stabilize degradation hotspots in Southern China? PhD Thesis. Paris, France: AgroParis Tech.68 p + 4 papers.
22. Ziemer RR, Swanston DN (1977) Root strength changes after logging in Southeast Alaska. Portland, Oregon: USDA Forest Service, Research Note. 10 p.
23. Phillips CJ, Watson AJ (1994) Structural tree root research in New Zealand: A review. Lincoln, Canterbury, New Zealand: Landcare Research Science. 71 p.
24. Hathaway RL, Penny D (1975) Root strength in some *Populus* and *Salix* clones. New Zealand Journal of Botany 13: 333–344.
25. Genet M, Stokes A, Salin F, Mickovski S, Fourcaud T, et al. (2005) The influence of cellulose content on tensile strength in tree roots. Plant and Soil 278: 1–9.
26. Wu TH, Watson A (1998) In situ shear tests of soil blocks with roots. Canada Geotechnical Journal 35: 579–590.
27. Nakamura H, Nghiem QM, Iwasa N (2007) Reinforcement of tree roots in slope stability: A case study from the Ozawa slope in Iwate Prefecture, Japan. In Stokes A, Spanos I, Norris JE, Cammeraat LH, editors. Eco- and ground bio-engineering: The use of vegetation to improve slope stability. Developments in Plant and Soil Sciences vol. 103. Dordrecht, Netherlands: Springer.pp. 81–90.
28. Danjon F, Barker DH, Drexhage M, Stokes A (2007) Using three-dimensional plant root architecture in models of shallow slope stability. Annals of Botany 101: 1281.
29. Hughes KA, Gandar PW, Silva HN (1995) Exploration and exploitation of soil by apple, kiwifruit, peach, Asian pear and grape roots. Plant and Soil 175: 301–309.
30. Oppelt AL, Kurth W, Jentschke G, Godbold D (2005) Contrasting rooting patterns of some arid-zone fruit tree species from Botswana – I. Fine root distribution. Agroforestry Systems 64: 1–11.
31. Oppelt AL, Kurth W, Jentschke G, Godbold D (2005) Contrasting rooting patterns of some arid-zone fruit tree species from Botswana – II. Coarse root distribution. Agroforestry Systems 64: 1–11.
32. Eissenstat DM, Wells CE, Yanai RD, Whitbeck JL (2000) Building roots in a changing environment: Implications for root longevity. New Phytologist 147: 33–42.
33. Yin Y (2008) Landslide mitigation strategy and implementation in China. In ISDR, ICL, editors. Parallel session volume of the first world landslide forum. pp. 693–696.
34. Sidle RC, Ziegler AD (2012) The dilemma of mountain roads. Nature Geoscience 5: 437–438.
35. Xu J, Yin R, Li Z, Liu C (2006) China's ecological rehabilitation: Unprecedented efforts, dramatic impacts, and requisite policies. Ecological Economics 57: 595–607.
36. Wang X, Lu C, Fang J, Shen Y (2007) Implications for development of Grain-for-Green policy based on cropland suitability evaluation in desertification-affected North China. Land Use Policy 24: 417–424.
37. Bennett MT (2008) China's sloping land conversion program: Institutional innovation or business as usual? Ecological Economics 65: 699–711.
38. Weyerhaeuser H, Wilkes A, Kahrl F (2005) Local impacts and responses to regional forest conservation and rehabilitation programs in China's Northwest Yunnan province. Agricultural Systems 85: 234–253.
39. Trac C, Harrell S, Hinckley T, Henck A (2007) Reforestation programs in Southwest China: Reported success, observed failure, and the reasons why. Journal of Mountain Science 4: 275–292.
40. Huang J, Deng X, Rozelle S (2004) Cultivated land conversion and bioproductivity in China. Pages 135–148 in Gao Wei, Shaw David R., editors. Proceedings of SPIE, Bellingham, WA.
41. Fattet M, Fu Y, Ghestem M, Ma W, Foulonneau M, et al. (2011) Effects of vegetation type on soil resistance to erosion: Relationship between aggregate stability and shear strength. Catena 87: 60–69.
42. Genet M, Kokutse N, Stokes A, Fourcaud T, Cai X, et al. (2008) Root reinforcement in plantations of *Cryptomeria japonica* D. Don: Effect of tree age and stand structure on slope stability. Forest Ecology and Management 256: 1517–1526.
43. Sidle RC, Hornbeck JW (1991) Cumulative effects: A broader approach to water quality research. Journal of Soil and Water Conservation 46: 268–271.

44. De Baets S, Poesen J, Reubens B, Muys B, De Baerdemaeker J, et al. (2009) Methodological framework to select plant species for controlling rill and gully erosion: application to a Mediterranean ecosystem. Earth Surface Processes and Landforms 34: 1374–1392.
45. Reubens B, Moeremans C, Poesen J, Nyssen J, Tewoldeberhan S, et al. (2011) Tree species selection for land rehabilitation in Ethiopia: from fragmented knowledge to an integrated multi-criteria decision approach. Agroforestry Systems 82: 303–330.
46. Zhou Y, Watts D, Li Y, Cheng X (1998) A case study of effect of lateral roots of *Pinus yunnanensis* on shallow soil reinforcement. Forest Ecology and Management 103: 107–120.
47. Munsell AH (1947) A color notation. 10th ed. Baltimore: Munsell Color Co., Incorp. 55 p.
48. Van Beek L, Wint J, Cammeraat L, Edwards J (2005) Observation and simulation of root reinforcement on abandoned mediterranean slopes. Plant and Soil 278: 55–74.
49. Ghestem M, Sidle RC, Stokes A (2011) The influence of plant root systems on subsurface flow: implications for slope stability. BioScience 61: 869–879.
50. Baize D, Girard MC (1995) Référentiel pédologique. Paris, France: INRA Editions. 332 p.
51. Association française de normalisation (1996) Qualité des sols: Recueil de normes françaises, 3rd ed. Paris-La Défense: AFNOR Editions. 534 p.
52. Kalra YP, Maynard DG (1991) Methods manual for forest soil and plant analysis. Information report NOR-X319. Edmonton, Alberta, Canada: Northern Forestry Centre. 110 p.
53. Baize D (2000) Guide des analyses en pédologie, 2nd ed. Paris, France: INRA Techniques et Pratiques. 266 p.
54. Schuppener B, Bohàč J, Dysli M (1999) Laboratory methods for direct shear tests. In Beuth Verlag GmbH editors. Recommendation of the ISSMGE for geotechnical testing. pp. 87–92.
55. Burylo M, Rey F, Roumet C, Buisson E, Dutoit T (2009) Linking plant morphological traits to uprooting resistance in eroded marly lands (Southern Alps, France). Plant and Soil 324: 31–42.
56. Genet M (2007) Variations spatio-temporelles des propriétés de renforcement racinaire de forêts naturelles et plantées dans le sud ouest de la Chine et leurs impacts sur la stabilité des pentes. PhD thesis. Bordeaux, France: University of Bordeaux I. 150 p.
57. Zobel RW (2008) Hardware and software efficacy in assessment of fine root diameter distributions. Computers and Electronics in Agriculture 60: 178–189.
58. Cofie P (2001) Mechanical properties of tree roots for soil reinforcement models. PhD thesis. Wageningen, The Netherlands: University of Wageningen. 165 p.
59. Isnard S, Rowe N, Speck T (2003) Growth habit and mechanical architecture of the sand dune-adapted climber *Clematis flammula* var. *maritima* L. Annals of Botany 91: 407–417.
60. Rowe N, Speck T (1996) Biomechanical characteristics of the ontogeny and growth habit of the tropical liana *Condylocarpon guianense* (Apocynaceae). International Journal of Plant Sciences 157: 406–416.
61. Jessome AP (1971) The bending strength of lumber in structural sizes. Department of the Environment, Canadian Forestry Service publication 1305: 1–33.
62. Van Soest PJ (1963) Use of detergents in the analysis of fibrous feeds: A rapid method for the determination of fibre and lignin. Journal of the Association of Official Agricultural Chemists 46: 829–835.
63. Packard GC (2011) Unanticipated consequences of logarithmic transformation in bivariate allometry. Journal of Comparative Physiology, Biochemical, Systemic, and Environmental Physiology 181: 841–849.
64. Hong DY (2011) Flora of China, Volume 10: Fabaceae. Missouri Botanical Garden Press. 680 p.
65. Irish G (2000) Agaves, yuccas, and related plants: A gardener's guide. Portland, Oregon: Timber Press. 398 p.
66. Heller J (1996) Physic Nut, *Jatropha curcas* L. Germany: International Plant Genetic Resources Institute. 66 p.
67. Reubens B, Achten WMJ, Maes WH, Danjon F, Aerts R, et al. (2011) More than biofuel? *Jatropha curcas* root system symmetry and potential for soil erosion control. Journal of Arid Environments 75: 201–205.
68. Petrone A, Preti F (2008) Suitability of soil bioengineering techniques in Central America: A case study in Nicaragua. Hydrology and Earth System Sciences Discussions 5: 379–403.
69. Achten WMJ, Reubens B, Maes W, Mathijs E, Verchot L, et al. (2007) Root architecture of the promising bio-diesel plant *Jatropha*. Communications in Agricultural and Applied Biological Sciences 72: 1–6.
70. Mao Z, Jourdan C, Bonis ML, Pailler F, Rey H, et al. (2013) Modelling root demography in heterogeneous mountain forests and applications for slope stability analysis. Plant and Soil 363: 357–382.
71. Polh M, Alig D, Körner C, Rixen D (2009) Higher plant diversity enhances soil stability in disturbed alpine ecosystems. Plant and Soil 324: 91–102.

6

Biomass Allocation Patterns across China's Terrestrial Biomes

Limei Wang[1], Longhui Li[1]*, Xi Chen[1], Xin Tian[2]*, Xiaoke Wang[3], Geping Luo[1]

1 State Key Laboratory of Desert and Oasis Ecology, Xinjiang Institute of Ecology and Geography, Chinese Academy of Sciences, Urumqi, China, **2** Institute of Forest Information Resource Techniques, Chinese Academy of Forestry, Beijing, China, **3** Beijing Urban Ecosystem Research Station, State Key Lab of Urban and Regional Ecology Research Center for Eco-Environmental Sciences, Chinese Academy of Sciences, Beijing, China

Abstract

Root to shoot ratio (RS) is commonly used to describe the biomass allocation between below- and aboveground parts of plants. Determining the key factors influencing RS and interpreting the relationship between RS and environmental factors is important for biological and ecological research. In this study, we compiled 2088 pairs of root and shoot biomass data across China's terrestrial biomes to examine variations in the RS and its responses to biotic and abiotic factors including vegetation type, soil texture, climatic variables, and stand age. The median value of RS (RSm) for grasslands, shrublands, and forests was 6.0, 0.73, and 0.23, respectively. The range of RS was considerably wide for each vegetation type. RS values for all three major vegetation types were found to be significantly correlated to mean annual precipitation (MAP) and potential water deficit index (PWDI). Mean annual temperature (MAT) also significantly affect the RS for forests and grasslands. Soil texture and forest origin altered the response of RS to climatic factors as well. An allometric formula could be used to well quantify the relationship between aboveground and belowground biomass, although each vegetation type had its own inherent allometric relationship.

Editor: Han Y.H. Chen, Lakehead University, Canada

Funding: This study was funded by the National Natural Science Foundation of China (grant number 41171094), the "Hundred Talent" Project of the Chinese Academy of Sciences (grant number Y174051001), one of National Basic Research Program of China (grant number 2013CB733404), and the National Natural Science Foundation of China (grant number 41101379). The funders had no role in study design, data collection and analysis, decision to publish, or preparation of the manuscript.

Competing Interests: The authors have declared that no competing interests exist.

* E-mail: lhli@ms.xjb.ac.cn (LHL); tianxin@caf.ac.cn (XT)

Introduction

Belowground biomass (BGB) is an important component of global terrestrial ecosystem carbon stocks and plays a critical role in global carbon cycling. Belowground biomass is more difficult and costly to measure and is a major source of uncertainties in large-scale biomass estimation and global carbon cycles [1–3]. The partitioning between aboveground and belowground biomass influences many of the functions performed by diverse terrestrial communities as well as the functions performed by individual plants (e.g., [4–7]). Root to shoot ratio (RS) is an effective parameter to describe the allocation between aboveground biomass (AGB) and BGB, and thus provides a practical tool to estimate BGB by relatively easily measured AGB. The RS reflects a plant's specific adaptive responses to its environment and has been widely used as a key descriptor for terrestrial ecosystem carbon modeling [8].

Biomass partitioning between belowground and aboveground parts can be predicted by plant allometric relationships [6,9]. At the level of individual plants, AGB scales nearly isometrically with respect to BGB, and this relationship has been validated across a broad spectrum of ecologically diverse vascular plants spanning several orders of magnitude in total body mass [10]. The allometric theory also estimates an isometric relationship between AGB and BGB at the community level [11], which has not been examined by adequate field measurements, especially its generality across diverse vegetation groups.

Recent studies have indicated that RS changes with environmental factors (e.g., climate, CO_2, soil texture, soil moisture, and nitrogen), biotic factors (e.g., plant type, stand age, and leaf traits), and forest origins [1,12–15]. Some improvements have been made in determining biomass allocation; however, many of the past studies focused on a particular ecosystem, and there have been few comprehensive investigations of RS across multiple ecosystems. Investigating biomass allocation patterns and their relationships with environmental factors crossing multiple ecosystems at regional scale may shed new light on this important topic.

China has rich vegetation resources and diverse vegetation types. China has the fifth highest total forest area and second highest total grassland area among all countries in the world, and thus the country plays an important role in global carbon stocks and cycling. The numerous vegetation types, diverse soil textures, complicated land use/cover patterns, and varying climatic zones offer a unique chance to examine the variation of RS and its influencing factors. Validating RS data from diverse biomass also provides a good opportunity to test the allometric relationships for ecosystems partitioning between AGB and BGB. Considerable work has been conducted to investigate the performance of allometric theory for a specific ecosystem and at sub-regional scales. For example, Wang et al. [16] examined RS relationships for major forest types in northeast China and pointed out that they differed significantly between natural and planted forests, and between broadleaf and coniferous forests. Yang et al. [17] focused their investigation on the Tibetan grasslands and suggested that the isometric relationships between AGB and BGB did not differ significantly between alpine steppe and alpine meadow. Further, Yang et al. [17] investigated relationships between AGB and BGB of grassland ecosystems across northern China and indicated that AGB was nearly proportional to BGB with a scaling exponent across various grassland types at the community level with no significant difference between temperate and alpine grasslands or between steppe and meadow. Comprehensive research on biomass allocation and its response to both climate and soil types for multiple ecosystems are essential to improving our understanding of the general rules of biomass allocation for biomes.

In this study, we compiled a comprehensive database of biomass and RS for the major terrestrial biomes across China. We used statistical approaches to address: 1) how RS varies across China's terrestrial biomes, 2) how RS responds to biotic and abiotic variables, and 3) how well the allometric theory performs in comparison with estimates of the empirical relationship between RS and environmental variables.

Materials and Methods

Datasets

We collected 2088 measurements of root and shoot biomass across China from hundreds of published papers (Fig. 1, Appendix S1). In the database, 276 records were extracted from Wang et al. [16], 377 records were extracted from Yang et al. [8], and 1139 records were extracted from Luo et al. [18]. The rest of the records in the dataset were compiled by the authors from related studies.

For each data set, we recorded the biomass data and whatever supporting information was available, including (1) shoot, root, and total biomass (dry weight of biomass per unit area); (2) site descriptions (including location, latitude, longitude, elevation, mean annual temperature, mean annual precipitation, mean annual potential evapotranspiration (PET), and soil texture); (3) vegetation descriptions (including vegetation type, dominate species, forest origin (primary, secondary, planted forest), age, height, density of stems, and mean diameter at breast height (DBH)); (4) descriptions of shoot and root biomass sampling methods (including sample size, dimensions of soil cores or soil pits, depth of sampling, whether root crowns were sampled, and whether existing allometric theories were applied).

The climatic variables mean annual precipitation (MAP) and temperature (MAT) at each biomass-sampling site were determined from the nearest surface meteorological stations across China (Fig. 1). Historically mean values of MAP and MAT during the period from 1961 to 2010 were used for analysis. Among 2088 sites, 77 sites have attached MAP and 88 sites have attached MAT. We assessed the accuracy of extracted MAP and MAT by comparing with site-attached values and found that extracted MAP was significantly correlated with the recorded measurements ($r^2 = 0.92$, $n = 77$, $p < 0.01$), and extracted MAT was in good agreement with the measurements ($r^2 = 0.9$, $n = 88$, $p < 0.01$) as well. We computed mean annual potential evapotranspiration (PET) using the FAO Penman–Monteith equation [19] from the meteorological variables recorded by the 664 surface stations, and then we computed potential water deficit index (PWDI) by the ratio of the mean annual precipitation to annual potential evapotranspiration (MAP/PET) [20].

Vegetation Types

The categorization of data into vegetation types relied primarily upon a soil texture classification scheme [17,18]. Different vegetation types had obviously different RS and soil texture may alter the value of RS for same vegetation types. Based on a soil texture map (Institute of Soil Science, Chinese Academy of Sciences, 1986), the dataset was divided into eight types: loam forest, sandy forest, clay forest, loam grassland, sandy grassland, clay grassland, loam shrubland, and sandy shrubland (Table 1). Recent studies have indicated that RS shows different physiolog-

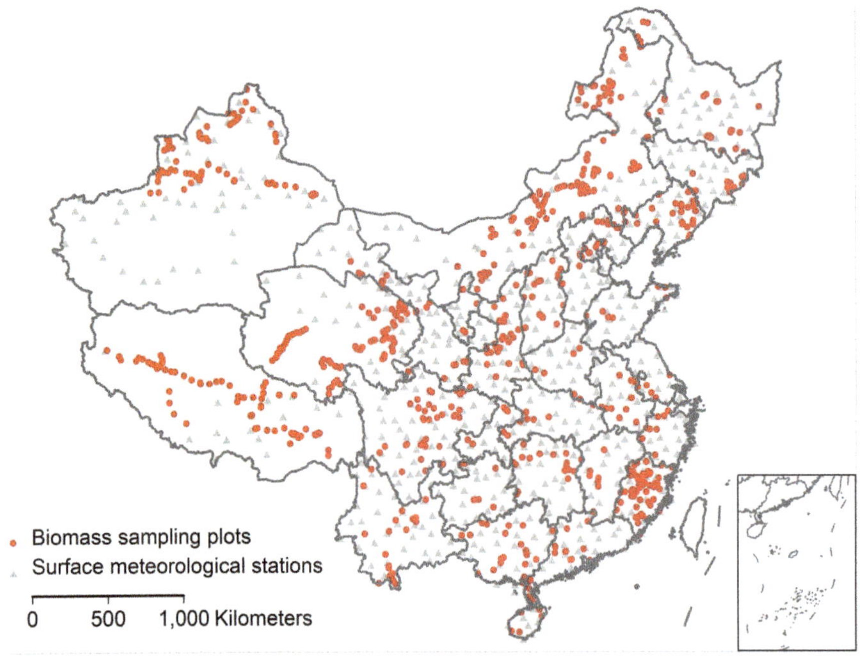

Figure 1. The geographical distribution of biomass sampling plots collected in this study. These data including forests, shrublands, and grasslands distributed across China.

Table 1. Median of above-ground biomass (AGB), below-ground biomass (BGB), root to shoot ratio (RS) and allometric relationships between AGB and BGB ($p<0.05$) across China's biomes.

Vegetation category	AGB(Mg ha^{-1})	BGB(Mg ha^{-1})	RS	Range	n	LogAGB–logBGB models		
						Slope	Intercept	r^2
Vegetation type								
Primary forest	99.1	24.0	0.24	0.03–1.20	454	0.87	−0.37	0.85
Secondary forest	93.7	22.6	0.23	0.08–0.58	97	0.95	−0.52	0.78
Planted forest	68.4	15.5	0.22	0.05–0.90	1002	0.81	−0.30	0.76
Temperate grassland	1.5	7.7	5.73	0.43–29.2	243	0.77	0.74	0.58
Alpine grassland	0.7	4.1	6.27	0.81–45.6	262	1.00	0.79	0.73
Soil texture								
Loam forest	76.5	16.7	0.22	0.02–1.20	967	0.90	−0.45	0.82
Sand forest	62.2	15.6	0.25	0.03–0.70	242	0.82	−0.30	0.79
Clay forest	91.1	19.3	0.22	0.05–0.50	345	0.87	−0.42	0.87
Loam grassland	1.0	5.5	6.10	0.43–29.2	336	0.78	0.74	0.66
Sand grassland	0.7	5.0	5.61	0.56–45.6	153	0.89	0.71	0.59
Clay grassland	1.1	7.2	7.60	2.0–12.72	10	0.96	0.87	0.74
Loam shrubland	9.5	11.3	0.71	0.26–1.56	12	1.00	−0.31	0.80
Sand shrubland	10.6	8.6	0.77	0.47–1.96	9	1.16	−0.44	0.95
Region								
China's forests	76.7	17.0	0.23	0.03–1.20	1551	0.86	−0.36	0.81
Global forests	81.0	22.5	0.25	0.09–1.16	207	0.88	−0.32	0.90
China's grasslands	1.0	5.2	6.00	0.43–45.6	502	0.82	0.74	0.63
Global grasslands	3.5	12.7	3.26	0.38–26.0	61	0.40	0.84	0.20
China's shrublands	10.6	10.6	0.73	0.26–1.96	22	1.05	−0.16	0.89
Global shrublands	16.6	14.0	1.84	0.34–4.25	8	–	–	–

n, number of samples; Slope, the scaling exponent of the log–log linear functions; Intercept, the allometric constant of log–log linear functions; *r*, the coefficient of determination.

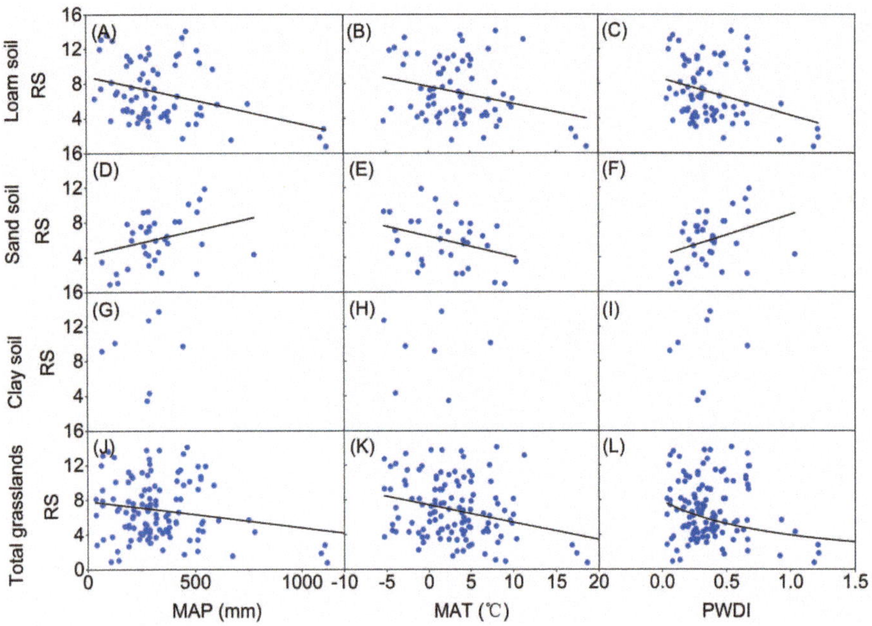

Figure 2. Root to shoot ratio (RS) for grasslands as a function of mean annual precipitation (MAP), temperature (MAT) and potential water deficit index (PWDI). Regression lines are given for loam, sandy, and claygrasslands and for all data together, if the relationships are significant at $p<0.05$.

ical and ecological responses to varied surroundings for a range of forest stand age and climate zones [8,12,16]. Summarizing these studies, the forests and grasslands were further investigated based on five groups divided by the forest origin and climatic zone: primary forest, secondary forest, planted forest, temperate grassland, and alpine grassland (Table 1).

Statistical Analysis

Descriptive analysis was used to determine the statistical characteristics (e.g., median value, range of data) of categorical data. Linear and nonlinear analyses were performed to develop regressions for RS dependent on the vegetation type, soil texture, and climatic variables. The significance of differences between slopes (scaling exponent) and y-intercepts (allometric constant) of the log-log transformed linear functions was evaluated by analyses of variance [21]. The standard error of the estimation (SEE) was used to measure the difference between predicted and actual values of root biomass.

Results

Variations of RS across China's Biomes

The biomass and RS values shown in Table 1 are referred to as the median rather than the mean values for the skewed distribution of data for many of the vegetation types. RS values ranged from 0.03 to 1.2, 0.43 to 45.6, and 0.26 to 1.96 for forests, grasslands, and shrublands, respectively (Table 1). The median RS tended to decrease from grasslands to shrublands to forests across China (Table 1).

The median RS of forests tended to increase from planted to secondary to primary stands. In addition, forests growing in sandy soil had a larger median RS than those in loam and clay soils. Further, loam soil forests had approximately the same median RS value as the clay soil forests (Table 1).

For grasslands, RS changed significantly with different grassland types. The median RS of alpine grassland was much higher than that of temperate grassland (Table 1). There was a general trend for RS to vary with soil texture changes across grasslands. The

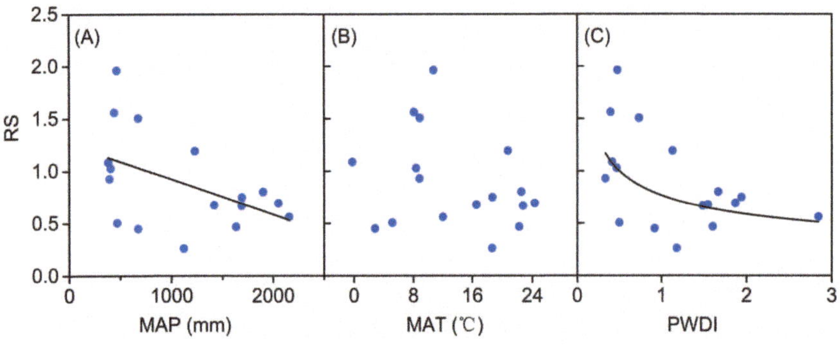

Figure 3. Root to shoot ratio (RS) for shrublands as a function of mean annual precipitation (MAP), temperature (MAT) and potential water deficit index (PWDI). Regression lines are given for all data together if the relationships are significant at $p<0.05$.

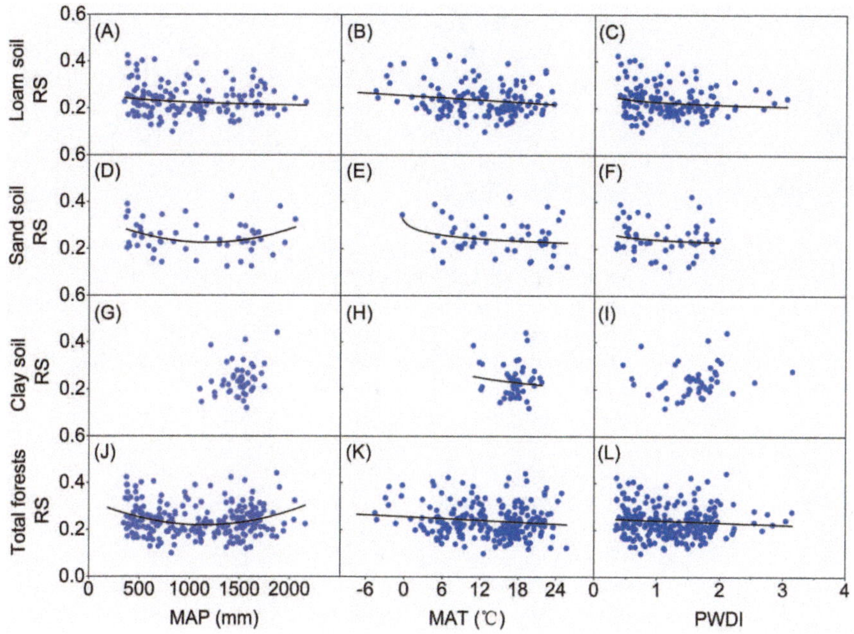

Figure 4. Root to shoot ratio (RS) for forests as a function of mean annual precipitation (MAP), temperature (MAT) and potential water deficit index (PWDI). Regression lines are given for loam, sandy, and clay forests and for all data together, if the relationships are significant at $p<0.05$.

median RS decreased from clay soil grassland to loam soil grassland to sandy soil grassland.

Responses of RS to Climatic Variables

China's forests, grasslands and shrublands responded differently to climatic variables. For both grasslands (Fig. 2J) and shrublands (Fig. 3A), RS decreased significantly ($p<0.01$) with increased MAP. A similar trend was found in RS for forests where MAP was less than 1250 mm·yr^{-1}; however, RS values tended to increase significantly ($p<0.01$) where MAP was over 1250 mm·yr^{-1} (Fig. 4J). Generally, soil nutrient availability dominated RS where precipitation was high (>1250 mm) [26] and more biomass was allocated into roots for uptaking nutrient and RS tend to increase with MAP. Although RS decreased significantly with increasing MAT for both forests ($p<0.05$, Fig. 4K) and grasslands ($p<0.01$, Fig. 2K), there was no significant relationship between MAT and

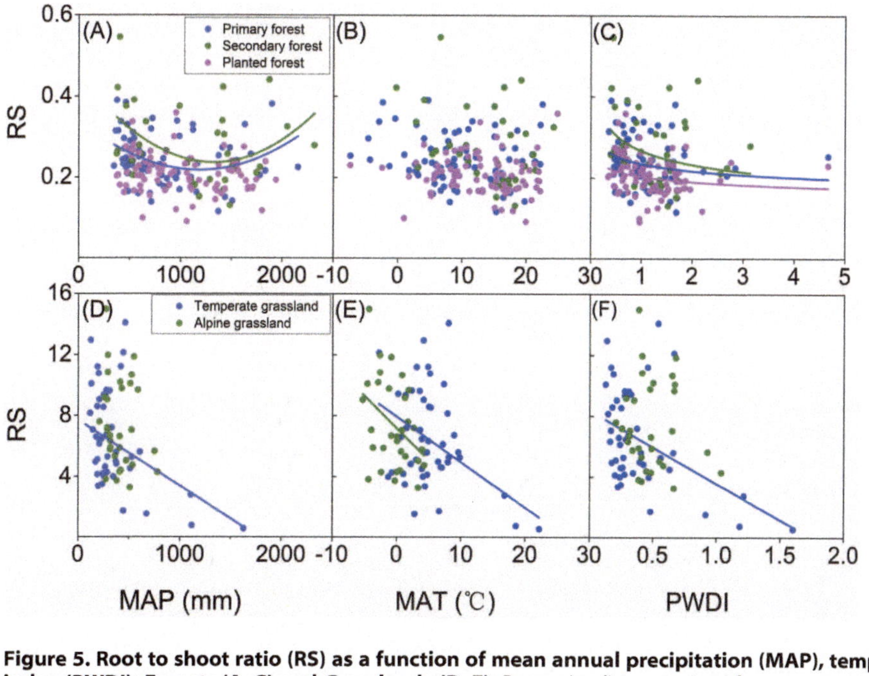

Figure 5. Root to shoot ratio (RS) as a function of mean annual precipitation (MAP), temperature (MAT) and potential water deficit index (PWDI): Forests (A–C) and Grasslands (D–F). Regression lines are given for every vegetation group if the relationships are significant at $p<0.05$.

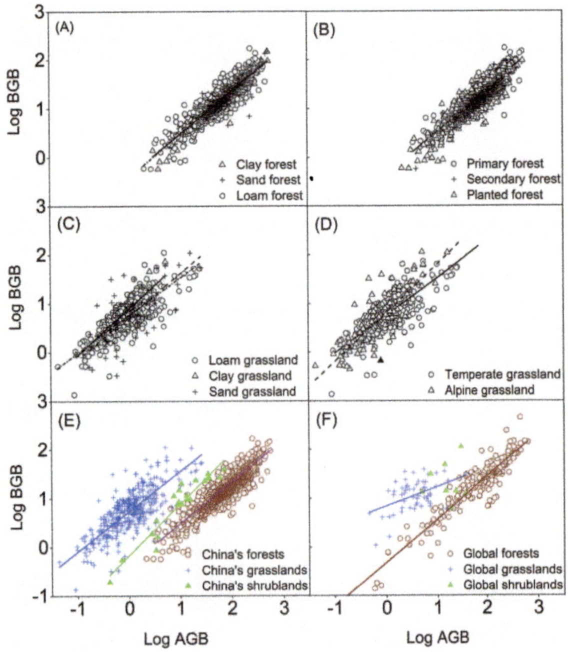

Figure 6. Allometric relationships between aboveground biomass (AGB) and belowground biomass (BGB) for different vegetation groups. (A) Comparison in between loam, sand and clay forest. (B) Comparison in between primary, secondary and planted forest. (C) Comparison in between loam, sand and clay grassland. (D) Comparison in between temperate and alpine grassland. (E) Comparison in between China's forest, grassland and shrubland. (F) Comparison in between global forest, grassland and shrubland. Regression lines and equations are given for every vegetation group if the relationships are significant at $p<0.05$. Global biomass data are from Mokany et al. [12].

RS for shrublands (Fig. 3B). RS values were negatively related to PWDI ($p<0.01$) for all forests (Fig. 4L), grasslands (Fig. 2L), and shrublands (Fig. 3C).

Soil texture affected RS's responses to climate. RS from both loam and sandy soils showed significant responses to climate variables for forests (Figs. 4A–F) and grasslands (Figs. 2A–F). However, no general trend was found between climate and RS from clay soil for either forests or grasslands (Figs. 2G–I). In addition to soil texture, the responses of RS to climate were influenced by forest origins. RS for natural forests showed a significant U-shaped trend ($p<0.05$) as MAP increased; however no trend existed between MAP and RS for planted forests (Fig. 5A). Neither natural nor planted forests showed a significant trend for RS as a function of MAT (Fig. 5B); however, RS decreased significantly ($p<0.01$) as PWDI increased for both natural and planted forests (Fig. 5C). Climatic zonation was an important factor influencing responses of grasslands' RS to climate. RS for temperate grasslands was negatively related to MAP, MAT, and PWDI (Fig. 5D–F, $p<0.01$). In contrast, no significant trend existed between RS and MAP (Fig. 5D) or PWDI (Fig. 5F) for alpine grassland, although RS for alpine grassland decreased significantly ($p<0.01$) with increasing MAT (Fig. 5E).

Allometric Relationships between Root and Shoot Biomass

The relationships between root biomass (y) and shoot biomass (x) across China's biomes fit well with an allometric power function, $y=ax^b$ (Fig. 6). The coefficients for a were 0.3, 8.02, and 2.35 and for b were 0.94, 0.72, and 0.66 for forests, grasslands, and shrublands, with corresponding r^2 of 0.75, 0.5, and 0.85, respectively. In general, the allometric relationships between root and shoot biomass were significantly ($p<0.01$) different from each other between China's major terrestrial ecosystems (forests, grasslands, and shrublands) and all slopes were significant ($p<0.01$) (Fig. 6; Table 1).

Across different forest types, none of the regressions between root and shoot biomass were significantly ($p<0.05$) different from each other and all slopes were significant ($p<0.01$) (Figs. 6A, B; Table 1). However, soil texture and climatic zonation altered grasslands' allometric relationships. The slope of the regression for clay grasslands was significantly ($p<0.01$) greater than that of both loam and sandy grasslands. The allometric relationship for loam soil is not significantly different from that of sandy soil ($p<0.01$) (Fig. 6C; Table 1). The slope for alpine grasslands was significantly greater ($p<0.01$) than that of temperate grasslands. The allometric relationship for alpine grasslands was significantly different from that of temperate grasslands ($p<0.01$) (Fig. 6D; Table 1). Meanwhile, the allometric relationship for China's grasslands was significantly different from the global average; however, the slope of the allometric equation for China's forests was similar to the global value (Figs. 6E, F).

Discussion

Variations in RS across China's Biomes

We reported that the overall median values of AGB were 76.7, 1.0 and 10.6, and the median of BGB were 17.0, 5.2 and 9.4 for China's forests, grasslands and shrublands, respectively (Table 2). As a consequence, the resulted median RS were 0.23, 0.60 and 0.74 for China's forests, grasslands and shrublands, respectively. Our estimated AGB, BGB and RS for grasslands were quite comparable with previous research using Northern China's dataset [8]. However, a remarkable difference existed in terrestrial root and shoot biomass and RS between China and global averages. Both root and shoot biomass in China were lower than global values for each vegetation type (Table 3). The median RS values for China's forests (0.23) and shrublands (0.73) were lower than global estimates (0.29 and 2.45, respectively [12]); however, the median RS for China's grasslands (6.0) was much higher than global estimates (3.3 [12] and 3.7 [22]). Such difference may come from several aspects. Firstly, sampling methods and the number of dataset used may be one of the main reasons. Makony et al. [24] pointed that data omitting was critical to the analysis of root and shoot ratio. Limited by axillary information for the dataset used in the current research, we ignored the selection and keep all data points we collected in analysis. Such treatment may avoid distorting the possible nature of biomass allocation pattern across wide biome, climate and soil conditions. Secondly, accuracy of climatic variables (mean annual precipitation and temperature) and soil properties (soil texture) may contribute the uncertainties of estimated RS values. Particularly for those data sites located in different altitude or slope of mountain areas, different climatic conditions should be expected but same climatic variables were used for analysis due to the lack of available meteorological observations. To overcome this drawback, we averaged all data points where shared meteorological measurements. However, the detailed response of RS to climatic variables may be also masked.

Factors Influencing Biomass Allocation

Biomass allocation is one of the most successful theories in modern ecology [36]. Although plant growth and biomass allocation are highly influenced by environmental and

Table 2. Statistical analysis for predicted root biomass (BGB) using median vegetation-specific root to shoot ratio (RS) and allometric equations.

Vegetation category		**Predicted root biomass by median vegetation-specific RS**			**Predicted root biomass by allometric equations**		
		Slope	r^2	**SEE**	**Slope**	r^2	**SEE**
Vegetation type							
Primary forest	443	0.95	0.77	12.1	1.09	0.76	12.1
Secondary forest	97	1.00	0.68	11.1	1.10	0.68	11.0
Planted forest	997	0.72	0.70	7.24	0.76	0.71	7.22
Temperate grassland	217	1.42	0.79	30.2	2.66	0.66	38.1
Alpine grassland	258	1.36	0.76	12.7	2.22	0.68	14.7
Soil texture							
Loam forest	951	0.97	0.72	9.68	1.01	0.72	9.66
Sand forest	242	0.82	0.78	8.50	0.97	0.78	8.63
Clay forest	342	0.89	0.81	7.85	0.93	0.80	7.92
Loam grassland	312	0.54	0.42	12.5	0.96	0.45	12.1
Sand grassland	152	0.73	0.56	11.8	1.18	0.60	11.3
Clay grassland	12	0.32	0.008	5.18	0.51	0.08	4.99
Loam shrubland	11	0.73	0.84	5.30	1.03	0.89	4.50
Sand shrubland	9	1.35	0.66	4.85	1.26	0.65	4.95
Region							
China's forests	1535	0.90	0.75	9.15	0.98	0.74	9.16
Global forests	207	1.02	0.56	30.1	0.98	0.56	30.1
China's grasslands	475	0.60	0.46	12.4	1.05	0.50	12.0
Global grasslands	57	0.13	0.04	9.66	0.71	0.04	9.66
China' s shrublands	21	0.76	0.79	5.26	1.04	0.85	4.53
Global shrublands	8	–	–	–	–	–	–

n is number of samples, Slope is the scaling exponent of the regressions, *r* is coefficient of determination and SEE is standard error of the estimation.

Table 3. Above-ground biomass (AGB), below-ground biomass (BGB) and root to shoot ratio (RS) for forests, grasslands and shrublands from different studies.

Vegetation category	AGB (Mg ha^{-1})	BGB (Mg ha^{-1})	RS	Reference
Forest				
China's forests	76.7	17.0	0.23	This study
China's forests	–	–	0.23	Luo et al. (2012)
Global forests	81.0	22.5	0.25	Mokany et al. (2006)
Grassland				
China grasslands	1.0	5.2	6.0	This study
China grasslands	1.0	5.7	5.7	Yang et al. (2009)
Global grasslands	3.5	12.7	3.3	Mokany et al. (2006)
Global grasslands	3.8	14.0	3.7	Jackson et al. (1996)
Shrubland				
China's shrublands	10.6	9.7	0.73	This study
Global shrublands	16.6	14.0	1.84	Mokany et al. (2006)

Arithmetic mean or median values are presented to compare with other studies.

non-environmental factors, a mass number of previous research have demonstrated that biomass allocation is impacted by climatic variables (particularly precipitation and temperature), soil type, nutrient availability and plant species [17,23–26,32,33].

Majority of previous studies reported that RS can be either positively or negatively and linearly correlated and even non-linearly related with the increase in annual precipitation at relatively larger regional or national scales [8,16,37,38,18,28] but the value of RS overall decreased with precipitation at global scale [3]. Similarly, the effect of temperature on RS can be different from region to region, vegetation type to type [18,38,27]. Various possible combinations of hydrothermal conditions (precipitation and temperature) may explain the variation of the diverse response of RS to individual factor of precipitation and temperature. Further, soil texture, vegetation type, composition and age, nutrient availability also induced variations in the RS across ecosystems [3]. Generally, there was an increasing trend for RS as soil texture changed from clay (0.22) to sand (0.25) for forests (Table 1). Lower water and nutrient availability from sandy soil may be responsible for the larger RS. In contrast, RS values of grasslands showed trend of decreasing when soil texture changed from clay to sand (Table 1). Our results also found that RS is more sensitive to MAT than MAP in grasslands. Lower temperatures in clay soil caused roots to gain more biomass to maintain energy and acquire water and nutrients [21,31], which may be responsible for the higher RS for clay grasslands.

One of key contributions of this study is that we comprehensively examined the responses of RS to multiply abiotic and biotic factors, including climatic variables (MAP, MAT, and PWDI), soil texture and vegetation type and age. The integrated analysis provided comprehensive understanding on the variations of RS and their response of RS to environmental factors. We found that RS could be significantly impacted by any of considered abiotic or biotic factors. Compared with any single factor from climate, a comprehensive index PWDI could be better to describe the dependency of RS on environmental factors. Biomass partition is commonly viewed as a result impacted by multiple environmental factors. Variation in RS and difference in their responses to individual or combined environmental factors supported the optimal partitioning hypothesis [29,30], suggesting an adaptation for plants in varied environments. It is widely recognized that RS varies with various biotic (e.g., stand age, vegetation type, and growing period) and abiotic factors (e.g., soil texture, soil nutrients, climate, and plant origin), as demonstrated in our analysis, although the mechanism is not well understood and inconsistencies between findings from existing studies are still quite large.

Differing from the analysis between RS and environmental factors, allometric equations have been also widely accepted as an effective method for investigating biomass partitioning between aboveground and belowground components (e.g., [1,34,35]). As an alternative, we attempt to examine whether the vegetation-specific RS presented in this study would provide a more accurate and easier means for estimating root biomass than allometric equations. We used the root and shoot biomass data collected in this study to compare the predictive accuracy of the two methods. We found that the application of median RS predicted root biomass with approximately the same accuracy as the allometric equations for 7 of the 13 vegetation groups (Table 2). Among the other 6 groups, we had a better prediction of root biomass using median RS for temperate and alpine grasslands, but lower predictive accuracy in loam, sandy, and clay grasslands and loam shrublands (Table 2). Moreover, the application of vegetation-specific RS for forests across soil textures had better predictive accuracy for root biomass than the single median RS for total forests in China. Similarly, the vegetation-specific RS for temperate and alpine grasslands also had better predictive accuracy for root biomass than the single median RS for total grasslands in China. This indicated that the accuracy of root biomass prediction varies significantly across vegetation types, particularly for grasslands, emphasizing that it is essential to advance our understanding of RS dynamics across vegetation types for improving the accuracy of root biomass predictions [12].

Conclusions

This paper systematically examined the variation in RS across China's terrestrial biomes and its influencing factors including vegetation type, soil properties, climatic variables, and stand age. We found that median RS tended to decrease from grasslands (6.0) to shrublands (0.73) to forests (0.23) in China. Although primarily decided by the plants' inherent isometric relationships, RS were significantly ($p<0.05$) influenced by vegetation type, soil texture, climatic variables, and stand age.

RS showed general trends with changing mean annual temperature, mean annual precipitation, and potential water deficit index. RS values were negatively related to mean annual precipitation and positively related to potential water deficit for forests, grasslands, and shrublands. For both forests and grasslands, RS were negatively related to mean annual temperature. Soil texture, forest origin, and climatic conditions affected the size of the RS and its relationships with climatic factors. Allometric equations performed better than empirical vegetation-specific RS equations in 6 out of 13 vegetation types.

Supporting Information

Appendix S1 Dataset on the root to shoot ratio, soil and vegetation types and climatic variables across China's terrestrial biomes.

Author Contributions

Conceived and designed the experiments: LHL. Performed the experiments: LMW XT. Analyzed the data: LMW. Contributed reagents/materials/analysis tools: XKW XC GPL. Wrote the paper: LHL XC GPL LMW. Compiled and formatted the data: LHL LMW.

References

1. Cairns MA, Brown S, Helmer EH, Baumgardner GA (1997) Root biomass allocation in the world's upland forests. Oecologia 111: 1–11.
2. Reich PB (2002) Root–shoot relations: optimality in acclimation and adaptation or the "Emperor's New Clothes"? In: Waisel Y, Eshel A, Kafkafi U, editors. Plant roots: The hidden half. Basel, Switzerland: Marcel Dekker. 205–220.
3. Mokany K, Raison RJ, Prokushkin AS (2005) Critical analysis of root:shoot ratios in terrestrial biomes. Glob Chang Biol 11: 1–13.
4. Poorter H, Nagel O (2000) The role of biomass allocation in the growth response of plants to different levels of light, CO_2, nutrients and water: A quantitative review. Aust J Plant Physiol 27: 595–607.
5. Binkley D, Stape JL, Ryan M (2004) Thinking about resource use efficiency in forests. For Ecol Manage 193: 5–16.
6. Niklas KJ (2005) Modeling below- and above-ground biomass for non-woody and woody plants. Ann Bot 95: 315–321.
7. Hui DF, Jackson RB (2006) Geographical and interannual variability in biomass partitioning in grassland ecosystems: a synthesis of field data. New Phytol 169: 85–93.
8. Yang YH, Fang JY, Ma WH, Guo DL, Mohammat A (2010) Large-scale pattern of biomass partitioning across China's grasslands. Global Ecol Biogeogr 19: 268–277.
9. Enquist BJ, Niklas KJ (2002) Global allocation rules for patterns of biomass partitioning across seed plants. Science 295: 1517–1520.
10. Niklas KJ (2006) A phyletic perspective on the allometry of plant biomass-partitioning patterns and functionally equivalent organ-categories. New Phytol 171: 27–40.
11. Cheng DL, Niklas KJ (2007) Above- and below-ground biomass relationships across 1534 forested communities. Ann Bot 99: 95–102.
12. Mokany K, Raison RJ, Prokushkin AS (2006) Critical analysis of root:shoot ratios in terrestrial biomes. Glob Chang Biol 12: 84–96.
13. Li LH, Yu Q, Zheng YF, Wang J, Fang QX (2006) Simulating the response of photosynthate partitioning during vegetative growth of winter wheat to environmental factors. Field Crops Res 96: 133–141.
14. Camila AC, Henrik S, Linda G, Annika N, Ulrika G, et al. (2011) Patterns of plant biomass partitioning depend on nitrogen source. PLoS One 6: e19211.
15. Yu Q, Wu H, He N, Lü X, Wang Z, et al. (2012) Testing the growth rate hypothesis in vascular plants with above- and below-ground biomass. PLoS One 7: e32162.
16. Wang XP, Fang JY, Zhu B (2008) Forest biomass and root–shoot allocation in northeast China. For Ecol Manage 255: 4007–4020.
17. Yang YH, Fang JY, Ji CJ, Han WX (2009) Above- and belowground biomass allocation in Tibetan grasslands. Journal of Vegetation Science 20: 177–184.
18. Luo YJ, Wang XK, Zhang XQ, Booth TH, Lu F (2012) Root:shoot ratios across China's forests: Forest type and climatic effects. For Ecol Manage 269: 19–25.
19. FAO (1998) Crop evapotranspiration–Guidelines for computing crop water requirements. Rome, Italy: Food and Agriculture Organization of the United Nations.
20. Brown S, Lugo AE (1982) The storage and production of organic matter in tropical forests and their role in the global carbon cycle. Biotropica 14: 161–187.
21. Crawley MJ (2002) Statistical computing–An introduction to data analysis using S-Plus. New York: John Wiley & Sons.
22. Jackson RB, Canadell J, Ehleringer JR, Mooney HA, Sala OE, et al. (1996) A global analysis of root distributions for terrestrial biomes. Oecologia 108: 389–411.
23. Ericsson T, Rytter L, Vapaavuori E (1996) Physiology of carbon allocation in trees. Biomass Bioenerg 11: 115–127.
24. Matsui Y, Fukuda Y, Inoue T, Matsushita T (2003) Effect of natural organic matter on powdered activated carbon adsorption of trace contaminants characteristics and mechanism of competitive adsorption. Water Res 37: 4413–4424.
25. Gholz HL, Vogel SA, Cropper WP Jr, McKelvey K, Ewel KC, et al. (1991) Dynamics of canopy structure and light interception in *Pinus elliottii* stands, north Florida. Ecol Monogr 61: 35–51.
26. Agren GI, Franklin O (2003) Root: shoot ratios, optimization and nitrogen productivity. Ann Bot 92: 795–800.
27. Read JJ, Morgan JA (1996) Growth and partitioning in *Pascopyrum smithii* (C3) and *Bouteloua gracilis* (C4) as influenced by carbon dioxide and temperature. Ann Bot 77: 487–496.
28. Li LH, Wang YP, Yu Q, Pak B, Eamus D, et al. (2012) Improving the responses of the Australian community land surface model (CABLE) to seasonal drought. J Geophys Res 117: G04002.
29. Friedlingstein P, Joel G, Field CB, Fung IY (1999) Toward an allocation scheme for global terrestrial carbon models. Glob Chang Biol 5: 755–770.
30. McConnaughay KDM, Coleman JS (1999) Biomass allocation in plants: ontogeny or optimality? A test along three resource gradients. Ecology 80: 2581–2593.
31. Gower ST, Vogt KA, Grier CC (1992) Carbon dynamics of Rocky Mountain Douglas scheme-fir: Influence of water and nutrient availability. Ecol Monogr 62: 43–65.
32. Schenk HJ, Jackson RB (2002) Rooting depths, lateral root spreads and below-ground/above-ground allometries of plants in water-limited ecosystems. J Ecol 90: 480–494.
33. Kellomäki S, Wang KY (2001) Photosynthetic responses to needle water potentials in Scots pine after a four-year exposure to elevated CO_2 and temperature. Tree Physiol 16: 765–772.
34. Vogt KA, Vogt DJ, Bloomfield J (1998) Analysis of some direct and indirect methods for estimating root biomass and production of forests at an ecosystem level. Plant Soil 200: 71–89.
35. Snowdon P, Eamus D, Gibbons P, Khanna PK, Keith H, et al. (2000) Synthesis of allometrics, review of root biomass and design of future woody biomass sampling strategies. NCAS Technical Report 17. Canberra: Australian Greenhouse Office.
36. Stearns SC (1992) The evolution of life histories. Oxford: Oxford University Press.
37. Kang M, Dai C, Ji W, Jiang Y, Yuan Z, et al. (2013) Biomass and its allocation in relation to temperature, precipitation, and soil nutrients in Inner Mongolia grasslands, China. PloS One, 8(7): e69561. doi:10.1371/journal.pone.0069561.
38. Luo YJ, Wang XK, Zhang XQ, Ren Y, Poorter H (2013) Variation in biomass expansion factors to forest type, climate and stand development. Annals of Forest Sciences, 70: 589–599.

7

Inorganic Nitrogen Supply and Dissolved Organic Nitrogen Abundance across the US Great Plains

Megan L. Mobley[1,2]*, Matthew J. Cleary[1], Ingrid C. Burke[1,2,3]

1 Department of Botany, University of Wyoming, Laramie, Wyoming, United States of America, 2 Helga Otto Haub School of Environment and Natural Resources, University of Wyoming, Laramie, Wyoming, United States of America, 3 Program in Ecology, University of Wyoming, Laramie, Wyoming, United States of America

Abstract

Across US Great Plains grasslands, a gradient of increasing mean annual precipitation from west to east corresponds to increasing aboveground net primary productivity (ANPP) and increasing N-limitation. Previous work has shown that there is no increase in net N mineralization rates across this gradient, leading to the question of where eastern prairie grasses obtain the nitrogen to support production. One as-yet unexamined source is soil organic N, despite abundant literature from other ecosystems showing that plants take up dissolved soil organic N. This study measured KCl-extractable dissolved organic N (DON) in surface soils across the grassland productivity gradient. We found that KCl-extractable DON pools increased from west to east. If available to and used by plants, this DON may help explain the high ANPP in the eastern Great Plains. These results suggest a need for future research to determine whether, in what quantities, and in what forms prairie grasses use organic N to support primary production.

Editor: Hormoz BassiriRad, University of Illinois at Chicago, United States of America

Funding: This research was funded by the University of Wyoming Excellence Funds. The funders had no role in study design, data collection and analysis, decision to publish, or preparation of the manuscript.

Competing Interests: The authors have declared that no competing interests exist.

* Email: meg.mobley@alumni.duke.edu

Introduction

It has in the past been widely accepted that nitrogen compounds must undergo mineralization – that is, transformation from organic to inorganic forms - to become available to plants [1]. Yet, in a variety of ecosystems, plants have been shown to take up dissolved organic nitrogen (DON) directly from the soil, as reviewed by [2,3]. These plant-accessible organic N compounds include amino acids, amino sugars, amino-polysaccharides, nucleic acids, and various purine and pyrimidine compounds [2]. A new paradigm concerning N cycling has gained acceptance [2], in which the rate of depolymerization of organic N polymers into monomers by microbial enzymes, i.e., the production of DON, regulates ecosystem N cycling rather than the rate of N mineralization.

The ecological importance of DON acquisition is still unclear in many ecosystems [4], including temperate grasslands. Jones et al. [5,6] suggest that temperate grassland species can take up DON directly from the soil, although it is unclear how important DON is relative to other sources. Knowledge about the magnitude and importance of plant uptake of DON is important for our understanding and predictions of N cycling using models, and for clarifying the dynamics of plant-microbial competition. Understanding the potential for grassland plant use of DON could lead to management strategies that reduce fertilizer use and associated nutrient runoff and pollution while still supporting productive and N-retentive cereal croplands and biofuel-producing grasslands.

For grasslands of the United States, the combined results of previous research do not frame a coherent understanding of grassland N cycling. A number of studies have shown grassland plant production to respond to experimental N additions, suggesting that these ecosystems are N–limited [7–10]; a literature review concluded that although water was an important co-limiter of the semi-arid western grasslands, N-limitation was the rule across the Great Plains [11]. Aboveground net primary production (ANPP) increases with mean annual precipitation from semiarid shortgrass steppe in the west through mixed grass prairie to the subhumid tallgrass prairie in the east [12]. Both the concentration of N in biomass and the litter decomposition rate decrease, and nutrient use efficiency increases, along this precipitation gradient [13]. Nevertheless, plant N uptake, as indicated by the quantity of N in aboveground production (ANPP-N), increases along this precipitation and production gradient [14,15]. These lines of evidence suggest that the severity of N-limitation increases from west to east.

To explain observed patterns of ANPP, simulations of the Century model predict a linear increase of net N mineralization rates with precipitation [16], yet field studies have reported no west-to-east trend in net N mineralization [14,15]. Total soil N, on the other hand, has been shown to be higher in tallgrass than shortgrass prairie [13–15,17], suggesting that organic N may play a role in supplying the additional N needed to support eastern tallgrass prairie production. This could explain the discrepancy between model predictions and field studies, as commonly used ecosystem models do not consider the rate of organic N depolymerization or plant uptake of organic N, focusing only on inorganic N mineralization [18–21].

This exploratory study examined the spatial pattern in abundances and rates of production of KCl-extractable dissolved inorganic N (DIN) and DON along the west to east precipitation gradient across US grasslands.

Materials and Methods

Site Description

The Nature Conservancy, Fort Hays State University, USDA-ARS, and the Long-Term Ecological Research programs at Konza Prairie and Shortgrass Steppe provided access to field sites spanning from eastern Colorado to eastern Kansas [14,15,17]. The entire gradient spans precipitation and temperature ranges of 500 mm and 4.4°C, respectively. The two westernmost sites, Shortgrass Steppe (SGS) Long-Term Ecological Research (LTER) and Arikaree River (ARI) Nature Conservancy site, are located in the drier shortgrass steppe. Two sites, Smokey River (SVR) Nature Conservancy site and a property of Ft. Hays State University (HAYS), were located in the northern mixed grass prairie. The easternmost site was located in the wetter tallgrass prairie, Konza Prairie LTER (KON) (Table 1). Each sampling location was located on native upland grasslands with moderate grazing by cattle in the summer and minimal disturbance. The soils at all sites ranged from loam to sandy loam. The vegetation of the shortgrass steppe was dominated by *Bouteloua gracilis* (Willd. ex Kunth) Lag. ex Griffiths and *Buchloë dactyloides* (Nutt.) J.T. Columbus, whereas *Andropogon gerardii* Vitman and *Schizachyrium scoparium* (Michx.) Nash dominated the tallgrass prairie. The mixed-grass prairie comprised a combination of all of the species above.

Field Sampling

At each site, we set up three 25 m long transects within a 100 m radius of each other in undisturbed areas representative of the surrounding vegetation. From June 6–15, 2011, we collected soil cores (10 cm length ×5 cm diameter) from two microsites (under plant canopies and interspaces between plants) and at two depths (0–5 cm and 5–10 cm) at three random locations on each transect, for a total of 180 soil samples (2 microsites ×2 depths ×3 locations ×3 transects ×5 sites). Soils were immediately sieved through a 2 mm mesh and homogenized in a plastic bag.

We extracted soils immediately for dissolved N pools, in order to minimize disturbance-induced N transformations associated with delays in processing, and to obtain the most realistic estimates of DON and DIN concentrations possible [22]. We extracted dissolved pools of N following the procedure outlined by Robertson et al. [23] and Bardgett et al. [24]. A 10 g subsample of field-fresh, sieved, homogenized soil was placed in an extraction cup with 100 mL 2M KCl solution and shaken for 10 minutes. The extraction cups were returned to the laboratory at the University of Wyoming, Laramie, WY, to be filtered and frozen within two days of sampling. The remaining soil was air-dried and stored in paper bags until further analysis.

We installed one pair of Plant Root Simulator (PRS) resin probes (Western Ag Innovations Inc., Saskatoon, Saskatchewan) per microsite at 10 random locations along each transect (2 microsites ×10 locations ×3 transects ×5 sites = 300 pairs), to estimate inorganic N supply rate. Each pair consisted of one anion and one cation probe to collect mobile NO_3^- and NH_4^+, respectively. Probes were inserted into the soil with the adsorption membrane extending from the surface to 5.5 cm depth, and were buried for 30 days from early May to early June 2011. At each site, the probes were removed on the same day as the soil sampling. Probes were sorted into two groups of five pairs of probes from each microsite type. This resulted in two replicate sets of probes per each microsite type per each transect for a total of 60 (2 replicates ×2 microsites ×3 transects ×5 sites) probe sets. Probes were washed in the laboratory with DI water within two days of unearthing and then sent to Western Ag Innovations, Inc. (Saskatoon, Saskatchewan) for inorganic N analysis.

Laboratory Analyses

The eluate from the extraction cups was shaken for 1 hour and filtered in the laboratory using Whatman #1 filter paper and frozen immediately thereafter. For measurement of total KCl-extractable dissolved total N and inorganic N, we thawed the eluate and diluted a 2.5 mL subsample with 17.5 mL DI water. We analyzed total dissolved N on a Shimadzu TOC-V Analyzer (Shimadzu Scientific Instruments, Wood Dale, IL, USA). Dissolved inorganic N was analyzed at the Natural Resource Ecology Laboratory (Colorado State University, Fort Collins, CO) using an Alpkem Flow Solution IV Automated wet chemistry system (O.I. Analytical, College

Table 1. Locations and attributes of grassland sites.

	SGS: Shortgrass Steppe LTER	ARI: Arikaree River TNC	SVR: Smokey River TNC	HAYS: Ft. Hays State Univ.	KON: Konza Prairie LTER
Latitude, Longitude	40°48.8′N, 104°42.8′W	39°45.4′N, 102°26.9′W	38°53.6′N, 100°58.7′W	38°52.4′N, 99°23.3′W	39°4.8′N, 96°33′W
MAP (mm)	320	450	502	575	835
MAT (°C)	8.6	9.7	10.8	11.9	13.0
Grassland Type	Shortgrass Steppe	Shortgrass Steppe	Mixed Grass Prairie	Mixed Grass Prairie	Tallgrass Prairie
ANPP-N [15] ($g\ N\ m^{-2}\ y^{-1}$)	1.1	1.7	1.5	1.6	2.4
ANPP-C [15] ($g\ C\ m^{-2}\ y^{-1}$)	143	123	221	211	337
Net N min. [14] ($g\ N\ m^{-2}\ y^{-1}$)	1.5 (0.49)	3.0 (1.04)	1.5 (0.53)	3.0 (0.63)	2.4 (0.26)
Soil C ($g\ C\ m^{-2}$)	1367 (67.9)	1440 (68.6)	1851 (52.9)	5204 (155.4)	2373 (151.2)
Soil N ($g\ N\ m^{-2}$)	287.2 (13.1)	274.2 (13.0)	214.6 (13.3)	285.1 (5.9)	210.7 (11.0)
Soil C:N	4.91 (0.31)	5.33 (0.25)	9.05 (0.50)	18.3 (0.44)	11.2 (0.32)
Volumetric Soil Moisture (%)	13.8 (0.40)	7.3 (0.53)	3.7 (0.17)	3.4 (0.36)	6.3 (0.30)

MAP = mean annual precipitation (mm), MAT = mean annual temperature (°C), TNC = The Nature Conservancy, LTER = National Science Foundation Long-Term Ecological Research site. Data in lower panel is from this study; values in parentheses represent one standard error of the mean.

Station, TX, USA). We estimated KCl-extractable DON as the difference between total dissolved N and dissolved inorganic N.

We mixed and subsampled remaining air-dried soils for analysis of total C and N. Subsamples were ground on a rotary soil grinder (Planetary Micro Mill Pulverisette 7, Fritsch GmbH, Idar-Oberstein, Germany), and 2 mg of ground soil were analyzed on a Perkin Elmer 2400 Elemental Analyzer (Perkin Elmer, Inc., Waltham, MA, USA) for total C and N. Volumetric water content was estimated after drying a 100 g subsample of field moist soil at 105°C for one day [23].

Statistical Analyses

We performed all analyses in the statistical program R, version 2.14.2 [25]. Extractable DIN and DON concentrations were multiplied by the soil sample depth (5 cm) and bulk density (mean = 0.877 g cm^{-3}) to obtain DIN and DON pools on a per-area basis. Data for the two soil depth increments (0–5 cm, 5–10 cm) were aggregated such that all analyses were of soil properties and constituents to 10 cm depth, with the exception of PRS probe data that characterized the top 5 cm of soil. We employed two-way ANOVAs coupled to Tukey multiple comparison tests [26] to test differences in the dependent variables total soil C (g C m^{-2}), N (g N m^{-2}), C:N (mass C:N), and 0–5 cm soil inorganic N supply rate (µg N 10 cm^{-2} $month^{-1}$) among sites and between microsites. Two-way ANCOVAs tested differences in the dependent variables total soil KCl-extractable DIN (g DIN m^{-2}) and KCl-extractable DON (g DON m^{-2}) among sites and between microsites with volumetric soil moisture as a continuous independent variable. One extreme outlier value for DON with strong influence on the ANCOVA (DON>17 g N m^{-2}; outlier test Bonferonni $p = 4.4\times10^{-26}$) was excluded from DON analyses. ANOVAs were called using aov(), with drop1(model, test = "F") to extract type III sums of squares. Multiple comparisons were performed using TukeyHSD(). We performed linear regressions using lm() to assess the effect of soil moisture on DIN and DON measurements, and to test for a relationship between KCl-extractable DON pools and MAP. A paired *t*-test tested differences in DON and DIN abundance (g N m^{-2}) across all sites and microsites. Significance was assessed at $\alpha = 0.05$. Data for this study can be found in the Supporting Information.

Results

Soil C differed significantly among sites (among-sites $F = 212$, d.f. = 4, $p<0.0001$) and was significantly higher in the tallgrass prairie than shortgrass steppe ($p<0.0001$, Table 1). Total soil N (g N m^{-2}) decreased from west to east across the grassland gradient (among-sites $F = 10.7$, d.f. = 4, $p<0.0001$, Table 1). Soil C:N increased significantly from west to east across the gradient (among-sites $F = 211$, d.f. = 4, $p<0.0001$). There were no significant differences in soil C ($p = 0.439$), N ($p = 0.537$), or C:N between under-plant and plant interspace microsites ($p = 0.101$).

The 30-day inorganic N supply rate estimated with PRS probes was highly variable and showed a slight decline across the gradient, with the supply rate at the easternmost site significantly lower than that of the westernmost site (among-sites $F = 8.07$, d.f. = 4, $p<0.0001$; Figure 1a). There were no significant differences in inorganic N supply rates between microsites.

The pool of KCl-extractable dissolved inorganic nitrogen (DIN) was greater in the eastern tallgrass prairie compared to the western shortgrass steppe (among-sites $F = 44.9$, d.f. = 4, $p<0.0001$; Figure 1b). KCl-extractable dissolved organic N was significantly more abundant in the tallgrass region compared to the shortgrass steppe (among-sites F = 10.6, d.f. = 4, p<0.0001; Figure 1b).

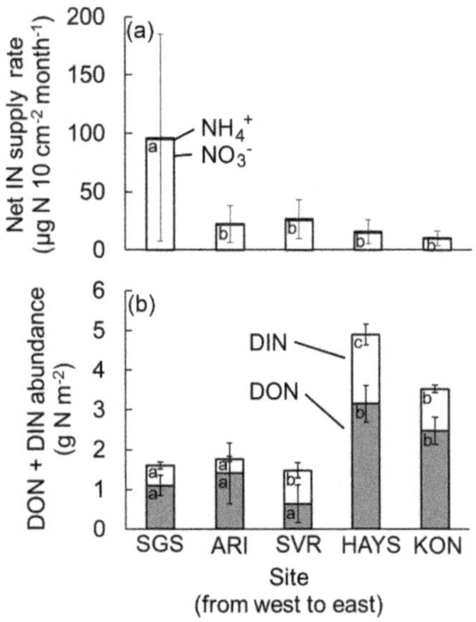

Figure 1. Inorganic N supply rates and extractable dissolved N pools at grassland sites. a) 0–5 cm soil net inorganic N supply rates (NH_4^+ in black bars, NO_3^- in white bars) at five grasslands of the US Great Plains, arranged from west to east. Error bars represent 95% confidence intervals on total (NH_4^++NO_3^-) inorganic N supply rate. Lowercase letters represent significant site differences at $\alpha = 0.05$. b) Total extractable dissolved soil N (DIN+DON) to 10 cm soil depth across the US Great Plains. Lowercase letters in the gray columns represent significant differences in DON among sites; lowercase letters in the white columns indicate significant differences in DIN among sites. Multiple comparisons were performed on output of ANOVAs that did not include soil moisture as a covariate. Error bars represent 95% confidence intervals. Significance assessed at $\alpha = 0.05$. One extreme outlier (DON>17 g N m^{-2} at SVR) was excluded from the analysis.

There were no significant differences in DIN ($p = 0.128$) or DON ($p = 0.461$) between microsites. Across all sites, abundances of KCl-extractable DON were significantly higher than DIN (paired $t = 6.648$, d.f. = 82, $p<0.0001$). Both DIN (F = 10.4, d.f. = 79, $R^2 = 0.106$, p = 0.0018) and DON (F = 21.1, d.f. = 78, $R^2 = 0.202$, p<0.0001) increased with soil moisture (Figure 2), but soil moisture was not significant as a covariate of DIN ($p = 0.277$) or DON ($p = 0.241$) in the ANCOVAs. There was a positive relationship between DON measurements and site-level ANPP (F = 6.257, d.f. = 42, $R^2 = 0.130$, $p = 0.0164$; Figure S1).

Discussion

This study tested the hypothesis that extractable dissolved organic N pools would increase along a precipitation and production gradient from west to east across US grasslands. Unlike two previous studies at the same sites [14,15], we did not find a west to east increase in total soil N; soil C and C:N did increase from west to east as in previous studies. Also like previous studies, we found no increase in inorganic N supply rates from west to east, even though plant uptake does increase (ANPP-N, Table 1). This supports previous conclusions that N becomes more limiting, and the N cycle grows tighter, from west to east. We did observe, however, that pools of KCl-extractable dissolved inorganic and organic N, representing exchangeable + free soil N [27], increased from west to east.

These data must be considered as suggestive, rather than conclusive, given a number of limitations to this small study. Our one-time measurements of KCl-extractable DIN and DON and

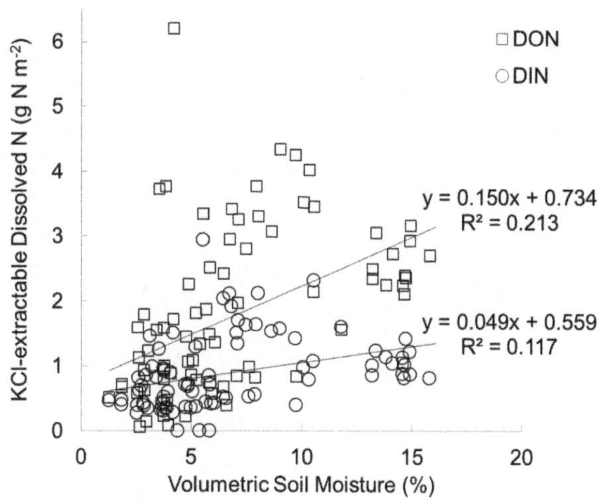

Figure 2. Simple linear regressions of KCl-extractable soil dissolved inorganic (DIN; circles) and organic nitrogen (DON; squares) against volumetric soil moisture for 0–10 cm soil samples from five sites across the Great Plains.

our one resin probe incubation give an incomplete characterization of the long-term mean pools and rates for these grassland sites. Pool size is not necessarily indicative of cycling rate or ecological importance, as a large pool might result from high production or low demand. We did not characterize the chemical composition or molecular weight distribution of DON, thus we cannot speculate on how much of the DON is potentially plant-available. There are artifacts associated with each of the field methods that could reduce the sensitivity of N pool and production estimates. The soil and root disturbance associated with soil sampling may release DON that would not naturally be available in the soil [4], and use of KCl as an extractant may result in a pool of extractable dissolved N that is quite different in composition, quantity, and quality than that found under natural conditions [28,29]. As evidenced by the significant relationships between KCl-extractable N pools and soil moisture (Figure 2), the KCl-extraction likely overestimates what is actually plant-accessible under normal dry soil conditions. Because less nitrate is likely to reach the resin probes when soil moisture is limiting, regardless of actual mineralization rates, we consider resin probes to be better indicators than extractions of the available N that is accessible by plants [30]. These methodological artifacts should not affect comparisons among sites within a given study if consistent methods are used at all sites, as was the case in this study and previous studies at these sites [14,15].

Alternative or additional sources of ANPP-N may include biological N fixation and atmospheric N deposition. N fixation, largely by free-living organisms, has been estimated at <0.1 g N m^{-2} y^{-1} in shortgrass steppe [31] and ~0.8 g N m^{-2} y^{-1} in tallgrass prairie [32], although it is unclear how much of that fixed N enters the plant-soil N cycle versus that volatilized or denitrified [33]. Total wet+dry inorganic N deposition increases from approximately 0.35 g N m^{-2} y^{-1} at SGS to approximately 0.71 g N m^{-2} y^{-1} at KON [34,35], and so could also provide a portion of the additional N required to support ANPP.

This study provides one piece of support for the broader hypothesis that organic N may serve as a nitrogen source for grassland production—a source that would explain how plant N uptake (ANPP-N) is able to increase along the precipitation gradient without a concomitant increase in N mineralization rates. Our results show that a sizeable pool of KCl-extractable dissolved organic N resides in these grassland soils, and that this N pool increases from west to east across the Great Plains precipitation gradient. The larger KCl-extractable DON pools at the eastern end of the precipitation gradient open the possibility that DON as a source of plant N may help resolve discrepancies in between model predictions and measurements of available N. If so, ecosystem models such as Biome-BGC, Century, PnET, and TEM may be missing a source of N for grasslands. These models represent the N required for primary productivity as "mineral N" generated by N mineralization [18–21]. If organic N, which these results indicate is abundant at these sites, is important for plant growth, then an important nutrient source is being left out of model simulations.

Conclusions

We found that the abundance of DON increased across the west-to-east along the precipitation and productivity gradient of US Great Plains grasslands, suggesting that it could be a significant source of N if plants are able to utilize it. If organic N is an important component of available N in US grasslands, it may explain why ecosystem models that do not include plant use of organic N are unable to predict N supply rates across the US grasslands. Additional studies must be conducted to determine if grass species dominating the US Great Plains assimilate biologically-relevant quantities of organic N.

Supporting Information

Figure S1 The relationship between KCl-extractable soil DON and ANPP. Simple linear regression of KCl-extractable soil dissolved organic nitrogen (DON; microsites and depths for each sample location mathematically combined to yield $n = 9$ replicate measurements per site) against site-level mean aboveground net primary production (ANPP; [15]) for five grassland sites across the US Great Plains.

Data S1 Bulk soil C and N, KCl-extractable N, and inorganic N supply rate data.

Acknowledgments

We wish to thank The Nature Conservancy, Fort Hays State University, USDA-ARS, and the Long-Term Ecological Research Sites at Konza Prairie and Shortgrass Steppe for access to field sites. We thank Colin Pinney (EcoCore Analytical Services, a Colorado State University Core Facility) and Leann Naughton for assistance with nitrogen analysis. This manuscript was improved by comments from William Lauenroth, Urszula Norton, David Bell, Daniel Schlaepfer, Otgonsuren Avirmed, Ariuna Lkhagva, Lynn Moore, Cait Rottler, and four anonymous reviewers.

Author Contributions

Conceived and designed the experiments: MJC ICB. Performed the experiments: MJC. Analyzed the data: MJC MLM. Contributed reagents/materials/analysis tools: ICB. Wrote the paper: MJC MLM ICB.

References

1. Stanford G, Smith S (1972) Nitrogen mineralization potentials of soils. Soil Science Society of America Journal 36: 465–472.
2. Schimel JP, Bennett J (2004) Nitrogen mineralization: challenges of a changing paradigm. Ecology 85: 591–602.
3. Neff JC, Chapin FS, Vitousek PM (2003) Breaks in the cycle: dissolved organic nitrogen in terrestrial ecosystems. Frontiers in Ecology and the Environment 1: 205–211.
4. Hobbie JE, Hobbie EA (2013) Microbes in nature are limited by carbon and energy: the starving-survival lifestyle in soil and consequences for estimating microbial rates. Frontiers in Microbiology 4: Article 324.
5. Jones DL, Healey JR, Willett VB, Farrar JF, Hodge A (2005) Dissolved organic nitrogen uptake by plants–an important N uptake pathway? Soil Biology and Biochemistry 37: 413–423.
6. Jones DL, Shannon D, Murphy DV, Farrar J (2004) Role of dissolved organic nitrogen (DON) in soil N cycling in grassland soils. Soil Biology and Biochemistry 36: 749–756.
7. Dodd JL, Lauenroth WK (1979) Analysis of the Response of a Grassland Ecosystem to Stress. In: French NR, editor.Perspectives in Grassland Ecology: Springer New York.pp. 43–58.
8. Lauenroth WK, Dodd JL, Sims PL (1978) The effects of water- and nitrogen-induced stresses on plant community structure in a semiarid grassland. Oecologia 36: 211–222.
9. Risser PG, Parton WJ (1982) Ecosystem Analysis of the Tallgrass Prairie: Nitrogen Cycle. Ecology 63: 1342–1351.
10. Clenton EO, Hyde RM, Anderson KL (1970) Effects of Clipping and Supplemental Nitrogen and Water on Loamy Upland Bluestem Range. Journal of Range Management 23: 341–346.
11. Hooper D, Johnson L (1999) Nitrogen limitation in dryland ecosystems: Responses to geographical and temporal variation in precipitation. Biogeochemistry 46: 247–293.
12. Sala OE, Parton W, Joyce L, Lauenroth W (1988) Primary production of the central grassland region of the United States. Ecology 69: 40–45.
13. Murphy KL, Burke IC, Vinton MA, Lauenroth WK, Aguiar MR, et al. (2002) Regional analysis of litter quality in the central grassland region of North America. Journal of Vegetation Science 13: 395–402.
14. McCulley RL, Burke IC, Lauenroth WK (2009) Conservation of nitrogen increases with precipitation across a major grassland gradient in the Central Great Plains of North America. Oecologia 159: 571–581.
15. Barrett J, McCulley R, Lane D, Burke I, Lauenroth W (2002) Influence of climate variability on plant production and N-mineralization in Central US grasslands. Journal of Vegetation Science 13: 383–394.
16. Burke IC, Lauenroth WK, Parton WJ (1997) Regional and temporal variation in net primary production and nitrogen mineralization in grasslands. Ecology 78: 1330–1340.
17. Vinton MA, Burke IC (1997) Contingent effects of plant species on soils along a regional moisture gradient in the Great Plains. Oecologia 110: 393–402.
18. Parton W, Schimel DS, Cole C, Ojima D (1987) Analysis of factors controlling soil organic matter levels in Great Plains grasslands. Soil Science Society of America Journal 51: 1173–1179.
19. McGuire AD, Melillo JM, Kicklighter DW, Joyce LA (1995) Equilibrium responses of soil carbon to climate change: empirical and process-based estimates. Journal of Biogeography 22: 785–796.
20. Running SW, Hunt ER Jr (1993) Generalization of a forest ecosystem process model for other biomes, BIOME-BGC, and an application for global-scale models. In: Ehleringer JR, Field CB, editors.Scaling Physiological Processes: Leaf to Globe.New York: Academic Press, Inc. pp. 141–158.
21. Aber JD, Ollinger SV, Driscoll CT (1997) Modeling nitrogen saturation in forest ecosystems in response to land use and atmospheric deposition. Ecological Modelling 101: 61–78.
22. Van Miegroet H (1995) Inorganic nitrogen determined by laboratory and field extractions of two forest soils. Soil Science Society of America Journal 59: 549–553.
23. Robertson GP, Wedin DA, Groffman PM, Blair JM, Holland EA, et al. (1999) Soil carbon and nitrogen availability. Nitrogen mineralization, nitrification and soil respiration potentials. In: Robertson GP, Coleman DC, Bledsoe CS, Sollins P, editors.Standard Soil Methods for Long-Term Ecological Research.New York: Oxford University Press. pp. 258–271.
24. Bardgett RD, Streeter TC, Bol R (2003) Soil microbes compete effectively with plants for organic-nitrogen inputs to temperate grasslands. Ecology 84: 1277–1287.
25. R Development Core Team (2012) R: A Language and Environment for Statistical Computing. 2.14.2 ed. Vienna, Austria: R Foundation for Statistical Computing.
26. Hsu JC (1996) Multiple comparisons: theory and methods. London: Chapman & Hall. xiv, 277 p., 274 p. of plates p.
27. Näsholm T, Kielland K, Ganeteg U (2009) Uptake of organic nitrogen by plants. New Phytologist 182: 31–48.
28. Inselsbacher E, Näsholm T (2012) The below-ground perspective of forest plants: soil provides mainly organic nitrogen for plants and mycorrhizal fungi. New Phytologist 195: 329–334.
29. Jones DL, Willett VB (2006) Experimental evaluation of methods to quantify dissolved organic nitrogen (DON) and dissolved organic carbon (DOC) in soil. Soil Biology and Biochemistry 38: 991–999.
30. Hook PB, Burke IC (1995) Evaluation of methods for estimating net nitrogen mineralization in a semiarid grassland. Soil Science Society of America Journal 59: 831–837.
31. Copley PW (1972) Evaluation of biological N_2 fixation in a grassland ecosystem. Fort Collins, CO: Colorado State University. 90 p.
32. DuBois JD, Kapustka LA (1983) Biological Nitrogen Influx in an Ohio Relict Prairie. American Journal of Botany 70: 8–16.
33. West NE, Skujins J (1977) The nitrogen cycle in North American cold-winter semi-desert ecosystems. Oecologia plantarum 12: 45–53.
34. National Atmospheric Deposition Program (NRSP-3) (2011) NADP Program Office, Illinois State Water Survey, 2204 Griffith Dr., Champaign, IL 61820.
35. U.S. Environmental Protection Agency Clean Air Markets Division (2011) Clean Air Status and Trends Network (CASTNET) Annual Total Deposition.

8

Assessment of the Soil Organic Carbon Sink in a Project for the Conversion of Farmland to Forestland: A Case Study in Zichang County, Shaanxi, China

Lan Mu[1,2], Yinli Liang[1,2]*, Ruilian Han[1,2]

1 Institute of Soil and Water Conservation, Northwest A&F University, Yangling Shaanxi Province, China, **2** Institute of Soil and Water Conservation, Chinese Academy of Sciences and Ministry of Water Resources, Yangling, Shaanxi Province, China

Abstract

The conversion of farmland to forestland not only changes the ecological environment but also enriches the soil with organic matter and affects the global carbon cycle. This paper reviews the influence of land use changes on the soil organic carbon sink to determine whether the Chinese "Grain-for-Green" (conversion of farmland to forestland) project increased the rate of SOC content during its implementation between 1999 and 2010 in the hilly and gully areas of the Loess Plateau in north-central China. The carbon sink was quantified, and the effects of the main species were assessed. The carbon sink increased from 2.26×10^6 kg in 1999 to 8.32×10^6 kg in 2010 with the sustainable growth of the converted areas. The black locust (*Robinia pseudoacacia L.*) and alfalfa (*Medicago sativa L.*) soil increased SOC content in the top soil (0–100 cm) in the initial 7-yr period, while the sequestration occurred later (>7 yr) in the 100–120 cm layer after the "Grain-for-Green" project was implemented. The carbon sink function measured for the afforested land provides evidence that the Grain-for-Green project has successfully excavated the carbon sink potential of the Shaanxi province and served as an important milestone for establishing an effective organic carbon management program.

Editor: Wen-Xiong Lin, Agroecological Institute, China

Funding: This work was financially supported by the National Science and Technology Support Program (2011BAD31B05) and by the National Science and Technology Support Program (2014BAD14B006). The funders had no role in study design, data collection and analysis, decision to publish, or preparation of the manuscript.

Competing Interests: The authors have declared that no competing interests exist.

* E-mail: liangyl@ms.iswc.ac.cn

Introduction

Soil organic carbon (SOC) is an important component of soil that plays a key role in the functions of both natural and agricultural ecosystems. In ecosystem services, SOC is critical for ensuring sustainable food production owing to its nutrient retention function and water-holding capacity [1,2]. The global SOC stock has been estimated to be 1400–1500 Pg C in the upper 100 cm soil layer [3–5], which is approximately twice the amount of C in the atmosphere and three times the amount stored in terrestrial vegetation [6]. Todd-Brown et al [7] reported that the present-day global SOC stocks range from 514 to 3046 Pg C among 11 earth system models (ESMs). Thus, slight reductions in SOC contents due to changes in land-use, soil management, or rates of soil erosion, could significantly raise the CO_2 in the atmosphere. Due to natural drought conditions, intensive human disturbance and severe soil erosion, the hilly-gully area of the Loess Plateau has the lowest soil organic carbon density (SOCD) in China [8]. However, it is possible to increase the organic carbon content and carbon sequestration capacity in the soils of this region through appropriate reforestation of degraded sloping croplands and other ecosystems, whose resilience capacity is intact [1,9].

Changes in land use may alter the land cover patterns in ways that can impact the biomass and soil carbon stocks [10] and alter the rate of C input and output to the soil, ultimately changing the soil C content [11]. Soil C sinks are not permanent and often persist only as long as appropriate management practices (conservation tillage and erosion control) are maintained. When a land-management or land-use change is reversed, the C accumulated as a result of the change is lost, usually more rapidly than it was accumulated [12]. And greater attention to the possibility of encouraging the growth of the forests as a means of removing carbon dioxide (CO_2) from the atmosphere [13]. Several studies have addressed the effects of the conversion of farmland to forestland on the SOC. In a long-term experiment, Guo and Gifford [14] showed that the conversion of forestland or grassland to farmland resulted in significant SOC reduction; the conversion of forestland to both farmland and grassland did not lead to SOC reduction in all cases, but always resulted in significant carbon sequestration from the atmosphere. Ostle et al [15] also found that soil carbon losses occur when grasslands, managed forest lands or native ecosystems are converted to croplands, and soil carbon gains are made when croplands are converted to grasslands, forest lands or native ecosystems. Carbon stored in forest ecosystems represents a substantial portion of the global C stock; worldwide, forests contain ~70% of all plant C and ~20% of all soil C [16]. The conversion of farmland to forestland is therefore an effective method for preventing soil erosion, which also has a large net sink effect on the SOC storage in the hilly-gully areas of China. However, the balance between inputs of organic matter, primarily from vegetation, and losses, as a result of decomposition, leaching and erosion, determines the magnitude of the carbon reservoir of the hilly-gully areas of China. The net CO_2 released via soil and water loss is calculated to be 8.4 g C m^{-2} yr^{-1} [17].

The hilly-gully areas of the Chinese Loess Plateau are known for their agricultural history and severe soil erosion [18]. Since the 1950s, the Chinese government has made substantially efforts to control the soil erosion and restore vegetation to the region. For example, an extensive tree-planting project was undertaken in the 1970s, and integrated soil erosion control was performed on the watershed scale in the 1980s and 1990s. Despite these efforts, the soil erosion remained unchecked, and vegetation had not been substantially restored to the region by the late 1990s [10]. In 1999, an extensive ecological rehabilitation program known as "Grain-for-Green" (conversion of farmland to forestland) was initiated by the Chinese government; this program is currently the largest land retirement project undertaken in the developing world. The Grain-for-Green program was initially intended to reduce soil erosion and desertification and increase the forest cover in China by retiring steep sloping and marginal lands from agricultural production in the hilly-gully area of the Loess Plateau [19]. The program has now been operating for about 10 years and the natural environment in parts of the Loess Plateau is improving as annual crops are replaced by perennial plants. However, limited attention has been paid to the effect of land use conversion on soil carbon sequestration in the Loess Plateau [20,21]. The Grain-for-Green program has potential for C gain through the improved land use conversions [[19] and we hypothesized that the conversion of farmland to forestland is one method that may increase SOC storage in soils in this study region in Loess Plateau.

Therefore, the objective of this paper was to assess the changes in the soil organic carbon content and carbon sink benefit following the implementation of the Grain-for-Green project on the Loess Plateau between 1999 and 2010. The paper also discussed the variation of the soil organic carbon content with the type of vegetation and soil age in the study area and assesses the effect of the land cover change project.

Material and Methods

Research region

Zichang County was located in the northern Shaanxi Province of the Loess Plateau in China within latitudes 36°59′–37°30′N and longitudes 109°11′–110°01′E. The region had a semiarid continental monsoon climate with a mean annual temperature of 9.1°C and mean annual precipitation of 514.7 mm. Over 70% of the precipitation occurred between June and September. The landform had the typical hilly-gully topology of the Loess Plateau, and the valleys and ravines accounted for approximately 94.6% of the total area, in which the gully density was 8.1 km km^{-2}. Soils in the research area originated from parent material of calcareous loess, with a low SOC concentration of less than 10 g kg^{-1}. These belong to the Calcic Cambisol group according to the Food and Agriculture Organization and United Nations Education Scientific and Cultural Organization (FAOUNESCO) soil classification system [10]. The soil type was loessal soil with a pH value of 8.6. The soil bulk density was 1.4 g cm^{-3} and total N content was 0.4 g kg^{-1}. The area of the croplands was 11.08×10^4 ha, accounting for 46.24% of the total land area. In this area, agricultural land occupied 39.5% of the area on slopes exceeding 25° (Table 1); this cultivated steep carried soil load of approximately 229.4×10^4 t yr^{-1} and a water loss of up to 857.8×10^4 m^3 yr^{-1}. The natural vegetation had been largely destroyed by deforestation and cultivation. Slope land reclamation and cultivation along the slope was the primary cause of soil and water loss.

Table 1. The area of the cultivated slope land classification of the ZiChang county.

slope (degree)	area	percentage
	hm^2	%
<5	17927.87	16.9
5~10	3050.33	2.9
10~15	9950.33	9.4
15~25	33247.4	31.3
25~35[a]	32532.93	30.7
>35	9362.93	8.8

[a]The slopes of greater than 25°were cultivated for agriculture occupied 39.5%.

Land degradation, desertification, soil erosion and declining soil fertility had degraded the environment and severely limited the crop yield. In addition, the SOC concentrations in these areas were less than 10 g C kg^{-1}. Consequently, an extensive ecological rehabilitation program known as "Grain-for-Green" was initiated in 1999 in the Loess Plateau region. Zichang was the demonstration county for the larger Shaanxi project. The program in the county had now been implemented for approximately 10 yr (since 1999), and the natural environment was improving as annual crops were replaced by perennial plants and as the vegetation cover increases. The vegetation coverage rate had increased from 6.16% to 32.8%, and the greening rate in the county had reached 62% (Figure 1). And the vegetation reduced the soil and water lost with the amount of approximately 27.5×10^4 t yr^{-1} and 102.9×10^4 m^3 yr^{-1}, respectively (the data come from the Bureau of Soil and Water conservation, Zichang county). These changes had significantly improved the ecological environment, reduced soil and water loss and produced a carbon sink effect in the area [22].

Land cover changes between 1998 and 2010

The main land use change in the Zichang region was the ongoing conversion of cropland to forestland between 1999 and 2010, with the afforestation area occupying up to 6.2×10^4 ha. Figure 2 showed the area over which the land cover was altered during the 10-yr period. 2 sites were selected for the study on the regional scale and a multi-year field study was conducted in the experimental sites at the Zichang experimental station, the institute of soil and water conservation, Chinese Academy of Science. The black locust (*Robinia pseudoacacia L.*) and alfalfa (*Medicago sativa L.*) were the main plantation species during this period.

Site A: alfalfa. Alfalfa is a perennial flowering plant in the pea family Fabaceae cultivated as an important forage crop in many countries around the world. Its root nodules contain bacteria, *Sinorhizobium meliloti* with the ability to fix nitrogen, producing a high-protein feed regardless of available nitrogen in the soil. It has been widely cultivated for restoration in Loess Plateau because of its nitrogen-fixing ability and livestock fodder. This site was situated in a hilly - gully slope which faced north, with a slope of 15°. The planting density was 150 plants /m^2. Three different ages of alfalfa (3, 7 and 12 yr old) were selected for detailed investigation to analyze the effect on the SOC content. The three afforestation chronosequences of alfalfa selected in the

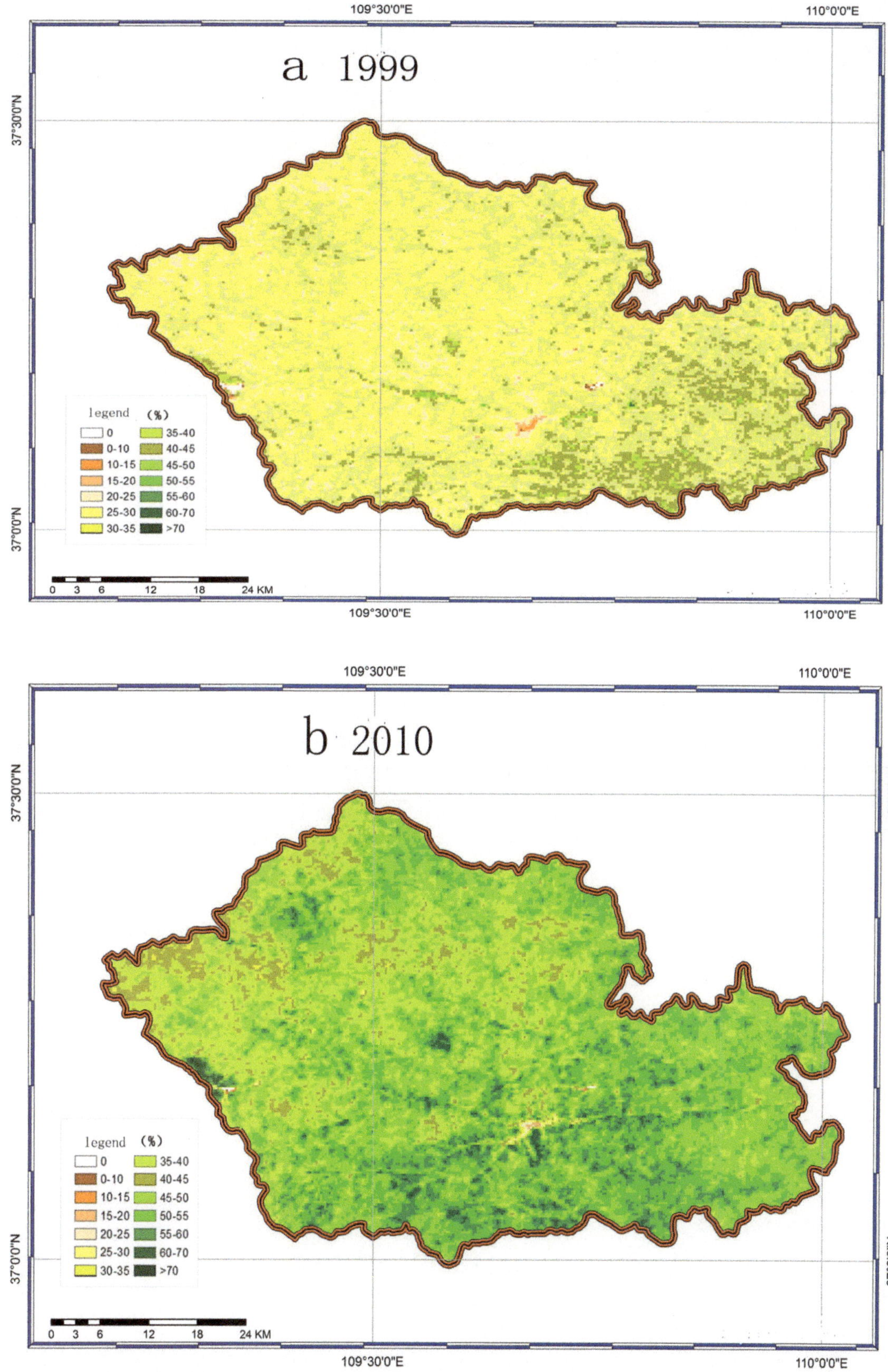

Figure 1. The images of the differences in the remote sensing vegetation coverage between 1999 and 2010 of ZiChang county.

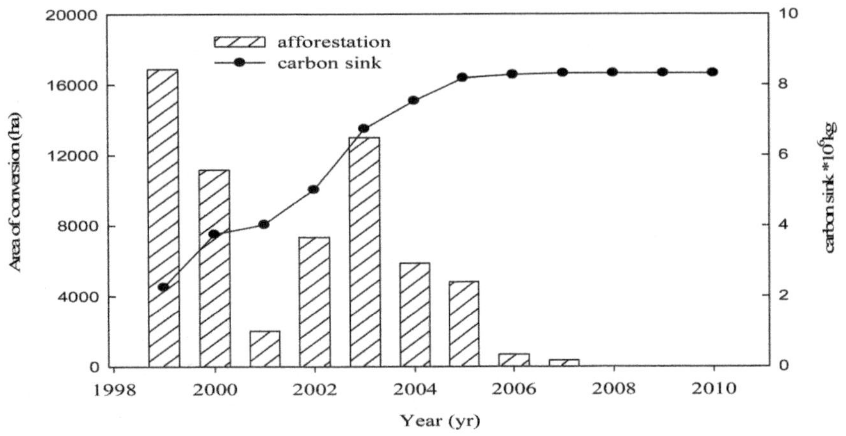

Figure 2. The areas of land cover change (farmland to forestland) and the carbon storage from 1998 to 2010.

same site and the only difference between them is the time when they had been afforested. Each chronosequences of alfalfa had three replications and the total were 3 ages×3 replications = 9 plots.

Site B: black locust. Black locust is a tree of the genus Robinia in the subfamily Faboideae of the pea family Fabaceae which is an exotic nitrogen-fixing tree native to Southeastern North America. It has been widely cultivated for restoration because of its drought resistance, high survival rate, ability to improve the soil nutrient status and remarkable growth rate [23]. At the present, it is the most widely cultivated species in the region. The black locust site was situated in a hilly - gully slope which faced north, with a slope of 19°. And the stand density of black locust was 1667 stands/hm^2. Four afforestation chronosequences of black locust (3, 7, 12 and 15 yr old) were selected for detailed investigation to analyze the effect on the SOC content. The four afforestation chronosequences of black locust selected in the same site and the only difference between them is the time when they had been afforested. Each chronosequences of black locust had three replications and the total were 4 ages×3 replications = 12 plots.

Calculation of forest carbon sink

Plenty of calculation methods for forest carbon sink have been evolved by experts and scholars such as carbon density method, carbon balance model F-CARBON and so on, which are precise but trivial and fall into the category of traditional nature science. In this research, the carbon sink was calculated on the basis of calculation method as forest storage extension suggested by Xi and Li [24] considering of the practicality and maneuverability [25].

Carbon sink was calculated using the following equation:

$$CF = \sum (Sx \times Cy) + \alpha \sum (Sx \times Cy) + \beta \sum (Sx \times Cy),\ Cy = V \times \delta \times \rho \times \gamma$$

Where S_x is the area of forest in the research region; C_y is the carbon density of the forest in the research region; V is the volume per unit area of forest; α is the carbon transfer coefficient of undergrowth plants, which is 0.195 in this research; β is the carbon transfer coefficient of forestland, which is 1.244 in this research; δ is the biomass expanding coefficient, which is 0.5 in this research; ρ is volumetric coefficient, which is 1.90 in this research; ϒ is carbon content rate, which is 0.5 in this research. Values of each conversion coefficient in calculations of forest carbon sink potential in this region are taken as default values prescribed by intergovernmental Panel on Climate Change [26].

Soil sample collection and analysis

In November 2010, the soil samples were collected with a 5-cm diameter soil auger and were extracted in 20-cm incremental subsamples, which were subsequently mixed by hand. Soil samples were collected from each soil layer (0–10 cm, 10–20 cm, 20–40 cm, 40–60 cm, 60–80 cm, 80–100 cm and 100–120 cm) at five different locations selected within a 10 m radius surrounding each plot to analyze the soil organic carbon content. The five samples collected for each layer were subsequently mixed by hand, yielding one representative sample for each layer at each site. The total carbon stock for multiple soil layers was calculated by summing the soil carbon stocks of the layers.

All of the soil samples were air-dried, sealed in airtight bags and transported to the Institute for Soil and Water Conservation of the Chinese Academy of Science in Yangling, Shaanxi province to determine the SOC contents. The SOC was determined using the oil bath-K_2CrO_7 titration method after digestion, and the soil bulk density (BD) was determined using the ring tube method suggested by the Chinese Editorial Committee for Soil Analysis [27]. Total N was measured by the Kjeldahl procedure and pH was measured by electronic pH-meter [28].

Statistical analysis

Each SOC contents in composite samples for each plantation species of various depths were averaged at the same ages following afforestation to perform statistical analysis. Analysis of variance was performed using the SPSS 16.0 software. Means were compared by least significant difference (LSD) at $p<0.05$ or $p<0.01$ level. A one-way analysis of variance (ANOVA) was used to analyze the effect of each factor (land use, soil depths, ages and species) on the SOC concentration.

Results

Carbon sink effects following conversion of farmland to forestland

The forest carbon storage increased continuously as the farmland to forestland conversion project was implemented between 1999 and 2010. The carbon sink increased from

2.3×10^6 kg in 1999 to 8.3×10^6 kg in 2010 with the sustainable growth of the converted areas (Figure 2). The total carbon sink increased by 268% compared to that in the first yr of implementation of the project. This result established that the sequestration of carbon (C) is another important environmental effect of afforestation. The land use policy affected the type, distribution, productivity and turnover of vegetation. This policy was therefore a key determinant of whether the land surface was a sink or source of carbon. Forestland had great carbon sequestration potential, especially compared to farmland. The carbon storage induced by the conversion of farmland to forestland in Zichang County was very large.

SOC content for alfalfa and black locust based on the afforestation chronosequence

In the alfalfa soil, there was a significant increase in the SOC content in the 0–20 cm and 40–100 cm layers with increasing age ($p<0.01$; Figure 3a) from 3 yr to 7 yr. Subsequently, the SOC content decreased slightly with increasing age (from 5 to 4.5 g kg^{-1} and from 3.2 to 2.4 g kg^{-1} in the last 5 yr). The SOC did not change appreciably in the 20–40 cm layer (Figure 3a). In contrast with the soil samples from <100 cm depth, the changes in the soil C in the 100–120 cm layer were significantly correlated with plantation age; the SOC decreased continuously with age from 2.7 to 2.0 g kg^{-1}.

The SOC exhibited a different trend with age in the black locust soil compared to the alfalfa soil (Figure 3b). The SOC content decreased in the first 7 yr at depths below 100 cm, from 3.0 to 1.7 g kg^{-1} (for yr 4–7, the concentration was consistently lower than that measured for the first 3 yr). Interestingly, the C content then exhibits a gradually increasing trend between 7 and 15 yr (Figure 3b) reaching 3.8 g kg^{-1} at an age of 15 yr. In contrast, the ecosystem C stores in the 100–120 cm layer of the old plantation were clearly higher than those in the younger afforestation sites, which had increased by 31.3% 15 yr after afforestation. The SOC content in the two afforestation plantations therefore provide evidence for a carbon sink once the farmland was converted to forestland.

A comparison of the SOC contents for plantation species of various depths at the same ages following afforestation

The statistical comparisons in the ANOVA demonstrate that the dependence of the SOC on depth in the areas converted to forestland is significant at a level of $p<0.05$ (Table 2). Between ages of 7 and 12 yr, the SOC content in the alfalfa soil decreases with increasing depth, and the SOC content in the 0–60 cm layer exceeds that at depths above 60 cm. The soil C content is greatest (5.5 and 4.5 g $kg^{-}1$) in the 0–10 cm layer. However, in the first 3 yr, the SOC decreases significantly ($p<0.01$, Table 2) with increasing depths of 0–80 cm and then exhibits a smaller increase at depths of 80–100 cm. For black locust, the SOC decreases continuously at depths of 0–100 cm and then increases slightly in the 100–120 cm layer in the 3-, 7-, 12- and 15-yr-old forests. Although the C concentration trend based on depth was different for the two plantations, the SOC content in the topsoil was significantly greater ($p<0.01$, Table 2) than that in the subsoil for both plantations.

To inform organic carbon management strategies, the dependence of the SOC content on the type of vegetation is also quantified (Figure 4). The SOC concentration in the alfalfa soil exceeded that for the black locust soil at a given age in the 0–100 cm and 100–120 cm layers. The soil C for the two plantations shows a highly significant difference between alfalfa and black locust ($p<0.05$ or $p<0.01$, Figure 4). The SOC in the 3-yr-old alfalfa was approximately 36.9% higher than that in the black locust in the 0–100 cm layer. The largest difference between the two species was measured for the 7-yr-old plantation; the SOC contents for this plantation exhibit highly significant differences between the two species at each depth ($p<0.01$, Figure 4). The soil C is similar for both species on the 12-yr-old plantation and exhibits a significant difference ($p<0.05$, Figure 4) compared to the 7-yr-old plantation at each depth.

Relationship between the SOC content, depth and other factors as a function of age following afforestation

A change in land use from farmland to forestland and grassland implies that the annual cycle of cultivating and harvesting crops is replaced by the much longer forest cycle. Therefore, many factors influence the SOC in the alfalfa and black locust soils after the land cover changes. The relationships between the SOC content, soil depth and age are summarized in Figure 5. For the 3-yr-old afforested land, there is a strong correlation ($R^2 = 0.93$, $R^2 = 0.99$, $p<0.01$; Figure 5a, b) between SOC content and depth. As shown in Fig 5, the same trend is obtained for 7- (Figure 5c, d) and 12-yr-old (Figure 5e, f) soil. The SOC is related to the afforestation yr, with the SOC decreasing gradually with increasing age (Figure 5).

Discussion

Effect of the farmland-to-forestland conversion project on the SOC stocks

Land use change can cause a change in land cover and an associated change in the carbon stocks [29]. Each soil has a given carbon carrying capacity, i.e., an equilibrium carbon content depending on the nature of the vegetation, precipitation and temperature [30]. The equilibrium carbon stock arises from a balance between the inflow and outflow to the carbon pool [31]. The equilibrium between the carbon inflow and outflow to the soil is disturbed by any land change until a new equilibrium is reached in the new ecosystem. In our study, we have found that in the farmland-to-forestland conversion project, the soil acts as a carbon sink (from 2.26×10^6 kg in 1999 to 8.32×10^6 kg in 2010). Other studied has got the same results. It has also been estimated that UK forestry and grassland sequester 110 ± 4 kg and 240 ± 200 kg of carbon per hectare yr, respectively, whereas croplands lose on average 140 ± 100 kg of carbon per hectare per yr [32]. Ostle et al [15] indicated that soil carbon accumulates more slowly (decadal) but gains can be made when croplands are converted to grasslands, plantation forest or native woodland. The results of this study partially support the hypothesis that the afforestation of formerly arable land leads to increased C storage in the soil over the short term (approximately 30 yr).

The balance between the inputs (primarily from vegetation) and losses of organic matter (as a result of decomposition, leaching and erosion) determines the magnitude of the area's land carbon reservoir. The net effect of the afforestation of cultivated slope land on the soil C content depends not only on the new C gained, but also on the C lost as a result of the previous management. Although the vegetation carbon stock in the hilly-gully areas are relatively small, plant matter is the single most important source of carbon input to the soil [33]. Therefore, the higher SOC content in the two plantations in our study could be a result of primarily the greater annual litter input and fine root biomass compared to the cropland.

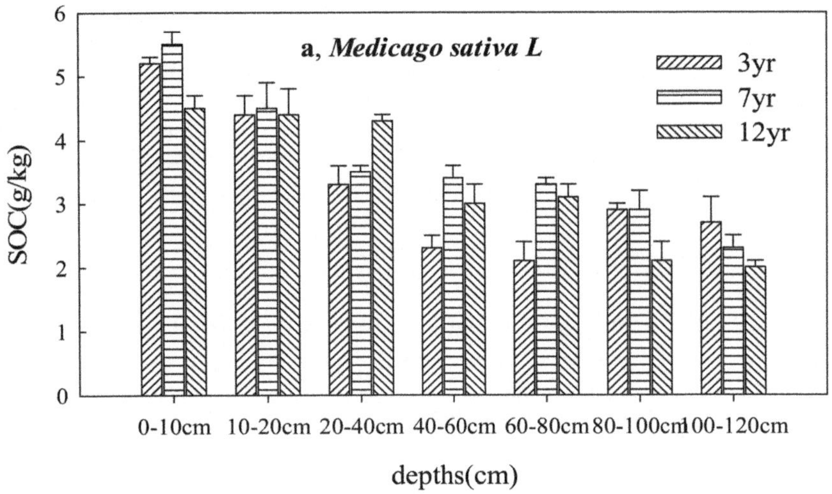

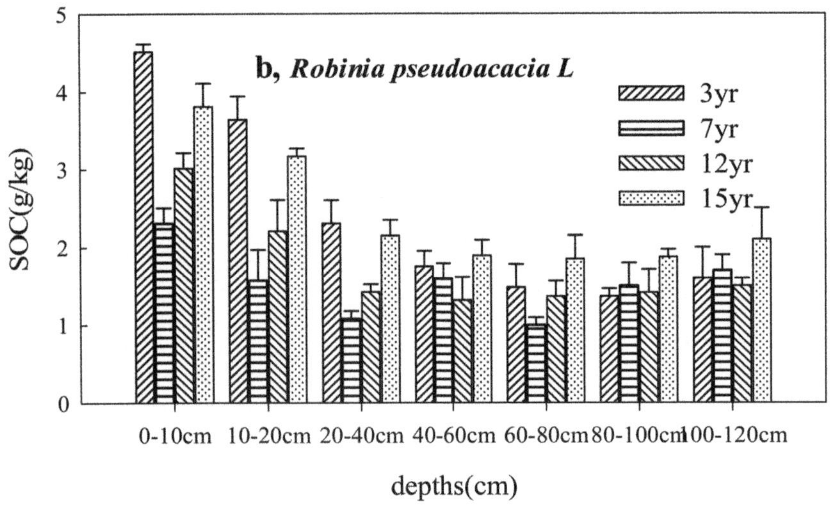

Figure 3. Changes of SOC sequestrations at different ages under alfalfa (*Medicago sativa L*) and black locust (*Robinia pseudoacacia L*) for each depth.

Table 2. SOC[a] at the different depth of the alfalfa (*Medicago sativa L*) and black locust (*Robinia pseudoacacia L*) following the afforestation.

Depth (cm)	alfalfa (*Medicago sativa L*) (g kg^{-1})				black locust (*Robinia pseudoacacia L*) (g kg^{-1})				
	3 yr	7 yr	12 yr	mean	3 yr	7 yr	12 yr	15 yr	mean
0–10 cm	5.2a	5.5a	4.5a	5.1a	4.5a	2.3a	3.0a	3.8a	3.0a
10–20 cm	4.4b	4.5b	4.4a	4.5a	3.6b	1.6b	2.2ab	3.2b	2.3b
20–40 cm	3.3c	3.5c	4.3a	3.8b	2.3c	1.1c	1.4bc	2.1c	1.5c
40–60 cm	2.3ef	3.4cd	3.0b	2.9c	1.8d	1.6b	1.3c	1.9c	1.6c
60–80 cm	2.01f	3.3cd	3.1b	2.8c	1.5d	1.0c	1.4bc	1.9c	1.4c
80–100 cm	2.9cd	2.9d	2.1c	2.6c	1.4d	1.5b	1.4bc	1.9c	1.6c
100–120 cm	2.7de	2.3e	2.0c	2.3c	1.6d	1.7b	1.5bc	2.1c	1.8cb

[a]Data in the column are mean values (n = 3 for alfalfa and black locust), which are compared among different depths within each ages and are not different at the 5% level of significance if followed by the same letter.

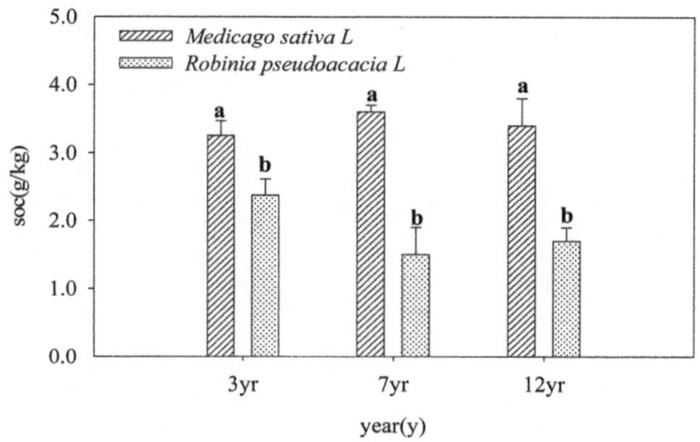

Figure 4. The difference of SOC between alfalfa (*Medicago sativa L*) and black locust (*Robinia pseudoacacia L*) at the same depths under the different ages at the 0–120 cm depths under the 3,7 and 12 ages after the conversion of farmland to forestland.

SOC accumulation for different depths and ages following the afforestation of the cropland

The soil organic carbon content increase in the topsoil (0–40 cm) for both plantations (from 3.8 to 5.1 g kg^{-1} and 1.5 to 3.0 g kg−1) after the conversion of cropland into the forest, is in consistent with other findings for the Loess Plateau [34,10] and other areas of the world [14,35–37] suggesting that the SOC may be highly sensitive to variations in the soil moisture, soil temperature and/or litter input (all of which depend strongly on depth). The increase in the topsoil SOC may be associated with the higher carbon input and lower rates of SOC decomposition and soil erosion associated with alfalfa plantations and black locust forest. It is well established that the top soil is the part of the soil profile that is most susceptible to land use change and disturbance, such as plowing and drainage. The increase in SOC decomposition resulting from tillage disturbance is a key reason for the decrease in SOC following the cultivation of perennial ecosystems [38]. Within limits, soil C increase with increasing soil water and decreasing temperature and the top soil of the increasing water within temperature zones can increase plant production and increase plant litter and root production [39]. In contrast, the cessation of tillage during the establishment of forest on cropland can reduce the SOC decomposition rate, thus increasing the SOC content. Furthermore, the addition of nitrogen derived from nitrogen-fixing species (in this case, alfalfa and black locust) has been well documented and can lower the decomposition of old and new carbon [40], hence favoring SOC sequestration. Moreover, the forest is widely found to be less susceptible to soil erosion compared to the cropland on the Loess Plateau [41], which may contribute to the higher topsoil SOC content following the afforestation of the cropland. And there is also an increase in the carbon input from aboveground litter and fine root biomass in the alfalfa and black locust, which may lead to SOC accumulation during the afforestation.

Understanding the dependence of the SOC content on the plantation age is important for the development of improved biological soil management practices on both local and national scales. This understanding will also help to identify different species that are ecologically compatible and economically useful. The SOC stores in the younger plantation were clearly higher than those in the old afforestation sites, which exhibited decreases of 15.6% and 50% for alfalfa and black locust 12 yr after afforestation. However, for black locust, the SOC continues to increase gradually 15 yr after the conversion of farmland to forestland. A fundamental assumption regarding the effect of land change on the carbon sink is that younger, growing forests sequester carbon more rapidly than older forests. For several yr following a disturbance, the land becomes a carbon source, but as it matures, it becomes a large carbon sink, which slowly declines with age during late succession [42,43]. This trend may owe to the past application of organic fertilizer when the land was used as cropland, producing a higher initial soil C content just before afforestation. When agricultural land is no longer used for cultivation and is allowed to revert to natural vegetation, the balance of C input and output in the agricultural system is broken, as no fertilizer is added to the afforested land.

Factors affecting the SOC sequestration in the "Grain-for-Green" project

The contribution of afforestation to the SOC improvement is controlled by several factors, including the severity of the land degradation, plant species and composition, climate, duration, land use history and management. The effect of the Grain for Green project on SOC sequestration was mainly influenced by ages and depths, which explained 43.2% and 35.6% (Table 3) of the variation of SOC. vegetation species explained 21.2% (Table 3) of the variation in SOC. Xu et al [44] found that the effect of the "Grain-for-Green" project on the SOC sequestration was influenced primarily by the land use type and age, which explained 55.6% and 24.1% of the variation in the SOC, respectively. The SOC content therefore varies across plantations because the C sequestration is affected by the varying land management. Our findings on the carbon sink induced by the project and the relationship between the SOC content and the ages following afforestation supported the conclusion of Xu et al [44]. Topographic factors such as the slope aspect and position were also found to explain 8.5% of the variation in the SOC [44]. Reay ea al [45] indicated that by increasing the productivity, nitrogen enrichment of some natural ecosystems may enhance their capacity to sequester carbon. Clearly, we demonstrate that afforestation has a positive influence on the SOC concentration in our study, depending on the species planted (alfalfa has a higher SOC content than black locust). The species-induced difference in the SOC content and other soil properties may owe to a difference in biomass production and nutrient cycling via litter fall and root turnover [46].

Conclusion

We examined the changes in the carbon sink and SOC content of the topsoil and subsoil following the "Grain-for-Green" (farmland-to-forestland conversion) project. The afforested land (alfalfa and black locust) soil displayed remarkable SOC sequestration compared to the sloping croplands. The two plantations studied exhibited SOC sequestration potential in the top soil (0–100 cm) in the initial 7-yr period, while the sequestration occurred later in the 100–120 cm layer (>7 yr after the "Grain-for-Green" project was implemented). Furthermore, our analyses provide evidence that SOC can be influenced by afforestation in semi-arid region such as the Loess Plateau, China. From a regional perspective, type of land use, age and species all had significant effects on the SOC sequestration in the "Grain-for-Green" project. The implementation of the program has further enhanced the vegetation restoration and ecological conservation and strengthened the management of the Loess Plateau ecosystem.

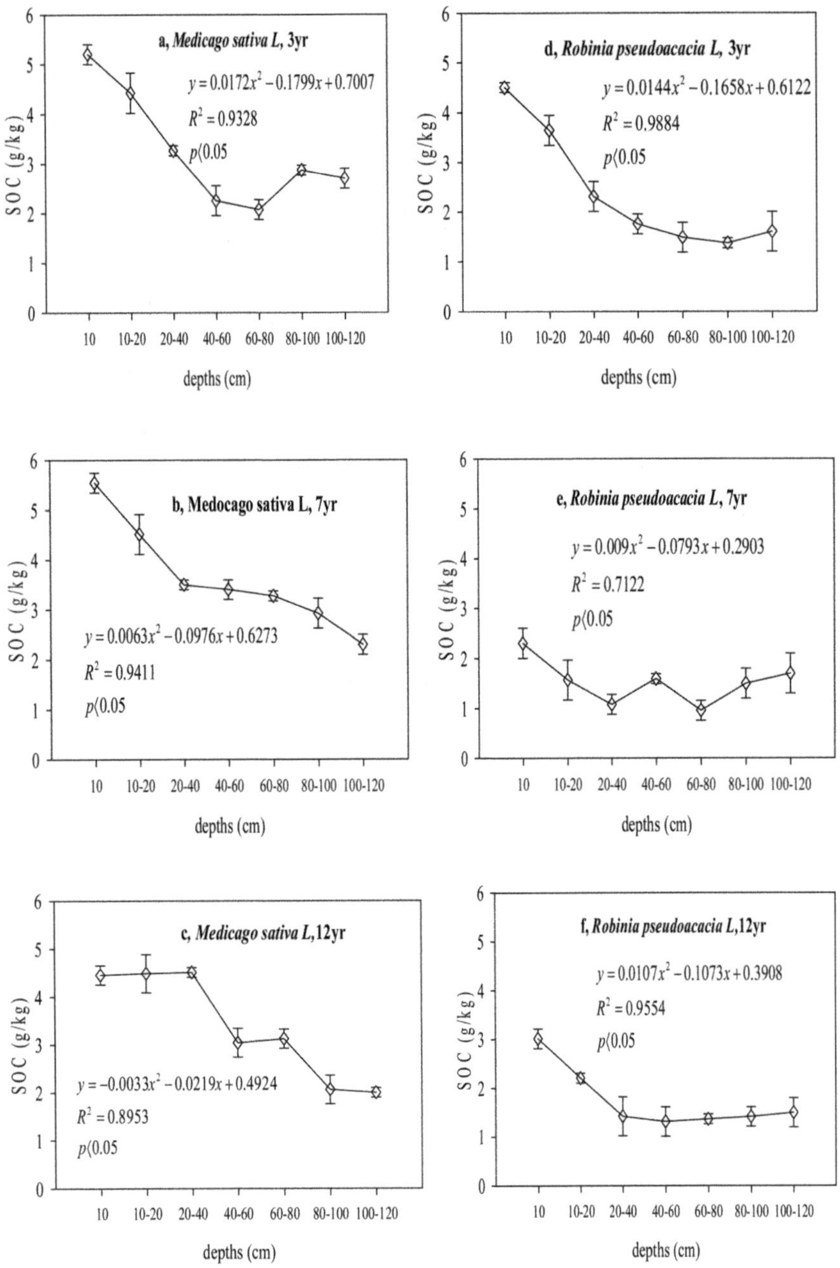

Figure 5. Relationship between the SOC and depths of the alfalfa (*Medicago sativa L*) and the black locust (*Robinia pseduoacacia L*) under 3, 7 and 12 years old following afforestation.

Table 3. Contribution of different factors to the variation of SOC content (variance components, n = 45).

	Variance Source		
SOC content	**Afforestation ages**	**Depths**	**Plantation Species**
Variance	16.62	13.69	8.15
Variance percentage to total variance (%)	43.20	35.60	21.20

Author Contributions

Conceived and designed the experiments: YLL LM. Performed the experiments: LM RLH. Analyzed the data: YLL LM. Contributed reagents/materials/analysis tools: YLL LM. Wrote the paper: LM YLL.

References

1. Lal R (2004) Soil carbon sequestration to mitigate climate change. Geoderma 123: 1–22.
2. Lal R (2010) Beyond Copenhagen: mitigating climate change and achieving food security through soil carbon sequestration. Food Secur 4: 1035–1064.
3. Post WM, Emanuel WR, Zinke PJ, Stangenberger AG (1982) Soil carbon pools and world life zones. Nature 298: 156–159.
4. Batjes NH (1996) Total carbon and nitrogen in the soils of the world. Eur.J. Soil Sci 47: 151–163.
5. Tarnocai C, Canadell JG, Schuur EAG, Kuhry P, Mazhitova G, et al. (2009) Soil organic carbon pools in the northern circumpolar permafrost region. Global Biogeochemical Cycles 23: 1–11.
6. Smith P (2004) Soils as carbon sinks- The global context. Soil Use Manage 20: 212–218.
7. Todd-Brown KEO, Randerson JT, Post WM, Hoffman FM, Tarnocai C, et al. (2013) Cause of variation in soil carbon simulations from CMIP5 Earth system models and comparison with observations. Biogeosciences 10: 1717–1736.
8. Yu DS, Shi XZ, Wang HJ, Sun WX, Chen JM, et al. (2007) Regional patterns of soil organic carbon stocks in China. J.Environ.Manage 85: 680–689.
9. Pan GX (2008) Soil organic carbon stock. Dynamics and climate change mitigation of China. Adv.Clim.Change Res 4: 282–289.
10. Chen LD, Gong J, Fu BJ, Huang ZL (2007) Effect of land use conversion on soil organic carbon sequestration in the loess hilly area, loess plateau of china. Ecol Res 22: 641–648.
11. Entry JA, Sojka RE (2002) Management of irrigated agriculture to increase organic carbon storage in soils. Soil Sci Soc Am. J 66: 1957–1964.
12. Smith P, Powlson DS, Glendining MJ (1996) Establishing a European soil organic matter network (SOMNET). In: Powlson DS, Smith P, Smith JU, editors. Evaluation of soil organic matter models using existing, long-term datasets. UK: Harpenden. pp: 429.
13. Lubowski RN, Plantinga AJ, Stavins RN (2005) Land-use change and carbon sinks: Econometric Estimation of the Carbon Sequestration Supply Function. Belfer Centerprograms or projects: Environment and Natural resources, Harvard Project on Climate Agreements 5: 135–152.
14. Guo LB, Gifford RM (2002) Soil carbon stocks and land use change: a meta analysis. Global Change Biol 8: 345–360.
15. Ostle NJ, Levy PE, Evans CD, Smith P (2009) UK land use and soil carbon sequestration. Land use polic 26s: S274–S283.
16. DeGryze S, Six J, Paustian K, Morris SJ, Merckx R, et al. (2004) Soil organic carbon pool changes following land-use conversions. Global Change Biology 10: 1120–1132.
17. Schlesinger WM (1999). Carbon sequestration in soils. Science 284: 2095.
18. Shi H, Shao MA (2000) Soil and Water loss from the Loess Plateau in China. J Arid Environment 45: 9–20.
19. Liu ZP, Shao MA, Wang YQ (2011) Effect of environment factors on regional soil organic carbon stocks across the Loess Plateau region, China. Agricultural, Ecosystemsand Environment 142: 184–194.
20. Xu XL, Zhang KL, Peng WY (2003) Spatial distribution and estimating of soil organic carbon on Loess Plateau. J Soil Water Conservation 17: 13–15.
21. Jia SW, He XB, Chen YM (2004) Effect of land abandonment on soil organic carbon sequestration in loess hilly areas. J Soil WaterConservation 18: 78–80.
22. Chang RY, Fu BJ, Liu GH, Wang SH, Yao XL (2012) The effects of afforestation on soil organic and inorganic carbon: A case study of the Loess Plateau of China. Catena 95:145–152.
23. Jin TT, Fu BJ, Liu GH, Wang Z (2011) Hydrologic feasibility of artificial forestation in the semi-arid Loess Plateau of China. Hydrol.Earth Syst.Sci 15:2519–2530.
24. Xi TT, Li SL (2006) Analysis of forestry carbon mitigation potential in Heilong jiang Province. Problems of Forestry Economics 26: 519–523.
25. Xu YR, Wang DT, Feng ZW (2003) Effect of carbon storage in several artificial vegetations on sea-beach salinity soil in Tianjin, China. Journal of Huazhong Agricultural University 22: 603–607.
26. IPCC (2000) Summary for Policymakers. In: Nakicenovic N, Swart R, editors. Special Report of the Intergovernmental Panel on Climate Change. Cambridge: Cambridge university. pp. 377
27. Chinese Editorial Committee of Soil Analysis (1996) Soil physical and chemical analysis and description of soil profile (in Chinese). China State Standards Press, Beijing.
28. ISS (1978) Physical and chemical analysis methods of soils. Institute of Soil Science. Shanghai: Shanghai Sci Technol Press. pp: 7–59.
29. Bolin B, Sukumar R (2000) Global perspective In: Watson RT, Noble IR, Bolin B, Racindranath NH, Verardo DJ, DokkenDJ, editors. Land use, land use change, and forestry. Cambridge: Cambridge university. pp. 23–51.
30. Gupta RK, Rao DLN (1994) Potential of wastelands for sequestering carbon by reforestation. Current science 66: 378–380.
31. Fearnside PM, Barbosa RI (1998). Soil carbon changes from conversion of forest to pasture in Brazilian Amazonia. Forest ecology and management 108: 147–166.
32. Dawson JJC, Smith P (2007) Carbon losses from soil and its consequences for land-use management. Science of the total environment 382: 165–190.
33. Leake MC, Jennifer HC, George HW, Fan B, Richard M, et al. (2006) Stoichiometry and turnover in single, functioning membrane protein complexes. Nature 443: 355–358.
34. Chang RY, Fu BJ, Liu GH, Liu SG (2011) Soil carbon sequestration potential for "grain for green" project in Loess plateau, China. Environmental management 48: 1158–1172.
35. Laganière J, Angers DA, Par D (2010) Carbon accumulation in agricultural soils after afforestation: a meta analysis. Global change biology 16: 439–453.
36. Paul KI, Polglase PJ, Nyakuengama JG, Khanna PK (2002) Change in soil carbon following afforestation. Forest ecology and management 168: 241–257.
37. Post WM, Kwon KC (2000) Soil carbon sequestration and land use change: processes and potential. Global change biology 6: 317–327.
38. Balesdent J, Chenu C, Balabane M (2000) Relationship of soil organic matter dynamics to physical protection and tillage. Soil and tillage research 53: 215–230.
39. Liski J, Iivesniemi A, Makela CJ (1999) CO_2 emission from soil in response to climate warming are overestimated-the decomposition of old soil organic matter is tolerantof temperature. Ambio 28: 171–174.
40. Binkley D (2005) How nitrogen-fixing trees change soil carbon. In: Binkley D, Menyailo O, editors. Tree species effects on soils: implications for global change. Dordrecht: Kluweracademic publishers. pp. 155–164.
41. Fu BJ, Wang YF, Lu YH, He CS (2009) The effects of land-use combinations on soil erosion: a case study in the loess plateau of China. Progress in physical geography 33: 793–804.
42. Law BE, Sun OJ, Campbell J, Vantuyl S, Thornton PE (2003) Changes in carbon storage and fluxes in a chronosequence of pondersosa pine. Global change biology 9: 510–524.
43. Turner DP, Guzy M (2003) Effects of land use and fine-scale environmental heterogeneity on net ecosystem production over a temperate coniferous forest landscape. Tellus 558: 657–668.
44. Xu MX, Wang Z, Zhang J, Liu GB (2012) Response of soil organic carbon sequestration to the "Grain for Green Project" in the hilly loess plateau regain. Acta Ecologicasinica 32: 5405–5415.
45. Reay DS, Dentener F, Smith P, Grace J, Feely R (2007) Global nitrogen deposition and carbon sinks. Nature Geoscience 1: 430–437.
46. Girmay G, Singh BR, Mitiku H (2008) Carbon stocks in Ethiopian soils in relation to land use and soil management. Land degradation and development 19: 351–367.

9

The Effect of Leaf Litter Cover on Surface Runoff and Soil Erosion in Northern China

Xiang Li, Jianzhi Niu*, Baoyuan Xie

Key Laboratory of Soil and Water Conservation and Desertification Combating of Education Ministry, Beijing Forestry University, Beijing, China

Abstract

The role of leaf litter in hydrological processes and soil erosion of forest ecosystems is poorly understood. A field experiment was conducted under simulated rainfall in runoff plots with a slope of 10%. Two common types of litter in North China (from *Quercus variabilis*, representing broadleaf litter, and *Pinus tabulaeformis*, representing needle leaf litter), four amounts of litter, and five rainfall intensities were tested. Results revealed that the litter reduced runoff and delayed the beginning of runoff, but significantly reduced soil loss ($p<0.05$). Average runoff yield was 29.5% and 31.3% less than bare-soil plot, and for *Q. variabilis* and *P. tabulaeformis*, respectively, and average sediment yield was 85.1% and 79.9% lower. Rainfall intensity significantly affected runoff ($R=0.99$, $p<0.05$), and the efficiency in runoff reduction by litter decreased considerably. Runoff yield and the runoff coefficient increased dramatically by 72.9 and 5.4 times, respectively. The period of time before runoff appeared decreased approximately 96.7% when rainfall intensity increased from 5.7 to 75.6 mm h^{-1}. Broadleaf and needle leaf litter showed similarly relevant effects on runoff and soil erosion control, since no significant differences ($p\leq0.05$) were observed in runoff and sediment variables between two litter-covered plots. In contrast, litter mass was probably not a main factor in determining runoff and sediment because a significant correlation was found only with sediment in *Q. variabilis* litter plot. Finally, runoff yield was significantly correlated ($p<0.05$) with sediment yield. These results suggest that the protective role of leaf litter in runoff and erosion processes was crucial, and both rainfall intensity and litter characteristics had an impact on these processes.

Editor: Ben Bond-Lamberty, DOE Pacific Northwest National Laboratory, United States of America

Funding: This study was supported by the Fundamental Research Funds for the Central Universities (No.BLYJ201406); National Natural Science Fund of China (41171028); National Advanced Project of the 12th Plan, China (2011BAD38B05); and National Forestry Public Welfare Industry Research Project (201104005). The funders had no role in study design, data collection and analysis, decision to publish, or preparation of the manuscript.

Competing Interests: The authors have declared that no competing interests exist.

* Email: nexk@bjfu.edu.cn

Introduction

Soil erosion has become a serious problem worldwide, causing to decrease productivity of agricultural and forest land, environmental and ecological degradations, and natural disasters such as mudflow that threaten to human safety and infrastructure [1]. Approximately five billion tons of soil is lost annually in China [2,3]. Soil erosion often damages forest ecosystems, including reduction of soil organic matter content and water-holding capacity [4], loss of valuable soil nutrients, and biota, also declines in biodiversity, and collectively lead to ecosystem instability [1,3]. While vegetation litter is often considered as an effective cover above soil surface that prevents soil erosion, it is often burned or removed for fuel by local population in many forests of Northern China. These activities may increase detachment of soil aggregates, so raising the sediment generation and transportation via runoff [5].

Although it is widely recognized that vegetation canopy is important in hindering soil erosion [6], by comparison, the role of vegetation litter layers in modulating surface runoff and soil erosion remains poorly understood. Litter layers are known to protect soil from raindrop splashes by intercepting rainfall, preventing surface sealing and crusting of soil, extend the time of soil infiltration, and enhance sediment deposition by increasing soil surface roughness [7–9]. Previous studies have mainly evaluated the effectiveness of various surface covers in reducing surface runoff and soil loss, including rock fragments [10–13], crop residues [14,15], grass [16–18], geo-textiles [19], post-fire ash and needle cover [20–22], and combined cover such as rock and litter [23]. Nonetheless, few leaf litter materials have been tested [24,25], with variable findings [26–34]. The effects of litter layers on surface runoff and soil loss were not consistent across different types of litter, soil, or different environments. In general, runoff volume and sediment yield were reduced in the presence of litter cover [26–30]. However, some disputable results have shown that runoff and soil erosion were accelerated by impervious plastic mulch covers [31–34].

Moreover, the early studies have mainly examined the percentage of litter cover as a dominant influential factor in surface runoff and soil erosion [20,23,28]. Rainfall intensity and litter characteristics such as litter type and areal litter mass (i.e., litter mass per unit area, kg m^{-2}), have rarely been regarded as key factors in hydrologic and erosion processes in forest ecosystems. These factors were proved to be effective in rainfall interception [35–38], thus possibly exerting strong effects on soil infiltration, surface runoff, and soil erosion.

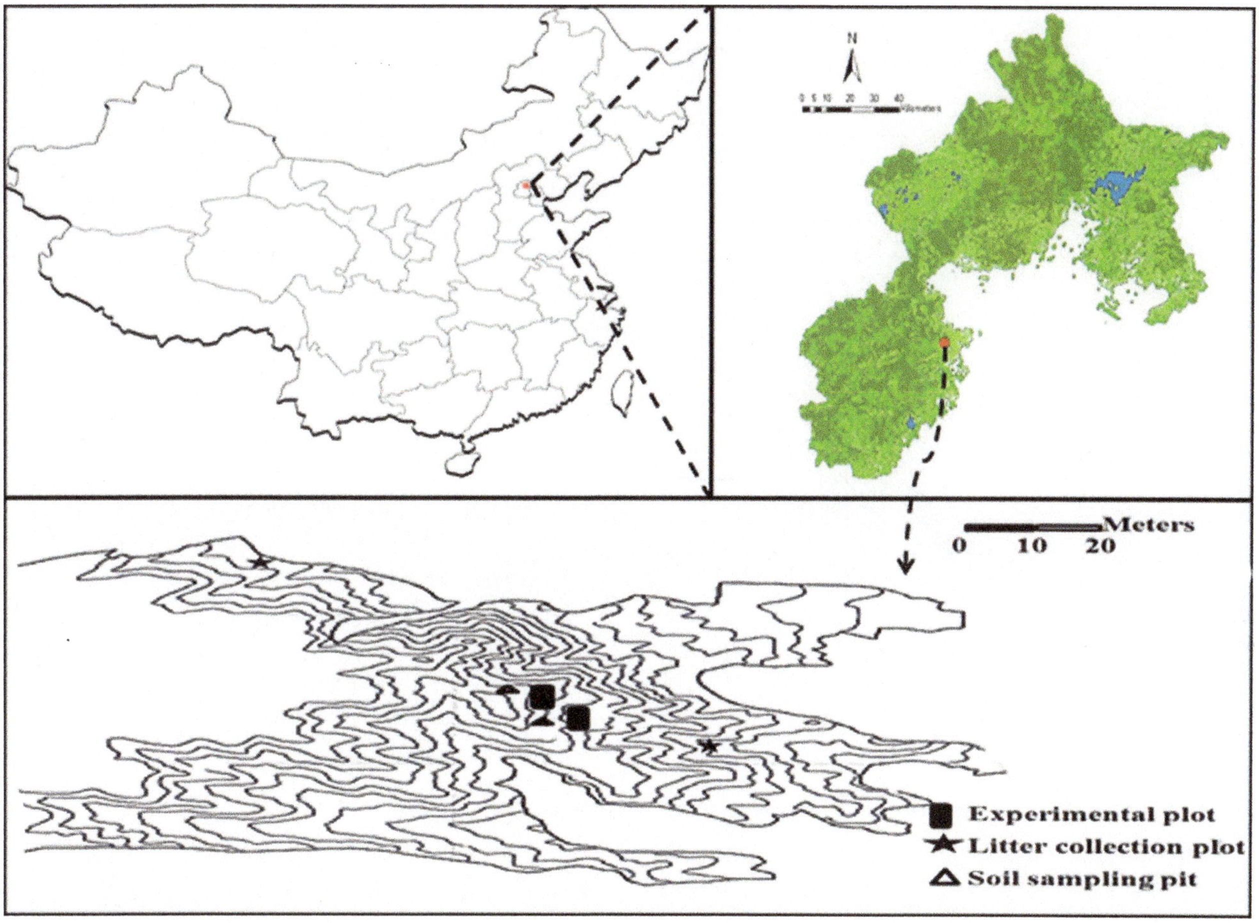

Figure 1. The location of the study area.

In this study, the effects of litter cover (i.e., little type and areal litter mass) and rainfall intensity on surface runoff and soil erosion were quantified through rainfall simulation experiments on sloped field plots with the presence or absence of litter covers. The purpose of the present work was to clarify the role of litter layers on surface runoff and soil erosion, which have caused serious environmental and ecological problems and sometimes even catastrophic events such as debris or mudslides in Northern China [39,40].

Materials and Methods

Ethics Statement

The experimental site, Jiu Feng National Forestry Park is managed by the Forestry Committee of Beijing Forestry University and is available for teaching and research of the university. This field study did not involve any endangered or protected species, and the tree species we selected were common in Northern China.

Study site

This study was performed at Jiu Feng National Forest Park, which is overseen by the Forestry Committee of Beijing Forestry University and available for the university's research and teaching activities. The park is located northwest of Beijing, China (116°28′E, 39°34′N) (Fig. 1). In its warm temperate climate, summers are hot and wet, and winters are cold and dry. Mean daily temperatures are in the range of 23–28°C from May to mid-October, and −5–15°C throughout the winter, resulting in mean annual temperature of 12°C. Mean annual precipitation was approximately 630 mm, most of which occurs as rainfall between June and September, i.e., the period with very intense and erosive rainfall events, usually following the summer drought period. In

Table 1. Forest stands and litter characteristics [38].

Litter species	Plot area (m^2)	Density (trees ha^{-1})	DBH* (cm)	Height (m)	Leaf litter length (cm)	Leaf litter width (cm)
Q. variabilis	25×15	1225	11.2	9.8	11.3–13.7	3.5–4.3
P. tabulaeformis	25×15	1748	7.6	5.2	9.5–12.5	3.3–5.1 (twigs + needles)

*Diameter at breast height.

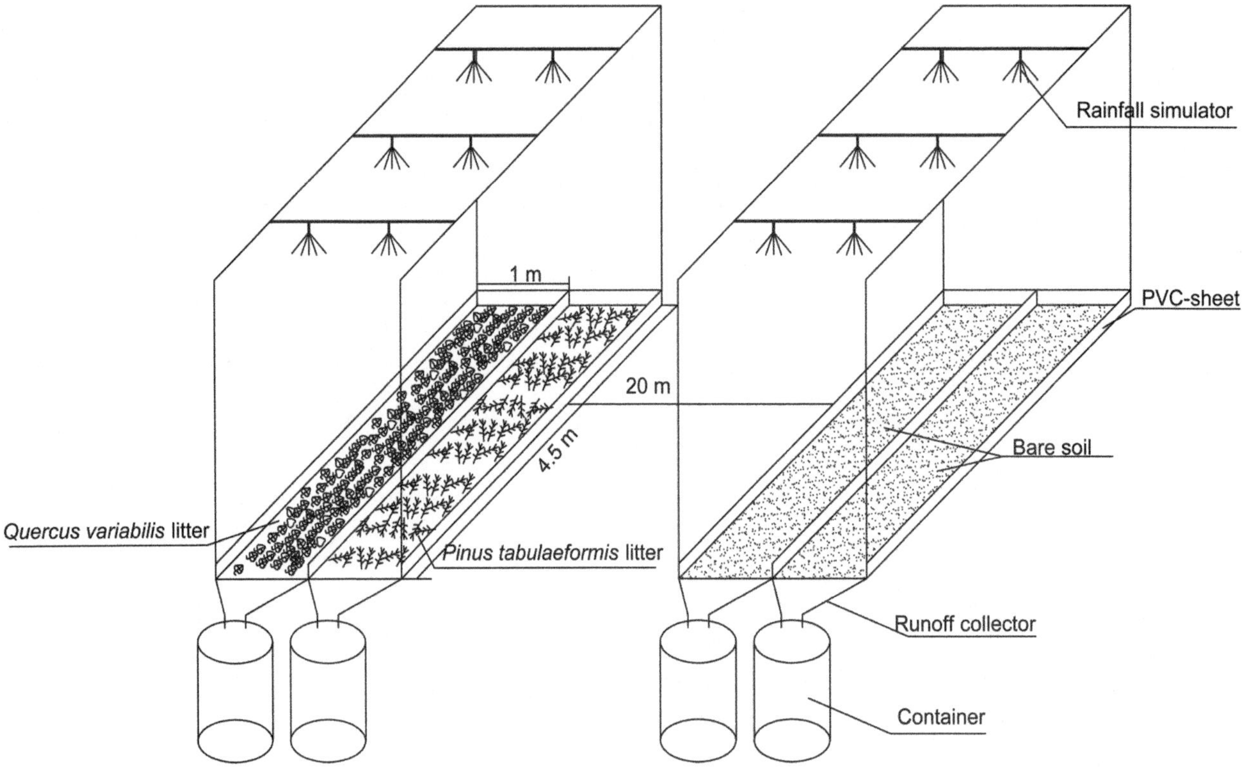

Figure 2. A single set of schematic diagram of experimental runoff plots.

the park Pinaceae *Pinus tabulaeformis*, Fagaceae *Quercus variabilis*, and Cupressaceae *Platycladus orientalis* are dominant tree species, most of which were planted in the 1950s and 1960s. At an elevation of 70–900 m, soils are mainly consisted of cinnamon soils, changing to brown soils at elevations above 900 m. The average slope is approximately 10% with a northeastern aspect.

Leaf litter

Collection and characterization of leaf litter used in this study have been described in detail in Li *et al.* [38]. Briefly, because the tree species of *Pinus tabulaeformis* and *Quercus variabilis* are broadly planted not only in the park but also across the Northern China to prevent wind and water erosion, two dominant and typical types of leaf litter was collected and tested: *Q. variabilis* litter (*QVL*), which here represents broadleaf litter, and *P. tabulaeformis* litter (*PTL*), which here represents needle-leaf litter. Regulators and cleaners remove and burn the litter for fuel almost every year. Prior to this study, the most recent removal and burning occurred in April 2010. In order to investigate the natural litter layer distribution, a 25×15 m^2 experimental stand for each type of litter, respectively. These were further divided into 1×1 m^2 sub-plots, where human activities were not allowed during the collection period from May 2010 to April 2013. Afterwards, in order to consistently investigate the variation of litter mass to ensure that the measured mass values were representative of field conditions, 10 sub-plots were randomly chosen and litter mass was measured and recorded every four months to cover seasonal variation. A 2-cm-high aluminum rectangular frame was placed over the ground to facilitate observation and collection of fallen leaves in each sub-plot. No cover was set in the sub-plots so that the litter was not isolated from the topsoil, also the biological decomposition of litter was not affected. Because the litter was moved around by wind, the mass of collected litter varied from one

Table 2. Soil properties (0–40 cm deep) in experimental and control plots.

	Bare-soil plot	Litter-covered plot
Soil moisture (%)	21.2	17.5
Porosity (%)	57.4	55.4
Bulk density (g cm^{-3})	1.57	1.43
pH	5.6	5.7
Sand (%)	79.5	73.6
Silt (%)	10.9	12.5
Clay (%)	9.6	13.9

Table 3. Runoff yield and runoff coefficient data corresponding to each treatment and simulated rainfall event.

Type of cover	Litter mass (kg m^{-2})	Rainfall intensity (mm h^{-1})				
		5.8	11.8	25.2	49.8	75.6
		Runoff yield (mm), runoff coefficient (%)				
Bare soil	0	2.09, 36.7	6.58, 56.2	21.3, 84.5	43.98, 88.3	68.42, 90.5
Q. variabilis	0.3	0.81, 14.2	4.64, 39.7	13.64, 54.1	26.71, 53.6	51.05, 67.5
	0.5	1.28, 22.5	3.66, 31.3	13.24, 52.5	26.53, 53.3	54.50, 72.1
	0.8	0.98, 17.2	3.66, 31.3	13.43, 53.3	33.26, 66.8	52.10, 68.9
	1.0	0.28, 4.9	2.75, 23.5	13.79, 54.7	33.14, 66.6	51.90, 68.7
P. tabulaeformis	0.3	0.71, 12.5	2.98, 25.5	12.31, 48.9	35.74, 71.8	52.06, 68.9
	0.5	0.65, 11.4	3.06, 26.2	11.58, 46.0	29.10, 58.4	43.54, 57.8
	0.8	0.69, 12.1	3.29, 28.1	9.60, 38.1	28.82, 57.9	54.24, 71.8
	1.0	0.35, 6.1	3.16, 27.0	12.76, 50.6	36.35, 73.0	50.30, 66.5

sub-plot to another. Because the interface between the litter and soil was clearly distinguishable, the uncomposed litter was selected as the experimental material instead of the fresh leaves, twigs and decomposed litter, only litter with intact shapes were collected. These samples were manually transferred into plastic bags, and transported to the laboratory, where they were allowed to air-dry.

The lengths and widths of the collected needle-leaf litter and broadleaf litter were measured (Table 1). Because the twigs and needles were difficult to separate, they were weighed together, and decomposed needles and broadleaves were excluded from the measurements. The areal litter mass in kg per m^2 was then calculated by dividing the weighed litter mass over the sub-plot areas. The undecomposed litter was chosen for two main reasons: first, it was the dominant component of the litter (approximately 85%) and was easy to identify; second, the undecomposed litter was located in the upper layer where it would directly intercept the raindrops. For this reason, that the undecomposed litter was assumed to play a more significant role in modulating surface runoff and soil erosion processes than decomposed litter and half-decomposed litter. Accordingly, the mass of this component of litter ranged from 0.33 to 1.24 kg m^{-2} for *QVL* and from 0.18 to 0.77 kg m^{-2} for *PTL* in the sub-plots (1×1 m^2). Based on the measurements mentioned above, to represent the litter variation characteristics and erosion in the slope accurately and to effectively illustrate and compare the hydrological and erosive responses of the two types of litter, four areal litter masses of 0.3, 0.5, 0.8, and 1.0 kg m^{-2} were selected for testing in rainfall simulation studies. This may explain the role of litter in runoff and sediment depletion.

Experimental plot design

Based on initial site survey, two northeastern-facing open runoff plots (4.5×2 m^2 each) with an approximate slope of 10% were selected for rainfall simulation experiments. These served as the bare plot and the litter-covered plot, respectively. The plots were approximately 20 m apart, the soil moisture and soil texture composition were almost the same, so they were not decisive factors in the infiltration and runoff generation in further analysis of runoff yields. All large stones were removed from both plots, and the remaining decomposed litter was removed from the litter-covered plot. No trees or other shrubs were present in either plot. Both plots were divided in half with each part measuring 4.5×1 m^2. The two halves of the divided plots were then covered with *Q. variabilis* litter (*QVL*) and *P. tabulaeformis* litter (*PTL*). These served as litter-covered plots in comparisons of the effects of the two different types of litter on runoff and erosion control (Fig. 2). Before each test run, leaf litter was brought out from plastic bags, then carefully distributed by hand in the litter-covered plot at litter masses of 0.3, 0.5, 0.8, and 1.0 kg m^{-2}. Litter thickness was measured in the upper, middle, and lower part of the plot to test whether the litter covered soil surface uniformly. Several adjustments were made if the litter thickness varied by more than 0.5 cm. The remaining two plots (4.5×1 m^2 each) were regarded as the control (or bare-soil) plot with no litter cover. Within almost 30 min of the cessation of rainfall, leaf litter was re-collected in plastic bags and then oven-dried at 80°C, and evenly returned to the plots before the next rainfall simulation.

Soil properties

The soil properties of each plot are presented in Table 2 (A common Chinese Soil Taxonomy was reported by CRGCST [41] and Shi *et al.* [42]). To comprehensively represent the soil characteristics of the experimental plot, bulk soil samples were collected (at a depth of 0–40 cm) from three different points outside each plot (2–5 m away from the plot) to determine soil properties from May to June, 2013. At each point, soil samples were separated into four sub-samples based on soil depth of 10 cm (0–10 cm, 10–20 cm, 20–30 cm, 30–40 cm) and stored in aluminum specimen boxes and trays. In order to determine soil bulk density (D_b) a weight basis of a 100 cm^3 soil core sample was collected under field-moisture conditions, oven-dried at 105°C for 24 h, and measured for the mass of oven-dried solids, V_t refer to the bulk volume of the soil, which includes the volume of the soil (m_s). Then D_b was calculated as m_s/V_t and the pore space between the soil particles [43], here $V_t = 100$ cm^3. Total porosity S_t can be calculated from the particle density (D_p) and bulk density (D_b) as follows:

$$S_t = 1 - \frac{D_b}{D_P}$$

Here, particle density ($D_p = \frac{m_s}{V_s}$) refers to the mass (m_s) of a unit volume of solid soil particles (V_s) [44]. Gravimetric soil water content (water mass/dry soil mass) was also measured after the

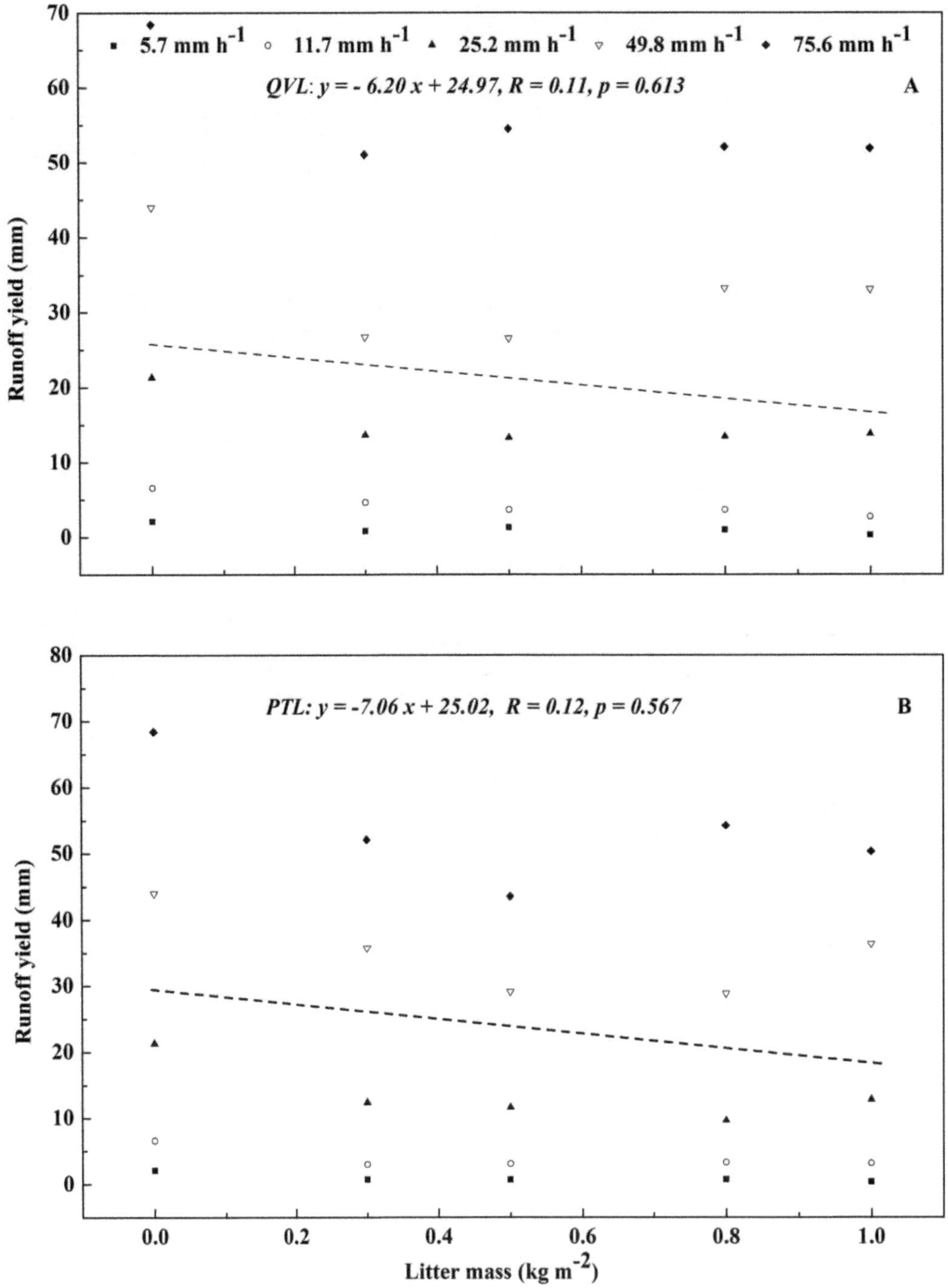

Figure 3. Runoff yield versus litter mass for five rainfall intensities for bare-soil plots (Litter mass equals zero represents 'bare'), *Quercus variabilis* litter (A), and *Pinus tabulaeformis* litter (B).

samples were oven dried [45]. Soil particle size distribution was measured after the soil samples were treated with H_2O_2 and dispersed in sodium hexametaphosphate solution using the Bouyoucos densimeter method [46,47]. Soil pH was determined with an electrode pH-meter on the saturated soil paste of 1:2.5 soil to distilled water ratio [48,49].

Rainfall simulation

The relatively high variability of the rainfall intensity in Northern China makes it difficult to examine its role in rainfall-runoff and soil erosion processes that take place under litter cover [50,51]. Here, simulated rainfall produced by an artificial rainfall simulator jointly developed by Beijing Normal University and Beijing Jiaotong University in 2006 was used [52,53]. A detailed description on the rainfall simulator was provided in Li *et al* [38]. The simulator delivered water at a height of 4.5 m and at varying speeds, spraying an area 2.2 m long ×1.5 m wide under a range of rainfall intensities. This rainfall simulator was designed to allow raindrops to reach terminal velocity and allow the flow rates to be easily controlled, resulting in simulated stable rainfall events with

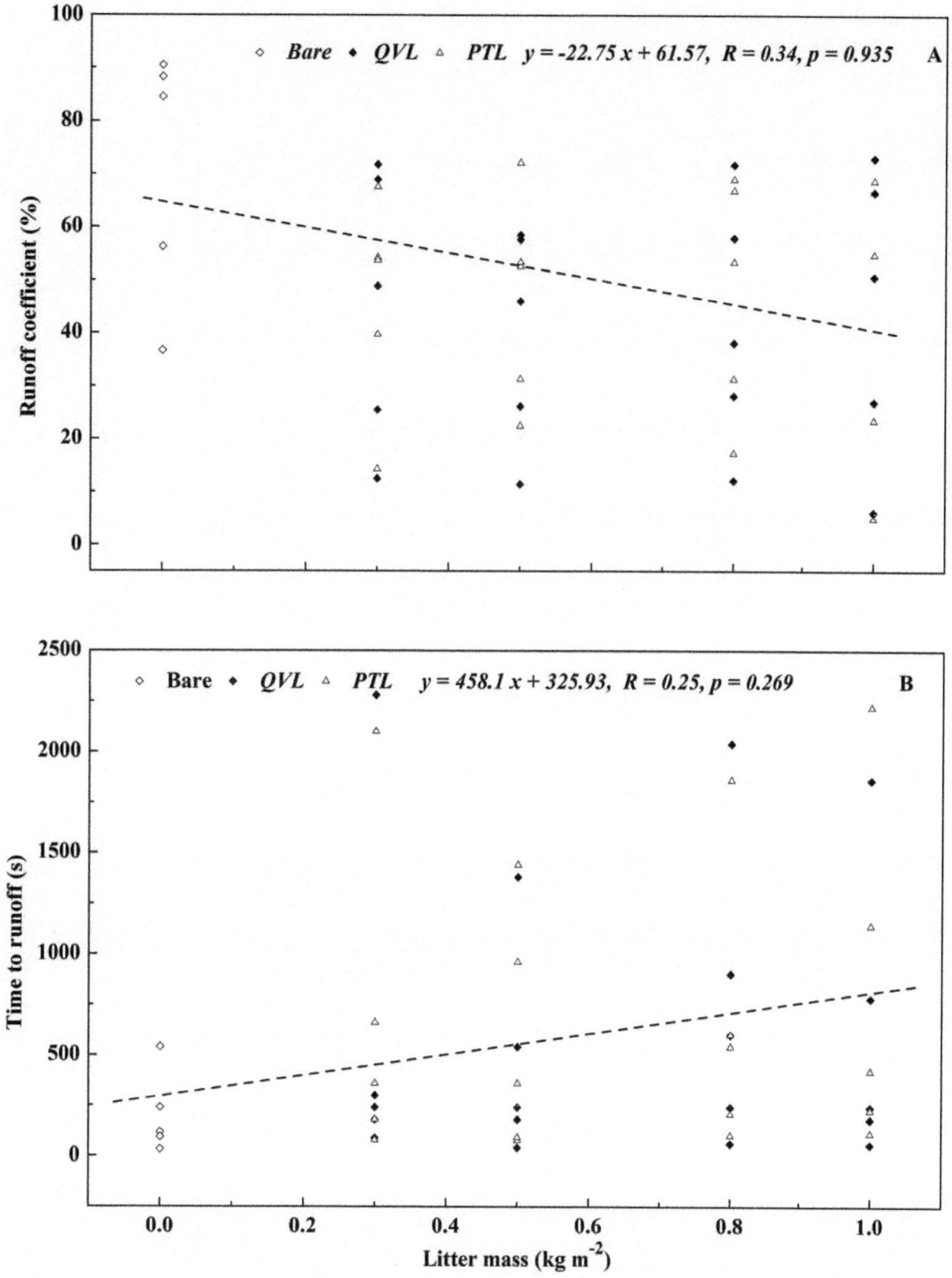

Figure 4. Runoff coefficient and time to runoff versus litter mass for bare-soil plots (open diamonds), *Quercus variabilis* litter (closed diamonds), and *Pinus tabulaeformis* litter (open triangles) for five rainfall intensities.

80% uniformity and raindrop sizes of 2.3±0.3 mm. Simulated rainfall chemistry is also an important factor to be considered in the simulation experiments. Water with high electrical conductivity tends to flocculate soil particles, and water with low electrical conductivity typical of natural rainfall may disperse and readily erode the same soil particles [54,55]. Distilled water was modified using NH_4Cl and NaCl to produce a solution with a pH of 5.5 and electrical conductivity ranged from 12.2 to 20.7 μS cm^{-1} (a detailed description of the electrical conductivity measurement was presented in the National Standard of China [56]), which was similar to the reference value of 14.8 μS cm^{-1} [57].

According to the precipitation data in the study region over 53 years from 1956 to 2008, the maximum rainfall intensity between June and September ranged from 2.6 to 82 mm h^{-1} [58], each plot was subjected to rainfall at intensities of 5.7, 11.7, 25.2, 49.8, and 75.6 mm h^{-1} for 1 h each to represent natural rainfall conditions and to measure the hydrological and erosive response to rainfall accurately. The duration of simulated rainfall was 1 h, which was consistent with the duration of the stable natural rainfall in the study region since 1984 [59,60]. Runoff and sediment samples were collected using a metal runoff collector, which was placed at the bottom of each plot to capture the runoff during the

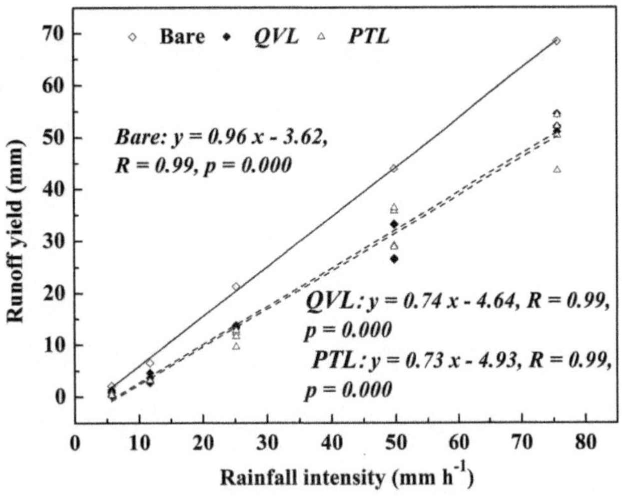

Figure 5. Runoff yield versus rainfall intensity for bare-soil plots (open diamonds), *Quercus variabilis* litter (closed diamonds), and *Pinus tabulaeformis* litter (open triangles) for four litter masses.

test run at 1-minute intervals, and the runoff sample was weighed using an electronic balance. Sediment was settled, separated from the water, and oven-dried at 105°C for 24 h, after which it was weighed to determine the sediment yield. Time to runoff (T_r) was measured as the time when runoff started to develop on the plot surface after the rainfall began [21,61]. Within 30 min after the cessation of rainfall, by which the gravitational water had drained out of the litter, the litter was removed from the soil manually and collected in two plastic bags.

Runoff coefficient (R_c) defined as the proportion of total rainfall that becomes runoff during a storm event, is used to describe the variation of runoff and water resource development [62]. R_c was calculated for each rainfall event. Sediment concentration (SC) was determined by the equation $SC = \frac{Y}{Q}$, where Y (g) represents sediment yield, and Q (L) represents runoff volume. In addition, infiltration and the antecedent soil water content affected runoff process very visibly [63]. Given the wide range of simulated rainfall intensities, the simulated rainfall intensity (RI) order 5.7, 25.2, 49.8, 11.7, and 75.6 mm h^{-1} was set to avoid that two high-intensity of rainfalls were simulated consecutively. To lessen the impact of infiltration on runoff and sediment generation, changes in soil water content in the soil profiles were measured by inserting time-domain reflectometry probes to a depth of 20 cm before and after the experiment at 2-day (48-h) intervals to determine whether the soil water content had reached 20%±2% (Table 2). If the soil water content reached or exceeded 25%, the rainfall simulation would be delayed for another 24 h until the initial water content before every experimental run was 20%±2%. All test runs were carried out from August to September 2013, during which no intensive rain or significant wind occurred. Each test was repeated once.

Statistical analysis

Linear, polynomial and non-linear regressions were used to determine the relationship between the litter mass (or RI) and runoff parameters (or sediment parameters). Correlation analysis among runoff (e.g. runoff yield, R_c, T_r), and sediment (e.g. sediment yield, SC) as dependent variables and litter mass and of RI as an independent variable was performed to evaluate possible relationships among them, and to facilitate understanding of the hydrological and erosion processes. Parameters were considered to be significantly correlated when they were at or above the 95% confidence level ($p \leq 0.05$). A test of normality was conducted by comparing the Sig. value in Kolmogorov-Smirnov and Shapiro-Wilk. Homoscedasticity was carried out to determine the whether the data was homogeneous. Sig>0.05 was defined as homoscedasticity in the Levene Statistic. One-way analysis of variance (ANOVA) was carried out to determine if there were any differences in runoff and sediment yield among three plot treatments (one bare-soil, two litter-covered plots), specifically differences between the broadleaf litter-covered plot and needle-leaf litter covered plot. The Fisher LSD (Least Significant Difference) test at $p \leq 0.05$ was used to test for significant differences. Data were grouped by treatment (bare, *Q. variabilis* litter cover and *P. tabulaeformis* litter cover). All statistical analyses were performed using IBM SPSS Statistics 20.0 software.

Results and Discussion

Runoff

Runoff yield and runoff coefficient (R_c) data are presented in Table 3. Generally, the most runoff took place in the control plot (bare soil). The mean amount of runoff in the QVL and PTL cover plots was lowered by 29.5% and 31.3% compared to the bare plot, respectively. In agreement, similar results were reported by Singer and Blackard [28], Pannkuk and Robichaud [20], Findeling *et al.* [15], and Cerdà and Doerr [21], with a reduction that ranged from 0.12–55.6% in contrast with the bare soil. This was likely due to the protection provided by the litter cover, which absorbed the energy of raindrops and also increased the roughness of soil surface to increase infiltration rate, and delayed and reduced runoff [49]. However, the opposite result was observed for straw mulch cover: runoff yield was slightly higher (by 0.09%) than that in a bare-soil plot [28,33]. Similar results were obtained for rock fragments cover [12,13,64] and plastic mulch cover [34], presumably because the physical shapes or impervious nature of these surface covers concentrated the flow of water to decrease the infiltration rate and increase the flow velocity [25,65]. However, results of LSD test in ANOVA revealed no significant differences ($p \leq 0.05$) in runoff yield between the control and two litter-covered plots ($p = 0.417$ and 0.390 respectively). In contrast, significant differences ($p \leq 0.05$) were found in R_c ($p = 0.000$ and 0.001) and T_r ($p = 0.028$ and 0.012) between bare and litter-covered plots. The results implied that litter cover had an impact on runoff, but other factors, such as precipitation and rainfall intensity should be taken into account as well.

The manner in which litter type (broadleaf *vs* needle-leaf), specially differences in physical leaf shape, might affect runoff was studied. However, no significant differences in runoff yield ($p = 0.696$), R_c ($p = 0.677$) and T_r ($p = 0.681$) were found between QVL and PTL plots, indicating that litter type was a the major influencing factor in controlling runoff in the present study as it was supposed to be. Very few studies concentrated on the role of litter type in runoff reduction, particularly on the physical differences in leaf shape. In a previous work, Neris *et al.* argued that runoff from the pine needle litter cover was twice of the rainforest litter cover, which does not match the current results [47]. The discrepancy suggests that further research is required.

Litter mass was treated as an influencing element in runoff process by covering the topsoil, but its effect on runoff remains unclear. In general, no significant linear correlations were found between litter mass and runoff yield for QVL ($R = 0.11$, $p = 0.923$)

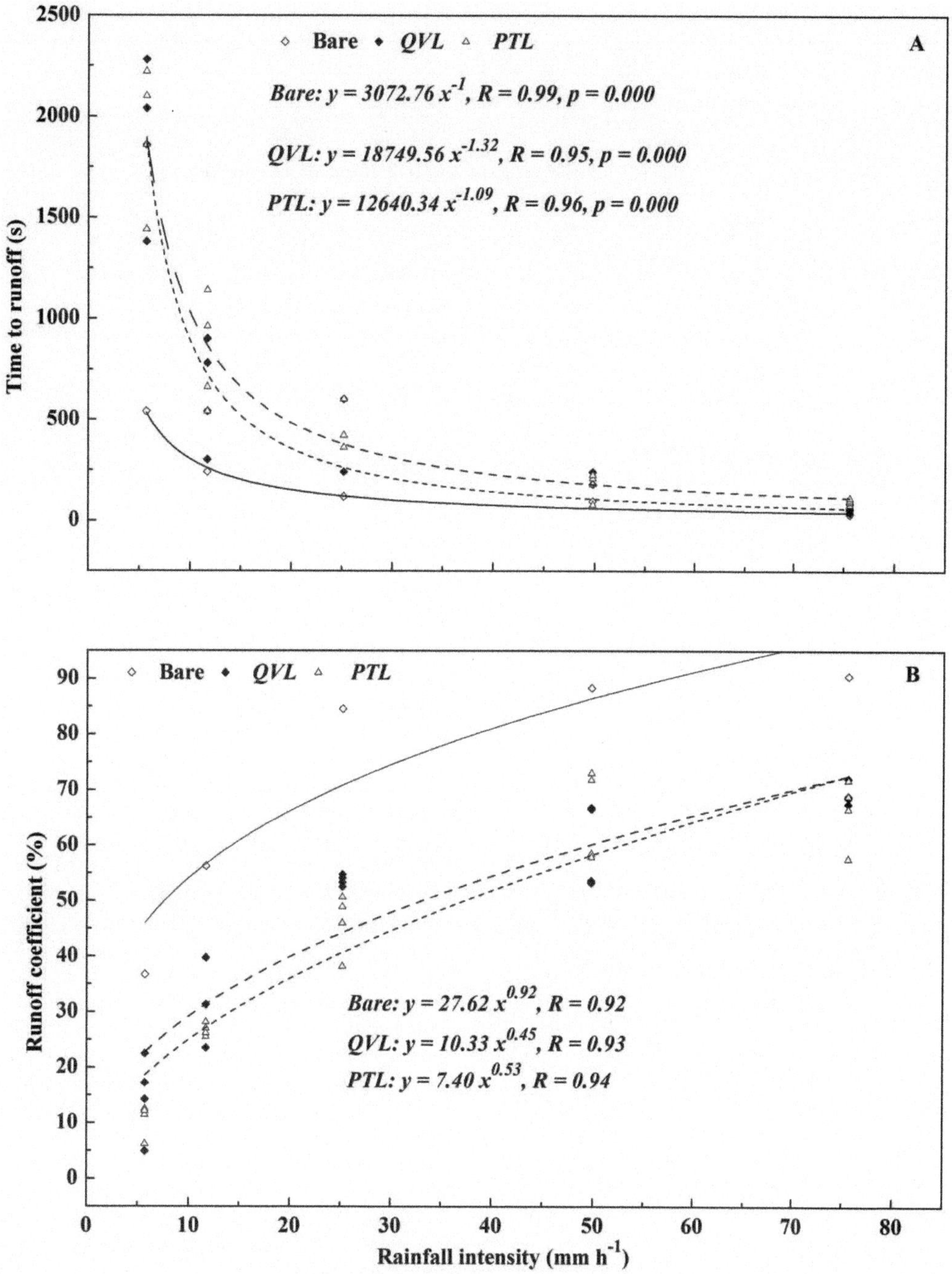

Figure 6. Rainfall intensity versus time to runoff and runoff coefficient for bare-soil plots (open diamonds), *Quercus variabilis* litter (closed diamonds), and *Pinus tabulaeformis* litter (open triangles) for four litter masses.

and *PTL* ($R = 0.12$, $p = 0.964$) (Fig. 3), and no non-linear correlations were observed. Runoff yield decreased dramatically when litter mass increased from 0 to 0.3 kg m^{-2} but then remained steady regardless of the increasing litter mass afterwards, which proved that litter cover had a strong impact on runoff reduction as stated above. R_c ($R = 0.34$, $p = 0.114$) and T_r ($R = 0.25$, $p = 0.078$) showed a similar correlations with litter mass (Fig. 4). One likely explanation was that the increased litter mass might intercept and store more rainwater. However, on the other hand, it may provide a flow channel for runoff by covering more bare topsoil so that runoff flowed on rather than through the leaf litter, particularly under rainfall intensities of 49.8 and 75.6 mm h^{-1}. For this reason, more runoff was observed with higher litter masses of 0.8 and 1.0 kg m^{-2}. While Findeling *et al.* compared the efficiency of different corn residue masses (0, 0.15 and 0.45 kg m^{-2}) in cutting runoff down, and reported that mean values of R_c were 0.44, 0.16, and 0.05, respectively [15]. They confirmed that surface cover contributed notably to the decrease of runoff in contrast with the bare soil, but because they tested only a narrow range of masses,, the slight increase in mass was not sufficient to facilitate extrapolation that runoff would decrease as the ground

Table 4. Sediment yield and sediment concentration data corresponding to each treatment and simulated rainfall event.

Type of cover	Litter mass (kg m^{-2})	Rainfall intensity (mm h^{-1})				
		5.8	11.8	25.2	49.8	75.6
		Sediment yield (kg), sediment concentration (g L^{-1})				
Bare soil	0	0.17, 18.29	0.23, 7.77	1.65, 17.26	4.70, 23.73	7.02, 22.79
Q. variabilis	0.3	0.07, 17.89	0.13, 6.19	0.79, 12.91	0.58, 4.79	2.30, 10.02
	0.5	0.07, 12.00	0.14, 8.42	0.51, 8.53	0.55, 4.64	1.24, 5.06
	0.8	0.05, 10.88	0.08, 4.98	0.18, 2.96	0.31, 2.09	0.42, 1.80
	1.0	0.06, 49.37	0.14, 11.39	0.16, 2.63	0.16, 1.10	0.22, 0.96
P. tabulaeformis	0.3	0.10, 30.14	0.12, 8.61	0.21, 3.77	0.85, 5.29	3.38, 14.44
	0.5	0.06, 19.97	0.15, 10.65	0.32, 6.14	0.68, 5.21	2.94, 15.03
	0.8	0.06, 17.97	0.09, 5.89	0.17, 3.92	0.50, 3.86	0.52, 2.11
	1.0	0.04, 24.44	0.13, 9.17	0.18, 3.08	0.26, 1.58	0.32, 1.40

cover mass increased, which suggested that a broader range of mass should be studied in further works.

Significant linear correlations were found between rainfall intensity (*RI*) and runoff yield for all the plots in Figure 5 ($R=0.99$, $p<0.05$). In all plots, runoff yield increased with increasing *RI*. Runoff yield at the maximum *RI* of 75.6 mm h^{-1} was 32.7, 62.4, and 83.4 times larger than that at the minimum *RI* of 5.7 mm h^{-1} in bare, *QVL* and *PTL* plots respectively. Based on regression analysis, R_c and time to runoff (T_r) showed a significant power ($R>0.9$, $p<0.05$) and reverse-power ($R>0.95$, $p<0.05$) relationships with *RI* (Figure 6A and 6B). As expected, mean R_c increased from 12.6% ($RI=5.7$ mm h^{-1}) to 67.8% ($RI=75.6$ mm h^{-1}) in litter-covered plots (Figure 6B), but T_r decreased apparently from nearly 30 min to 1~1.5 min with increasing *RI*. The findings were in accordance with those reported by Gholami *et al.* [65], who noted that mulch cover was more effective in controlling runoff at a low *RI* of 30 mm h^{-1} ($R_c=49.9\%$, $T_r=81.83$ s) than at a high *RI* of 90 mm h^{-1} ($R_c=66.6\%$, $T_r=55.22$ s). The phenomena in the present study also proved that *RI* (rather than litter mass) was a dominant factor in determining runoff irrespective of litter cover. This was probably because when *RI* was high, leaf litter was pressed close to the soil surface, blocking the channels between leaves. The accumulative precipitation was assumed to surpass infiltration rate, which resulted in the generation of more runoff.

Sediment

The magnitude of the sediment yield and sediment concentration (*SC*) are shown in Table 4. Generally, bare-soil plot generated 2.75 kg sediment, which was almost 6.7 and 5 times larger than that in *QVL* (0.41 kg) and *PTL* (0.55 kg) plots, respectively. Further, ANOVA analysis indicated there were significant differences ($p=0.000$ and $p=0.001$) between bare plot and two litter-covered plots. Similar results were obtained in terms of *SC* in bare and litter-covered plots ($p=0.001$ and 0.003). In this way, litter cover was able to retard erosion. The observed outcomes were similar to those presented in previous studies. For example, Pannkuk and Robichaud reported that rill and inter-rill erosion were reduced by 70% when plots were covered with Douglas fir needles and 30% when they were covered with Ponderosa pine needles, as compared with the erosion in a bare-soil plot [20]. Cerdà and Doerr reached similar conclusions in their study on soil erosion from the plots covered with ash and pine needles [21]. Benkobi *et al.* suggested that soil loss was reduced by about 87% in 100% litter cover relative to bare soil [23]. Miyata *et al.* reported that annual soil erosion in uncovered plots was 3.7 times greater than in covered plots [30]. However, Singer and Blackard found that more soil loss was observed when mulch covers <40% in contrast with bare soil [28], this was mainly because they excluded the splash erosion off the inter-rill plot, and overall sediment was reduced under mulch cover.

In the present study, because the litter cover prevented raindrops from hitting the soil surface directly by intercepting and storing the rainfall, the detachment of soil aggregates and splash erosion was lessened [66,67]. Less soil was lost in the litter-covered plots than in the bare plot.

Additionally, mean sediment yield and *SC* in the *QVL* plot were slightly lower than in the *PTL* plot, but no statistical differences were found in sediment yield ($p=0.544$) and *SC* ($p=0.810$) between the two plots. The findings indicated that litter type (broadleaf or needle-leaf) was may not be a defining predictor in reducing soil loss. This was probably because even though more topsoil was exposed and soil aggregations were detached under needle-leaf litter cover compared with broadleaf litter, needle-leaf litter tended to form mini-debris dams to trap down the soil particles in the runoff and return them to the soil surface [20]. The results were consistent with Neris *et al.*, who also reported that no significant differences were observed between pine forest floor (including needle leaf litter) and a rainforest floor (including broadleaf litter) [47].

Linear relationships between litter mass and sediment yield in *QVL* ($R=0.50$, $p=0.05$) and *PTL* plot ($R=0.52$, $p=0.112$) are shown in Figure 7. In general, total sediment yield decreased from 13.77 to 0.75 kg (*QVL*) and 0.92 kg (*PTL*) when the litter mass increased from 0 (bare) to 1 kg m^{-2}, but the largest decreases of 71.9% and 66.2% were revealed when litter mass increased from 0 to 0.3 kg m^{-2}. Similar relationships are presented in Figure 8 in terms of litter mass and *SC*. *SC* decreased as litter mass increased. Although no significant correlations were observed, the effect of litter mass on reducing sediment was found to be relevant, as the increasing litter mass covered the remaining bare soil, which indirectly protected soil from raindrop impact, and reduced the velocity of runoff [15,67]. Similarly, Lal reported that soil loss decreased from 52.4 to 0.04 kg m^{-2} when the maize grain yields increased from 0.06 to 0.68 kg m^{-2} [68].

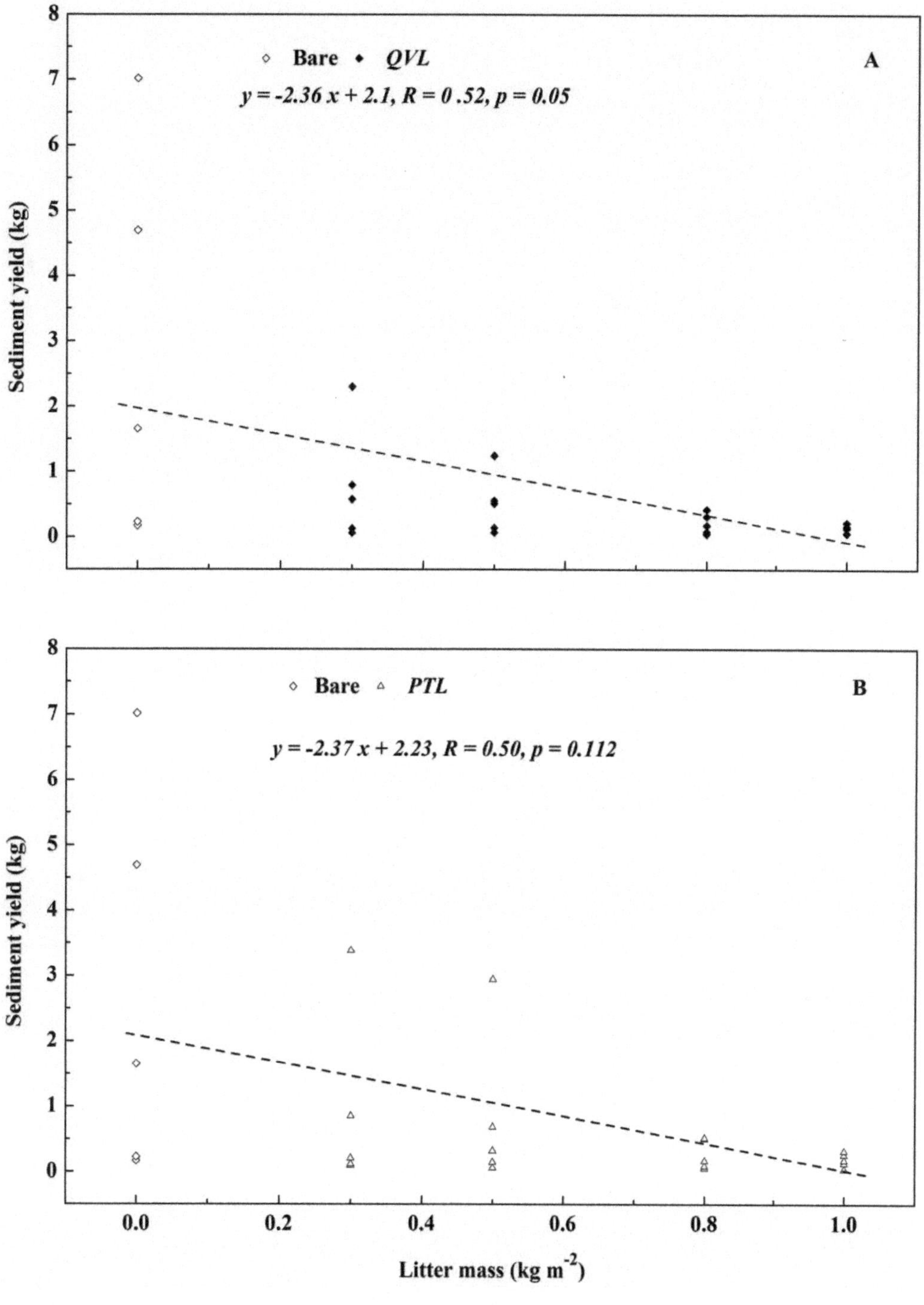

Figure 7. Relationship between litter mass and sediment yield for bare-soil plots (open diamonds), ***Quercus variabilis*** **litter (closed diamonds), and** ***Pinus tabulaeformis*** **litter (open triangles) for five rainfall intensities.**

RI also had an impact on sediment generation. Both linear (bare soil: $R = 0.99$) and polynomial relationships (*QVL*: $R = 0.63$, $p = 0.055$; *PTL*: $R = 0.71$, $p = 0.022$) between *RI* and sediment yield are presented in Figure 9A. When *RI* increased from 5.7 to 75.6 mm h^{-1}, sediment yield increased 41.3, 17.5, and 29.8 times in bare, *QVL* and *PTL* plots, respectively. *SC* showed a reverse-power decrease trend with *RI* in *QVL* and *PTL* but increased in bare plots. These results verified the protective role of the leaf litter layer, which was in line with the previous studies [24,25,46]. This was mainly because on one hand, leaf litter reduced the detachment of soil aggregation by covering on the topsoil, particularly, high-intensity rainfall (49.8 and 75.6 mm h^{-1}) with intense striking force pressed leaf litter closer to the soil surface [17]; on the other hand, the coverage by leaf litter increased the surface roughness, which facilitated sediment deposition in the runoff [30,67].

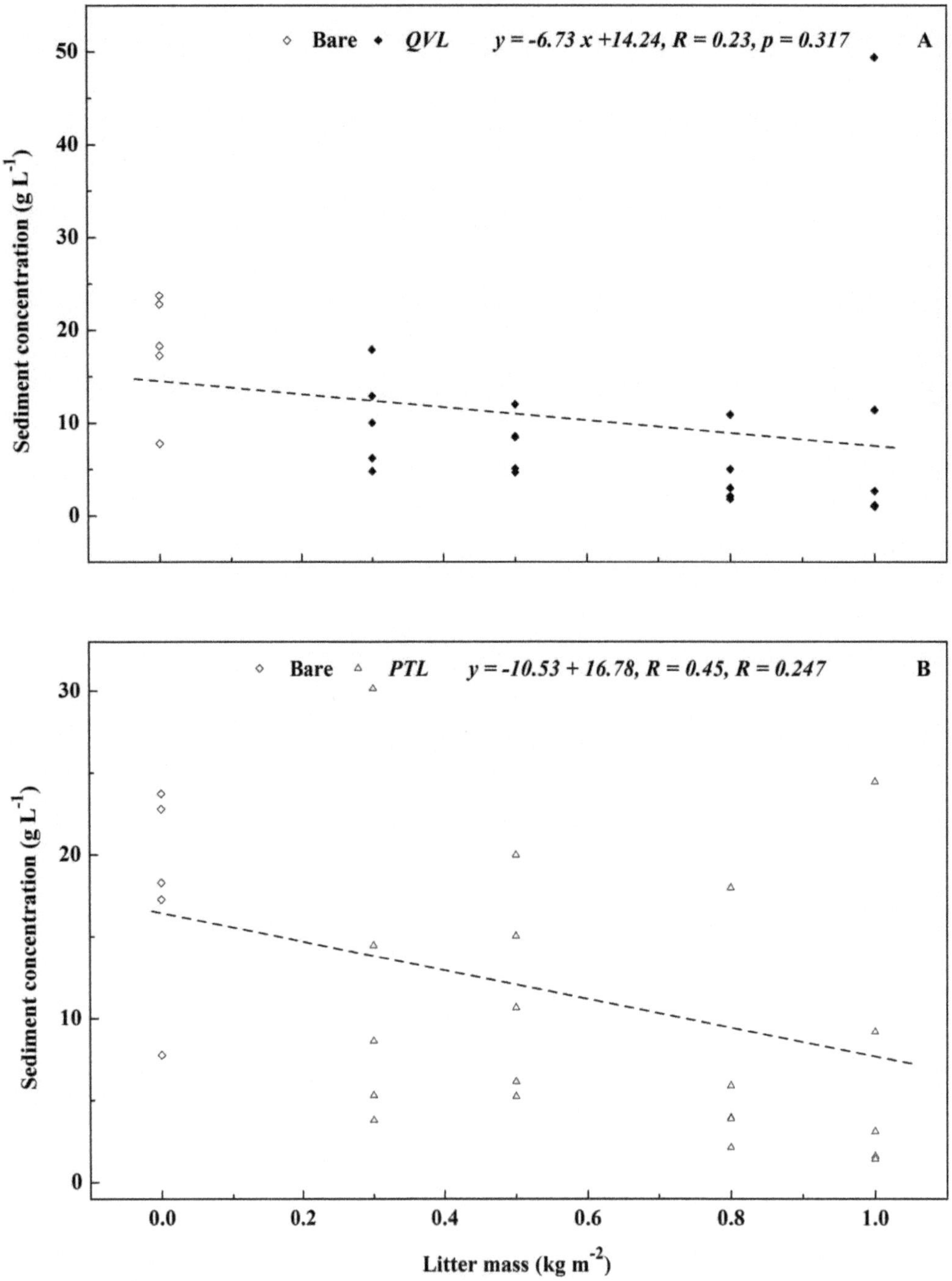

Figure 8. Relationship between litter mass and sediment concentration for bare-soil plots (open diamonds), *Quercus variabilis* litter (closed diamonds), and *Pinus tabulaeformis* litter (open triangles) for four litter masses.

Relationship between runoff and sediment

As analyzed above, runoff yield in the litter-covered plots was not significantly different ($p = 0.226$ and 0.220) from that in the bare plot, but significant differences ($p = 0.000$ and 0.000) were shown in sediment between the bare plot and litter-covered plots. This can be explained briefly by the different forms of interaction between litter type and runoff-erosion process. For broadleaf litter, runoff generated and flowed on the leaf surface, while the presence of needle-leaf litter increased the soil surface roughness and trapped down soil particles in the flow as a barrier [15,21,67,69]. Significant correlations ($p \leq 0.05$) between runoff and sediment were observed for plots in Figure 10A and 10B. Apparently, sediment yield increased with the increasing runoff. This is similar with the previous studies (e.g. [30,65,70]). In addition, an excellent linear regression was observed in bare plot ($R = 0.99$). The greatest runoff, which showed large amounts of energy, was capable of detaching soil aggregation and moving the soil particles. However, leaf litter reduced runoff and sediment by covering soil to prevent splash erosion and largely increasing surface roughness which in turn increased infiltration. The phenomenon confirmed again that litter cover played an irreplaceable role in runoff and soil erosion control.

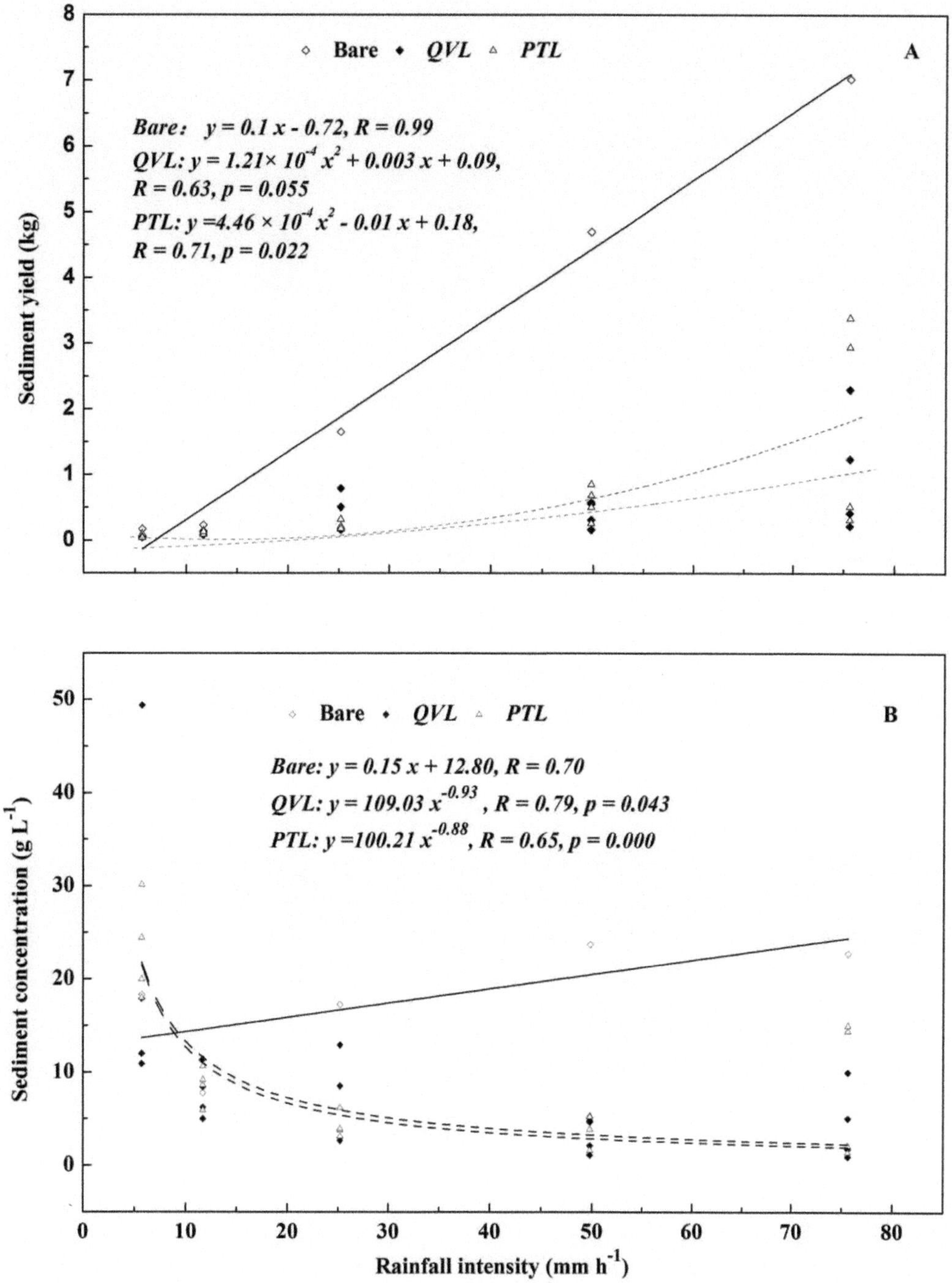

Figure 9. Relationships between sediment yield (A) and sediment concentration (B) versus rainfall intensity for bare-soil plots (open diamonds), *Quercus variabilis* litter (closed diamonds), and *Pinus tabulaeformis* litter (open triangles) for four litter masses.

Conclusion

The research effort was conducted to examine the role of leaf litter in runoff and soil erosion. Results showed that runoff and sediment yield were lower in litter-covered plots than in bare plot, litter cover had a relevant impact on runoff and sediment control. A detailed analysis of how runoff and sediment interacted with litter type, litter mass and rainfall intensity showed that rainfall intensity rather than litter type (broadleaf or needle-leaf) or litter mass was the decisive factor in generating runoff. The correlation between runoff yield and sediment yield was significantly positive ($p<0.05$) in all the plots, which confirmed that increasing runoff with larger energy was able to detach soil aggregation and cause severe erosion afterwards.

These results improved our understanding of litter as a protective layer and rainfall re-distributor. The data may be used to establish hydrological models to predict changes in runoff and soil erosion, and to provide scientific support for managing water resources in Northern China. Future studies should focus on the effects of slope, plot length, other litter types, and hydraulic characteristics (e.g., flow speed, erosive power, and shearing force)

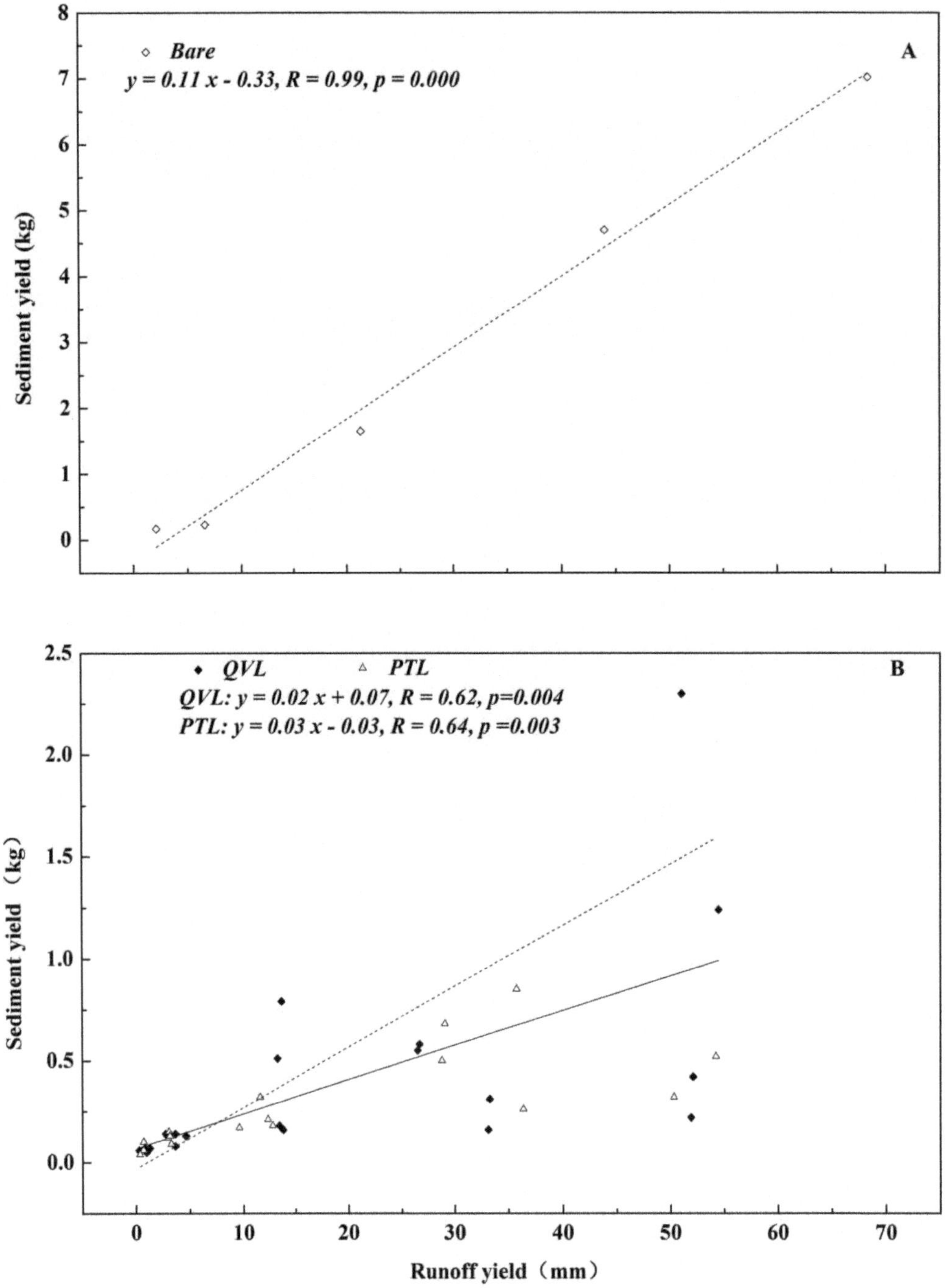

Figure 10. Relationships between runoff yield and sediment yield for bare-soil plots (open diamonds), *Quercus variabilis* litter (closed diamonds), and *Pinus tabulaeformis* litter (open triangles).

on runoff and soil erosion under litter cover to draw comprehensive conclusions.

Acknowledgments

We would like to thank the Jiu Feng National Forest Park Administration for their help and support. Further, we deeply appreciate the efficient and effective assistance provided by graduate students Shangjie Chen, Xiaoqing Du, Mingxing Wang, Bingpeng Qu, and Zhentai Pan during the field experiment.

Author Contributions

Conceived and designed the experiments: XL JN BX. Performed the experiments: XL. Analyzed the data: XL JN BX. Contributed reagents/materials/analysis tools: JN. Wrote the paper: XL JN.

References

1. Morgan RPC (2005) Soil and Conservation, 3rd edition. London: Blackwell Publishing Ltd. 304 p.
2. Hugo V, Pleguezuelo C (2008) Soil-erosion and runoff prevention by plant covers. A review. Agron J 28: 65–86. doi: 10.1007/978-90-481-2666-8_48.
3. Li ZG, Cao Y, Liu BZ, Luo ZD (2008) Current status and developing trend of soil erosion in China. Science of Soil and Water Conservation 6(1): 57–62. (In Chinese with English abstract)
4. Zhang HJ (2008) Soil erosion theory, 2nd edition. Beijing: China Forestry Publishing House. 355 p.
5. Kimoto A, Uchida T, Mizuyama T, Li CH (2002) Influences of human activities on sediment discharge from devastated weathered granite hills of southern China: effects of 4-year elimination of human activities. Catena 48: 217–233. doi: 10.1016/S0341-8162(02)00029-2.
6. Bochet E, Poesen J, Rubio JL (2006) Runoff and soil loss under individual plants of a semi-arid Mediterranean shrubland: influence of plant morphology and rainfall intensity. Earth Surf Proc Land 31: 536–549. doi: 10.1002/esp.1351.
7. Walsh RPD, Voight PJ (1977) Vegetation litter: an underestimated variable in hydrology and geomorphology. J Bio 4: 253–277.
8. Geddes N, Dunkerley D (1999) The influence of organic litter on the erosive effects of raindrops and of gravity drops released from desert shrubs. Catena 36: 303–313. Doi: 10.1016/S0341-8162(99)00050-8.
9. Sayer EJ (2006) Using experimental manipulation to assess the roles of leaf litter in the functioning of forest ecosystems. Biol Rev 81: 1–31. Doi: 10.1017/S1464793105006846.
10. Abraham AD, Parsons AJ (1991) Relation between infiltration and stone cover on a semiarid hillslope Arizona. J Hydrol 122: 45–59. doi: 10.1016/0022-1694(91)90171-D.
11. Valentin C, Casenave A (1992) Infiltration into sealed soils as influenced by gravel cover. Soil Sci Soc Am J 56: 1667–1673. doi:10.2136/sssaj1992.03615995005600060002x.
12. Peosen JW, Torri D, Bunte K (1994) Effects of rock fragments on soil erosion by water at different spatial scales: a review. Catena 23: 141–166. Doi: 10.1016/0341-8162(94)90058-2.
13. Jean JS, Ai KF, Shih K, Hung CC (2000) Stone cover and slope factors influencing hillside surface runoff and infiltration: laboratory in investigation. Hydrol Process 14: 1829–1849. doi: 10.1002/1099-1085(200007)14:10<1829::AID-HYP66>3.0.CO;2-#.
14. Dickey EC, Shelton DP, Jasa PJ, Perterson TR (1985) Soil erosion from tillage systems used in soybean and corn residues. T ASAE 28: 1124–1129.
15. Findering A, Ruy S, Scopel E (2003) Modeling the effects of a partial residue mulch on runoff using a physically based approach. J Hydrol 275: 49–66. doi: 10.1016/S0022-1694(03)00021-0.
16. Pan CZ, Shangguan ZP (2006) Runoff hydraulic characteristics and sediment generation in sloped grassplots under simulated rainfall conditions. J Hydrol 331: 178–185. doi: 10.1016/j.jhydrol.2006.05.011.
17. Pan CZ, Ma L, Shangguan ZP (2010) Effectiveness of grass strips in trapping suspended sediments from runoff. Earth Surf Proc Land 35: 1006–1013. doi: 10.1002/esp.1997.
18. Adekalu KO, Olorunfemi IA, Osunbitan JA (2007) Grass mulching effect on infiltration, surface runoff and soil loss of three agricultural soils in Nigeria. Bioresource Technol 98(4): 912–917. doi: 10.1016/j.biortech.2006.02.044.
19. Bhattacharyya R, Smets T, Fullen MA, Poesen J, Booth CA (2010) Effectiveness of geotextiles in reducing runoff and soil loss. A ynthesis. Catena 81: 184–195. doi: 10.1016/j.catena.2010.03.003.
20. Pannkuk CD, Robichaud PR (2003) Effectiveness of needle cast at reducing erosion after forest fires. Water Resour Res 39 (12): 1–9. doi: 10.1029/2003WR002318.
21. Cerdà A, Doerr SH (2008) The effect of ash and needle cover on surface runoff and erosion in the immediate post-fire period. Catena 74: 256–263. doi: 10.1016/j.catena.2008.03.010.
22. Robichaud PR, Lewis SA, Wagenbrenner JW, Ashmun EL, Brown RE (2013) Post-fire mulching for runoff and erosion mitigation Part 1: Effectiveness at reducing hillslope erosion rates. Catena 105: 75–92. doi: 10.1016/j.catena.2012.11.015.
23. Benkobi L, Trlica MJ, Smith JL (1993) Soil loss as affected by different combinations of surface litter and rock. J Environ Qual 4: 657–661. doi:10.2134/jeq1993.00472425002200040003x.
24. Smets T, Poesen J, Knapen (2008) Spatial scale effects on the effectiveness of organic mulches in reducing soil erosion by water. Earth-Sci Rev 89: 1–12. doi: 10.1016/j.earscirev.2008.04.001.
25. Smets T, Poesen J, Bochet E (2008) Impact of plot length on the effectiveness of different soil-surface covers in reducing runoff and soil loss by water. Prog Phys Geog 32 (6): 654–677.
26. Lowdermilk WC (1930) Influence of forest litter on runoff, percolation, and erosion. J Forest 28: 474–491.
27. Adams JE (1966) Influence of mulches on runoff, erosion, and soil moisture depletion. Soil Sci Soc Am J 30 (1): 110–114. doi:10.2136/sssaj1966.03615995003000010036x.
28. Singer MJ, Blackard J (1978) Effect of mulching on sediment in runoff from simulated rainfall. Soil Sci Soc Am J 42 (3): 481–486. doi: 10.2136/sssaj1978.03615995004200030022x.
29. Netto ALC (1987) Overlandflow production in a tropical rainforest catchment: The role of litter cover. Catena 14: 213–231. doi: 10.1016/S0341-8162(87)80019-x.
30. Miyata S, Kosugi K, Gomi T, Mizuyama T (2009) Effect of forest floor coverage on overland flow and soil erosion on hillslopes in Japanese cypress plantation forests. Water Resour Res 45: 1–17. doi: 10.1029/2008WR007270.
31. Pierce SP (1967) Evidence of overland flow on forest watersheds. Inter Symp Forest Hydrol. 25: 247–253.
32. Ruxton BP (1967) Slopewash under primary rainforest in Northern Papua. In: Jennings JN, Mabbutt JA, editors. Landform Studies from Australia and New Guinea. Cambridge University Press. pp. 85–94.
33. McGregor KC, Bengtson RL, Muchler CK (1988) Effects of surface straw on interrill runoff and erosion of Grenada Silt Loam soil. T ASAE 31: 111–116.
34. Y . Wan, SA . El-Swaify (1999) Runoff and soil erosion as affected by plastic mulch in a Hawaiian pineapple field. Soil Till Res 52: 29–35. doi: 10.1016/S0167-1987(99)00055-0.
35. Putuhena W, Cordery I (1996) Estimation of interception capacity of the forest floor. J Hydrol 180: 283–299. doi: 10.1016/0022-1694(95)02883-8.
36. Sato Y, Kumagai T, Kume A, Otsuki K, Ogawa S (2004) Experimental analysis of moisture dynamics of litter layers – the effect of rainfall conditions and leaf shapes. Hydrol Process18: 3007–3018. doi: 10.1002/hyp.5746.
37. Guevara-Escobar, Gonzalez-Sosa, Ramos-Salinas, Hernandez-Delgado (2007) Experimental analysis of drainage and water storage of litter layers. Hydro Earth Syst Sci 11 (5): 1703–1716. doi:10.5194/hess-11-1703-2007.
38. Li X, Niu JZ, Xie BY (2013) Study on Hydrological Functions of Litter Layers in North China. Plos One 8 (7): 1–11. doi: 10.1371/journal.pone.0070328.
39. Wu ZH (2001) The mud-rock flow disaster and their touch off condition by rainfall in Beijing Area. Research Soil Water Conserv 8 (1): 67–72. (In Chinese with English abstract).
40. Wang HZ (2008) A study of the critical rainfall volume for mudflows based on historical data in the mountainous area of Beijing. Analysis Research 3 (1): 18–21. (In Chinese with English abstract).
41. Cooperative Research Group on Chinese Soil Taxonomy (CRGCST) (2001) Chinese Soil Taxonomy, Beijing: Beijing Science Press.
42. Shi XZ, Yu DS, Yang GX, Yang GX, Wang HJ, et al. (2006) Cross-Reference Benchmarks for Translating the Genetic Soil Classification of China into the Chinese Soil Taxonomy. Pedosphere 16(2): 147–153.
43. Blake GR, Hartge KH (1986) Bulk density. In: Klute AK, editor. Methods of soil analysis. Part I: Physical and mineralogical method. American Society of Agronomy-Soil Science Society of America, Madison, pp. 363–375.
44. Carter MC, Gregorich EG (2008) Soil Samping and Methods of Analysis. Canadian Society of Soil Science, pp. 743–760.
45. Carter MC, Gregorich EG (2008) Soil Samping and Methods of Analysis. Canadian Society of Soil Science, pp. 911–938.
46. Gee GW, Bauder JW (1986) In: Klute AK, editor. Methods of soil analysis. Part I: Physical and mineralogical method. American Society of Agronomy-Soil Science Society of America, Madison, pp. 383–411.
47. Neris J, Tejedor M, Rodriguez M, Fuentes J, Jimenez C (2013) Effect of forest floor characteristics on water repellency, infiltration, runoff and soil loss in Andisols of Tenerife (Canary Islands, Spain). Catena 108: 50–57. doi: 10.1016/j.catena.2012.04.011.
48. Carter MC, Gregorich EG (2008) Soil Sampling and Methods of Analysis. Canadian Society of Soil Science, pp. 173–175.
49. Ayed GM, Mohammad AA (2010) The impact of vegetative cover type on runoff and soil erosion under different land uses. Catena 81: 97–103. doi: 10.1016/j.catena.2010.01.008.
50. Jiali W, Renhe Z, Yingchun W (2012) Characteristics of precipitation in Beijing and the precipitation representativeness of Beijing Weather Observatory. Journal of Applied Meteo Sci 23: 265–273. (In Chinese with English abstract).
51. Ma JJ, Li SY, Wang J (2012) The study of Beijing's rainstorm division and its returning periods. Meteo Monthly 38: 569–576. (In Chinese with English abstract).
52. Zhang GH, Liu BY, Li PK (2007) Principles and properties of artificial trough rainfall simulator. Bulletin of Soil and Water Conservation 27 (6): 56–60. (In Chinese with English abstract).
53. Xie Y, Lin XJ, Liu YN, Zhen YZ, Liu BY, et al. (2008) Calibration of simulated rainfall intensity and its special distribution for trough rainfall simulator. Bulletin of Soil and Water Conservation 28 (4): 1–6. (In Chinese with English abstract).
54. Battany MC, Grismer ME (2000) Rainfall runoff and erosion in Napa Vally vineyards: effects of slope, cover and surface roughness. Hydrol Process 14: 1289–1304. doi: 10.1002/(SICI)1099-1085(200005)14:7<1289::AID-HYP43>3.0.CO;2-R.
55. Kim KH, Miller WP (1996) Effect of rainfall electrolyte concentration and slope on infiltration and erosion. J Soil Tech 9: 173–185. doi: 10.1016/S0933-3630(96)00011-6.
56. Ministry of Environmental Protection of the P.R. China (1993) Determination of specific conductance in the wet precipitation. http://www.mep.gov.cn/image20010518/1881.pdf. Accessed 2014 Aug 29.
57. Tang J, Xue HS, Yu XL Cheng HB, Xu XB, et al. (2000) The preliminary study on chemical characteristics of precipitation at Mt. Waliguan. Acta Sci Circums 20(4): 420–425. (In Chinese with English abstract).

58. Zhong YD, Jia YW, Li ZW (2013) Spatial and Temporal Changes of Maximum 1h Precipitation Intensity in Beijing Region in Last 53 Years. J China Hydrol 33 (1): 32–37. (In Chinese with English Abstract).
59. Jiali W, Renhe Z, Yingchun W (2012) Characteristics of precipitation in Beijing and the precipitation representativeness of Beijing Weather Observatory. J Applied Meteo Sci 23: 265–273. (In Chinese with English abstract).
60. Ma JJ, Li SY, Wang J (2012) The study of Beijing's rainstorm division and its returning periods. Meteo Monthly 38: 569–576. (In Chinese with English abstract).
61. Imeson AC (1983) Studies of erosion thresholds in semi-arid areas:field measurements of soil loss and infiltration in northern Morocco. Catena Sup 4: 79–89.
62. Savenije HHG (1996) The runoff coefficient as the key to moisture recycling. J Hydrol 176: 219–225. doi: 10.1016/0022-1694(95)02776-9.
63. Uttam KM, Rao KV, Mishra PK, Vittal KPR, Sharma KL, et al. (2005) Soil infiltration, runoff and sediment yield from a shallow soil with varied stone cover and intensity of rain. Eur J Soil Sci 56: 435–443. doi: 10.1111/j.1365-2389.2004.00687.x.
64. Abrahams AD, Parsons AJ (1991) Relation between infiltration and stone cover on a semiarid hillslope, southern Arizona. J Hydrol 122: 45–59. doi: 10.1016/0022-1694(91)90171-D.
65. Gholami L, Sadeghi SH, Homaee M (2013) Straw mulching effect on splash erosion, runoff, and sediment yield from eroded plots. Soil Sci Soc Am J 77: 268–278. doi:10.2136/sssaj2012.0271.
66. Ghahramani A, Ishikawa Y, Gomi T (2011) Downslope soil detachment–transport on steep slopes via rain splash. Hydrol. Processes 25: 2471–2480. doi: 10.1002/hyp.8086.
67. Ghahramani A, Ishikawa Y, Gomi T (2011) Effect of ground cover on splash and sheetwash erosion over a steep forested hillslope: A plot-scale study. Catena 85: 34–47. doi: 10.1016/j.catena.2010.11.005.
68. Lal R (1981) Soil erosion problems on alfisols in Western Nigeria, VI. Effects of erosion on experimental plots. Geoderma 25: 215–230. doi: 10.1016/0016-7061(81)90037-9.
69. Miura S, Yoshinaga S, Yamada T (2003) Protective effect of floor cover against soil erosion on steep slopes forested with Chamaecyparis obtusa (hinoki) and other species. J Forest Res-Japan 8: 27–35. doi: 10.1007/s103100300003.
70. Dunjó G, Pardini G, Gispert M (2004) The role of land use–land cover on runoff generation and sediment yield at a microplot scale, in a small Mediterranean catchment. Journal of Arid Environments 57: 99–116. doi: 10.1016/S0140-1963(03)00097-1.

10

Salinity and Bacterial Diversity: To What Extent Does the Concentration of Salt Affect the Bacterial Community in a Saline Soil?

Loredana Canfora[1]*, Giovanni Bacci[1,3], Flavia Pinzari[1], Giuseppe Lo Papa[2], Carmelo Dazzi[2], Anna Benedetti[1]

1 Consiglio per la Ricerca e la Sperimentazione in Agricoltura, Centro di Ricerca per lo studio delle relazioni tra Pianta e Suolo, Roma, Italy, **2** Dipartimento di Scienze Agrarie e Forestali, Università degli Studi di Palermo, Palermo, Italy, **3** Department of Biology, University of Florence, Florence, Italy

Abstract

In this study, the evaluation of soil characteristics was coupled with a pyrosequencing analysis of the V2-V3 16S rRNA gene region in order to investigate the bacterial community structure and diversity in the A horizon of a natural saline soil located in Sicily (Italy). The main aim of the research was to assess the organisation and diversity of microbial taxa using a spatial scale that revealed physical and chemical heterogeneity of the habitat under investigation. The results provided information on the type of distribution of different bacterial groups as a function of spatial gradients of soil salinity and pH. The analysis of bacterial 16S rRNA showed differences in bacterial composition and diversity due to a variable salt concentration in the soil. The bacterial community showed a statistically significant spatial variability. Some bacterial phyla appeared spread in the whole area, whatever the salinity gradient. It emerged therefore that a patchy saline soil can not contain just a single microbial community selected to withstand extreme osmotic phenomena, but many communities that can be variously correlated to one or more environmental parameters. Sequences have been deposited to the SRA database and can be accessed on ID Project PRJNA241061.

Editor: A. Mark Ibekwe, U. S. Salinity Lab, United States of America

Funding: The authors have no funding or support to report.

Competing Interests: The authors have declared that no competing interests exist.

* Email: loredana.canfora@entecra.it

Introduction

Saline soils are environments characterised by high concentrations of salts and by an uneven temporal and spatial water distribution. According to an early classification, a soil is considered to be saline when the Electrical Conductivity (ECe) of a saturated paste is greater than 4 dS m^{-1} [1]. More recently, the two international soil classification systems introduced higher minimum thresholds of ECe to classify a soil horizon as saline. In fact, the WRB (World Soil Resources Reports) [2] considers the reference value 15 dS m^{-1} of ECe in defining the saline horizon, while the Soil Taxonomy [3] fixes the threshold at 30 dS m^{-1}. A high concentration of salt in soil changes the availability of water and nutrients for both plants and microorganisms, and it has direct and indirect influences on soil organic matter, content, and stability [4]. Salinity, in fact, has been found to influence the size and the activity of soil microbial biomass [5,6], which in turns plays a key role in biogeochemical cycles.

A basic distinction must be made between primary and secondary salinisation processes. Primary salinisation consists of salt accumulation through natural processes, such as a high salt content of the parent material or in groundwater. Secondary salinisation is usually caused by human interventions such as inappropriate irrigation practices, i.e. after the use of salt-rich irrigation water, and/or insufficient drainage. A natural secondary soil salinisation mechanism is represented by the long-term effects of tsunami waves, which can deposit salty seawater on large flooded areas with dramatic consequences for agriculture. Depending on the climatic conditions, secondary soil salinisation can also be temporary, and the soils may recover by washing out the infiltrated salt deposits through rainfall; but this is not the case for Mediterranean and arid environments, which can hardly spontaneously recover from a secondary salinisation event. Microorganisms that occur in naturally saline habitats are supposed to share a strategy for resisting high salt concentrations, and to have developed multiple adaptations for maintaining their population active while coping with such extreme environmental conditions. From the genetic point of view, these species display an under- or over-expression of peculiar genes and metabolites, which confer them the capability of coping with an osmotic stress [7].

A naturally saline soil is also a mutable environment where rain and water movements can strongly change the distribution of salts, and create an evolving patchy landscape. The researcher's perception of environmental variability, and, consequently, the scale at which specific properties (like salinity) are measured, can misrepresent the spatial scale at which microbial groups differentiate their structure. Measurements of microbial community's structure and function are often based on broad-scale characterisations, and rarely consider the real spatial scale within which individuals and populations interact [8,9]. Space and scale in population, community, and ecosystem processes are increasingly recognised as fundamental factors in the study of microbial functions and activities in soil [8]. Understanding the spatial pattern in the abundance and structure of microbial communities occurring in saline soils represents a crucial target in ecology [9,10] as it sheds light on the selection mechanisms exerted by the environment on bacterial groups with specific functions and properties. The choice between a niche higher in salinity and instability, where instead of developing metabolic tools for resistance the bacterial communities wait for more favorable conditions, can represent the key mechanism that shapes microbial heterogeneity and taxa spatial composition in problematic soils.

In this framework, a series of surveys on "extreme" environments where salinity is the main determinant addressed the issue of some microbial taxa's specific ability to resist osmotic stress or other limiting factors associated with the presence of high salt concentrations. On the basis of these studies [11,12,13,14,15,16,17,18,19,20,21,22,23,24,25,26,27,28], we have attempted a rough discrimination between taxa frequently listed in these extreme environments (salinity related), and taxa less represented or even absent in salt affected sites (salinity unrelated). Clearly, this approach can lead to a dangerous generalisation, especially for the taxa that contain genera and species with strongly different physiologies and very broad geographical distributions. This is the case of the phylum *Proteobacteria*, which is probably the biggest group of bacteria associated to practically any environment.

A recent meta-analysis of soil microbial communities reported that the global microbial composition in a saline soil is influenced more by salinity than by any other extreme chemical factors such as temperature or pH [11,12]. Ma and Gong [12] recently examined the bacterial and archaeal diversity in saline soils using a meta-analysis approach indicating that approximately 50% of the archaeal diversity and less than 25% of the total bacterial diversity has been recorded from saline soil habitats. Ma and Gong [12] updated the available information on DNA sequences gained from a wide array of studies on soil microorganisms in saline environments, but also showed that there is a significant gap in the published information on the relevant soil properties where microbial communities had been sampled. Despite the useful and unique information collected, the authors could not address the beta diversity of the microbial species according to the different salt concentrations [12], thus failing to relate bacterial taxa composition with such drastic soil physical-chemical factors.

In the present study, a thorough evaluation of soil characteristics was supported by a pyrosequencing of the V2–V3 16S rRNA gene region in order to investigate the bacterial structure and diversity in the A horizon of a natural saline soil habitat located in Sicily (Italy). The application of metagenomic strategies has been recognised by several authors as a valid instrument to exploit the microbial biodiversity within soil habitats [29,30,31], but up to now saline habitats have still been largely unexplored by the metagenomic approach. The main aim of the research is to evaluate the organisation and diversity of microbial taxa by means of a spatial scale that reveals heterogeneity in the distribution of physical and chemical properties of the environment under investigation. As salinity is one of the most widespread soil degradation processes on earth affecting an estimated 1 million hectares in the European Union, mainly in the Mediterranean countries, and as it is a major cause of desertification, the understanding of soil microbial resilience and resistance in primary salinisation processes represents a cognitive platform for any application of bioremediation in secondary salinisation events.

Materials and Methods

Site description

The study was performed in Sicily, Italy, in an abandoned natural area, *Piana del Signore* -a semiarid Mediterranean environment- characterised by an alluvial flat land where the geomorphology has been modeled by the river *Gela* (Fig. S1). The area falls within the Natura 2000 network as a site of Community Importance (SCI), with the number SIC ITA050012- Gela. In the basin of the river *Gela* the prevalent lithology is made up by Messinian evaporites belonging to the *Gessoso-Solfifera* geological formation, among which many types of saline rocks crop out: Gypsum, Carbonates and Marls (with frequent chloride and sulfide rock inclusions). The area we surveyed is 12.3 hectares wide, and lies about 1 km far from the coastline and 1.2 km from the river estuary. The vegetation is a patchy mosaic plant association defined *Junco subulati-Salicornietum fruticosae* belonging to the *Thero-Salicornietea* class, in which the most plants are salt pioneer swards typical of salt marshes. The vegetation pattern consists of *Salicornia fruticosa* (L.), *Suaeda fruticosa* (L.), *Juncus subulatus* (Forsskal), *Juncus bufonius* (L.), *Phragmites australis* (Cav.), *Aster squamatus* (Spreng.), *Polypogon monspeliensis* (L.), *Hainardia cylindrica* (Wild.). The vegetation distribution is discontinuous: plant covered zones alternate with bare zones with visible salt crusts deposited above the soil surface. The whole area is temporarily flooded in the autumn and winter seasons, with a long permanence of water in those zones where salt crusts have formed on the soil surface after the water has dried out. Mean daily air temperature ranges from a maximum of 26.6°C in August to a minimum of 4.9°C in January. The average annual rainfall is 383 mm according to the nearest meteorological station (37°4′48″N, 14°13′12″E).

Soil Survey and sampling

Nine soil sites from A horizons (with a mean depth of 0–10 cm) were collected following a random simple sampling scheme (Fig. S2a) in Summer 2011. No specific permission was required for this location, as no endangered or protected species live in the area (Table S1). Sampling sites were positioned with a minimum distance between points of 50 m. Sites were accurately recorded with a GPS. In each site, three soil samples were collected in the vertices of 1-meter side equilateral triangle (Fig. S2b) and mixed together in a unique representative analytical sample [32]. Vegetation, salt crusts, and other features of the soil surface were described and recorded for each site. Soil samples from each site were subdivided in two representative subsamples: the first one was air dried, 2 mm sieved, then chemically and physically analysed, while the second one was stored at −80°C and later processed for the 454 pyrosequencing analysis reported below.

Soil physical and chemical analysis

Air-dried and sieved soil subsamples were analysed for the following physical and chemical properties: texture; reaction (pH), electrical conductivity (EC), and organic carbon (C_{org}). Texture was determined by the pipette method, without carbonate and organic matter removal, and after complete removal of soluble salts by using distilled water [33]; pH was measured on 1:2.5 (w/v) soil to water mixtures; $EC_{1:5}$ was measured on 1:5 (w/v) soil to water mixtures at 25°C; C_{org} was obtained using the Walkley and Black method; $EC_{1:5}$ was converted in electrical conductivity of the saturation paste extract (EC_e) using the correlation model proposed by Khorsandi and Yazdi [34] for arid and semiarid environments.

DNA extraction

DNA was extracted from the fresh soil sub-sample with the MoBio Power soil DNA extraction kit following the manufacturer's instructions. The samples were then purified from excess impurities with GeneReleaser [35]. DNA crude extract concentrations were measured using Qubit 2.0 Fluorometer by means of the kit Quanti HS assay Invitrogen, following the manufacturer's instructions. DNA was extracted in duplicates from each soil site, and then pooled and used for the following analytical steps.

Amplification of 16S rRNA genes and pyrosequencing

The V2–V3 region of the 16S rRNA bacterial gene was amplified by PCR. The PCR reaction mixture (50 μl) contained 10 μl 10-fold reaction buffer (Fusion GC buffer, FINNZYMES, Espoo, Finland), 800 mM of each of the four deoxynucleoside triphosphates, 3% DMSO, 1.2 mM of each of the primers, V2 For and V3 Rev, 0.5 U of Phusion hot start high- fidelity DNA Polymerase (FINNZYMES), and 20 ng of isolated DNA as template. The V2–V3 region was amplified with the following set of primers containing the Roche 454 pyrosequencing adaptors (underlined): V2for 5′-CGTATC**GCCTCCCTCGCGCCA**TCAG**ACGACTGCGTAGTGGCGGACGGGTGAGTAA**-3′ and V3rev 5′- CTATGC**GCCTTGCCAGCCCGC**TCAG**AGACGCACTCATTACCGCGGCTG**C-3′ ([31], modified following the instructions of BMR Genomics to better adapt the primers to the 454 project and overcome possible PCR biases. (In bold, the reader can find the starting bp number for each primer modified, adding the tag CGTATC and CTATGC at 5′ and a barcode in the middle of each primer TCAG). The following thermal cycling scheme was used: initial denaturation at 98°C for 5 min, 25 cycles of denaturation at 98°C for 45 s, annealing at 68°C for 45 s, and extension at 72°C for 25 s, followed by a final extension period at 72°C for 5 min. All samples were amplified in two series of triplicates, pooled in equal amounts, and then purified using the Qiaquick PCR purification kit (Qiagen Inc. Chatsworth, CA, USA). Six independent PCR products for each site were combined to minimise the impact of PCR errors. Quantification of the PCR products was performed using the Quant-iT dsDNA BR assay kit (Invitrogen GmbH, Karlsruhe, Germany) and a Qubit fluorometer (Invitrogen GmbH, Karlsruhe, Germany) as recommended by the manufacturer. The samples were stored at −20°C and sent to BMR Genomics s.r.l. (Padova, Italy) for pyrosequencing by means of a Genome Sequencer FLX System platform (454 Life Science Branford, CT, USA).

Analysis of pyrosequencing data: dataset clean up and taxonomic assignments

The obtained sequences were assigned to each site using a custom script developed by BMR Genomics. 9 different sequence files, with an average sequence number of 6108, and an average length of 492 nucleotides per sequence were obtained (Table S2, Table S3). In order to ensure a correct nucleotide assignment in the raw sequence files, 2 control steps were performed. First, the nucleotide distribution along each sequence was analysed. This analysis showed that each file had an unbalanced nucleotide distribution in the first 5–10 bases of each sequence (Fig. S4). Next, the quality distribution along each sequence was analysed in order to identify possible low quality segments. As a result, low quality segments of 50–100 bases were identified in the terminal region of each sequence. Finally, a trimming step was performed using StreamingTrim software [36]. An offset of 10 nucleotides was set in order to remove the first 10 bases of each sequence. After the trimming step, 6062 sequences were collected with an average length of 425 bases.

The last control step performed consisted in the identification of chimeric sequences in the dataset [37]. In order to detect all possible chimeric sequences, a dataset was constructed that contained all the 16s rRNA available genes in the NCBI (National center for Biotechnology Information). This database was used as reference for the UCHIME algorithm [38]. After UCHIME analysis 1936 chimeric sequences were detected and removed from the dataset. As a result, 9 different sequence files (one for each site), containing an average sequence number of 5847, were recovered (for additional details, see Fig. S3).

In order to construct a community data matrix, the cleaned files were analysed using the standalone version of RDP multiclassifier [37]. As the average length of the sequences to be analysed was bigger than 250 nucleotides, an assignment's confidence cutoff of 0.8 was set to perform a much stringent analysis (according to the RDP multiclassifier pipeline: http://rdp.cme.msu.edu/tutorials/classifier/RDPtutorial_MULTICLASSIFIER.html).

Statistical analysis of community data

The community data matrix obtained was used in a series of statistical and ecological evaluations carried out using the R software (http://cran.r-project.org/) [39] and the *vegan* package (http://cran.r-project.org/web/packages/vegan/index.html) [40]. A Rarefaction Analysis [41], based on *genus*-level data (all taxonomic assignments that reached the *genus* level), was performed in order to inspect the different grade of diversity explained in each site. *Richness* [42], *inverse Simpson* [43] and *evenness* [44] indexes were calculated on the same data set. *Richness* index was calculated on the basis of the number of the taxonomic assignments at the *genus* level gained for each site, while *evenness* and *inverse Simpson* were calculated as follows:

$$Evenness = \frac{-\sum_{i=1}^{S} p_i \log p_i}{\log S}$$

$$Inverse\ Simpson = \frac{1}{\sum_{i=1}^{S} {p_i}^2}$$

Where, S is the site *richness* (defined as the number of genera in each site) and p_i is the proportion of genera in the site [42]. In order to inspect the putative number of unseen *genera* present in the sites, Chao [45] has calculated the index using the following equation:

$$Chao = S + \frac{{f_1}^2}{2f_2}$$

Where, S is the site *richness* (defined above), f_1 and f_2 are the numbers of genera observed once and twice respectively. The community data matrix has then been transformed into a relative abundance matrix in order to highlight the differences in the community composition correlated to the salinity and pH levels of each site. The relative abundance values have been calculated dividing the number of reads assigned to each *genus* by the total number of reads present in each sequence file, in order to compare these values with each others. Moreover, a Canonical Correlation Analysis (CCA) was run with soil salinity and pH fitted onto the ordination analysis obtained, using the *envfit* function of the *vegan* R package, with a number of permutations of 10000. To emphasise the differences in the composition of the different *taxa* inside the community data matrix, a relative abundance matrix was generated, as described above, considering just the assignments at the Phylum level. This matrix was used to generate a heatmap plot of the ordered sites, following the soil degree of salinity (from the low salinity level, at the bottom, to the high salinity level, at the top). Above the generated heatmap, a coloured bar was placed to indicate the phyla correlated to a high level of salinity and the phyla not strictly correlated to high salinity levels. In addition, a dendrogram showing a cluster distribution was plotted and added to the map, using the Bray-Curtis distance and the UPGMA algorithm.

Furthermore, data on soil properties and on the abundance of microbial phyla were combined as variables in a principal component analysis (PCA), used as exploratory analysis, assuming that microbial groups (absolute abundance values) have a linear response to environmental gradients [46]. The PCA was performed on autoscaled data, and based on Spearman's rank correlation matrix; XLStat 7.0 (Addinsoft, Paris) statistical software was used for the purpose. Spearman's rank correlation test was used to define the degree of dependence among the variables. The advantage of using Spearman's rank correlation coefficient is the independence of the population's distribution, so that the data can be collected over regular spaced intervals. The Spearman's correlation coefficient (rho) was used as a measure of the correlation (dependence) between the variables, giving a value between +1 and −1, where 1 stands for the total positive correlation, 0 stands for no correlation, and −1 stands for the total negative correlation. The rho coefficient is based on the ranks of the observations; the Spearman's rank correlation coefficient does not assume that the relationship among the variables is linear [47].

Results

The 9 soil sites represented nine plots distributed in an area characterised by a great spatial variability, which encompassed different levels of salinity, and a significant variation in soil pH, organic carbon, vegetation type and cover percentage, saline crust percentage, and texture class (Table 1). Electrical conductivity ranged between 169.96 dS m^{-1} to 5.37 dS m^{-1} with values of pH between 6.4 and 8. Total organic Carbon ranged between 4.26 g kg^{-1} and 0.38 g kg^{-1}.

Microbial DNA extraction yielded between 4.5 ng/µl to 10.2 ng/µl, showing variability in DNA recovery. Given the very standardised methodology applied to all the samples during the extraction step, the reason for the yield variability can be searched in the different soil characteristics among the nine sites. A different salt concentration and the presence of a variable amount of organic carbon can modify the degree of interaction between microbial DNA and the used extracts [48]. In order to check whether possible interferences occur between the DNA extraction procedure here adopted and some of the main soil properties, namely Organic Carbon content, pH and Salt concentration, an evaluation of the degree of correlation (Spearman's correlation) between DNA recovery (ng/µl) and measured soil parameters has been carried out. The results (Table S5) excluded a statistically significant positive or negative role of soil organic carbon, pH and salt concentration in DNA recovery during extraction.

Table 1. Main physical and chemical properties of the soil sites and percentage of vegetation and salt crust cover of the sampling sites estimated in the field.

Plot	pH	EC_e	C_{org}	Clay	Silt	Sand	Texture class[1]	Vegetation cover	Salt crust cover
		dS m^{-1}	g kg^{-1}					%	
1	7.9	50.45	2.27	189	177	634	loam	90	70
2	7.0	15.06	1.32	80	98	822	loamy sand	85	5
3	7.2	92.73	0.67	76	74	850	loamy sand	40	100
4	7.4	28.00	4.26	245	294	461	silty loam	100	0
5	7.8	5.82	1.4	83	145	772	loamy sand	100	0
6	6.9	63.50	1.46	369	263	368	silty clay loam	90	0
7	7.2	169.96	2.35	188	217	595	loam	3	100
8	8.0	5.37	0.38	54	38	908	sand	70	0
9	7.5	54.88	1.27	107	105	788	sandy loam	90	100

[1] Classification according to the international scheme (Leeper and Uren, 1993).

General analyses of the pyrosequencing- derived dataset

The average length of the sequences analysed was bigger than 250 nucleotides, more than 75% of the reads analysed was assigned at the Phylum level, and just 25–30% of the reads was assigned to the *genus* level. The description of the assignments' number at each taxonomic level and for each site is reported in Figure 1. A total of 16342 sequences for the nine soil sites was used for bacterial diversity analysis. Figure S4 displays the nucleotide's relative frequency distribution along the sequences, where the first 10 bases of each sequence file have an unbalanced nucleotide distribution, suggesting an uneven distribution of sequences among the sites in the soil. The number of sequences per site ranged from 1968 to 15614, except for site 1. Figure 1 displays the number of sequences assigned by RDP classifier for each site above and below the domain of Bacteria 52617 sequences were assigned to the domain of Bacteria and 42459 of these sequences were classified below the domain level.

Bacterial diversity and richness

Figure 2 shows the rarefaction curves obtained for each soil site. Rarefaction curves were created by randomly re-sampling the pool of N samples multiple times, and then plotting the average number of species found in each site. This method generated the expected number of species in a small collection of n samples drawn at random from the large pool of N samples. Rarefaction curves grow rapidly at first, as the most common species are found, then the curves reach a plateau when the rare species remain to be sampled. In sites 1, 7 and 8 the curves didn't reach saturation, suggesting that taxonomic diversity was not fully exploited. In contrast, the remaining six sites showed rarefaction curves that reached a plateau. In particular, site 9 reached the point of saturation faster than the other sites.

A comparison of the rarefaction analyses with the Chao1 (Fig. 3) index revealed that a substantial fraction of soil sites (6 out of 9) showed a relevant number of putative "unseen genera", in agreement with the rarefaction analysis.

In table 2, the values of biodiversity indexes calculated on the nine soil sites were compared; site 9, despite showing a rarefaction curve that reached the plateau, was characterised by low values of the *Invsimpson* and *evenness* indexes, when compared with the other soil sites. The values of the *Invsimpson* index (Table 2) ranged between 1.6 and 22.7 while the *richness* index ranged between 34 and 148. The *evenness* index ranged from 0.2 to 0.8, indicating an uneven distribution of bacterial genera between the sites.

The comparison of the soil sites based on the values of the *richness* index showed the greatest bacterial richness in site 3, followed by sites 4, 6, and 2. Both *Invsimpson* and *richness* indexes (Table 2) discriminated between the different soil sites, but didn't succeed in emphasising peculiar behaviours of some sites. In the case of site 9, for example, which showed a curve of rarefaction very different from the other sites, the *evenness* index was the only indicator that showed sensitivity to this particular behaviour.

Bacterial phyla distribution along a salinity gradient

The 52623 sequences classified below the domain level were affiliated to 15 bacterial phyla. The dominant phyla across all sites were: *Acidobacteria*, *Actinobacteria*, *Bacteroidetes*, *BRC1*, *Chlor-*

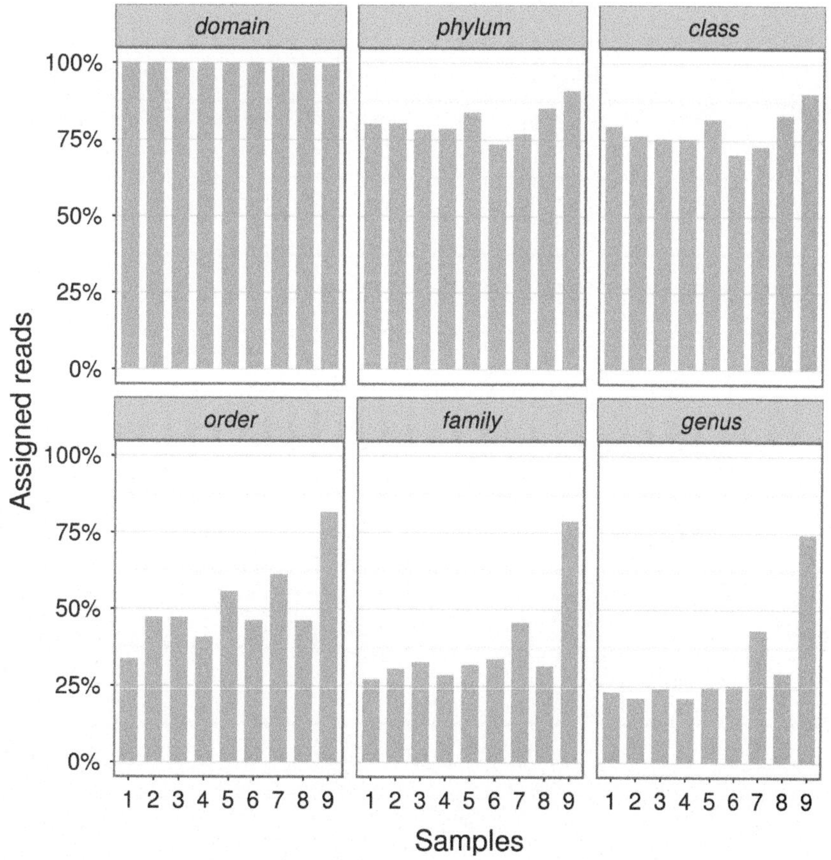

Figure 1. Number of sequences assigned by RDP multiclassifier for each site at each taxonomic level.

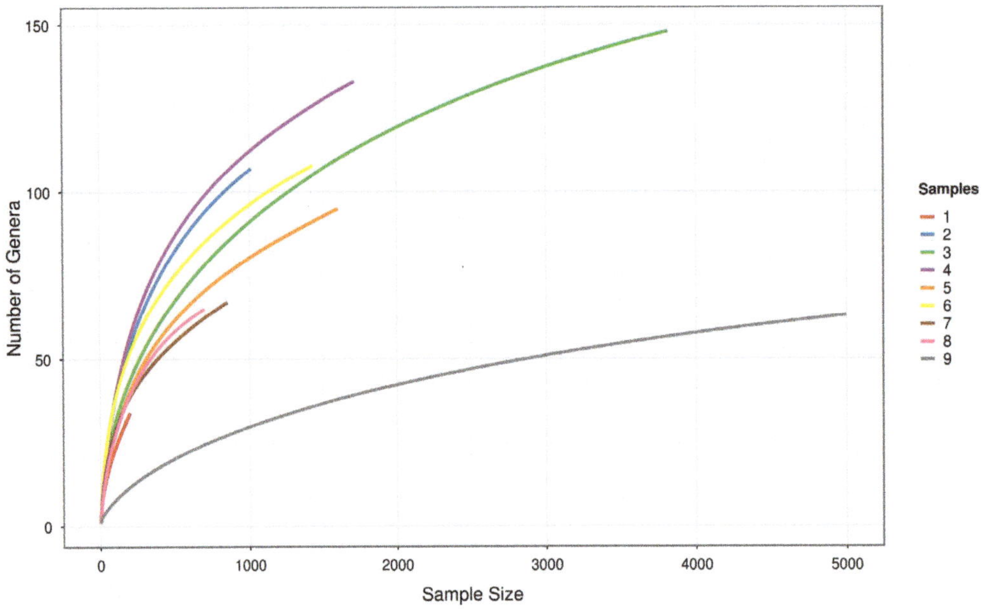

Figure 2. Rarefaction curves of the sites. These curves were obtained considering only the taxonomic assignments that reached the *genus* level in the RDP multiclassifier analysis.

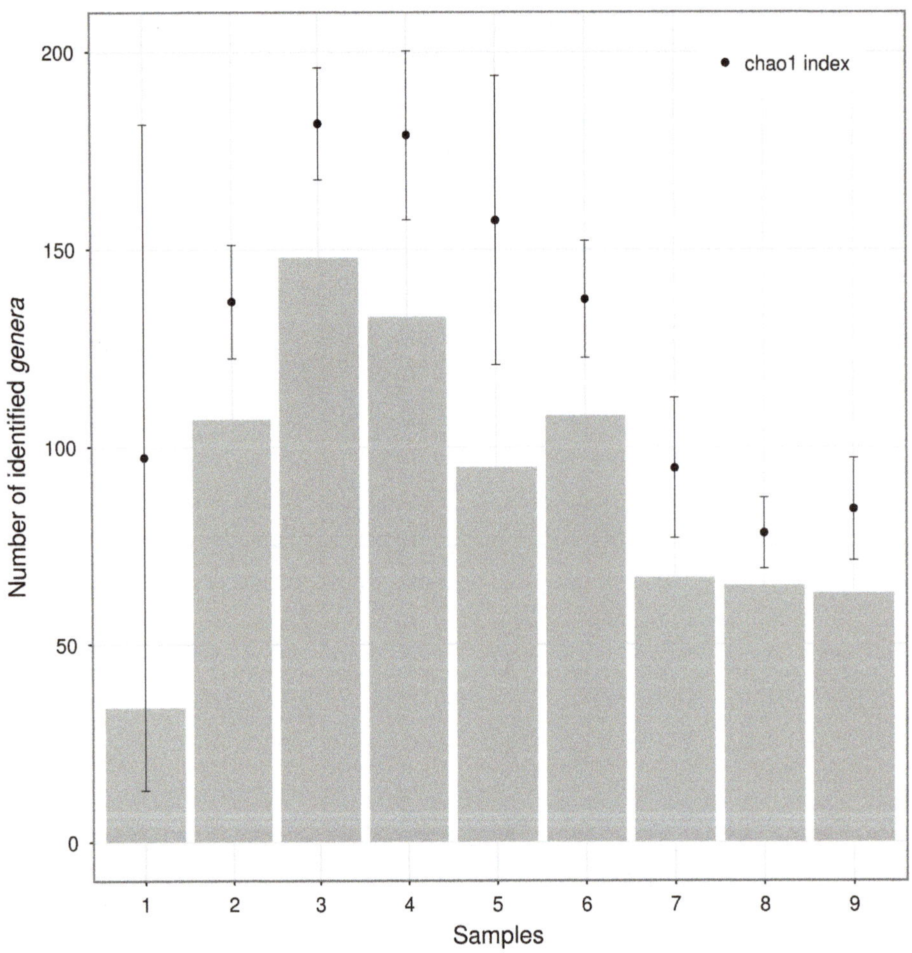

Figure 3. Richness values compared to Chao1 indexes. As shown in this plot almost all sites have a relevant number of putative "unseen *genera*" according to the Rarefaction analysis (Fig. 3).

Table 2. Diversity indexes. Each index has been calculated as reported in the section: "Statistical analysis of community data" of the "Material and Methods" chapter.

	value of biodiversity indices		
Plot	*Invsimpson*	*Richness*	*Evennes*
1	6.67	34	0.70
2	14.78	107	0.74
3	13.72	148	0.66
4	11.04	133	0.70
5	5.17	95	0.58
6	22.71	108	0.77
7	14.22	67	0.74
8	6.41	65	0.65
9	1.57	63	0.21

obi, Chloroflexi, Cyanobacteria, Deferribacteres, Firmicutes, Gemmatimonadates, Nitrospira, Planctomycetes, Proteobacteria, Spirochaetes, Tenericutes, Verrucomicrobia, WS3. The dominant taxa in the analysed soil sites were: *Proteobacteria* (95.95%), *Actinobacteria* (85.39%), *Acidobacteria* (72.12%), *Verrucomicrobia* (70.60%), *Firmicutes* (64.14%), followed by a second group showing a lower but still important percentage of distribution across all the soil sites (*Bacteroidetes, Chloroflexi, Chlorobi, Gemmatomonadates*, with a percentage between 70 and 50%). The taxa with a lower relative distribution were: *Planctomycetes, Tenericutes, Deferribacteres, Cyanobacteria, Spirochaetes, Nitrospira*, and the uncultured candidate bacterium divisions *WS3* and *BRC1*.

The relative distribution of the groups of bacteria varied among soil sites and along a salinity gradient, as showed in the heatmap of Figure 4, where the relative distribution of *phyla* assignments is clustered and plotted with respect to the different values of salinity of soil sites. Each row (sites) of the heatmap shows the phyla's relative abundance in a soil site. The rows of the heatmap are ordered according to the degree of salinity in the soil sites (from the lowest values, at the *bottom*, to the higher, at the *top*). The heatmap of Figure 4 reports the following annotations: "salinity related" and " salinity unrelated", referring to taxa frequently listed in saline environments and taxa less represented or even absent in salt affected sites, as roughly determined on the basis of what reported in previous studies [12,13,14,15,16,17,18,19,20,21,22,23,24,25,26,27,28]. Moreover, the cluster structure shows five main groups of phyla which share a peculiar composition and abundance among the sites. For example, site 3 presented a very peculiar abundance of three phyla (*Cyanobacteria, Deferribacteres, Nitrospira*), which resulted unrelated to salinity, while a large group of phyla, that appeared grouped together and uniformly distributed across the other sites, resulted strongly correlated to salinity, although showing a low abundance. In fact, the phyla *Acidobacteria, Actinobacteria, Firmicutes, Proteobacteria*, and *Bacteriodetes* resulted related to all the 9 sites, with almost the same abundance degree. *Spirochaetes* and *Tenericutes* resulted related just to site 7, and the BRC1 candidate Phylum was only found in site 5.

Salinity and pH values of each soil site were used as variables in the CCA ordination analysis, run in order to highlight the influence of these two soil properties on the bacterial community structure, and to highlight the phyla's relative abundance among sites (Fig. 5). The Canonical Correlation Analysis (CCA) (Fig. 5) summarises the joint variation of the two sets of variables, namely soil pH and salinity, in relation to the bacterial phyla ordination (obtained using the *envfit* function of the *vegan* R package, as described in the Materials and Methods Section). The CCA method combines a correspondence analysis with a multivariate regression analysis, taking into account the underlying model that assumes chi-squared dissimilarities among the sites. The plot obtained by CCA allowed us to extract synthetic environmental gradients from the metagenomic data-sets. The gradients are fundamental for succinctly describing and visualising the different habitat preferences (niches) of the phyla; in order to do so we used an ordination diagram that is shown in Figure 5. The graph obtained on the basis of permutations gave a clear picture of the variability expressed by the different sites, showing different ecological niches for both salinity and pH, but also with respect to the inhabiting microbial communities. In particular, Figure 5 showed a different spatial position within the plot of the sites 7, 9 and 1, due to a high salinity level in the first one, and due to a significantly lower pH in the other two sites. It also emerged that the bacterial communities of sites 5 and 8 were characterised by a tendency to low values of salinity, and that there was a core of sites with closer bacterial assemblages and soil characteristics (sites 2, 3, 4, 5, 6 and 8).

Spearman's rank correlation results (Table 3) show the dependence between a relative abundance of bacterial phyla and the soil properties. *Proteobacteria, Spirochaetes, Tenericutes, WS3 Plantomycetes* and *Bacteroidetes* showed a significant positive correlation with salinity. The *Acidobacteria* showed a statistically significant negative correlation with salinity (-0.717 for $p<0.01$). The only bacterial group that showed a relative abundance significantly correlated with soil pH was the *Plantomycetes*, which resulted distributed among sites with a negative correlation to the variable's values (a higher abundance of bacteria was related to lower pH values). Two phyla showed a statistically significant dependence on the organic carbon content (Corg) of soil sites: *Verrucomicrobia* and *Chlorobi*. Interestingly, the Spearman's rank correlation between the degree of salt crust coverage of the sites and the composition of bacterial phyla did not follow the trend showed by the variable ECe (salinity) and, while not showing statistically significant values, it divided the groups of bacteria into two categories, positively dependent and negatively dependent on the presence of a salt crust in the site.

Figure 6 shows the scatter plots of sites obtained by plotting the variables "abundance" and "soil salinity (ECe) for each bacterial phyla. The plots allow us to check for linearity, that is, for a monotonic relationship between the two variables, and to check

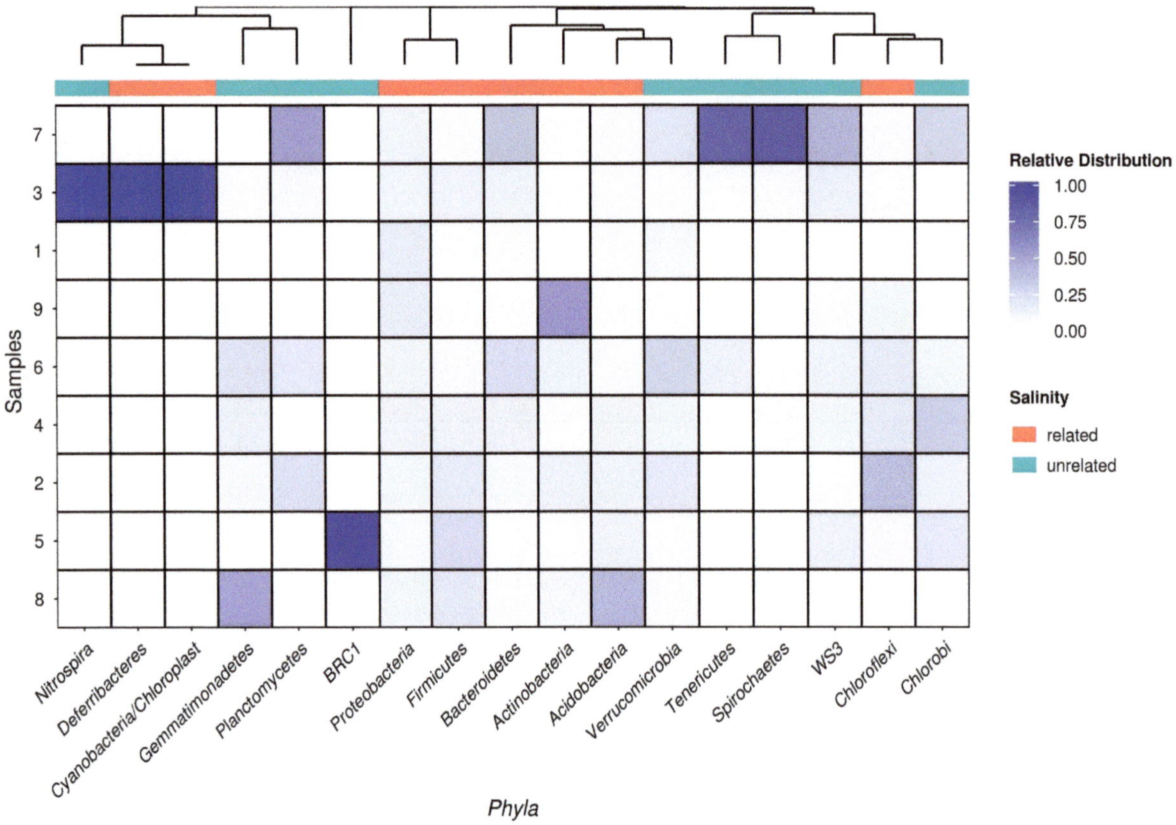

Figure 4. Heatmap of *Phyla* assignments. The heatmap reports the normalized values of the taxonomic assignments at *phylum* level. Each value has been normalized following this criterion:

$$X_{ij}^{norm} = \frac{X_{ij}}{\sum_{k=1}^{N} X_{ik}}$$

Where X_{ij} is the occurrence of the phyla 'j' in the site 'i' and N is the number of site in the dataset (in this case 9). Using this transformation each *phyla* assignment can be compared in all sites independently from its order of magnitude.

whether either the variables increase in value together, or, as one variable value increases, the other variable value decreases. The plots showed in Figure 6 relate to the bacterial phyla showing a significant rho value in the Spearman's rank correlation test. With regards to *Bacteroidetes*, *Spirochaetes* and *Tenericutes,* their relative abundances in the different sites increase positively with the salinity degree in the soil (from site 8, showing one of the lowest values of salinity, through to sites 5, 2, 4, 1, 9, until the sites 6,3, and 7 showing the highest values of salinity). *Proteobacteria* and *WS3* also showed a monotonic relationship between abundances and soil salinity, although some sites (i.e. site 1 for *Proteobacteria*) showed a non-linear behaviour. Furthermore, *Acidobacteria* showed an opposite trend, as relative abundances decrease with the increase of soil salinity, as also indicated by the Spearman's rank correlation test (Table 3).

A principal component analysis is reported as biplot in Figure 7. The bacterial phyla abundances in each sample were used as variables, together with soil chemical properties (C_{org}, pH, vegetation cover, salt crust, ECe), while the soil sites were showed as observations.

The biplot, obtained using the first two components, which together explained about 52.31% of the total variability of data, showed that salinity (ECe variable) is correlated with salt crust presence, and that both variables are positively correlated with the values of PC1 axes. Soil pH appeared correlated to the vegetation cover, and both variables are negatively correlated to the PC1 axes. Soil organic matter content (C_{org}) is positively correlated to PC2 axes. The loading of the sites on the principal components axes confirms the main patterns that were delineated by both Spearman's rank correlation analysis (Fig. 6; Table 3) and the cluster analysis (Fig. 4). The variability of soil properties between the 9 sites corresponded to a diverse relative abundance of bacterial phyla, which actually showed a different degree of correlation with soil characteristics. In some cases, there is an inverse relationship between bacterial phyla, thus showing a kind of vicariousness, suggesting that the presence of some groups of bacteria in a site could be the reason for the absence of others. Rather the opposite behaviour is found in some phyla that are strongly associated in the biplot, and seem to occur in the same sites and when the same micro-environmental conditions occur.

The percentages of the different bacterial phyla present in each soil site, and the percentages observed in the whole study site (values consisting of an average of what obtained in the nine sites) are reported in Figure 8, where, for each site, a pie-plot is showed, representing the bacterial community as emerged by the pyrosequencing analysis of the V2–V3 16S rRNA bacterial gene

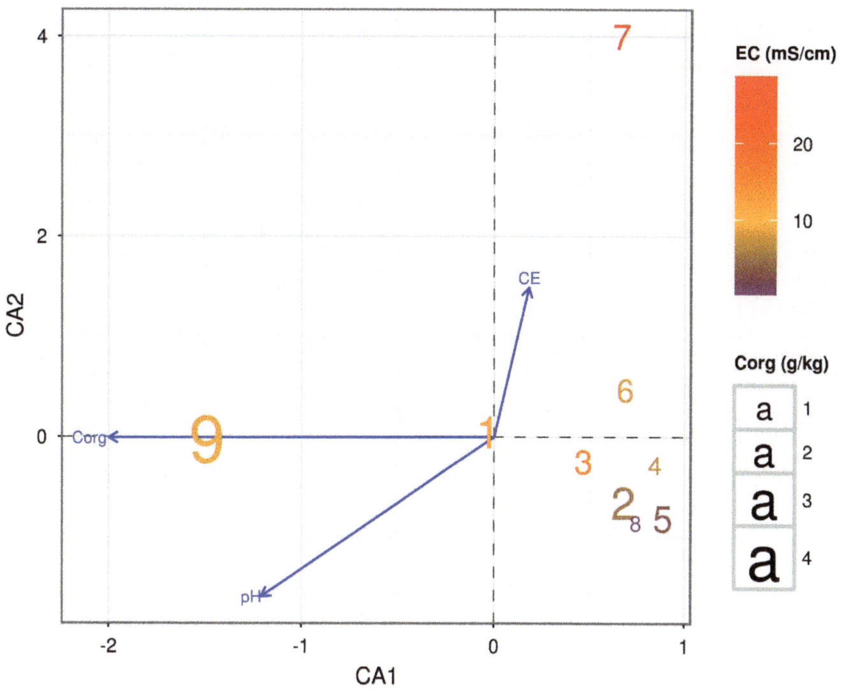

Figure 5. Canonical Correlation Analysis (CCA) based on community data matrix. The salinity levels and the pH levels of each site have been fitted onto the CCA ordination analysis in order to inspect the relevance of these two factors in relation to the bacterial communities distribution.

region; each slice of the pie represents a bacterial phyla. The large pie represents the average of the entire microbial community of the saline soil, and shows the labels for each slice/bacterial phyla that are also valid for the pies of each site. At a glance, if one compares the average composition of the saline soil under study and the composition of the bacterial community in each sampled

Table 3. Spearman's rank correlations between the relative abundances of the six most abundant bacterial phyla and the soil properties across the nine soil sites.

Taxonomic group	**Correlation values**				
	ECe dS m−1	**pH**	**Corg**	**Salt Crust cover**	**Vegetation Cover**
Proteobacteria	0.500	0,142	0.317	0.541	0.034
Actinobacteria	−0.317	−0.226	−0.467	−0.213	0.196
Acidobacteria	**−0.717****	0.084	−0.333	−0.638	0.009
Verrucomicrobia	0.217	−0.603	**0.667****	−0.301	−0.17
Firmicutes	−0.650	0.109	−0.367	−0.585	0.051
Bacteroidetes	0.533	−0.661	0.467	−0.027	−0.400
Chloroflexi	−0.033	−0.655	0.301	−0.236	0.393
Chlorobi	0	−0.345	**0.722***	−0.319	0.276
Gemmatimonadetes	−0.331	−0.179	−0.235	−0.625	−0.049
WS3	0.536	−0.445	0.301	0.183	−0.103
Plantomyccetes	0.566	**−0.816****	0.146	0.330	−0.634
Tenericutes	0.688*	−0.586	0.444	0.079	−0.231
Defferibacteres	0.411	−0.206	−0.411	0.437	−0.420
Cyanobacteria	0.411	−0.206	−0.411	0.437	−0.420
Spirochaetes	**0.712***	−0.468	0.390	0.271	−0.208
Nitrospira	0.411	−0.206	−0.411	0.437	−0.420
BRC1	−0.411	0.275	0	−0.364	0.490

In the table is reported the rho value. The significant correlation values are indicated as follows: * P<0.05; ** P 0.01.

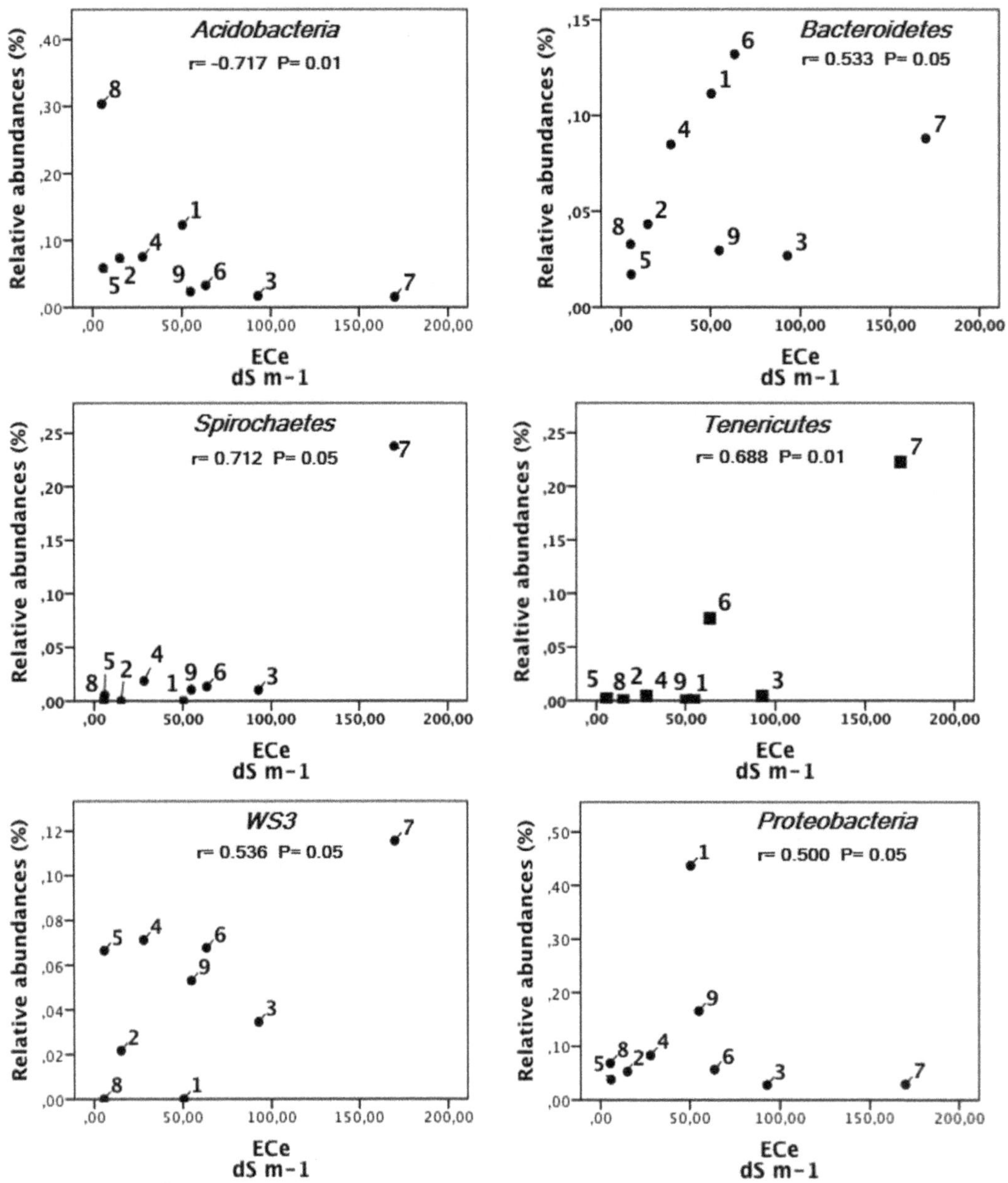

Figure 6. Correlations between relative abundances of different taxonomic groups and soil salinity. Circles represent the soil sites. Spearman rank correlation coefficient (r) with the related P values are shown for each taxonomic group.

site, a considerable spatial variability is noted. What is visually evident from Figure 8 is the way the microbial communities differ greatly both qualitatively and quantitatively from site to site, although the sites are only fifty meters away. Although some groups (like *Proteobacteria* and *Actinobacteria*) were dominant in almost all the sites, a very different composition of the microbial communities among the sites appeared. The graphs also show, from site to site, the relationships among the major microbial groups, and that those who dominate in one site can be absent or just visible in the next one. In each site, given the spatial proximity, it is likely that certain species of bacteria prevail better than others, but this likelihood is apparently expressed in a different way from point to point, probably depending on local environmental forces, and also on positive and/or negative interactions between the different microbial groups.

Discussion

Relative abundances of the phyla found in soils naturally affected by salt, and spatial variability

Some of the bacterial phyla found in the salt-affected soil under examination were also reported by Ma and Gong [12]. Using meta-analysis, these authors retrieved 10,082 sequences longer than 250 bp from the two databases GenBank and RDP, using the search terms "saline" OR "hyper saline", AND "soil" AND "16S". Ma and Gong [12] reported that 90% of the bacterial sequences they enumerated belonged to six phyla, namely *Proteobacteria*, *Actinobacteria*, *Firmicutes*, *Acidobacteria*, *Bacteroidetes*, and *Chloroflexi*, which were also detected in our study.

All the bacterial phyla reported as "salinity related" in previous studies were enumerated in the present survey, and, in addition,

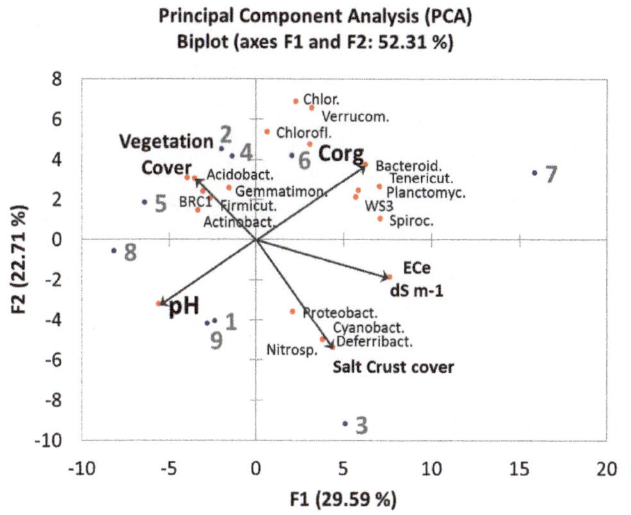

Figure 7. Principal component analysis of bacterial communities as affected by soil properties, based on the abundance of bacterial phyla. Every vector points to the direction of increase for a given variable so that soil sites with similar bacterial communities, are localized in similar positions in the diagram.

some phyla that, according to different authors [15,22], cannot be classified as "salinity related" were also found. This wide group contains the following phyla: *Nitrospira, Deferribacteres, Cyanobacteria/Chloroplast, Gemmatimonadetes, Planctomycetes, BRC1, Verrucomicrobia, Tenericutes, Spirochaetes, WS3* and *Chlorobi*. What is more, the following bacterial phyla were significantly related to soil salinity for the first time: *Nitrospira, Deferribacteres, Cyanobacteria/Chloroplast, Tenericutes* and *Spirochaete*., These phyla were abundant in the two sites showing the highest salinity grade (sites 7 and 3). On the other hand, in the present survey, some phyla classified as "salinity related" by some authors [12] showed an equal distribution all over the sites, apparently uninfluenced by the degree of salinity.

Proteobacteria seemed to be one of the most common bacterial taxon in saline soils [12,49]. In the present study, the occurrence of *Proteobacteria* is actually the highest, with 95.95% of frequencies in the sites, followed by the *Actinobacteria* that represented the second most spread taxon, as it was present in 83.39% of the sites. In contrast with what reported by other meta-analysis studies [12,50], the third largest phylum was *Acidobacteria* (72.12%), followed by *Verrucomicrobia* (70.60%), *Gemmatimonodates* (66.14%), *Firmicutes* (64.14%), *Chloroflexi* (62.69%), *Bacteroidetes* (56.62%), and *Chlorobi* (54.09%). In addition to the nine phyla reported above, 8 phyla characterised by a relative patchy abundance were observed. These taxa with a leopard-spot distribution among the sites had been related to hypersaline soil [12,13,17,23,24,25,27] by other authors. This is the case of groups like *Cyanobacteria* and *Deferribacteres* which characterised site 3, or the *BRC1* group which represented an isolated "spot" in the study area, being present just in site 5. These findings indicated that testing the variability of microbial community in a saline soil using a spatial scale was a successful operation, since it was effective in providing a picture of the subdivision of the microbial community according to a micro-environmental gradient. Although we examined bacterial communities at coarse levels of taxonomic resolution, we expect that a distribution of the bacterial groups based on a mosaic-like scheme would also apply at finer levels of taxonomic resolution.

We noticed that *Proteobacteria* and *Actinobacteria* were common denominators among the investigated sites, being dominant in soils with both high and low salinity. Both phyla contain representatives of the most abundant halophilic bacteria occurring in saline soils [12]. The *Chloroflexi* group, which was found by some authors in hypersaline wastewater [12,49], and is known to be a potential phototroph, was fairly well distributed between the different sites. *Firmicutes* can also be considered common denominators in all the nine sites, because they resulted homogeneously spread in the soil sites. This phylum was apparently absent in a number of hypersaline environments previously investigated [15,46], but it was found to be abundant in low salinity environments such as deep-sea sediments [21]. Among the dominant genera assigned to *Firmicutes*, the *Bacillus* outstands, as it proves to be an important resource for exploring halophilic enzymes and metabolic pathways for pollutant remediation in saline soil [22].

The phyla *Gemmatimonodates* is known for having members showing active roles in biogeochemical transformations, especially in hypersaline soils, where it was undeservingly described as a minor phylum [26]. The assignments attributed to this phylum showed a patchy distribution in the studied area, with a relatively high abundance in site 8 and in other sites characterised by a low salinity level. Another phylum showing a discontinuous distribution was *Planctomycetes,* found in previous studies as symbionts of marine algae and sponges [16,20]. *Nitrospira* dominated site 3 which was the site most affected by the salt crust coverage. Not surprisingly, *Nitrospira* phylum is represented by nitrite-oxidising bacterial species with a marked chemolithoautrophic nature. *Cyanobacteria* presented an identical distribution as that of *Nitrospira* and *Deferribactereres*, which were found as dominant groups in site 3. *Cyanobacteria* is a phylum represented by oxygen evolving and chlorophyll containing photosynthetic bacteria [17,24,25,46]. While *Nitrospira* and *Deferribactereres,* have an important role in biogeochemical cycles, being ammonia-oxidising bacteria (AOB) in saline soils [23] and sulphur oxidising, respectively. The phylum *Deferribacteres* comprehends also chemoorganotrophic heterotrophs that breath anaerobically with terminal electron acceptors including Fe(II), Mn(IV), SO, Co(III), and nitrate [23]. The phylum *Verrucomicrobia* showed, on the one hand, a uniform distribution among the considered sites, with no correlation to salinity, but also, on the other hand, a highly significant dependence on organic matter, with a Spearman's rho value for C_{org} of 0.667, for $p>0.01$. The bacteria belonging to this group is likely to overcome the selection imposed by the high salinity by means of an intimate association with the organic matter of the soil, and perhaps with a direct involvement in the carbon cycle. *Spirochaetes* showed a particular distribution, as they were present with low abundance in almost all sites, but also showing a particular association with the site 7, where the phylum exhibited a very high abundance and a correlation with the presence of another bacterial phylum, the *Tenericutes*. *Spirochetes* are widely distributed in nature; presumably they play an important role as free-living microbes in environments such as soil. The *Spirochaeta* phylum also contains moderately halophilic bacteria, such as species of the *Halomonas* and *Deleya* genera, being members of the gamma subclass of *Proteobacteria* [28,51]. Facultative aerobic halophilic *Spirochaeta* bacteria were isolated by various authors, close to salted lakes [18,28], although the phylum had never been recovered from saline soils, even according to meta-analysis based studies. *Tenericutes* is a phylum of bacteria that contains the Class *Mollicutes* and that, as reported above, presented almost the same distribution of *Spirochaeta*. *Tenericutes* comprehend denitrifying bacteria, and, due to the lack

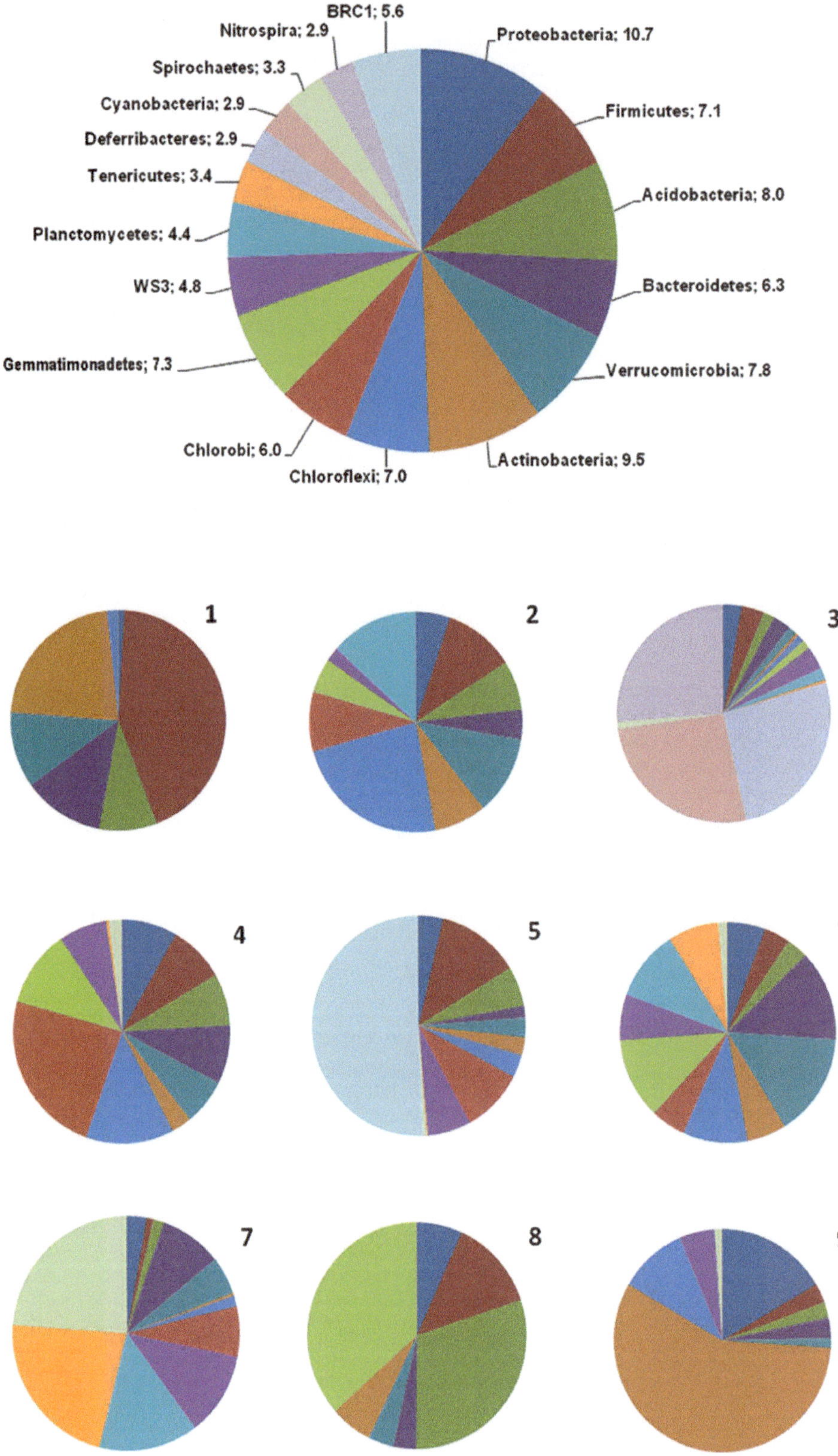

Figure 8. Piecharts based on the abundance (%) of bacterial phyla. Every pie shows the percent of the abundance (%) of the bacteria phyla.

of a cell wall, they are more sensitive to osmotic stress [28]. The correlation with *Spirochaetes* suggests a symbiotic or parasitic role of *Tenericutes* with respect to *Spirochaetes*.

A further interpretation key for understanding the relationship between the relative abundances of these bacterial phyla and salinity and the other soil properties, is given by the illuminating work done by Noah Fierer and colleagues [52] who suggested that certain bacterial phyla can be differentiated into copiotrophic and oligotrophic categories that correspond to the r- and K-selected categories used to describe the ecological attributes of plants and animals. By applying the copiotroph–oligotroph concept to saline soil phyla, we can further understand the structure and function of soil bacterial communities in extreme conditions, and in discontinuous/patchy environments. Copiotrophic bacteria that have higher growth rates, a greater degree of variability in population size, and lower substrate affinities than oligotrophic bacteria, should be dominating when there is abundance of nutritive substrates and, in general, a non-limiting situation. The oligotrophs should increase in relative abundance, as substrate quality and/or quantity declines over time and harsh environmental conditions prevail. Fierer et al. [52] found that bacteria belonging to the *Acidobacteria* phylum showed an oligotrophic behaviour while β-Proteobacteria, and *Bacteroidetes* exhibited copiotrophic attributes, changing their abundances in a predictable manner to changes in soil C availability. Moreover, Fierer et al. [52] found that, across 71 different samples, the *Acidobacteria* were less abundant, while the beta-Proteobacteria and the *Bacteroidetes* were more abundant. This study confirms the abundance of *Proteobacteria* all over the sites, and also shows a considerable presence of *Actinobacteria* as the second most spread taxon, and of *Acidobacteria* as the third largest phylum. Both *Acidobacteria* and *Actinobacteria* frequencies were not correlated to soil organic carbon contents (C_{org}), while a strong dependence to C_{org} is exhibited by *Verrucomicrobia* and *Chlorobi* phyla.

Conclusions

In the latest years, saline soils received a great attention because of the general shortage of arable land, and of the increasing demand for ecological restoration of areas affected by secondary salinisation processes. This is due to the fact that naturally salt-affected soils have a biotechnological potential in their microbial communities, which represent not only a gene reserve for future exploitation in biotechnological applications, assuming they could be used in some kind of restoration or conservation techniques of saline environments, but they can also serve as model systems for exploring the relationships between diversity and activity at the soil level in selective/limiting situations. As outlined in the introduction, very few studies succeeded in addressing the beta diversity of the microbial species in soils, according to the different salt concentrations and, at a different scale, to bacterial taxa distribution in relation to salinity gradients [12].

Although some of the enumerated phyla related to saline soils have already been found by other authors, this study complements the limited information available on these extreme habitats by providing specific information on the type of distribution of different bacterial groups as a function of spatial gradients in salinity and pH. The analysis of bacterial 16S rRNA-based datasets obtained from a naturally saline soil revealed significant differences in bacterial community composition and diversity, along an increasing salinity level, which underlies a multi-scale spatial variability. What is more, a spatial heterogeneity of microbial communities at a relatively small scale has emerged from this study, especially with respect to the macro-scale environmental scheme in terms of geography and soil. The soil of the study showed a patchy distribution of the vegetation structure and of chemical properties, which coincided with an heterogeneous distribution of many bacterial groups. Some bacterial phyla appeared, however, spread in the whole study area.

It is possible to make some assumptions that could be the basis for future in-depth studies on the association between groups of bacteria, or on their variance in certain extreme environments. The first assumption is that spatial autocorrelation in terms of microbial diversity can hardly be found at the soil scales used for physical-chemical studies. According to some authors [53], spatial autocorrelation in soil ranges from 30 cm to more than 6 m, depending on the sampling extent considered. In some locations, Franklin et al. [53] found up to four different correlation length scales. The presence of nested scales of variability suggests that the environmental factors regulating the development of the communities in the saline soil of the present study may have operated at different scales. The presence of spatial patterns in the distribution of bacteria was demonstrated at the microscale by Nunan et al. [54] who showed ranges of spatial autocorrelation of 1 mm and below. The second assumption is that an environment in which some limiting factors favour some microbial groups and not others is in fact compared to a set of islands that allow the formation of different communities, separated from each other by the discontinuity of the chemical-physical factors and by the availability of nutrients. One could imagine that in spite of the same element of "noise" (salinity), the spatial discontinuity allows the formation of more possible microbial assortments. Therefore, a patchy saline environment can contain not just a single microbial community selected to withstand extreme osmotic phenomena, but many different though efficient communities.The occurrence of a significant number of "salinity unrelated" phyla (e.g. *Nitrospira*, *Spirochaetes*) captures our interest, therefore we strongly believe a further analysis, and a further step in metatrascriptomic of functional genes, are needed.

Responding to the initial question on the role of salt concentration in defining the diversity of the bacterial community in a saline soil, we can say that salinity had the strongest effect on bacterial community structure, as revealed by the study of the correlation between soil properties and bacterial phyla occurrence. Soil pH and other chemical properties seemed to have a minor impact on bacterial group distribution when analysed at the considered spatial scale. The relative abundances of a number of taxonomic groups, as a matter of fact, changed significantly between soil sites according to differences in soil salt content. Nevertheless, the abundance of some other taxa resulted almost unaffected by the salinity level (e.g. *BRC1*, *Gemmatimonodates*). This may indicate, on the one hand, a high plasticity of bacterial phyla that evidently possess genera and species adaptable to different conditions, while on the other hand that the sensitivity to salinity of some groups is poor or, in any case, less dependent on other factors, such as the presence of organic matter, plant roots, etc.

Furthermore, it is not certain that bacterial phyla co-occurring at a given site occupy the same ecological niche; rather, the spatial variability can indicate the existence of different scales in the distribution of some major environmental factors, just as the salinity factor. In any case it is evident that the correlation of some groups (*Nitrospira*, *Deferribacteres*, *Spirochaetes*) to the degree of salinity seems to be a necessary condition for the proliferation of the species belonging to those particular groupings.

In conclusion, we feel the need to deepen the scale at which we analyse the bacterial communities in extreme environments. To go back to the more general discussion on saline system ecology, and to the measurement of the "extent of species replacement or biotic change along environmental gradients", which corresponds to the beta diversity *sensu* Whittaker [55], one should distinguish between two rather antithetical phenomena: nestedness and turnover. In the saline soil here studied, we have seen that nestedness occurred only for some taxa, when the biota of a site with a lower number of representatives was a subset of a biota with a greater number of elements of the same taxa (i.e. *Bacteroidetes*, *Chloroflexi*, *Chlorobi*, *Gemmatomonadates*). In this case, the dissimilarity between two sites is related to the difference in specific *richness*, and it occurs even in the absence of a real turnover of species. In contrast, the spatial turnover implies that he replacement of some species by others can easily occur in a mutable environment, where rain and water movements can strongly change the distribution of salts, although it requires a different experimental scheme, with time-related samplings.

It appears evident that the assortment and distribution of microorganisms in a heavily fragmented environment depend on very complex dynamics of colonisation and dispersion, and that the analysis of the correlation between the population of microorganisms and environmental parameters, such as the organic matter, pH, and salinity, adds important information that can help to unravel the mechanisms of formation and structure of the bacterial communities.

Supporting Information

Figure S1 Location of the study area.

Figure S2 Sampling scheme.

Figure S3 Sequences lost during "quality refinement" steps. The piecharts report the fraction of the sequences maintained and the fraction of sequences lost during quality refinement steps. The green portion of each piechart is the maintained portion of sequences (approximately more than 90% of the total sequences) while the other two portions (the red and the blue ones) are the portion of sequences lost during the trimming and the chimera check steps, respectively.

Figure S4 Nucleotide relative frequency distribution along the sequences. The first 10 bases of each sequences file showed an unbalanced nucleotide distribution.

Author Contributions

Conceived and designed the experiments: LC GB AB. Performed the experiments: LC GB GLP. Analyzed the data: LC GB FP GLP. Contributed reagents/materials/analysis tools: CD AB GLP. Contributed to the writing of the manuscript: LC FP GB.

References

1. Richards LA (1954) Diagnosis and improvement of saline and alkali soils. US Salinity Lab., US Department of Agriculture Handbook 60 California, USA.
2. IUSS Working Group WRB (2007) World Reference Base for Soil Resources 2006, first update. 2007. World Soil Resources Reports No. 103. FAO, Rome.
3. Soil Survey Staff (2010) Keys to Soil Taxonomy, 11th ed. USDA-Natural Resources Conservation Service, Washington, DC.
4. Mavi MS, Sandarman J, Chittleborough DJ, Cox JW, Marchner P (2012) Sorption of dissolved organic matter in salt-affected soils: Effect of salinity, sodicity and texture. Sci. Total Environ. 435–436.
5. Rietz DN, Haynes RJ (2003) Effects of irrigation induced salinity and sodicity on soil microbial activity. Soil Biol. Biochem. 35, 845–854.
6. Tripathi S, Kumari S, Chakraborty A, Gupta A, Chakrabarti K, et al. (2006) Microbial biomass and its activities in salt-affected coastal soils. Biol. Fert. Soils. 42: 273–277.
7. Dion P, Nautiyal CS (2008) Microbiology of Extreme Soils. Soil Biology 13 3© Springer-Verlag Berlin Heidelberg.
8. Ettema CH, Wardle DA (2002) Spatial soil ecology. Trends Ecol Evol 17: 177–183.
9. Parkin TB (1993) Spatial variability of microbial processes in soil- a review. J. Environ. Qual. 22: 409–417.
10. Pereira MC S, Cavalcante AF D, van Elsas JD, Salles JF (2012) Spatial and temporal variation of archaeal, bacterial and fungal communities in agricultural soils. Plos One, 7, 12.
11. Lozupone CA, Knight R (2007) Global patterns in bacterial diversity. P. Natl. Acad. Sci. USA. 104: 11436–11440.
12. Ma B, Gong J (2013) A meta- analysis of the publicly available bacterial and archaeal sequence diversity in saline soils. World J Microbiol Biotechnol. DOI 10.1007/s11274-013-1399-9.
13. Benlloch S, Lopez-Lopez A, Casamayor EO, Ovreas L, Goddard V, et al. (2002) Prokaryotic genetic diversity throughout the salinity gradient of a coastal solar saltern. Environ Microbiol 4(6), 349–360.
14. de los Rios A, Valea S, Ascaso C, Davila A, Kastovsky J, et al. (2010) Comparative analysis of the microbial communities inhabiting halite evaporates of the Atacama deserts. International Microbiol 13: 79–89.
15. Demergasso C, Casamayor EO, Chong G, Galleguillos P, Escudero L, et al. (2004) Distribution of prokaryotic genetic diversity in athalassohaline lakes of the Atacama Desert, Northern Chile. FEMS Micro- biol Ecol 48: 57–69.
16. Fukunaga Y, Kurahashi M, Sakiyama Y, Ohuchi M, Yokota A, et al. (2009) Phycisphaera mikurensis gen. nov., sp. nov., isolated from a marine alga, and proposal of Phycisphaeraceae fam. nov., Phycisphaerales ord. nov. and Phycisphaerae classis nov. in the phylum Planctomycetes. J Gen Appl Microbiol 55: 267–275.
17. Garcia-Pichel F, Cortes AL, Nubel U (2001) Phylogenetic and morphological diversity of cyanobacteria in soil desert crusts from the Colorado plateau. Appl. Environ. Microbiol. 67: 1902–1910.
18. Greenberg EP, Canale-Parola E (1976) *Spirochaeta halophila* sp. nov., a fermentative anaerobic bacterium from a high-salinity pond. Arch. Microbiol. 110: 185–194.
19. Hollister EB, Engledow AS, Hammett A, Provin JM, Wilkinson TL, et al. (2010) Shifts in microbial community structure along an ecological gradient of hypersaline soils and sediments. The ISME Journal 4, 829–838.
20. Jogler C, Glockner FO, Kolter R (2011) Characterization of *Planctomyces limnophilus* and development of genetic tools for its manipulation establish it as a model species for the phylum *Planctomycetes*. Appl Environ Microbiol 77: 5826–5829.
21. Li LN, Kato C, Horikoshi K (1999) Bacterial diversity in deep-sea sediments from different depths. Biodivers Conserv 8(5): 659–677
22. Liszka M, Clark M, Schneider E, Clark DS (2012) Nature versus nurture: developing enzymes that function under extreme conditions. Ann Rev Chem Biomol Eng 3: 77–102
23. Mohamed NM, Saito K, Tal Y, Hill RT (2010) Diversity of aerobic and anaerobic ammonia-oxidizing bacteria in marine sponges. ISME J 4: 38–48.
24. Nagy ML, Pérez A, Garcia- Pichel F (2005) The prokaryotic diversity of biological soil crusts in the Sonoran Desert (Organ pice cactus national monument, AZ). FEMS Microbiol Ecol 233–245.
25. Srivastava AK, Bhargava P, Kumar A, Rai Lal C, Neilan BA (2009) Molecular characterization and the effect of salinity on cyanobacterial diversity in the rice fields of Eastern Uttar India. Saline Systems 5: 4.
26. Zhang H, Sekiguchi Y, Hanada S, Hugenholtz P, Kim H, et al. (2003) *Gemmatimonas aurantiaca* gen. nov., sp nov., a gram-negative, aerobic, polyphosphate-accumulating micro-organism, the first cultured representative of the new bacterial phylum Gemmatimonadetes phyl. nov. Int J Syst Evol Microbiol 53: 1155–1163.
27. Yousuf B, Sanadhya P, Keshri J, Jha B (2012) Comparative molecular analysis of chemolithoautitrophic bacterial diversity and community structure from coastal saline soils, Gujarat, India. BMC Microbiol 12: 150.
28. Ventosa A, Nieto JJ, Oren A (1998) Biology of moderately halophilic aerobic bacteria. Microbiol. Mol. Biol. Rev. 62: 504–544.
29. Roesch LFW, Fulthorpe RR, Riva A, Casella G, Hadwin AKM, et al. (2007) Pyrosequencing enumerates and contrasts soil microbial diversity. ISME J 1: 283–290.
30. Acosta-Martinez V, Dowd S, Sun Y, Allen V (2008) Tag-encoded pyrosequencing analysis of bacterial diversity in a single soil type as affected by management and land use. Soil Biol Biochem 40: 2762–2770.
31. Nacke H, Thürmer A, Wollherr A, Will C, Hodac L, et al. (2011) Pyrosequencing- based assessment of bacterial community structure along a different management types in German forest and grassland soils. 6, 2 e17000.
32. FAO (2006) Guidelines for Soil Description (Fourth edition). Food and Agriculture Organisation of the United States, Rome.
33. U.S. Salinity Lab Staff (1954) Diagnosis and improvement of saline and alkali soils. Agr. Handbook 60, USDA, Washington, DC, 122–124.

34. Khorsandi F, Yazdi FA (2011) Estimation of Saturated Paste Extracts' Electrical Conductivity from 1: 5 Soil/Water Suspension and Gypsum. Communications in Soil Science and Plant Analysis 42(3), 315–321.
35. Zidan WMH, Radwan SMA, El-Khawas H, Zahra MK, Badr El-Din SM (2005) Extraction of Microbial Community DNA from Soil for Polymerase Chain Reaction. Egyptian Journal of Applied Science 20(10B): 430–441.
36. Bacci G, Bazzicalupo M, Benedetti A, Mengoni A (2013) StreamingTrim 1.0: a Java software for dynamic trimming of 16S rRNA sequence data from metagenetic studies. Molecular ecology resources.
37. Wang Q, Garrity GM, Tiedje JM, Cole JR (2007) Naive Bayesian classifier for rapid assignment of rRNA sequences into the new bacterial taxonomy. Applied and environmental microbiology, 73(16), 5261–5267.
38. Edgar RC, Haas BJ, Clemente JC, Quince C, Knight R (2001) UCHIME improves sensitivity and speed of chimera detection. Bioinformatics, 27(16), 2194–2200.
39. R Core Team (2013) R: A language and environment for statistical computing. R Foundation for Statistical Computing, Vienna, Austria. URL http://www.R-project.org/.
40. Oksanen J, Guillaume Blanchet F, Kindt R, Legendre P, Minchin PR, et al. (2013) vegan: Community Ecology Package. R package version 2.0–10. http://CRAN.R-project.org/package = vegan
41. Gotelli NJ, Colwell RK (2001) Quantifying biodiversity: procedures and pitfalls in the measurement and comparison of species richness. Ecology letters 4(4): 379–391.
42. Tuomisto H (2010) A diversity of beta diversities: straightening up a concept gone awry. Part 1. Defining beta diversity as a function of alpha and gamma diversity. Ecography, 33(1): 2–22.
43. Simpson EH (1949) Measurement of diversity. Nature.
44. Magurran AE (2004) Measuring biological diversity. Blackwell Science, Oxford.
45. Chao A, Gotelli NJ, Hsieh TC, Sander EL, Ma KH, et al. (2014) Rarefaction and extrapolation with Hill numbers: a framework for sampling and estimation in species diversity studies." Ecological Monographs 84(1): 45–67.
46. Ramette A (2007) Multivariate analyses in microbial ecology, Minireview. FEMS Microbiol Ecol 62: 142–16.
47. Gautier TD (2001) Detecting trends using Spearman's rank correlation coefficient. Environ Forensics 2: 359–362.
48. Purohit MK, Singh SP (2009) Assessment of various methods for extraction of metagenomic DNA from saline habitats of coastal Gujarat (India) to explore molecular diversity. Letters in Applied Microbiology 49: 338–34.
49. Valenzuela-Encinas C, Neria-Gonzalez I, Alcantara-Hernandez RJ, Estrada-Alvarado I, Zavala-Diaz de la Serna FJ, et al. (2009) Changes in the bacterial populations of the highly alkaline saline soil of the former lake Texcoco (Mexico) following flooding. Extremophiles 13: 609–621.
50. Siddikee MA, Tipayno SC, Kim K, Chung JB, Sa T (2011) Influence of varying degree of salinity-sodicity stress on enzyme activities and bacterial populations of coastal soils of Yellow Sea, South Korea. J Microbiol Biotechnol 21: 341–346.
51. DasSarma S, Arora P (2001) Halophiles. Encyclopedia of Life sciences.
52. Fierer N, Bradford MA, Jackson RB (2007) Toward and ecological classification of soil bacteria. Ecology 88 (6) 1354–1364.
53. Franklin RB, Mills AL (2003) Multi-scale variation in spatial heterogeneity for microbial community structure in an eastern Virginia agricultural field. FEMS Microbiology Ecology 44: 335–346.
54. Nunan N, Kejian Wu, Young IM, Crawford JW, Ritz K (2003) Spatial distribution of bacterial communities and their relationships with the micro-architecture of soil. FEMS Microbiology Ecology 44 203–215.
55. Whittaker RH (1972) Evolution and measurement of species diversity. Taxon, 21: 213–251.

11

Bioinformatic Approaches Reveal Metagenomic Characterization of Soil Microbial Community

Zhuofei Xu, Martin Asser Hansen, Lars H. Hansen, Samuel Jacquiod, Søren J. Sørensen*

Section of Microbiology, Department of Biology, University of Copenhagen, Copenhagen, Denmark

Abstract

As is well known, soil is a complex ecosystem harboring the most prokaryotic biodiversity on the Earth. In recent years, the advent of high-throughput sequencing techniques has greatly facilitated the progress of soil ecological studies. However, how to effectively understand the underlying biological features of large-scale sequencing data is a new challenge. In the present study, we used 33 publicly available metagenomes from diverse soil sites (i.e. grassland, forest soil, desert, Arctic soil, and mangrove sediment) and integrated some state-of-the-art computational tools to explore the phylogenetic and functional characterizations of the microbial communities in soil. Microbial composition and metabolic potential in soils were comprehensively illustrated at the metagenomic level. A spectrum of metagenomic biomarkers containing 46 taxa and 33 metabolic modules were detected to be significantly differential that could be used as indicators to distinguish at least one of five soil communities. The co-occurrence associations between complex microbial compositions and functions were inferred by network-based approaches. Our results together with the established bioinformatic pipelines should provide a foundation for future research into the relation between soil biodiversity and ecosystem function.

Editor: Raffaella Balestrini, Institute for Plant Protection (IPP), CNR, Italy

Funding: This work was partly funded by the EU ITN project TRAINBIODIVERSE and the Center for Environmental and Agricultural Microbiology (CREAM) funded by The Villum Foundation. The funders had no role in study design, data collection and analysis, decision to publish, or preparation of the manuscript.

Competing Interests: The authors have declared that no competing interests exist.

* E-mail: SJS@bio.ku.dk

Introduction

Soil is considered to be the most diverse natural environment on the Earth [1,2]. The soil microbial communities harbor thousands of different prokaryotic organisms that contain a substantial number of genetic information, ranging from 2,000 to 18,000 different genomes estimated in one gram of soil [3]. One of the most important issues in the field of soil ecology is to uncover the complex relationships between microbial compositions and functional diversity in soil.

Based on traditional approaches for cultivating and isolating soil microorganisms, early studies have focused on culturable bacteria which only account for less than 1% of soil microbial populations [4]. These studies have already discovered many novel genes encoding interesting enzymes and antimicrobials in soils via functional screens and clone-based Sanger sequencing [1,5,6]. Due to the recent advent of High-Throughput Sequencing (HTS) technologies, metagenomic sequencing approaches have been applied to investigate characterizations of diverse soil microbial communities, including target sequencing of the phylogenetic marker gene encoding 16S rRNA [7,8] and whole-metagenome shotgun sequencing [9–13]. However, the majority of 16S rRNA gene-based studies are committed to the interpretation of community composition but poorly focus on the functional and metabolic properties in a microbial community [14]. In addition, integrated bioinformatic analyses for microbial community-level taxonomic affiliation, metabolic reconstruction, and interaction network, seems to be less studied for the highly diverse soil ecosystems. Currently, MG-RAST [15], IMG/M [16], and CAMERA [17] are the major databases that can support deposition and analysis of metagenomic datasets. Uploading large sequencing data and the subsequent analysis jobs on these web servers sometimes take long waiting time and even weeks. The computational pipelines implemented by these prominent platforms are capable of processing many analysis tasks, but some approaches for special biological inference and graphical visualization still need to be complemented [18].

Recently, together with the rapid development of the Human Microbiome Project, numerous computational tools and methodologies have been developed for effective interpretation and visualization of taxonomic and metabolic profiling of complex microbial communities [19,20] and could be applied to the analysis of the soil microbiota. Particularly, some outstanding computational techniques that could better explain the complexity and heterogeneity of microbial communities are still less applied in the study of the soil microbiota, e.g. prediction of metagenomic biomarkers and network-based correlation analyses [21,22]. In this study, we aim to explore the characterizations of the soil microbiota through integrating the current state-of-the-art bioinformatics tools. A collection dataset of 33 publicly available soil metagenomes was investigated in a custom metagenomic data mining pipeline for explaining and visualizing microbial compositions and metabolic potential. A full spectrum of metagenomic biomarkers and a network of taxon co-occurrence patterns were inferred to hopefully provide some new insights into the underlying mechanisms of complex ecological relationships in the soil microbial community.

Materials and Methods

Ethics statement

No specific permissions were required for the described field studies. The study locations are not privately owned and the field studies did not involve endangered or protected species.

Collection and quality control of metagenomic datasets

Thirty-three metagenomes sampled from five natural soil environments were publicly available and collected in the present study: 14 from grassland, seven from forest soil, nine from desert, two from Arctic soil, and one from mangrove sediment. The metagenomic datasets used can be downloadable according to the list of sequence accession numbers or web links shown in Table S1. All datasets have been produced by whole-metagenome shotgun sequencing using the Roche 454 or Illumina platforms. More reference information about these chosen metagenomes was listed in Table 1. For the datasets of FASTQ formatted sequence reads without quality control, we performed a quality check of bases by using the package Biopieces (http://www.biopieces.org). Low quality ends per read were trimmed by *trim_seq*. Trimming progressed until all bases in a 3-bp stretch with minimum quality score of 20. High quality reads were retained if satisfying the following criteria: minimum average quality score of 15 in a sliding window of 20 bp; minimum read length of 50 bp.

Estimation of microbial composition

MetaPhlAn v1.7 [24] and BLAST v2.2.22 [25] were employed for profiling the taxonomic clades in the metagenomic datasets. Briefly, metagenomic reads were firstly mapped to the MetaPhlAn reference database composed of unique clade-specific marker genes using BLASTN. The non-default parameters used for BLASTN sequencing similarity searching were as follows: E-value cutoff of 1e-10, word size of 12, and minimum alignment length of 75 nt. Relative abundance scores at all taxonomic levels from the domain level to the species level were then estimated by MetaPhlAn. In the text, mean values of abundances were shown for the mentioned taxon. To assess the compositional similarity among soil samples from different microbial communities, the Bray-Curtis measure of beta diversity [26] was employed to compare all pairwise taxonomic abundances between each sample-pair using a R function *vegdist* in the package vegan [27]. The permutation-based multivariate analysis of variance (PERMANOVA) and 2D stress value were then estimated. Based on the resultant Bray-Curtis similarity distance matrix, non-metric multidimensional scaling (NMDS) was adopted to visualize the dispersion of community structure. Multivariate analysis was carried out using vegan [27] and R (http://www.R-project.org) [28].

Metabolic reconstruction of metagenomes

Metabolic reconstruction was carried out using the HUMAnN methodology designed for the functional analyses of meta'omics [29]. High quality reads were initially mapped to the characterized protein functional database KEGG Orthology v54 [30] using the accelerated translated BLAST program USEARCH v6.0.307 [31]. The cutoff E-value was set to 1e-6 and best hits were then used to estimate relative abundances of KEGG orthologous (KO) gene families by HUMAnN v0.98. Base on the resulting KO information, MinPath was used to calculate the coverage and relative abundances of KEGG modules that are manually defined functional units [32]. Circular cladograms representing microbial taxonomic compositions and metabolic modules were implemented by using a standalone graphical tool GraPhlAn v0.9.5 (http://huttenhower.sph.harvard.edu/GraPhlAn).

Detection of metagenomic biomarkers

In order to further test whether some taxa/metabolic modules are significantly overrepresented in the individual soil habitat, statistical analyses were performed according to the inferred relative abundances. Differentially abundant features were identified by the approach of the linear discriminant analysis (LDA) effect size (LEfSe) and could be used as metagenomic biomarkers [21]. As the sample size is not very large in this test, the significance threshold of the alpha parameter for the Krushkal-Wallis (KW) test among classes was set to 0.01 and the cut-off logarithmic LDA score was 2.0. These analyses were performed through the Galaxy server [33]. Additionally, a non-parametric test of Spearman rank correlation between the relative abundances of each KO entry and taxonomic unit was employed to estimate co-variation of community composition and functional features using the R function *cor.test*.

Detection of microbial interactions

A recently developed computational methodology was used to investigate microbial co-occurrence and co-exclusion relationships within and between soil sites [22]. The microbial network of significant co-occurrence and co-exclusion interactions was built by a Cytoscape plugin CoNet 1.0b2 (http://psbweb05.psb.ugent.be/conet/). The taxonomic abundances estimated by MetaPhlAn were used to prepare an input matrix consisting of data from three sites (grasslands, deserts, and forest soils). The analysis was carried out with the non-default parameters listed below: 50 initial top and bottom edges; four similarity measures (Spearman, Pearson, Kullbackleibler, and Bray Curtis); edgeScores for the randomization routine; 1000 permutations and bootstraps. The resulting networks were merged based on the Simes method [34] and Benjamini-Hochberg false discovery rate (FDR) correction [35]. The FDR cutoff was set to 0.05. The ensemble co-occurrence network was visualized by Cytoscape 2.8 [36].

Results and Discussion

General characterization of soil community composition

To explore comprehensive characterizations of taxonomic compositions in the soil microbiota, 33 metagenomes sampled from five soil habitats (i.e. grassland, forest soil, desert, Arctic soil, and mangrove sediment) were included in this analysis (Table 1). Based on the assessment by MetaPhlAn, a total of 63 clades (11 phyla and 53 genera) were identified at ≥0.5% abundance in at least one sample (Table S2). Proteobacteria was the most dominated phylum in the microbial community of soil, ≥70% abundance detected in all soil sites except for the microbiota in the desert samples (Figure 1A). In desert, both phyla Proteobacteria and Actinobacteria exhibit almost identical abundance: 30% for Proteobacteria and 29% for Actinobacteria. In addition, Firmicutes and Bacteroidetes, which are the two major phyla dominating the human microbiome [37,38], were not frequently present in the soil microbial communities. Particularly, bacterial species within the Firmicutes rarely occurred in soil. As the taxonomic distribution of environmental metagenomic sequences are greatly affected by distinct reference databases [23], the 16S amplicon approach should provide more accurate taxonomic profiling than metagenome shotgun sequencing [12]. Previous amplicon surveys of 16S rRNA gene have pointed out that bacterial phyla Acidobacteria, Actinobacteria, Bacteroidetes, Proteobacteria, and Verrucomicorbia are often abundant and

Table 1. Summary of 33 soil metagenomes used in this study.

Biome type	ID	Sequencer	Original name	Location	Coordinates	No. of sequences	No. of bases (Mb)	Reference
Forest soil	F1	454	NA	Puerto Rican	18°18′N, 65°50′W	782,404	322	[9]
Forest soil	F2	454	NA	Massachusetts, USA	42°54′N, 72°18′W	1,439,445	742	[11]
Forest soil	F3	Illumina	AR3	Misiones, Argentina	26°44′S, 54°41′W	5,235,352	524	[12]
Forest soil	F4	Illumina	PE6	Manu National Park, Peru	12°38′S, 71°14′W	9,206,662	921	[12]
Forest soil	F5	Illumina	BZ1	Bonanza Creek, Alaska, USA	64°48′N, 148°15′W	6,543,903	654	[12]
Forest soil	F6	Illumina	CL1	South Carolina, USA	34°37′N, 81°40′W	6,402,940	640	[12]
Forest soil	F7	Illumina	DF1	North Carolina, USA	35°58′N, 79°5′W	3,890,044	389	[12]
Arctic soil	A1	454	Control	Alert, Canada	82°31′N, 62°17′W	495,998	254	[13]
Arctic soil	A2	Illumina	TL1	Toolik Lake LTER, Alaska, USA	68°38′N, 149°35′W	6,011,971	601	[12]
Grassland	G1	454	F1	Hertfordshire, UK	51°48′N, 0°14′E	976,268	358	[10]
Grassland	G2	454	F2a	Hertfordshire, UK	51°48′N, 0°14′E	1,094,883	471	[10]
Grassland	G3	454	F2b	Hertfordshire, UK	51°48′N, 0°14′E	890,966	321	[10]
Grassland	G4	454	F3	Hertfordshire, UK	51°48′N, 0°14′E	754,829	311	[10]
Grassland	G5	454	F4	Hertfordshire, UK	51°48′N, 0°14′E	946,839	391	[10]
Grassland	G6	454	F5	Hertfordshire, UK	51°48′N, 0°14′E	754,135	256	[10]
Grassland	G7	454	F6	Hertfordshire, UK	51°48′N, 0°14′E	782,342	306	[10]
Grassland	G8	454	J1	Hertfordshire, UK	51°48′N, 0°14′E	1,130,719	466	[10]
Grassland	G9	454	J1a	Hertfordshire, UK	51°48′N, 0°14′E	1,137,813	433	[10]
Grassland	G10	454	J1b	Hertfordshire, UK	51°48′N, 0°14′E	919,406	343	[10]
Grassland	G11	454	J1rhizo	Hertfordshire, UK	51°48′N, 0°14′E	1,025,699	369	[10]
Grassland	G12	454	J4	Hertfordshire, UK	51°48′N, 0°14′E	1,135,084	506	[10]
Grassland	G13	454	J7	Hertfordshire, UK	51°48′N, 0°14′E	938,860	339	[10]
Grassland	G14	Illumina	KP1	Kansas, USA	39°6′N, 96°36′W	5,348,832	535	[12]
Desert	D1	Illumina	EB017	Garwood Valley, Antarctica	78°2′S, 163°52′E	7,947,086	795	[12]
Desert	D2	Illumina	EB019	Lake Bonney Valley, Antarctica	77°44′S, 162°18′E	5,454,640	545	[12]
Desert	D3	Illumina	EB020	Lake Fryxell Valley, Antarctica	77°36′S, 163°15′E	9,446,684	945	[12]
Desert	D4	Illumina	EB021	Lake Hoare Valley, Antarctica	77°38′S, 162°53′E	6,543,681	654	[12]
Desert	D5	Illumina	EB024	Wright Valley, Antarctica	77°32′S, 161°42′E	10,863,646	1,086	[12]
Desert	D6	Illumina	EB026	Lake Bonney Valley, Antarctica	77°44′S, 162°19′E	5,951,684	595	[12]
Desert	D7	Illumina	MD3	Mojave Desert, California, USA	34°54′N, 115°39′W	5,899,497	590	[12]
Desert	D8	Illumina	SF2	Chihuahuan Desert, New Mexico, USA	35°23′N, 105°56′W	6,805,456	681	[12]
Desert	D9	Illumina	SV1	Chihuahuan Desert, New Mexico, USA	34°20′N, 106°44′W	11,122,546	1,112	[12]
Mangrove sediment	M1	454	NA	Bertioga, Brazil	23°53′S, 46°12′W	913,752	216	[23]

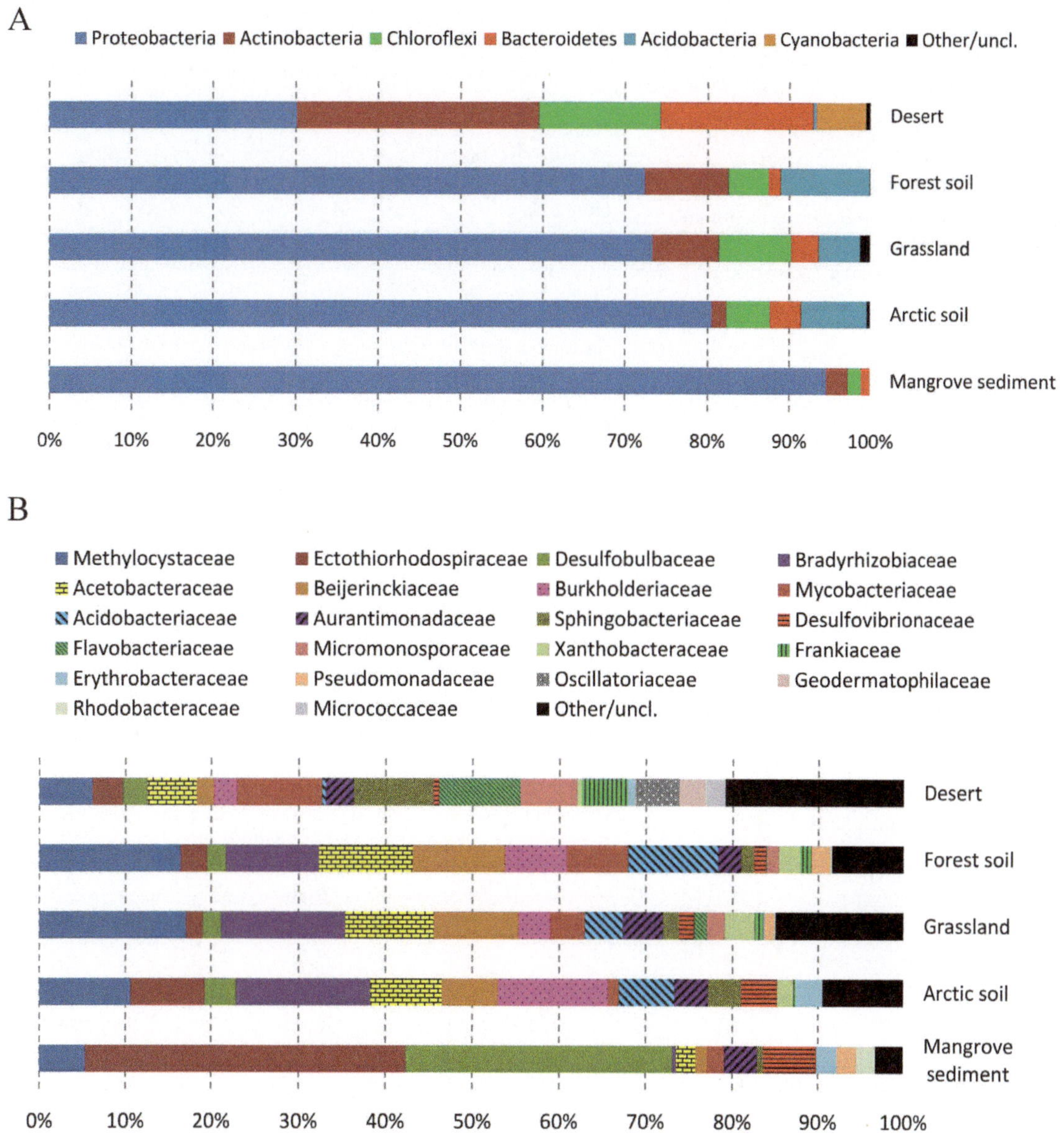

Figure 1. Taxonomic distribution of the soil microbial communities. A) Distribution at the phylum level; B) Distribution at the family level. Labels show the taxonomic units with average relative abundance >2% in at least one of five soil habitats: desert, forest soil, grassland, Arctic soil, and mangrove sediment.

ubiquitous in soil [12,39]. Although the clade-specific marker gene database in MetaPhlAn has successfully validated the composition of the human microbiome [24], it still needs to be updated with more genomes sequenced recently from various environments.

At the family level, several families were observed to be evidently more prevalent in and specific to one soil habitat or closely related habitats (Figure 1B). For instance, the family *Methylocystaceae* was dominated and almost equally present in both sites of forest soil (16.4% abundance) and grassland (17.1%) comparing with the other three soil habitats. Additionally, among five soil habitats, both families *Ectothiorhodospiraceae* (37.0%) and *Desulfobulbaceae* (30.8%) were found to be extremely abundant in the microbiota of mangrove sediment. The enrichment of these families could be reasonably explained by the selective pressures acting on certain ecological sites. For example, organisms within *Methylocystaceae* are usually methanotrophs that can metabolize methane as their only carbon source and involved in methane oxidation [40,41]. The DNA-level evidence identified herein may support the oxidation of methane observed in forest soil and grasslands [42,43]. In addition, the microbiota of mangrove sediments is known to be sampled from anaerobic and hyperhaline seawater [23]. The corresponding environmental features should be beneficial for the dominance of *Ectothiorhodospiraceae* and *Desulfobulbaceae* in this particular habitat. The former comprises the most halophilic eubacteria [44] and bacteria in the latter family are strictly anaerobe sulphate reducers [45]. However, it is worth mentioning that taxonomic profiling of individual metagenome is visually distinguished from those of the other metagenomes within the same soil habitat (Figure S1). This is probably because publicly available soil metagenomes were generated by different research groups and varied in sampling strategies as well as sequencing methods. Thus, analysis of more soil metagenomes newly sequenced or coming soon is still required to statistically support the findings of soil biodiversity in the present study.

Structure similarity and taxonomic biomarkers of soil microbial communities

For a glimpse of structural similarity of soil microbial communities, ecological dissimilarity indices Bray-Curtis similarity scores were inferred and summarized in Table 2. The PERMANOVA test demonstrated that taxonomic compositions of microbial communities were significantly varied among soil habitats ($p = 0.001$). Meanwhile, the NMDS plot in Figure 2 further illustrated the compositional similarity among 33 samples from five soil sites. These results demonstrated that the microbiota from the same soil habitat should be more similar to each other. The community structure similarity is also influenced by varied geographical locations. E.g., the soil samples from grasslands were intensively clustered together and the corresponding similarity score (Bray-Curtis index 0.80 ± 0.07) is indeed the highest among all inter- or intra-group comparisons (Table 2). On the contrary, the Bray-Curtis similarity score between nine desert samples is the lowest (0.58 ± 0.16) among all intra-group comparisons, perhaps due to their sampling environments: three samples from hot deserts but the remaining ones from cold deserts [12]. Likewise, A2 sampled from the edge of the Arctic Circle is distant from A1 from high Arctic soil (Table 1). In addition, it was observed that the distances of most samples between forest soil and grasslands were closely clustered (Figure 2) and the Bray-Curtis similarity score was consistently high (0.76 ± 0.07). Whereas, the microbiota from two extreme conditions, desert and mangrove soil, respectively, exhibited the greatest compositional dissimilarity (0.37 ± 0.07) among all inter-group comparisons.

To further investigate the taxonomic distribution and differentially abundant clades of diverse soil ecosystems, we compared the abundances of microbial compositions at each taxonomic level. Figure 3A shows a cladogram visualizing all detected microbial compositions (≥0.5% abundance) from domain to species, respectively. Based on the inferred taxonomic profiling of all samples, a statistical strategy for discovering metagenomic biomarkers was carried out by LEfSe and determined 46 differentially abundant taxa (Table S2). Among these differentially abundant taxa, 10 and 12 were found to be family- and genus-level biomarkers, respectively (Figure 3A). These detected taxonomic biomarkers could be used as candidate indicators to distinguish at least one microbial community of five individual soil habitats. E.g., two families *Beijerinckiaceae* and *Methylocystaceae* that consist of methanotrophic taxa [46] were detected to be family-level biomarkers (*P* value <0.01) that were most abundant in the forest soil and grassland, respectively. The abundances of both families were found to be significantly decreased in desert and mangrove sediment (Figure 3B). The abundance differences of these methanotrophs might be positively associated with the expected capability of methane oxidation among distinct soil ecosystems. Although the organisms within the *Alphaproteobacteria* class were most differentially abundant in the grassland community, a genus-level biomarker within *Alphaproteobacteria* was specially enriched in the communities of forest soil and Arctic soil, respectively (Figure 3A). Intriguingly, the desert community had two phylum-level markers, Cyanobacteria and Chloroflexi, both of which showed the highest abundance in deserts comparing with other soil sites (Figure 3A). Bacteria in both phyla can produce their energy through photosynthesis [47]. It was worth noting that the family Oscillatoriaceae within Cyanobacteria was significantly enriched in the desert microbial community (Figure 3B). The enrichment of these bacterial groups should be consistent with the

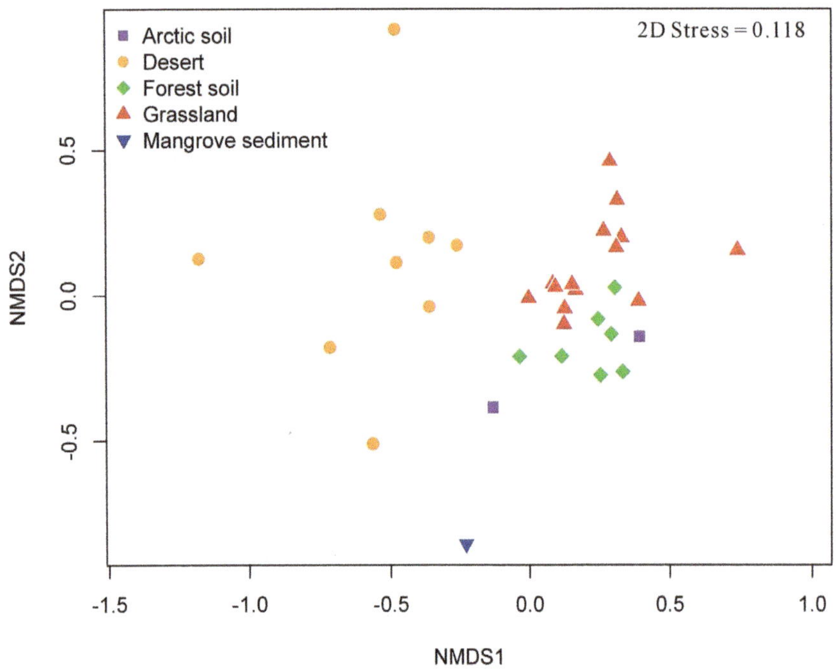

Figure 2. A nonmetric multidimensional scaling (NMDS) plot showing diversity of soil ecosystems. A Bray-Curtis distance similarity matrix was calculated based on the pairwise taxonomic profiles of 33 soil samples and used to generate NMDS coordinates of each sample. The distance linking two samples is shorter, indicating higher similarity between these samples. Samples from five soil sites were illustrated by different symbols and colors.

Table 2. Community structure similarity of the soil metagenomes within a habitat or between habitat pair.

Biome type	Grassland	Forest soil	Arctic soil	Desert	Mangrove soil
Grassland	**0.80±0.07**	0.76±0.07	0.68±0.09	0.50±0.12	0.45±0.02
Forest soil	0.76±0.07	**0.77±0.05**	0.68±0.10	0.48±0.12	0.45±0.05
Arctic soil	0.68±0.09	0.68±0.10	NA	0.45±0.11	0.52±0.15
Desert	0.50±0.12	0.48±0.12	0.45±0.11	**0.58±0.16**	0.37±0.07
Mangrove soil	0.45±0.02	0.45±0.05	0.52±0.15	0.37±0.07	NA

Mean and standard deviation of Bray-cutis similarities of all pairwise samples between any pair of soil habitats were shown herein. The number of sample combination less than 2 was denoted by NA.

following environmental features of deserts: extreme arid, strong light, and poor nutrient conditions. In addition, six species were found to be differentially abundant, some of which were uniquely present in the individual soil habitat. E.g., *Rubrobacter xylanophilus* only occurs in the microbiota of deserts. *R. xylanophilus* is the most thermophilic actinobacterium known and bears extreme tolerance to desiccation [48]. *Bradyrhizobium japonicum*, an agriculturally important species of legume-root nodulating [49], was found to be the most abundant in grassland and second in forest soil.

Metabolic potential and functional biomarkers of soil microbial communities

Besides microbial composition, metabolic potential of soil microbial communities was also investigated. In this study, we focused on the KEGG modules that are tight functional units composed of approximately 5 to 20 genes and beneficial for biological interpretation of metagenomes [29,30,38]. To further enhance the performance of statistical inference on the functional analysis, two soil sites with limited samples (two samples from Arctic soil and one sample from mangrove sediment) were excluded. After translated BLAST searching against the database of KO gene families, we found an average of ~33.6% of reads mapped to at least one KO entry (Table S3). Based on the metabolic reconstruction of 30 metagenomic datasets using HUMAnN, Figure 4 shows 119 functional modules detected in the microbial communities of grasslands, forest soils, and deserts (Table S4). Of these functional modules, we found 20 core metabolic modules that were almost entirely present at >90% coverage in all soil metagenomes tested (Table 3). Some of these core modules were essential for basic life activities of prokaryotic cells in soil, such as central carbon metabolism (M00002-3, M00007, M00009, M00011-12), nucleotide and amino acid metabolism (M00016, M00018, M00048, M00115, M00125), translation (M00178, M00359-360), and ATP synthesis (M00144). In addition, all the remaining core modules were involved in certain transport systems, three (M00207, M00222, M00239) of which are also detected in the core modules of the human microbiome [29]. On the other hand, three functional modules (M00026, M00133, M00319) were differentially covered among three soil sites (Figure 4; Table S5). It was worth noting that structural complex module manganese/zinc/iron transport system (M00319) was completely present only in the deserts but appeared to be absent in both grasslands and forest soils. It indicates that deserts microbiota is well-equipped with metal acquisition systems that play potential roles in the maintenance of metal homeostasis [50].

Furthermore, 33 functional modules were detected to be differentially abundant in at least one of three soil sites (Figure 4 and Table S4). Interestingly, two thirds of these modules were significantly enriched in the microbiota of deserts in comparison to the microbiota of grasslands and forest soils. Of them, three metabolic modules (M00165-167) are involved in the reductive pentose phosphate cycle (Benson-Calvin cycle), which is the main pathway for the conversion of atmospheric CO_2 to organic compounds [51]. It was worth noting that these overrepresented modules involved in carbon fixation might be consistent with high abundance of photosynthetic organisms Cyanobacteria present in the microbiota of desert. Additionally, eight structural complex modules detected to be functional biomarkers in deserts are responsible for the transport of metallic cation (M00317, M00319), mineral and organic ion (M00321, M00299), saccharide and polyol (M00201, M00199), glutamate (M00233), and urea (M00323). On the other hand, we found that two metabolic modules (M00022: Shikimate pathway and M00237: Branched chain amino acid transport system) were significantly overrepresented in grasslands and forest soils comparing with plant-free deserts (Figure 4 and Table S4). Both modules are associated with plant-derived metabolites [52,53]. These results showed that some modular metabolic activities are likely to be associated with the individual soil ecosystem. However, more metagenome samples from different sites are needed for accurately statistical validation of these characterized modules as promising biomarkers for diverse soil communities.

Correlation between microbial compositions and functions

Similar to the approach presented by Segata et al. [38], we assessed the correlations between microbial compositional and functional enrichment. The results showed that some significant associations between taxonomic clades and functional gene families were detected in the soil microbial communities (Spearman non-parametric test; Benjamini-Hochberg corrected *p*-value <0.01) (Figure S2). Notably, several taxonomic biomarkers possessed by individual microbial community mentioned above were further confirmed by the related strong associations between gene families and taxonomic clades. E.g. the gene *petA* (K02634) encoding apocytochrome f protein involved in photosynthesis was positively associated with the members of Cyanobacteria (Spearman test; *q*-value <0.001), one of the earliest prokaryotic organisms which can carry out oxygenic photosynthesis on Earth [47]. In addition, a significantly positive correlation (Spearman test; *q*-value <0.001) between methanotrophs *Methylocystaceae* and the gene *mcl* (K08691) coding for malyl-CoA lyase was observed in the microbial community of grassland. The enrichment of protein Mcl involved in both pathways of methane metabolism and carbon fixation, should be consistent with the featured metabolic activities of these methanotrophs.

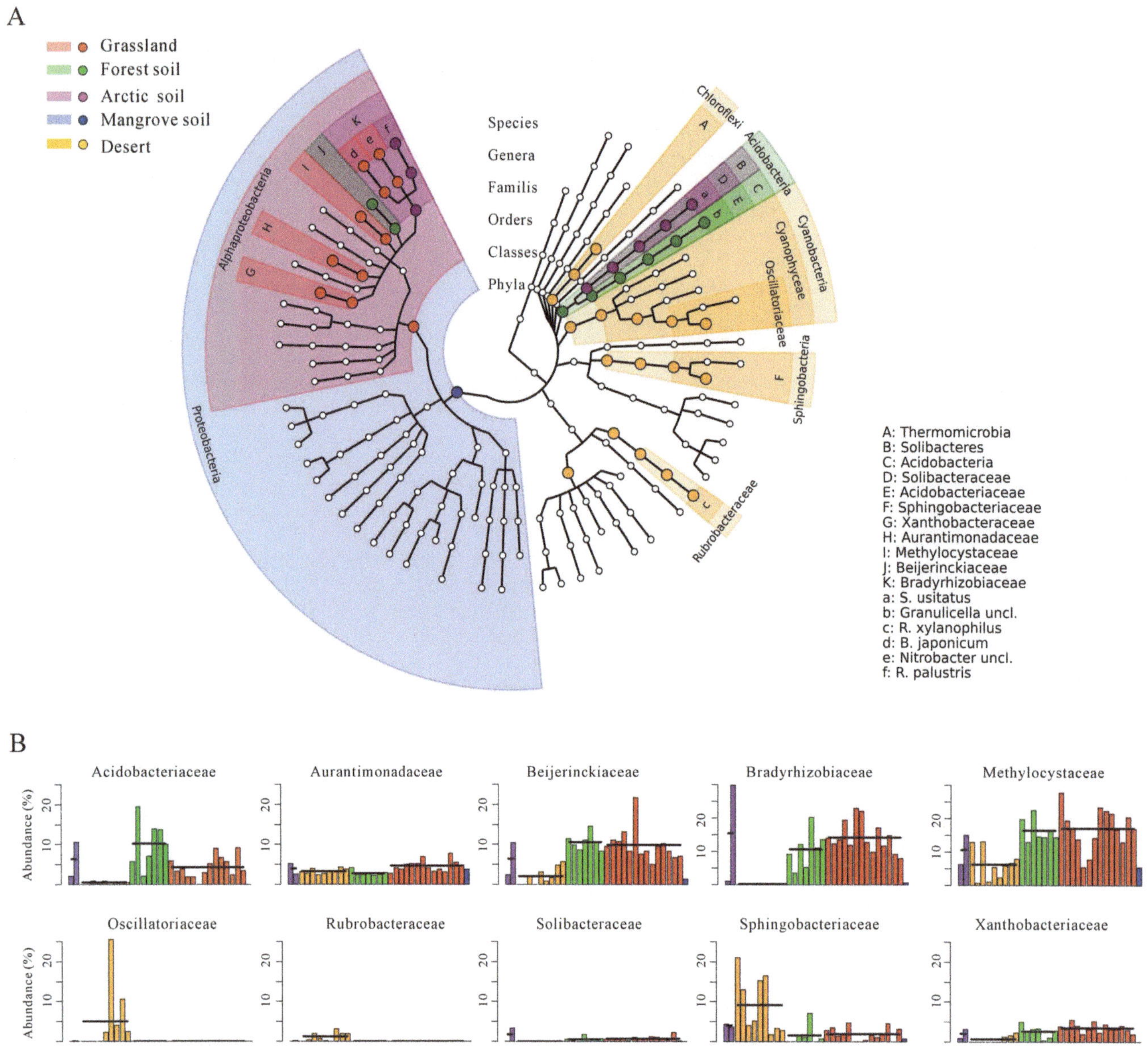

Figure 3. Taxonomic composition of soil microbial community based on the metagenomes from five soil habitats. A) Taxonomic cladogram showing all detected taxa (relative abundance ≥0.5%) in at least one sample. Taxonomic clades with more than five samples ≥0.5% abundance were used as inputs for LEfSe. Seven rings of the cladogram stand for domain (innermost), phylum, class, order, family, genus, and species (outermost), respectively. Enlarged circles in color are the differentially abundant taxa identified to be metagenomic biomarkers and the circle color is corresponding to the individual soil habitat in which the taxon is the most abundant among 5 soil ecosystems (Green for forest soil, red for grassland, purple for Arctic soil, blue for mangrove sediment, and orange for desert). B) The histograms of relative abundances of family-level biomarkers in each sample. Bacterial families significantly differential among all pairwise comparisons were illustrated. The average abundance of each family in the individual soil habitat was denoted by the horizontal line.

Soil microbial interaction network

To further decipher complex ecological relationships in the individual soil microbial community, microbial association networks were inferred based on the estimated taxonomic profiling. In this case, we intended to focus on the microbial associations within the single soil habitat, i.e. forest soil, grassland, and desert. The resultant metagenome-wide networks comprised 126 significant associations among 66 phylotypes at or above the genus level (Benjamini-Hochberg corrected p-value <0.05) (Figure 5). Of these significant phylotype correlations, 54% was detected to be co-present and the remaining was mutually excluded. Interestingly, we found that three quarters (~74%) of co-occurrence patterns were constituted by the taxa within the same phyla; whereas nearly all co-exclusion patterns (~90%) consisted of the taxa from the distinct phyla. The evidence presented herein can again support the previous notion that phylotypes with closely evolutionary relationships usually tend to co-occur [8]. E.g., three families (*Bifidobacteriaceae*, *Mycobacteriaceae*, and *Frankiaceae*) belonging to the same class Actinobacteria showed pairwise positive correlation in the microbiota of desert (Figure 5). Similar taxon co-occurrence pattern was also found between *Bifidobacteriaceae* and *Frankiaceae* in the microbiota of grassland. Additionally, two genera within the family *Bradyrhizobiaceae* co-occurred in the grassland community: one is nitrogen-fixing bacteria *Nitrobacter* and the other is

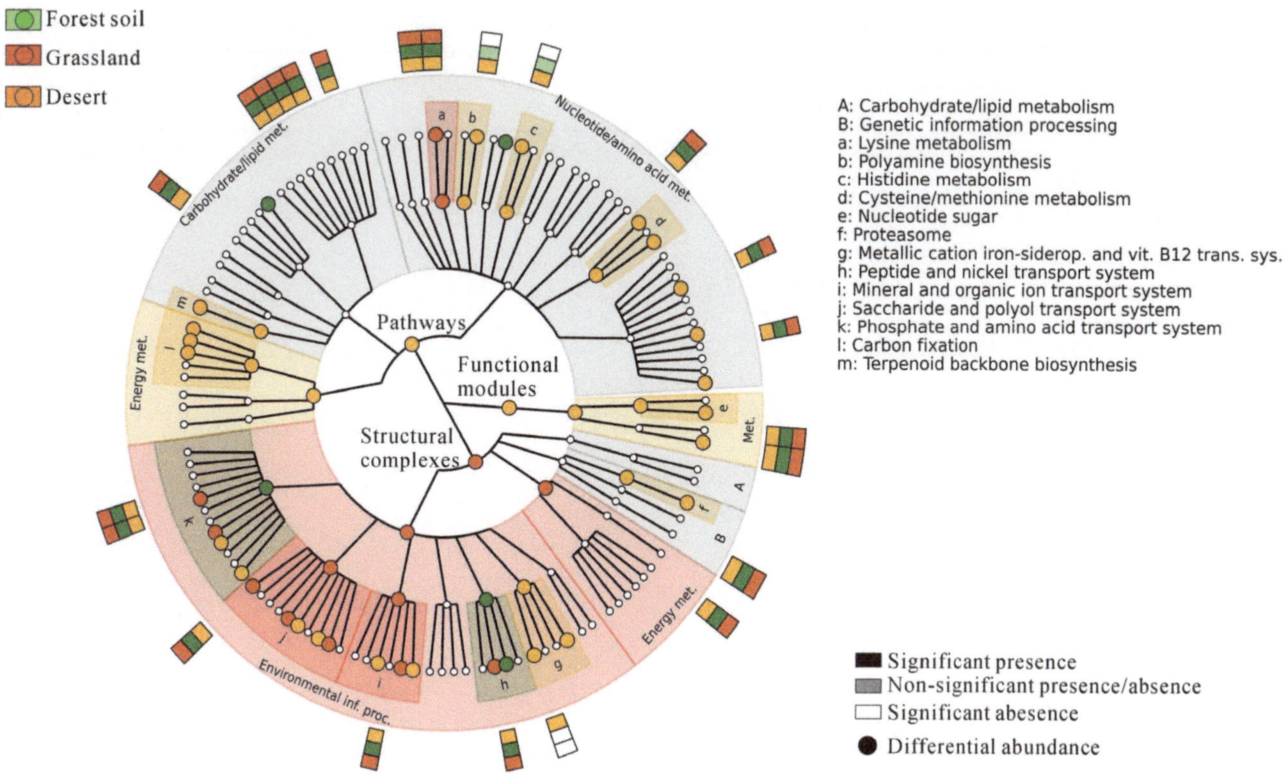

Figure 4. Metagenome-level metabolic reconstruction of the soil microbial community. KEGG BRITE hierarchical structures that are illustrated by the innermost four rings were used to cluster metabolic modules. The outermost ring composed of circles denotes KEGG functional modules detected in at least one of 30 metagenomes from three soil sites. Differentially abundant modules were inferred by LEfSe and illustrated by the enlarged circles in distinct colors: green stands for the modules most abundant in the forest soil, red for the grassland, and orange for the desert. The outermost rectangles denote core and differentially covered modules among three soil sites: ≥90% coverage stands for presence and ≤10% coverage for absence.

phototrophic bacteria *Rhodopseudomonas*. On the other hand, those mutually excluded bacteria were found to be evolutionarily unrelated. E.g., *Sphingobacteriaceae* belonging to the Bacteroidetes were negatively associated with *Desulfovibrionaceae* from the Proteobacteria and *Rubrobacteraceae* from the Actinobacteria in the microbiota of desert (Figure 5). Although most phylotype associations in the network lack empirical evidence to support their natural presence, it provides some promising targets at least to shed light on the complex cooperative or competitive mechanisms among soil microorganisms.

Conclusions

In this study, comparative metagenomic characterizations of divergent soil microbial communities were described in details by an integrated bioinformatics analysis pipeline. Complicated phylogenetic and metabolic networks with a spectrum of taxonomic and functional biomarkers were comprehensively illustrated at the metagenome level for soil. Cooperative or competitive associations among microbes from diverse soil ecosystems were also inferred to understand complex microbial interactions in the soil metagenome. This study provides new insights into the relation between soil biodiversity and ecosystem function, and provides applicable analysis and visualization approaches for studying soil microbial communities.

Supporting Information

Figure S1 Taxonomic distribution of 33 metagenomes from soil microbial communities. A) Distribution at the phylum level; B) Distribution at the family level. Labels show the taxonomic units with average relative abundance >2% in at least one of 33 samples.

Figure S2 Co-variation of bacterial clades and KEGG orthologous gene families in the desert microbiome. The spearman non-parametric correlation of each KEGG gene family against each taxonomic clade was assessed. After multiple testing corrections based on the Benjamini-Hochberg procedure, a network of significant correlations between gene families and taxonomic clades was shown herein (q-value <0.01). Ellipses denote taxa and rectangles stand for KEGG gene families. The edge linking taxonomic clade and gene family indicates that strong correlation was detected in the individual microbial community: green for forest soil, red for grassland, and orange for desert.

Table S1 Sequence data accession numbers and/or web links of soil metagenomes used in this study.

Table S2 Taxonomic profiling of the soil metagenomes estimated in this study. Relative abundances of taxa were inferred by MetaPhlAn. Differentially abundant clades among five soil habitats were detected by LEfSe and labeled by the soil site with

Table 3. Core metabolic modules shared by grasslands, deserts, and forest soils.

Module ID	Definition of modules in KEGG
M00002	Glycolysis, core module involving three-carbon compounds
M00003	Gluconeogenesis, oxaloacetate => fructose-6P
M00007	Pentose phosphate pathway, non-oxidative phase, fructose 6P => ribose 5P
M00009	Citrate cycle (TCA cycle, Krebs cycle)
M00011	Citrate cycle, second carbon oxidation
M00012	Glyoxylate cycle
M00016	Lysine biosynthesis, aspartate => lysine
M00018	Threonine biosynthesis, apartate => homoserine => threonine
M00048	Inosine monophosphate biosynthesis, PRPP + glutamine => IMP
M00115	NAD biosynthesis, aspartate => NAD
M00125	Riboflavin biosynthesis, GTP => riboflavin/FMN/FAD
M00144	Complex I (NADH dehydrogenase), NADH dehydrogenase I
M00178	Ribosome, bacteria
M00185	Sulfate transport system
M00207	Multiple sugar transport system
M00222	Phosphate transport system
M00237	Branched-chain amino acid transport system
M00239	Peptides/nickel transport system
M00359	Aminoacyl-tRNA biosynthesis, eukaryotes
M00360	Aminoacyl-tRNA biosynthesis, prokaryotes

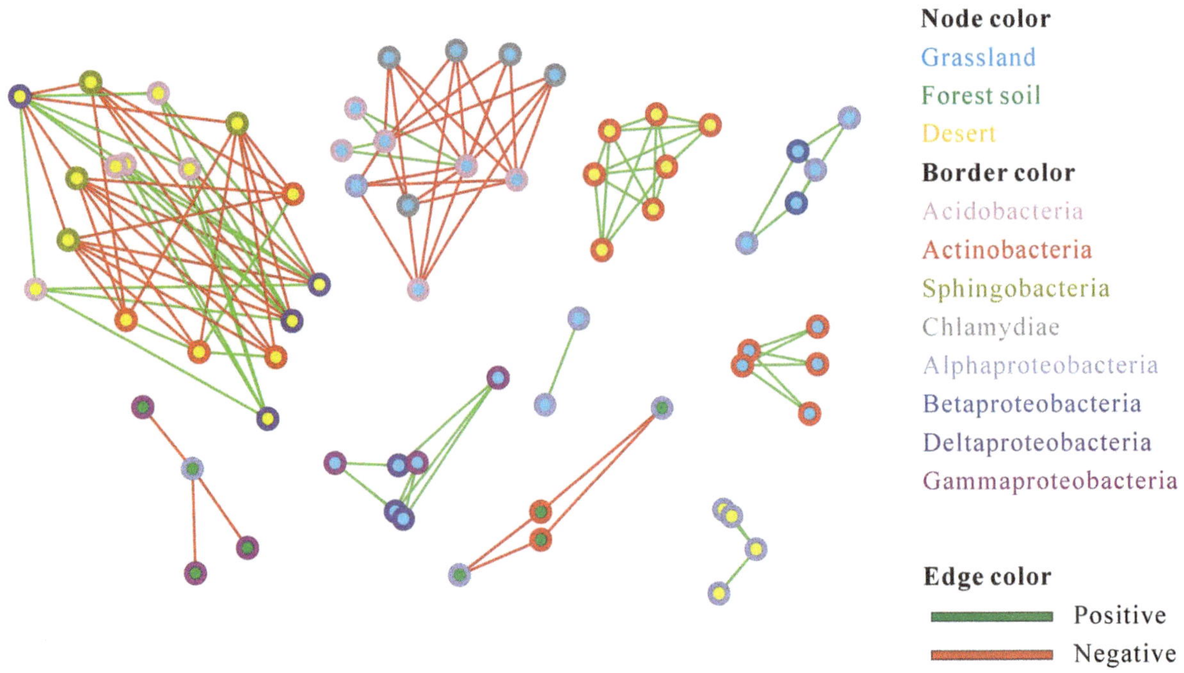

Figure 5. A global microbial interaction network of the soil microbial community. The network captured all significant associations (multiple corrected *p*-value <0.05) among the abundances of phylotypes at or above the genus level in the soil microbial community within and across the three soil sites. Phylotypes were illustrated by nodes (light blue for grasslands, blue for forest soils, and yellow for deserts) and edges denote significant correlations between phylotypes: positive correlation colored in green means co-occurrence whereas negative correlation in red means mutual exclusion. The border of nodes was colored according to taxonomic affiliations at the class level.

the highest LDA score among pairwise comparisons of all sites. According to the non-strict and strict statistical strategy, 46 taxonomic biomarkers were detected to be significantly differential in at least one of five soil habitats.

Table S3 The proportion of reads mapped to MetaPhlAn clade-specific marker genes and KEGG orthologous gene families.

Table S4 Estimated values for relative abundances of KEGG functional modules in the soil microbial community. Differentially abundant modules were detected by LEfSe and labeled by the soil habitat with the highest LDA score among pairwise comparisons of all habitats.

Table S5 Estimated values represented by percentage of the coverage of KEGG functional modules in the soil microbial community. The presence/absence of modules was defined as follows: the median of coverage estimates of the samples per site > 0.9 stands for presence; the median <0.1 for absence.

Author Contributions

Conceived and designed the experiments: ZX SJS. Performed the experiments: ZX. Analyzed the data: ZX MAH SJ. Contributed reagents/materials/analysis tools: ZX LHH SJS. Wrote the paper: ZX MAH LHH SJ SJS.

References

1. Daniel R (2005) The metagenomics of soil. Nat Rev Microbiol 3: 470–478.
2. Vogel TM, Simonet P, Jansson JK, Hirsch PR, Tiedje JM, et al. (2009) TerraGenome: a consortium for the sequencing of a soil metagenome. Nat Rev Microbiol 7: 252.
3. Delmont TO, Robe P, Cecillon S, Clark IM, Constancias F, et al. (2011) Accessing the soil metagenome for studies of microbial diversity. Appl Environ Microbiol 77: 1315–1324.
4. Torsvik V, Øvreås L (2002) Microbial diversity and function in soil: from genes to ecosystems. Curr Opin Microbiol 5: 240–245.
5. Henne A, Daniel R, Schmitz RA, Gottschalk G (1999) Construction of environmental DNA libraries in *Escherichia coli* and screening for the presence of genes conferring utilization of 4-hydroxybutyrate. Appl Environ Microbiol 65: 3901–3907.
6. Rondon MR, August PR, Bettermann AD, Brady SF, Grossman TH, et al. (2000) Cloning the soil metagenome: a strategy for accessing the genetic and functional diversity of uncultured microorganisms. Appl Environ Microbiol 66: 2541–2547.
7. Fierer N, Lauber CL, Ramirez KS, Zaneveld J, Bradford MA, et al. (2012) Comparative metagenomic, phylogenetic and physiological analyses of soil microbial communities across nitrogen gradients. ISME J 6: 1007–1017.
8. Barberán A, Bates ST, Casamayor EO, Fierer N (2012) Using network analysis to explore co-occurrence patterns in soil microbial communities. ISME J. 6: 343–351.
9. De Angelis KM, Gladden JM, Allgaier M, Dhaeseleer P, Fortney JL, et al. (2010) Strategies for enhancing the effectiveness of metagenomic based enzyme discovery in lignocellulolytic microbial communities. Bioenerg Res 3: 146–158.
10. Delmont TO, Prestat E, Keegan KP, Faubladier M, Robe P, et al. (2012) Structure, fluctuation and magnitude of a natural grassland soil metagenome. ISME J 6: 1677–1687.
11. Stewart FJ, Sharma AK, Bryant JA, Eppley JM, DeLong EF. (2011) Community transcriptomics reveals universal patterns of protein sequence conservation in natural microbial communities. Genome Biol 12: R26.
12. Fierer N, Leff JW, Adams BJ, Nielsen UN, Bates ST, et al. (2012) Cross-biome metagenomic analyses of soil microbial communities and their functional attributes. Proc Natl Acad Sci U S A 109: 21390–21395.
13. Yergeau E, Sanschagrin S, Beaumier D, Greer CW (2012) Metagenomic analysis of the bioremediation of diesel-contaminated Canadian high arctic soils. PLoS One. 7: e30058.
14. Langille MG, Zaneveld J, Caporaso JG, McDonald D, Knights D, et al. (2013) Predictive functional profiling of microbial communities using 16S rRNA marker gene sequences. Nat Biotechnol 31: 814–821.
15. Meyer F, Paarmann D, DSouza M, Olson R, Glass EM, et al. (2008) The metagenomics RAST server - a public resource for the automatic phylogenetic and functional analysis of metagenomes. BMC Bioinformatics 9: 386.
16. Markowitz VM, Chen IM, Chu K, Szeto E, Palaniappan K, et al. (2012) IMG/M: the integrated metagenome data management and comparative analysis system. Nucleic Acids Res 40: D123–129.
17. Sun S, Chen J, Li W, Altintas I, Lin A, et al. (2011) Community cyberinfrastructure for Advanced Microbial Ecology Research and Analysis: the CAMERA resource. Nucleic Acids Res 39: D546–551.
18. Wu S, Zhu Z, Fu L, Niu B, Li W (2011) WebMGA: a customizable web server for fast metagenomic sequence analysis. BMC Genomics 12: 444.
19. Gevers D, Knight R, Petrosino JF, Huang K, McGuire AL, et al. (2012) The Human Microbiome Project: a community resource for the healthy human microbiome. PLoS Biol 10: e1001377.
20. Segata N, Boernigen D, Tickle TL, Morgan XC, Garrett WS, et al. (2013) Computational meta'omics for microbial community studies. Mol Syst Biol 9: 666.
21. Segata N, Izard J, Waldron L, Gevers D, Miropolsky L, et al. (2011) Metagenomic biomarker discovery and explanation. Genome Biol 12: R60.
22. Faust K, Sathirapongsasuti JF, Izard J, Segata N, Gevers D, et al. (2012) Microbial co-occurrence relationships in the human microbiome. PLoS Comput Biol 8: e1002606.
23. Andreote FD, Jiménez DJ, Chaves D, Dias AC, Luvizotto DM, et al. (2012) The microbiome of Brazilian mangrove sediments as revealed by metagenomics. PLoS One 7: e38600.
24. Segata N, Waldron L, Ballarini A, Narasimhan V, Jousson O, et al. (2012) Metagenomic microbial community profiling using unique clade-specific marker genes. Nat Methods 9: 811–814.
25. Altschul SF, Gish W, Miller W, Myers EW, Lipman DJ (1990) Basic local alignment search tool. J Mol Biol 215: 403–410.
26. Bray JR, Curtis JT (1957) An ordination of the upland forest communities of southern Wisconsin. Ecol Monographs, 27: 325–349.
27. Oksanen J (2011) Multivariate analysis of ecological communities in R: vegan tutorial. URL: http://cc.oulu.fi/~jarioksa/opetus/metodi/vegantutor.pdf.
28. R Development Core Team (2007) R: A language and environment for statistical computing [http://www.R-project.org].R Foundation for Statistical Computing, Vienna, Austria.
29. Abubucker S, Segata N, Goll J, Schubert AM, Izard J, et al. (2012) Metabolic reconstruction for metagenomic data and its application to the human microbiome. PLoS Comput Biol 8: e1002358.
30. Kanehisa M, Goto S, Furumichi M, Tanabe M, Hirakawa M. (2010) KEGG for representation and analysis of molecular networks involving diseases and drugs. Nucleic Acids Res 38: D355–360.
31. Edgar RC (2010) Search and clustering orders of magnitude faster than BLAST. Bioinformatics 26: 2460–2461.
32. Ye Y, Doak TG (2009) A parsimony approach to biological pathway reconstruction/inference for genomes and metagenomes. PLoS Comput Biol 5: e1000465.
33. Goecks J, Nekrutenko A, Taylor J, Galaxy Team (2010) Galaxy: a comprehensive approach for supporting accessible, reproducible, and transparent computational research in the life sciences. Genome Biol 11: R86.
34. Sarkar SK, Chang CK (1997) The Simes method for multiple hypothesis testing with positively dependent test statistics. J Am Stat Assoc 92: 1601–1608.
35. Benjamini Y, Hochberg Y (1995) Controlling the false discovery rate: a practical and powerful approach to multiple testing. J Royal Statis Soc B 57: 289–300.
36. Smoot ME, Ono K, Ruscheinski J, Wang PL, Ideker T. (2011) Cytoscape 2.8: new features for data integration and network visualization. Bioinformatics 27: 431–432.
37. Qin J, Li R, Raes J, Arumugam M, Burgdorf KS, et al. (2010) A human gut microbial gene catalogue established by metagenomic sequencing. Nature 464: 59–65.
38. Segata N, Haake SK, Mannon P, Lemon KP, Waldron L, et al. (2012) Composition of the adult digestive tract bacterial microbiome based on seven mouth surfaces, tonsils, throat and stool samples. Genome Biol 13: R42.
39. Janssen PH (2006) Identifying the dominant soil bacterial taxa in libraries of 16S rRNA and 16S rRNA genes. Appl Environ Microbiol 72: 1719–1728.
40. Dedysh SN (2009) Exploring methanotrophs diversity in acidic northern wetlands: molecular and cultivation-based studies. Microbiology 78: 655–669.
41. Dedysh SN (2011) Cultivating uncultured bacteria from northern wetlands: knowledge gained and remaining gaps. Front Microbiol 2: 184.
42. Ojima DS, Valentine DW, Mosier AR, Parton WJ, Schimel (1993) Effect of land use change on methane oxidation in temperate forest and grassland soils. Chemosphere 26: 675–685.
43. Jang I, Lee S, Hong JH, Kang H (2006) Methane oxidation rates in forest soils and their controlling variables, a review and a case study in Korea. Ecol Res 21: 849–854.
44. Imhoff JF (2005) Family II. *Ectothiorhodospiraceae*. In: Brenner DJ, Krieg NR, Staley JT, Garrity GM, editors. *Bergey's manual of systematic bacteriology*, 2nd. Springer, New York, NY. pp. 41–52.

45. Kuever J, Rainey FA, Widdel F (2005) Family II. *Desulfobulbaceae* fam. nov. In: Brenner DJ, Krieg NR, Staley JT, Garrity GM, editors. p.*Bergey's manual of systematic bacteriology*, 2nd. New York: Springer, p.988.
46. Lau E, Fisher MC, Steudler PA, Cavanaugh CM (2013) The methanol dehydrogenase gene, mxaF, as a functional and phylogenetic marker for proteobacterial methanotrophs in natural environments. PLoS One 8: e56993.
47. Mulkidjanian AY, Koonin EV, Makarova KS, Mekhedov SL, Sorokin A, et al. (2006) The cyanobacterial genome core and the origin of photosynthesis. Proc Natl Acad Sci U S A 103: 13126–13131.
48. Empadinhas N, da Costa MS (2011) Diversity, biological roles and biosynthetic pathways for sugar-glycerate containing compatible solutes in bacteria and archaea. Environ Microbiol 13: 2056–2077.
49. Kaneko T, Nakamura Y, Sato S, Minamisawa K, Uchiumi T, et al. (2002) Complete genomic sequence of nitrogen-fixing symbiotic bacterium *Bradyrhizobium japonicum* USDA110. DNA Res 9: 189–197.
50. Desrosiers DC, Sun YC, Zaidi AA, Eggers CH, Cox DL, et al. (2007) The general transition metal (Tro) and Zn2+ (Znu) transporters in Treponema pallidum: analysis of metal specificities and expression profiles. Mol Microbiol 65: 137–152.
51. Wolosiuk RA, Ballicora MA, Hagelin K (1993) The reductive pentose phosphate cycle for photosynthetic CO2 assimilation: enzyme modulation. FASEB J 7: 622–637.
52. Maloney GS, Kochevenko A, Tieman DM, Tohge T, Krieger U, et al. (2010) Characterization of the branched-chain amino acid aminotransferase enzyme family in tomato. Plant Physiol 153: 925–936.
53. Maeda H, Dudareva N (2012) The Shikimate pathway and aromatic amino acid biosynthesis in plants. Annu Rev Plant Biol 63: 73–105.

12

A Thesaurus for Soil Invertebrate Trait-Based Approaches

Benjamin Pey[1,2]◑, Marie-Angélique Laporte[3,4*]◑, Johanne Nahmani[5], Apolline Auclerc[6], Yvan Capowiez[7], Gaël Caro[1], Daniel Cluzeau[8], Jérôme Cortet[9], Thibaud Decaëns[10], Florence Dubs[11], Sophie Joimel[6], Muriel Guernion[8], Charlène Briard[8], Fabien Grumiaux[12,13], Baptiste Laporte[2], Alain Pasquet[14], Céline Pelosi[1], Céline Pernin[12], Jean-François Ponge[15], Sandrine Salmon[15], Lucia Santorufo[6,16], Mickaël Hedde[1]*

1 INRA, UR251 « PESSAC », Versailles, France, **2** CESAB/FRB, Aix-en-Provence, France, **3** German Centre for Integrative Biodiversity Research (iDiv), Leipzig, Germany, **4** Bioversity International, Montpellier, France, **5** Centre d'Ecologie Fonctionnelle et Evolutive (CEFE), CNRS, Université of Montpellier II, Montpellier, France, **6** Laboratoire Sols et Environnement, UMR 1120 INRA, Vandoeuvre-lès-Nancy, France, **7** INRA, UR1115 « Plantes et Systèmes Horticoles », Avignon, France, **8** Université de Rennes 1, UMR CNRS 6553 « EcoBio », Paimpont, France, **9** Université Paul Valéry Montpellier III, Centre d'Ecologie Fonctionnelle et Evolutive, Laboratoire de Zoogéographie, UMR 5175 CEFE, Montpellier, France, **10** UFR Sciences et Techniques, EA 1293 « ECODIV », Université de Rouen, Mont Saint Aignan, France, **11** IRD, UMR IEES-P, Bondy, France, **12** Université de Lille 1, EA 4515 « Laboratoire Génie Civil & géo Environnement », Lille Nord de France, Ecologie Numérique et Ecotoxicologie, Villeneuve d'Ascq, France, **13** Université Lille Nord de France, Ecole Supérieure du Professorat et de l'Education (ESPE), Arras, France, **14** UR AFPA, Faculté des Sciences et Technologies, Université de Lorraine, Vandœuvre-lès-Nancy, France, **15** CNRS, UMR 7179, Muséum National d'Histoire Naturelle, Brunoy, France, **16** Department of Structural and Functional Biology, University of Naples Federico II, Naples, Italy

Abstract

Soil invertebrates are known to be much involved in soil behaviour and therefore in the provision of ecosystem services. Functional trait-based approaches are methodologies which can be used to understand soil invertebrates' responses to their environment. They (i) improve the predictions and (ii) are less dependent on space and time. The way traits have been used recently has led to misunderstandings in the integration and interpretation of data. Trait semantics are especially concerned. The aim of this paper is to propose a thesaurus for soil invertebrate trait-based approaches. T-SITA, an Internet platform, is the first initiative to deal with the semantics of traits and ecological preferences for soil invertebrates. It reflects the agreement of a scientific expert community to fix semantic properties (*e.g.* definition) of approximately 100 traits and ecological preferences. In addition, T-SITA has been successfully linked with a fully operational database of soil invertebrate traits. Such a link enhances data integration and improves the scientific integrity of data.

Editor: Raymond Schuch, Rockefeller University, United States of America

Funding: Support was provided by the Centre de synthèse et d'analyse sur la biodiversité/ Fondation pour la Recherche sur la Biodiversité (http://cesab.org/index.php?lang=fr;http://www.fondationbiodiversite.fr/). All authors benefit from the funding. One of them, BL is employed by the CESAB. The funders had no role in study design, data collection and analysis, decision to publish, or preparation of the manuscript.

Competing Interests: The authors have declared that no competing interests exist.

* Email: mickael.hedde@versailles.inra.fr (MH); marieangelique.laporte@gmail.com (M-AL)

◑ These authors contributed equally to this work.

The Need for Semantic Data Integration for Soil Invertebrate Traits

The soil fauna consists of a variety of animals which may represent as much as a quarter of all currently described biodiversity [1]. Of this, soil invertebrates are known to be highly involved in soil behaviour (*e.g.* carbon transformation and sequestration, soil aggregation) and therefore in the provision of ecosystem services [2–4]. As a consequence, soil ecologists aim to understand the interactions between soil invertebrates and their environment. Functional trait-based approaches are methodologies which can help us to understand soil invertebrates' response to their environment through their traits. In this paper, we consider functional traits as being characteristics of individuals that affect their fitness and govern their responses to their surrounding environment [5–8]. The main advantages of trait-based approaches are that they (i) improve the prediction of the relationship between soil invertebrates and environmental changes and (ii) reduce the dependence of such predictions on time and space [8]. Trait-based approaches have confirmed the existence of environmental filters which filter a sub-set of individuals from the regional pool to form local communities [9]. Furthermore, trait-based approaches have been shown to be reliable over eco-regions and for whatever kind of environmental change is considered [10].

The current use of traits for soil invertebrates resulted from isolated initiatives which produced large amounts of unconnected heterogeneous data [8]. Without efforts to integrate such data, the emergence of new knowledge from combining, reusing or sharing it will remain scarce and time-consuming. Our main aim is to provide soil invertebrate scientists with tools which allow data identification, availability and interoperability [11]. The semantic web offers such kinds of tools by being based on the key principles of metadata, controlled vocabularies and ontologies [11].

The integration of the trait data on soil invertebrates is a key issue which can be resolved through semantic data integration. It deals with the variation of the terms employed (terminology) over time. It preserves their meanings (scientific concepts) and also captures their interrelationships [12,13]. In the following, trait scientific concepts will be called "concepts" and trait terms which pertain to trait scientific concepts will be called "terms". As has been stressed by some authors, semantic inconsistencies can not only impede data integration [14] but could also lead to ambiguous scientific data interpretation [13]. For instance, concerning the problem of data integration, some authors employed either the term "body length" [15] or "body size" [16] to describe the same concept, *i.e.* the length of the body, for two soil invertebrate taxonomic groups respectively. Without a semantic link between these two terms, data integration is impossible as data were described by two different terms. Is the concept associated with these two terms the same? Only a semantic structure would remove any doubt and identify these two terms as synonyms. Such semantic inconsistencies also exist within a given soil invertebrate taxonomic group. For instance, the development of ground beetle wings has been called "wing morphology" [17,18], "wing form" [10] or "wing type" [19]. Otherwise, concerning examples of misunderstanding scientific data interpretation, the type of food eaten by soil invertebrates (*e.g.* carnivorous which means that they eat animals, usually alive), the way they feed on them (*e.g.* predators, which means that they feed by killing their live prey) or finally their trophic position in the food chain (*e.g.* tertiary consumers which eat animals feeding on herbivores) refer to different concepts. Nevertheless, the literature contained several categorical traits whose attributes described several of the above concepts simultaneously. For instance, the term "food of the adult" [20,21] referred both to the type of materials ingested (*e.g.* plants, springtails) but also to the way the materials were eaten (*e.g.* generalist predators). The terms "feeding guilds" [22] or "food strategy" [23] are other terms used. Another example is that, to refer to the body colour, some authors employed the term "body colour" for carabid beetles [21,24] while others used "body pigmentation" for earthworms [15,25]. However, the concept of "coloration" is different from the concept of "pigmentation" since pigmentation does not necessarily imply the presence of colour. As soon as the traits are not clearly defined, confusion will emerge from comparisons between trait data from multiple literature sources.

As far as we know there has been no attempt to deal with these shortcomings for soil invertebrates. One solution is to build a thesaurus, which is a list of terms used in a particular topic, with some of their properties, organized into a hierarchy according to their meanings, i.e. their concepts. The aim of this paper is to present a first thesaurus for soil invertebrate trait-based approaches, called T-SITA.

The Thesaurus Construction

The thesaurus for soil invertebrate trait-based approaches (T-SITA) was constructed through a web-based tool, designed for the collaborative construction of thesauri in ecology: the Thesauform [13].

1. The tool: Thesauform features

The Thesauform allows a thesaurus to be created which resulted in a hierarchy of terms organized according to their meaning, i.e. their concept. Each term of the hierarchy is described by a defined number of its properties: preferred label, definition, bibliographic reference of the definition, abbreviation, synonym(s), related term(s) and preferred unit. The building procedure is performed collaboratively by a scientific expert community. It is divided into three successive steps: editing, validation and supervision. The editing step consists of the opportunity for each scientific expert to (i) modify and enrich the properties of a term, (ii) modify the hierarchy, (iii) add or delete a term or (iv) add a comment. The validation phase consists of a voting procedure within the scientific expert community on the different amendments produced during the editing phase. At each of these first two steps, several scientific experts can access the Thesauform simultaneously. The supervision phase aims to control the semantic consistency of the votes. It is mainly done by the editors of the thesaurus before the release of the final version. The whole procedure described above can be repeated indefinitely to continually improve the semantic content of the thesaurus.

2. The thesaurus for soil invertebrate traits: method

Before starting the editing phase, twenty-one experts in soil invertebrate ecology provided a list of approximately 80 well-known terms of traits and ecological preferences (see definitions in [8]). These trait/ecological preference terms were first chosen because they are commonly used for at least four notable invertebrate taxonomic groups with different biological strategies: earthworms, ground beetles, spiders and springtails. Nevertheless, the thesaurus design is not limited to such soil invertebrate groups. It is possible to input trait terms for all soil invertebrates and/or specific trait terms for a single soil invertebrate group (*e.g.* collembolan ocelli number).

Some of the properties of these selected trait and ecological preference terms (*e.g.* definition, unit, preferred label) were given as an input to the Thesauform system. They were organized in a conceptual hierarchical tree with their mother and daughter terms. Each term is conceptually included in its mother term. For instance, the "Reproduction type" trait term was included in the mother term "Physiology". This means that the concept linked to the "Reproduction type" term is included in the concept linked to the "Physiology" term (Fig. 1). This last term is also included in the term: "Trait". All the terms which have a position above a trait/ecological preference term, in the conceptual hierarchy tree, are called "multi-level mothers" of the trait/ecological preference term. For instance, the multi-level mothers of "Reproduction type" are: "Physiology" and "Trait" (Fig. 1). "Reproduction type" has two daughter terms: "Asexual reproduction" and "Sexual reproduction" (Fig. 1). We call "multi-level daughters" of a trait/ecological preference term all the terms which have a position below it in the conceptual hierarchy tree. For instance, the multi-level daughters of "Reproduction type" are: "Asexual reproduction", "Sexual reproduction", "Arrhenotokous", "Deuterotokous" and "Thelytokous" (Fig. 1).

T-SITA has been coupled with a soil invertebrate trait database to improve data management and enrich trait/ecological preference information (see Section 4). To achieve this, a necessary condition is that each term must be typified by a unit. Trait and ecological preference terms are identified by having either a numerical unit or by being "categorical". Quantitative traits require numerical values and therefore have numerical units. For instance, the unit of the "body length" trait term is mm. Otherwise, qualitative traits are described by textual data. To be usable, they need to be categorized into attributes, *e.g.* by fuzzy coding procedures [8,26], so their units are described as being "categorical". For instance, the unit of the "habitat" preference term is "categorical". "Habitat" is categorized into several attributes, such as "agricultural area" or "wetland". Such

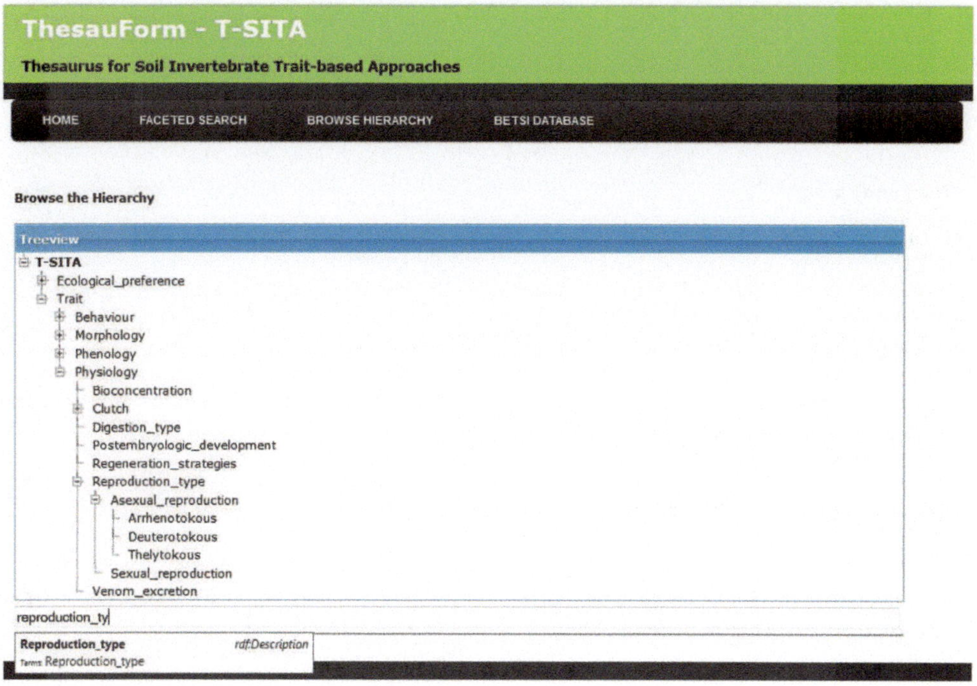

Figure 1. Auto-completed field and navigation tree searches in T-SITA.

attributes appear as daughter terms of the habitat preference in the thesaurus hierarchy. Attributes are identified in the thesaurus by having a unit specified as being an attribute. For instance, the unit of the "Wetland" term is "attribute". In addition, a categorical trait or ecological preference can have multi-level daughters. For instance, the habitat preference is the mother of "Agricultural area" which is itself the mother of "Arable land", "Fallow" and "Perennial crop". This was done to take into account the variable accuracy of the textual literature describing the categorical traits or ecological preferences.

To deal with a trait which is described by using a different data format, for instance from one soil invertebrate taxonomic group to the other, two different traits must be created. They must have the same definition but with different terms and units. For instance, there are in the thesaurus an "Antenna length" term (unit: mm) which represents an antenna length trait described by numerical data and an "Antenna categorical length" term (unit: categorical) which represents an antenna length trait described by textual data. Finally multi-level mothers of traits/ecological preferences have an empty unit. For instance, "Nutrition", which is the mother of the "Mouthpart type" trait, has an empty unit.

The initial hierarchy was then inserted into the Thesauform and was made available to experts on the web at a URL address (no longer available).

From October 2011 to October 2012, experts carried out the editing phase. From October 2012 to April 2013, they carried out the validation phase. From April 2013 to October 2013, editors of the thesaurus checked the consistency of the thesaurus before its first available on-line version, as presented in this paper.

The Thesaurus Content and Browsing

T-SITA is freely available at the following URL address: http://t-sita.cesab.org/Thesauform/BETSI_vizIndex.jsp. It contains 71 traits and 24 ecological preferences.

Two types of semantic search engines are offered to access the T-SITA content. The first one is a classic semantic search engine which allows thesaurus terms to be found through an auto-completed search field and/or through a navigation tree (Fig. 1). It reflects how the experts of soil invertebrate ecology first organized the terms into a conceptual hierarchical tree and then amended it during the editing phase. To have access to information on a given term, it is necessary to click on it in the tree. Then a new web page appears with the properties of the term (Fig. 2).

The second search engine available is a faceted search engine (Fig. 3). It is defined as being "a technique for accessing a collection of information allowing users to explore by filtering available information. It allows the assignment of multiple classifications to an object" [27]. It enables users to filter thesaurus terms on characteristics they share that are called 'facets'. At the moment in T-SITA, terms can be filtered according to several facets which gather terms either by their expression basis, the organ concerned, the main biological function concerned, their nature by distinguishing the traits from ecological preferences [8] and finally by the environmental property concerned. Each facet is divided into several categories that the user can select (when selected, the categories are coloured green). For instance, the "expression basis" was divided into four categories (area, length, mass, time) so that the user can select one/several of them. A dynamically updated list of terms appears then in the result part according to the selected category(ies) (Fig. 3). Simultaneously multiple selection of facet categories is possible. For instance, the user can select simultaneously the category "growth and development" from the "biological function" facet and the category "by mass" from the "expression basis" facet. He will find three terms: the "body weight", the "fresh body weight" and the "dry body weight" in the result part since they correspond to both filters (Fig. 3). To access the information on a given term, the user has to click on it in the result part. Then a new web page appears displaying the complete information on the term.

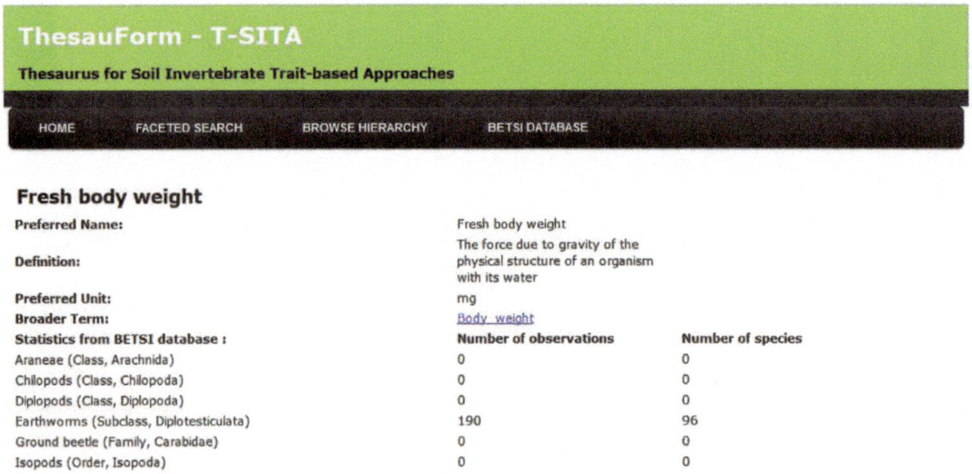

Figure 2. Panel of properties of the "fresh body weight" trait term in T-SITA.

A Useful Tool for Data Integration, Data Scientific Integrity and Navigation among Multiple Data Sources

The web of data focuses on the sharing of data on the web. SKOS (Simple Knowledge Organization System) (Isaac 2008) is the W3C standard dedicated to thesaurus representation, based on RDF triples [28]. SKOS was designed to provide an existing thesaurus standard (ANSI 2005) by providing a common format to adequately manage not only thesauri, but also any knowledge organizational system. T-SITA, built using Thesauform [13], is defined through the SKOS format and makes full use of it. As a consequence, T-SITA serves as a stable reference resource, specifically when available as linked data on the web. Additionally, T-SITA has been coupled with a soil invertebrate trait database to enhance data integration and data scientific integrity. The BETSI database will be soon in production. It is a relational database defined under the database management system PostgreSQL (http://www.postgresql.org/docs/9.1/static/reference.html), which contains soil invertebrate data on traits and ecological preferences. The linkage of the thesaurus and the database presents huge advantages.

First, each insertion of trait data in the BETSI database is under the control of T-SITA, which improves data quality in the database. Indeed, a trait term must be present in T-SITA in order to insert related data in the BETSI database, so data contributors to the BETSI database must consult T-SITA before inputting data. It guides the data integration and limits scientific misunderstanding. Concerning the data integration, it allows integrating trait data from two trait terms that represent the same concept (see section 1). For instance, sclerotization of the body is called "integument sclerotization" or "cuticule sclerotization" in the literature. As both terms are defined as synonyms in T-SITA, data inserted from these two terms will be identified in the database as belonging to the same trait concept. As a consequence, a data query concerning the sclerotization of the body will return data from both terms. Concerning the data interpretation, contributors

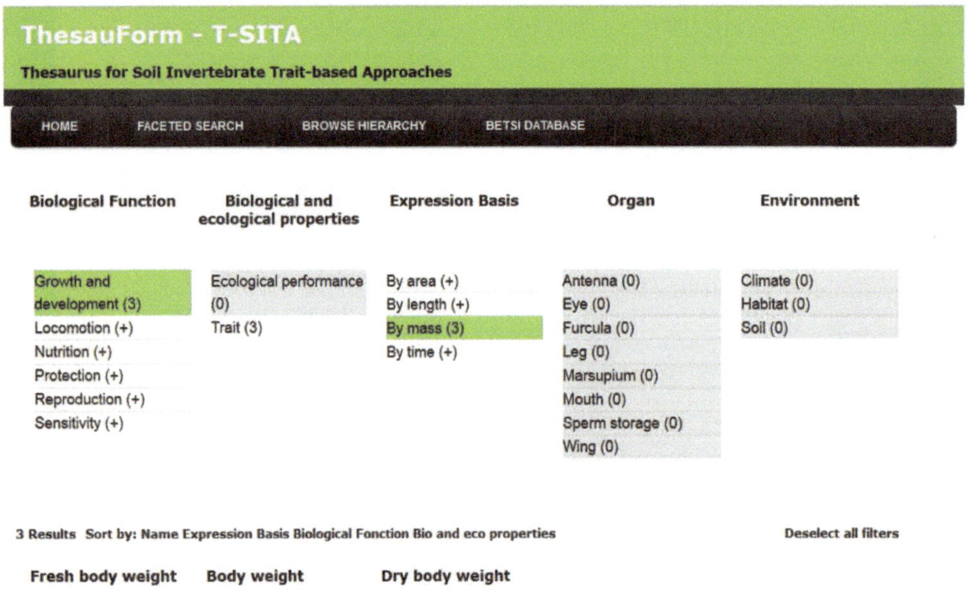

Figure 3. Faceted search system implemented to access T-SITA.

to the database have to associate their data with a term coming from a limited set of trait terms. Each of them represents a trait concept which does not overlap with other trait concepts in the thesaurus. Two trait terms representing the same concept are synonyms. This can resolve the problems of scientific misunderstanding mentioned in section 1. Data contributors who want to insert raw data describing the "body colour" trait for carabid beetles and the "body pigmentation" trait for earthworms will realize that both data apply to the same concept. As a consequence, they will insert the data under the same trait term "body colour" in T-SITA and not under the trait term "body pigmentation" which refers to a different concept.

Second, in the web page describing the trait/ecological preference information in T-SITA, the dynamic link with the database enriches information about a given trait/ecological preference by providing statistics coming from the BETSI database (Fig. 2). It displays for each trait or ecological preference (as soon as a unit has been allocated to them, see section 2), the number of raw data observations input into the database and how many species they concern (Fig. 2). In addition, the statistics are aggregated according to the level of the tree hierarchy. When clicking on an above term (no unit, see Section 2), the number of raw data observations input into the database and how many species they concern are aggregated from its trait/ecological preference daughter terms. For instance, the "body dimension" term displays the aggregated statistics coming from its daughter terms, i.e. the "body length" and "body width" trait terms.

Interoperability between T-SITA and the database is dynamic. Therefore, each time the content is modified either via the database or the Thesauform, the modification is instantaneously updated in the other tool without any human intervention.

Conclusion

Harmonization of trait data for soil invertebrates requires a handbook to answer questions such as: what is really to be understood by this trait term, how can I measure it? T-SITA forms the first step in this huge task by being, to our knowledge, the first initiative to deal with the semantics of traits and ecological preferences for soil invertebrates. It reflects the agreement of a scientific expert community to fix the semantic properties (*e.g.* definition) of approximately 100 traits and ecological preferences. In addition, T-SITA has been successfully linked with a fully operational database on soil invertebrate traits. Such a link enhances data integration and improves data scientific integrity.

The future of T-SITA depends on the tool used to build it (Thesauform), which allows improvements in the current content by (i) performing other complete procedures (edit, validate, supervise), (ii) including other scientific experts and (iii) including new trait/ecological preference properties such as methods of measurement. Finally, a more long-term prospect for T-SITA could be its use for the construction of a soil invertebrate trait ontology.

Author Contributions

Conceived and designed the experiments: BP M-AL BL MH. Performed the experiments: BP M-AL JN AA YC GC DC JC TD FD SJ MG CB FG BL AP C. Pelosi C. Pernin JFP SS LS MH. Analyzed the data: BP M-AL BL MH. Contributed reagents/materials/analysis tools: BP M-AL BL MH. Contributed to the writing of the manuscript: BP M-AL.

References

1. Decaëns T, Jimenez JJ, Gioia C, Measey GJ, Lavelle P (2006) The values of soil animals for conservation biology. Eur J Soil Biol 42: 23–38.
2. Barrios E (2007) Soil biota, ecosystem services and land productivity. Ecol Econ 64: 269–285.
3. Kibblewhite MG, Ritz K, Swift MJ (2008) Soil health in agricultural systems. Philos Trans R Soc B-Biol Sci 363: 685–701.
4. Lavelle P, Decaëns T, Aubert M, Barot S, Blouin M, et al. (2006) Soil invertebrates and ecosystem services. Eur J Soil Biol 42: 3–15.
5. Lavorel S, Garnier E (2002) Predicting changes in community composition and ecosystem functioning from plant traits: revisiting the Holy Grail. Funct Ecol 16: 545–556.
6. Violle C, Navas ML, Vile D, Kazakou E, Fortunel C, et al. (2007) Let the concept of trait be functional! Oikos 116: 882–892.
7. Statzner B, Hildrew AG, Resh VH (2001) Species traits and environmental constraints: Entomological research and the history of ecological theory. Annu Rev Entomol. 291–316.
8. Pey B, Nahmani J, Auclerc A, Capowiez Y, Cluzeau D, et al. (2014) Current use of and future needs for soil fauna response traits. Basic Appl Ecol 15: 194–206.
9. Decaëns T, Margerie P, Aubert M, Hedde M, Bureau F (2008) Assembly rules within earthworm communities in North-Western France–A regional analysis. Appl Soil Ecol 39: 321–335.
10. Vandewalle M, de Bello F, Berg MP, Bolger T, Dolédec S, et al. (2010) Functional traits as indicators of biodiversity response to land use changes across ecosystems and organisms. Biodivers Conserv 19: 2921–2947.
11. Laporte MA, Garnier E, Mougenot I (2013) A faceted search system for facilitating discovery-driven scientic activities: a use case from functional ecology. Semantics for Biodiversity (S4BioDiv 2013), 25.
12. Madin J, Bowers S, Schildhauer M, Krivov S, Pennington D, et al. (2007) An ontology for describing and synthesizing ecological observation data. Ecol Inform 2: 279–296.
13. Laporte M-A, Mougenot I, Garnier E (2012) ThesauForm–Traits: A web based collaborative tool to develop a thesaurus for plant functional diversity research. Ecol Inform 11: 34–44.
14. Jones MB, Schildhauer MP, Reichman OJ, Bowers S (2006) The New Bioinformatics: Integrating Ecological Data from the Gene to the Biosphere. Annu Rev Ecol Evol Syst 37: 519–544.
15. Decaëns T, Margerie P, Renault J, Bureau F, Aubert M, et al. (2011) Niche overlap and species assemblage dynamics in an ageing pasture gradient in north-western France. Acta Oecol 37: 212–219.
16. Makkonen M, Berg MP, van Hal JR, Callaghan TV, Press MC, et al. (2011) Traits explain the responses of a sub-arctic Collembola community to climate manipulation. Soil Biol Biochem 43: 377–384.
17. Gerisch M, Agostinelli V, Henle K, Dziock F (2012) More species, but all do the same: contrasting effects of flood disturbance on ground beetle functional and species diversity. Oikos 121: 508–515.
18. Gobbi M, Fontaneto D (2008) Biodiversity of ground beetles (Coleoptera: Carabidae) in different habitats of the Italian Po lowland. Agric Ecosyst Environ 127: 273–276.
19. Karen M, O'Halloran J, Breen J, Giller P, Pithon J, et al. (2008) Distribution and composition of carabid beetle (Coleoptera, Carabidae) communities across the plantation forest cycle–Implications for management. For Ecol Manag 256: 624–632.
20. Cole LJ, McCracken DI, Dennis P, Downie IS, Griffin AL, et al. (2002) Relationships between agricultural management and ecological groups of ground beetles (Coleoptera: Carabidae) on Scottish farmland. Agric Ecosyst Environ 93: 323–336.
21. Ribera I, Doledec S, Downie IS, Foster GN (2001) Effect of land disturbance and stress on species traits of ground beetle assemblages. Ecol 82: 1112–1129.
22. Grimbacher PS, Stork NE (2009) Seasonality of a diverse beetle assemblage inhabiting lowland tropical rain forest in Australia. Biotrop 41: 328–337.
23. Gerisch M (2011) Habitat disturbance and hydrological parameters determine the body size and reproductive strategy of alluvial ground beetles. ZooKeys: 353–370.
24. Barbaro L, van Halder I (2009) Linking bird, carabid beetle and butterfly life-history traits to habitat fragmentation in mosaic landscapes. Ecogr 32: 321–333.
25. Pérès G, Vandenbulcke F, Guernion M, Hedde M, Beguiristain T, et al. (2011) Earthworm indicators as tools for soil monitoring, characterization and risk assessment. An example from the national Bioindicator programme (France). Pedobiol 54: 77–87.
26. Chevenet F, Dolédec S, Chessel D (1994) A fuzzy coding approach for the analysis of long-term ecological data. Freshw Biol 31: 295–309.
27. MUMIA. Available: http://www.mumia-network.eu/index.php/working-groups/wg4. Accessed: 2014 Sep 10.
28. Manola F, Miller E (2004) RDF Primer, W3C Recommendation.

13

Gypsophile Chemistry Unveiled: Fourier Transform Infrared (FTIR) Spectroscopy Provides New Insight into Plant Adaptations to Gypsum Soils

Sara Palacio[1]*, Matt Aitkenhead[2], Adrián Escudero[3], Gabriel Montserrat-Martí[4], Melchor Maestro[4], A. H. Jean Robertson[2]

1 Instituto Pirenaico de Ecología (IPE-CSIC), Jaca, Huesca, Spain, **2** The James Hutton Institute, Aberdeen, Scotland, United Kingdom, **3** Biodiversity and Conservation Group, E.S.C.E.T., Universidad Rey Juan Carlos, Móstoles, Madrid, Spain, **4** Instituto Pirenaico de Ecología (IPE-CSIC), Zaragoza, Spain

Abstract

Gypsum soils are among the most restrictive and widespread substrates for plant life. Plants living on gypsum are classified as gypsophiles (exclusive to gypsum) and gypsovags (non-exclusive to gypsum). The former have been separated into wide and narrow gypsophiles, each with a putative different ecological strategy. Mechanisms displayed by gypsum plants to compete and survive on gypsum are still not fully understood. The aim of this study was to compare the main chemical groups in the leaves of plants with different specificity to gypsum soils and to explore the ability of Fourier transform infrared (FTIR) spectra analyzed with neural network (NN) modelling to discriminate groups of gypsum plants. Leaf samples of 14 species with different specificity to gypsum soils were analysed with FTIR spectroscopy coupled to neural network (NN) modelling. Spectral data were further related to the N, C, S, P, K, Na, Ca, Mg and ash concentrations of samples. The FTIR spectra of the three groups analyzed showed distinct features that enabled their discrimination through NN models. Wide gypsophiles stood out for the strong presence of inorganic compounds in their leaves, particularly gypsum and, in some species, also calcium oxalate crystals. The spectra of gypsovags had less inorganic chemical species, while those of narrow gypsum endemisms had low inorganics but shared with wide gypsophiles the presence of oxalate. Gypsum and calcium oxalate crystals seem to be widespread amongst gypsum specialist plants, possibly as a way to tolerate excess Ca and sulphate. However, other mechanisms such as the accumulation of sulphates in organic molecules are also compatible with plant specialization to gypsum. While gypsovags seem to be stress tolerant plants that tightly regulate the uptake of S and Ca, the ability of narrow gypsum endemisms to accumulate excess Ca as oxalate may indicate their incipient specialization to gypsum.

Editor: Heidar-Ali Tajmir-Riahi, University of Quebect at Trois-Rivieres, Canada

Funding: SP was funded by a Juan de la Cierva contract (Ministerio de Ciencia e Innovación, Spain) and projects [CGL2011-26654] (Ministerio de Economía y Competitividad, Spain) and [ARBALMONT/786-2012] (Organismo Autónomo de Parques Nacionales, Ministerio de Agricultura, Alimentación y Medio Ambiente, Spain). The funders had no role in study design, data collection and analysis, decision to publish, or preparation of the manuscript.

Competing Interests: The authors have declared that no competing interests exist.

* Email: s.palacio@ipe.csic.es

Introduction

The ability of plants to survive in substrates with limiting conditions for plant life has intrigued biologists since early times [1,2]. Gypsum soils are amongst the most widespread special substrates, extending over 100 million hectares [3,4], but yet they have received comparatively less attention than other substrates such as serpentines, saline or calcicolous soils [5]. Gypsum soils develop from gypsic rocks in arid and semi-arid areas where low precipitation prevents gypsum from being leached [6]. Together with the arid conditions, gypsum soils have particularly stressful physical and chemical properties for plant life including the presence of hard soil crusts, high mechanical instability, low soil porosity, extreme nutritional deficits, high concentration of sulphates and moderate salinity [7,8,9,10]. As a consequence, they are among the most restrictive soils for plants [6]. Nevertheless, the adverse conditions of gypsum soils contrast with the rich and specialized flora they shelter, comprising diverse arrays of narrow endemic and rare plants in arid and semiarid regions, many of which are threatened or endangered and constitute a global conservation biodiversity concern [6,11].

Depending on their specificity to gypsum soils, plants can be classified as gypsophiles, i.e. plants growing exclusively on gypsum substrates, or gypsovags, i.e. plants growing both in and out of gypsum [11]. Despite the recent efforts devoted to understand plant life on gypsum, the mechanisms displayed by plants to become competitive on gypsum soils are still not fully understood [5]. Early investigations showed that the chemical composition of gypsophiles and gypsovags differs, with the former showing higher concentration of certain nutrients (N, P, Ca, S) and total ashes than the latter [12,13,14]. Previous studies also showed that gypsophile seedlings (like *H. squamatum* and *L. subulatum*) show a higher ability than gypsovags to surpass the physical soil crust characteristic of gypsum soils, and hence are better adapted to cope with the physical limitations of gypsum soils [15]. More recently, however, it was found that gypsophiles could be segregated in two different groups of plants, with distinct chemical composition and different putative ecological strategies [16]. These authors found differences in the composition of narrowly and widely distributed gypsophiles, the former being more similar to the chemical composition of gypsovags. According to Gankin and Major's interpretation of edaphism origin [17], narrowly distributed gypsophiles (similarly to gypsovags) would fit the *refuge* model, being stress tolerant species not specifically adapted to gypsum soils that avoid competition in marginal soils, while widely distributed gypsophiles would fit the *specialist* model, being specifically adapted to gypsum and dramatically lowering their performance in other soils [16]. The increased concentration of Ca and S of widely distributed gypsophiles suggests the existence of certain (still unexplored) physiological adaptations to cope with the excess of calcium sulphate in gypsum soils [13,16]. The high concentrations of certain nutrients (such as N or P) of wide gypsophiles are also intriguing, as gypsum soils are inherently nutrient poor [18,19].

The results of the study by Palacio *et al.* [16] are crucial since they show that various chemical strategies are feasible for plants living exclusively on gypsum soils. However, they are solely based on the elemental composition of plant species. To know if the generality of their conclusions stands with more comprehensive analyses of plant biochemistry, biochemical fingerprinting techniques allowing for the identification of the functional chemical groups of plants are needed. Such approaches could also shed light on the biochemical and physiological adaptations of different groups of plants to survive on gypsum soils. For example, they could inform on the biochemical mechanisms of wide gypsophiles to accumulate S, Ca, N or P. Such information is critical to understand plant life on gypsum substrates.

Fourier transform infrared (FTIR) spectroscopy is a powerful tool for the chemical analysis of biological samples [20]. This technique offers a fast, cost-effective, non-destructive way of obtaining a biochemical fingerprint of samples, where the main functional groups and bonds can be identified, thus giving structural information on the chemical compounds present rather than just elemental information. Crucially, both organic and inorganic compounds have features in the FTIR spectrum, allowing not only identification of the main organic constituents of plant material but also characterisation of the forms of inorganic compounds present in the plants [21]. In addition, by comparing the replication of FTIR spectra for different plants of the same species grown in the same conditions and localities, the technique provides a means of assessing the variability of the chemical profile among individuals.

FTIR spectroscopy is a widely used tool in plant biological studies [20]. It has been used to detect changes in plant chemistry in relation to fertilization [22,23,24] or heavy metal addition [25,26,27]. FTIR spectra analysed through artificial neural networks (NN) show a high potential to discriminate biological samples. For instance, [28], [29] and [30] applied NN to FTIR spectral data in order to distinguish between plant species, using a range of different kinds of source material. Despite the huge potential and increasing use of FTIR in plant studies, it has never been applied to the study of edaphic endemisms before.

The aim of this study was to compare the identity of the main chemical groups present in the leaves of plants with different specificity to gypsum soils and to explore the ability of FTIR spectra analyzed with NN to discriminate groups of gypsum plants. We hypothesize that: 1) data obtained by FTIR spectroscopy will show a high correlation with the elemental composition of plants. Consequently, owing to the observed differences in the mineral composition of gypsum plants in previous studies, we further hypothesize that 2) narrow gypsum endemisms will show a similar FTIR spectral composition to gypsovags, while widely distributed gypsophiles will show distinct spectral features.

Materials and Methods

Species and study sites

Species and study sites generally followed [16], except for the addition of the gypsovag *Thymelaea tinctoria*, and included fourteen species for analysis: five gypsovags and nine gypsophiles. The latter comprised most woody gypsophile species of study areas, including five widely distributed gypsophiles and four narrow gypsum endemics (Table 1). The distinction between both types of gypsophiles was made according to the extent of their distribution area: widely distributed gypsophiles were species that showed a wide distribution range in the Iberian Peninsula, comprising almost every gypsum outcrop in this large territory, whereas narrow gypsum endemics were species that showed a limited number of populations, growing only in one of the gypsum areas of the Iberian Peninsula. In order to minimize bias due to variations in plant functional strategies, all study species were shrubs or sub-shrubs, which are prevalent growth forms in gypsum outcrops [6,31], and had a similar branch morphology and architecture (for further details see [16]). None of the species included in this study is endangered or protected [32].

Plant species were collected from the two more massive and distinctive gypsum outcrops of the Iberian Peninsula: Central Spain (Middle Tajo Basin, near Madrid) and NE Spain (Middle Ebro Basin, near Zaragoza). Samples were not replicated across both areas owing to the absence of qualitative differences in the spectra of samples of *Ononis tridentata* from different sites (Fig. S1) and the previously reported similar elemental composition of samples collected from both areas [16]. Study species and sampling sites are shown in Table 1, while climatic and edaphic features of each sampling site are in Table 2. All permits required for plant collection at Madrid Community were requested by AE. Permit for sampling of plant material in Aragón was issued by the Government of Aragón (Diputación General de Aragón, DGA) to GMM and SP.

Plant and soil sampling and processing

To avoid variability in the chemical composition of plants due to phenology, all plant material was collected during winter, when chemical concentrations of nutrients in the leaves of study plants are steady [16]. Five adult individuals were harvested in each studied population (Table 1). Once in the laboratory, a 15 g sample of the leaves of the short branches of each individual was collected and oven-dried to a constant weight at 60°C. Dry and damaged leaves were excluded from the analyses.

Chemical analyses

Samples were ground in a ball mill (Retsch Mixer MM301, Leeds, UK) to a fine powder. N and C concentrations were analyzed with an elemental analyzer (Elementar VarioMax N/CN, Hanau, Germany). Subsamples were burnt at 550°C for 4 hours and ash was dissolved in HNO3-HCl-H20 (1:3:9) and filtered. Concentrations of Na and K were measured in the soluble (silica-free) ash by flame photometry, Ca and Mg concentrations were determined by complexometry [33] and P concentration was assessed by vanado-molybdate colorimetry [34]. Total sulphur was analyzed by a turbidimetric method with barium chloride [33].

Table 1. Location and characteristics of study species and sites.

Species	Species code	Species type[a]	N[b]	Sampling sites[c]	UTM	Altitude (m a.s.l.)	Date of sampling	Subst.[d]
Centaurea hyssopifolia Vahl.	Ch	NE	4	Chinchón (M)	VK6247	640	15-Mar-10	G
Gypsophila struthium L. subsp. *hispanica* (Wilk.) G. López	Gh	G	5	Villamayor (Z)	XM8820	300	19-Feb-10	G
Helianthemum marifolium (L.) Mill. subsp. *conquense* Borja and Rivas Goday	Hco	NE	7	Tendilla (Cu)	WK4214	780	3-Mar-03	G
Helianthemum squamatum (L.) Pers.	Hsq	G	5	Villamayor (Z)	XM8718	290	21-Jan-03	G
Herniaria fruticosa L.	Hf	G	C	Villamayor (Z)	XM8820	300	20-Jan-03	G
							19-Feb-10	
					XM8718	290	21-Jan-03	G
Lepidium subulatum L.	Lsb	G	5	Villamayor (Z)	XM8820	320	13-Jan-03	G
							19-Feb-10	
Linum suffruticosum L.	Lsf	GV	5	Villamayor (Z)	XM8820	320	7-Jan-03	G
Ononis tridentata L.	Ot	G	5	Villamayor (Z)	XM8820	300	20-Jan-03	G
							19-Feb-10	
Rosmarinus officinalis L.	Ro	GV	5	Peñaflor (Z)	XM8626	300	24-Jan-03	G
Salvia lavandulifolia Vahl.	Sl	GV	5	Villamayor (Z)	XM8820	320	21-Jan-03	G
Teucrium polium L. subsp. *capitatum* (L.) Arcangelli	Tc	GV	5	Chinchón (M)	VK6247	640	6-Feb-03	G
Teucrium pumilum L.	Tp	NE	6	Chinchón (M)	VK6247	640	6-Feb-03	G
Thymelaea tinctoria (Pourr.) Endl. subsp. *tinctoria*				Villamayor (Z)			19-Feb-10	
Thymus lacaitae Pau	Tl	NE	5	Alcalá de Henares (M)	VK8282	710	28-Jan-03	G

[a]NE = Narrow endemism, G = gypsophile, GV = gypsovag.
[b]N = number of sampled individuals, C = composite sample.
[c]M = Madrid (Central Spain, Middle Tajo Basin), Cu = Cuenca (Central Spain, Middle Tajo Basin), Z = Zaragoza (NE Spain, Middle Ebro Basin).

Table 2. Climatic and edaphic characteristics of study sites.

Study sites	Area[a]	Climate		Soil				
		T (°C)	P (mm)	pH	SOM (%)	N (%)	$CaCO_3$ (%)	Gypsum (%)
Villamayor (Z)	Z	14.2	370	7.7	1.9	0.14	20.0	55.5
Peñaflor (Z)	Z	14.2	370	8.00	0.88	0.051	13.04	66.30
Alcalá de Henares (M)	M	13.2	350	7.8	1.8	0.15	20.6	32.0
Chinchón (M)	M	13.8	422	7.7	2.1	0.12	9.7	59.3
Tendilla (Cu)	M	13.0	410	7.8	2.4	0.17	22.3	29.3

Abbreviations: T = Mean annual temperature; P = total annual rainfall; SOM = soil organic matter; N = nitrogen.
[a]M = Madrid (Central Spain, Middle Tajo Basin), Z = Zaragoza (NE Spain, Middle Ebro Basin).

FTIR Spectroscopic Analysis

As for the other chemical analyses, dried samples for FTIR analysis were prepared by grinding in a ball mill (Retsch Mixer MM301, Leeds, UK) to a fine powder. FTIR spectra were recorded on a Bruker Vertex 70 FTIR spectrometer (Bruker, Ettlingen, Germany) fitted with a potassium bromide beam splitter and a deutroglycine sulphate detector from three replicates of each species and, where the replicates did not match closely, spectra were recorded for two further replicates. A Diamond Attenuated Total Reflectance (DATR) sampling accessory, with a single reflectance system, was used to produce "transmission-like" spectra. Samples were placed directly on a DATR/KRS-5 crystal, and a flat tip powder press was used to achieve even distribution and contact. Spectra were acquired by averaging 200 scans at 4 cm^{-1} resolution over the range 4000 – 370 cm^{-1}. A correction was made to the ATR spectra to allow for differences in depth of beam penetration at different wavelengths, using OPUS software (Bruker, Ettlingen, Germany, version 6.0). The spectra were also baseline corrected. No correction was required for water vapour and CO_2 as the spectrometer is continuously flushed with dry air.

Statistical analyses

Predictions of chemical properties from the FTIR spectra. The relationship between FTIR spectra and the results of the chemical analyses was evaluated through Partial Least Squares (PLS) regression, using the OPUS QUANT version 6.0 software (Bruker, Ettlingen, Germany). Data were available for C (%), N (mg/g), C/N, Na (mg/g), K (mg/g), Ca (%), Mg (mg/g), P (mg/g), S (%) and Ash (%) (Table S1). The validation was done by cross validation, which allows the maximum representation of the samples, essential for such a small dataset. Correlation analyses between "predicted" and "true" values were used to assess the relationships between spectral information and elemental composition of plant samples (R^2, RMSECV (Root Mean Square Error of Cross Validation), bias and RPD (SD/SECV), Regression Point Displacement, which equals the standard deviation in the field data values divided by the standard error in the Cross Validation error values). There was insufficient representation of all types of species to enable robust calibrations to be developed for prediction of unknown samples, but the data should give a good indication of the extent to which these parameters are related to the IR spectral features.

Neural network analysis. The ability of FTIR spectra to discriminate groups of plants with different specificity to gypsum was evaluated by neural network (NN) analyses. Neural networks are particularly useful when the relationships between the input data (in this case the FTIR spectra) and the output parameters (plant species classification) are complex and/or unknown (e.g. [35]). They are also useful when applied to datasets with a large number of input parameters, such as infrared spectra [36]. In order to train the neural network model, the FTIR spectral data was pre-processed to make it more acceptable for NN training. This was carried out in two stages. Firstly, a moving window averaging process was applied to each spectrum, with the moving average at each point subtracted from the actual value. This had the effect of 'flattening' the spectra and setting the value of flatter sections of the curve to zero (Fig. S2). The moving window width was set at 101 (averaged over a window that extended 50 values to the left and 50 to the right from each value), following trial and error to optimise the performance of the neural network model. Secondly, the maximum absolute value within each derived spectrum was determined, and the value at each point was divided by this. This had the effect of normalising the spectra within the range [−1, 1] (Fig. S2).

The neural network architecture had four layers, with one input layer of 1882 nodes, two hidden layers each of 100 nodes and one output layer of 3 nodes. The neural network was fully connected (i.e. each node in the input layer was connected to each node in the first hidden layer, and so on). The training rate of the network was set at 0.05, and a total of 10,000 training epochs were used to train the network. These parameter values were selected using trial and error to optimise the model performance. The training algorithm used was the feed-forward back-propagation, a standard method that has been applied in many cases. For further details on this algorithm as applied to environmental data, see [37].

The training data consisted of 45 spectra, which is a relatively small number of data points for training a neural network. To ensure that overfitting of the data was not an issue, a separate neural network was trained for each of the data points, using all of the data except that point and using the one left out for testing purposes. This 'leave-one-out' approach ensures that training and testing data do not overlap, and provides more accurate information about how well the model is performing under conditions of low training data numbers. For each data point, the network was trained to predict a value of 1 for output node corresponding to the plant type being presented, and a value of 0 for the other output nodes.

Results

Description of FTIR spectra of the different groups

Although the spectra of each species were different and were interpreted individually, this provided a large amount of potentially useful information. Therefore, the results have been summarised for each of the three groups of species analyzed, i.e. "wide gypsophiles", "narrow gypsum endemisms" and "gypsovags", and in Table 3 the frequencies and assignments of the main peaks identified are presented.

Overall, the major difference between the spectra of wide gypsophiles and the other two groups is the presence of appreciable gypsum in their spectra (with one exception - see below) and often considerable oxalate too (Figs. 1, 2, and 3, Table 3). In general, the spectra of gypsovags had a lowest proportion of inorganic compounds present in them, while the spectra of narrow gypsum endemisms had a higher proportion, and mostly show the presence of oxalate (Figs. 1, 2, and 3, Table 3). However, for a number of species there was very little difference between the gypsovag and narrow gypsum endemism categories (see below for details).

Gypsovags. Replication of the FTIR spectra obtained from gypsovags was generally good (data not shown) with subtle variations in intensities of bands, indicating similar overall chemical profiles within each species. They all exhibited sharp CH_2 stretching bands, (indicating the presence of long carbon chains) and the same basic spectral patterns with features relating to protein (amide), polysaccharide (cellulose, pectin), lignin, lipids and other long chain hydrocarbons, as might be expected in plant spectra (Fig. 1, Table 3). A feature of the spectra of *R. officinalis* and *S. lavandulifolia* was a distinct band at 1686 cm^{-1}, which is possibly an aromatic ester (Fig. 1, Table 3). Notably, in comparison with the spectra of the species in the other two groups, the spectra of gypsovags had the least bands attributable to inorganic components (although all species may have traces). Among the few exceptions were the evidence for calcite, and possibly ammonium sulphate, in the spectrum of *L. suffruticosum*, and oxalate in the spectrum of *T. polium* subsp. *capitatum* (Fig. 1, Table 3).

Narrow gypsum endemisms. Replication of the FTIR spectra of narrow gypsophiles was also good, again implying consistent chemical composition within each species. The spectra of this group of species generally appeared to have greater evidence for the presence of inorganic compounds than those of the gypsovag group, although the spectra still showed predominantly features expected from plant material (Fig. 2, Table 3). Apart from *C. hyssopifolia*, where it was not readily detected, the spectra of narrowly distributed gypsophile species all appeared to show the presence of some oxalate. Despite this common feature, there was variation among narrow gypsum endemism in the intensity of the CH_2 stretching bands in their spectra (indicative of long carbon chains and sometimes referred to as a "waxy" character, [38]). Two of the species (*C. hyssopifolia* and *H. marifolium* subsp. *conquense*) had much weaker CH_2 absorption in their spectra compared with the spectra of the other two (*T. pumilum* and *T. lacaitae*) which appeared to have stronger absorption, much more similar to the spectra of species in the gypsovag group (Figs. 1–2, Table 3). In fact, *T. pumilum* and *T. lacaitae* had spectra generally quite closely related to those of gypsovags; in particular the spectrum of *T. pumilum* quite closely resembled that of its congener *T. polium* subsp. *capitatum* (including the presence of oxalate). The spectrum of *T. lacaitae* also resembled that of *T. polium* subsp. *capitatum*, and it had also evidence for a band at 1686 cm^{-1}(sh), compatible with an aromatic ester, as indicated above for the spectra of other Lamiaceae (*R. officinalis* and *S. lavandulifolia*) of the gypsovag group (Figs. 1–2, Table 3). *C. hyssopifolia* had a spectrum more reminiscent of that of *L. suffruticosum* than the spectra of the other narrow gypsophile species (with a high proportion of protein/amide bands and other bands present which could be assigned to ammonium sulphate or carboxylate). Finally, *H. marifolium* subsp. *conquense* had a spectrum that generally differed from the other narrow gypsophiles, and spectral searches suggested that the presence of tannic acid is a possibility. However it had the oxalate present and relatively strong protein/amide bands, as seemed to be common in the rest of plants from this group (Figs. 1–2, Table 3).

Wide gypsophiles. The replication of spectra within this group was noticeably less tight, with variations in intensity of bands, particularly those arising from inorganic components, rather than different bands being present (i.e. same chemistry but different proportions). A distinct feature of the spectra of this group of plants was the relatively high dominance of bands related to inorganic compounds (Fig. 3, Table 3), with clear bands for gypsum in all but one of them (*L. subulatum*). The presence of crystalline gypsum could be confirmed, even in the presence of other sulphates, by shoulders on the broad OH stretch at 3518 and 3398 cm^{-1} (Fig. 3, Table 3). These bands were similar to those detected in the spectra of soil samples collected at the sampling site (data not shown) and had not been identifiable in any of the previous spectra of the gypsovag and narrow gypsophile groups. Oxalate was also strongly present in the spectra of two of the wide gypsophiles analyzed (*H. fruticosa* and *G. struthium* subsp. *hispanica*) and some small traces were also visible in *H. squamatum* and *O. tridentata*, with the most diagnostic band being at 1315 cm^{-1} (Fig. 3, Table 3). Although *L. subulatum*, had a spectrum where inorganic compounds were not so immediately obvious, a broad absorption band in its spectrum at $\sim$1600 cm^{-1} could possibly be ammonium sulphate or carboxylate, and sulphates other than gypsum may also be present (Fig. 3, Table 3). This spectrum also had the strongest bands relating to amide/protein.

Bands expected from plant material (such as cellulose, lignin, pectin, protein etc) were also present in spectra of wide gypsophiles, but were often obscured by the inorganic compounds

Table 3. Frequencies and assignments of the main peaks identified in the analysis of FTIR spectra.

Wave number, cm^{-1}	Assignment	Characterisation
3522,3400	O-H stretching	Gypsum[1, 2]
3340	O-H stretching	Cellulose, in samples with defined 3340 peak[3]
2920	antisymmetric CH_2 stretching	Fats, wax, lipids[3]
2850	symmetric CH_2 stretching	Fats, wax, lipids[3]
1740-1720	C=O stretch of COOR	Esters[3, 4]
1707-1710	C=O stretch of COOH	Carboxylic acids[3, 4]
1653	C=O of amide I	Proteinaceous origin[3]
1615	C-O stretching	Calcium oxalate[5]
1600-1650 (1610)	Aromatic C=C stretching and/or asymmetric C-O stretch in COO-	Lignin and other aromatics, or aromatic or aliphatic carboxylates[3]
1550	N-H in plane (amide-II)	Proteinaceous origin[3]
1505-1515	Aromatic C=C stretching	Lignin/Phenolic backbone[3, 4]
1450-1410	C-O stretching	Calcium carbonate[4, 6]
1426	Symmetric C-O stretch from COO- or stretch and OH deformation (COOH)	Carboxylate/Carboxylic structures (humic acids)[3]
1371, 1450	C-H deformations	Phenolic (lignin) and aliphatic structures[3]
1312	C-O stretching	Calcium oxalate[5]
1265 (approx.)	C-O stretching of phenolic OH and/or arylmethylethers	Indicative of lignin backbone[3]
1265 -1240	C-O-C stretching	Esters
	C-N stretching	Amide III[4]
1140-1080	S-O stretching	Sulphates[4]
1100-1000	P-O stretching	Phosphates[6]
1100-950	Si-O stretching	Silicates[4, 6]
1050 (1030-1080)	Combination of C-O stretching and O-H deformation	Polysaccharides[3]
874	C-O in plane bending	Calcium carbonate[4, 6]
835	Aromatic CH out of plane	Lignin[3]
780	COO bending	Calcium oxalate[5]
720	CH_2 wag	Long chain (> C4) alkanes[3]
715	C-O in plane bending	Calcium carbonate[4, 6]
680-610	S-O bending	Sulphates[4]
669,597	S-O bending	Gypsum[1, 2]

[1][39]; [2] [56]; [3] [38]; [4] [21]; [5] [57]; [6] [58].

(Fig. 3, Table 3). Absorption in the region between 1200 - 1000 cm^{-1} could arise from a range of inorganic compounds rather than polysaccharide but this region is very hard to interpret as sulphate, phosphate, polysaccharide and silicate groups all absorb in this frequency range and often have overlapping bands [21,38,39].

Correlation between FTIR spectral data and chemical analyses

In general, FTIR data adjusted reasonably well (very well for some cases) to the concentration of most nutrients in plants (Table 4). Correlation coefficients were above 90.0 for ashes, carbon and sulphur, and above 79.0 for Mg, P, Ca, Na and N. The different forms of N in different species, particularly those in the wide gypsophile group, may result in poorer correlation overall than might be the case if they are separated out, but there were not enough cases to represent each type of N form separately and not enough replicates to separate out each group (Table S1). The main exception was K, which implies that there was not such a clear cut relationship between spectral features and the variance in values of K for the samples analysed.

Suitability of FTIR spectra to discriminate ecological strategies of gypsum plants

The neural network (NN) approach was successful in correctly assigning the majority (39 out of 44) of the cases analyzed (Table 5). Predictive accuracies were assessed in terms of the error between target and actual values at each output node (RMSE), and are given in Table 5. RMSE values for gypsovag and narrow endemisms were high. However, they were not as high as would be obtained through purely random output values, for which RMSE values greater than 0.4 would be expected. The RMSE value for gypsophiles was much lower. In general, the NN models built were able to discriminate between the three groups of plants (Table 5). The wide gypsophile category was most accurately predicted with 15 data points correctly predicted out of 15 points analysed. It was followed by gypsovags, with 15 data points correctly predicted out of 17, and narrow gypsum endemisms or narrow gypsophiles, with

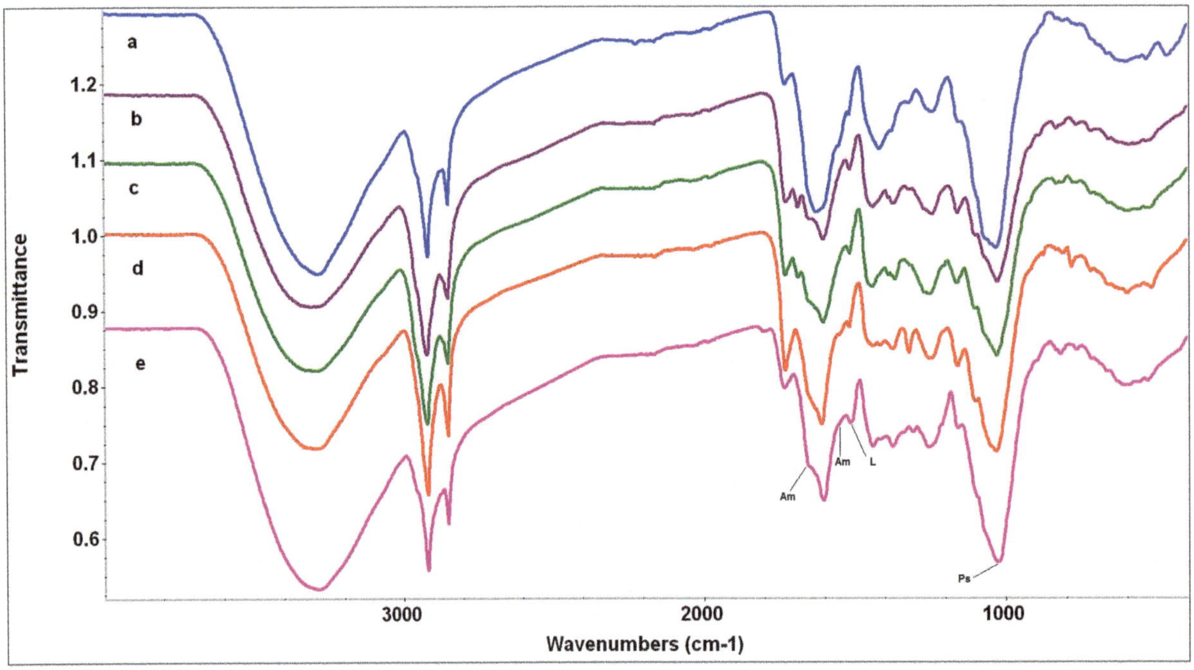

Figure 1. FTIR spectra of gypsovag species. The scale of the Y axis is offset for illustration purposes. a) *Linum suffruticosum* (blue), b) *Rosmarinus officinalis* (purple), c) *Salvia lavandulifolia* (green), d) *Teucrium polium* subsp. *capitatum* (red), e) *Thymelaea tinctoria* (pink). Abbreviations of compound peaks: Am: Amide, L: Lignin/Phenolic backbone, Ps: Polysaccharides.

9 data points correctly predicted out of 12, and therefore the least accurately predicted group.

Of the five data points that were incorrectly identified, none were from the wide gypsophile category and no other species from either of the other two categories was incorrectly classified as a

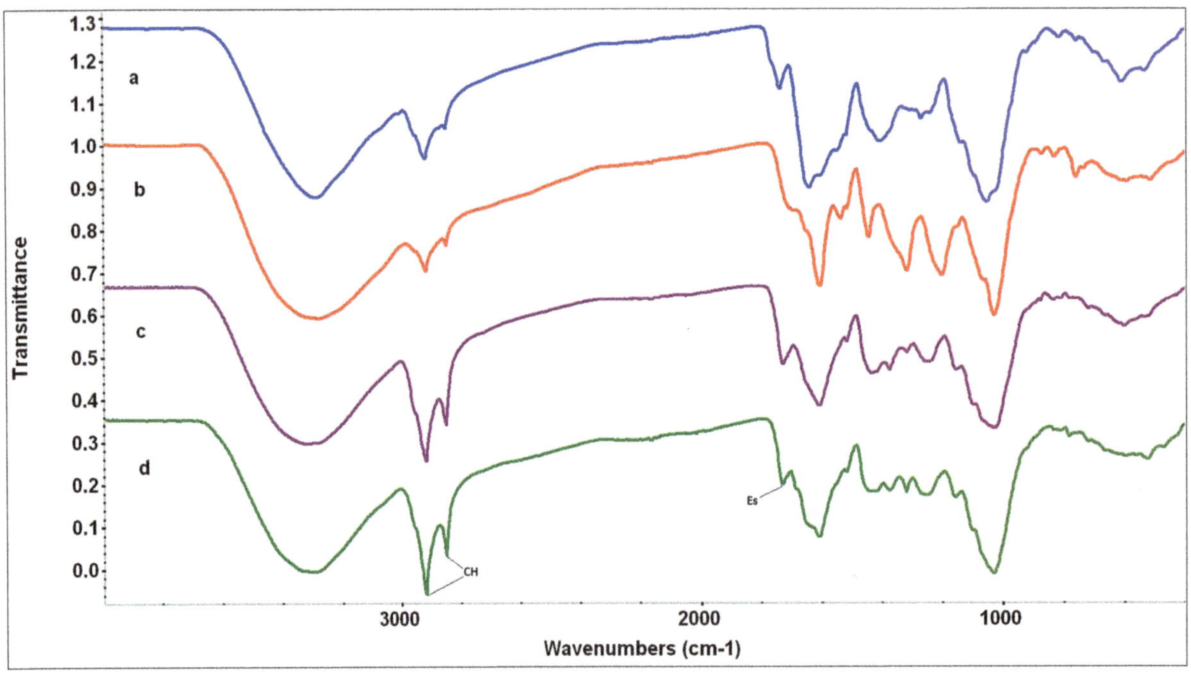

Figure 2. FTIR spectra of narrow gypsum endemisms. The scale of the Y axis is offset for illustration purposes. a) *Centarurea hyssopifolia* (blue), b) *Helianthemum marifolium* subsp. *conquense* (red), c) *Teucrium pumilum* (purple), d) *Thymus lacaitae* (green). Abbreviations of compound peaks: CH: CH-stretching, Es: Esters.

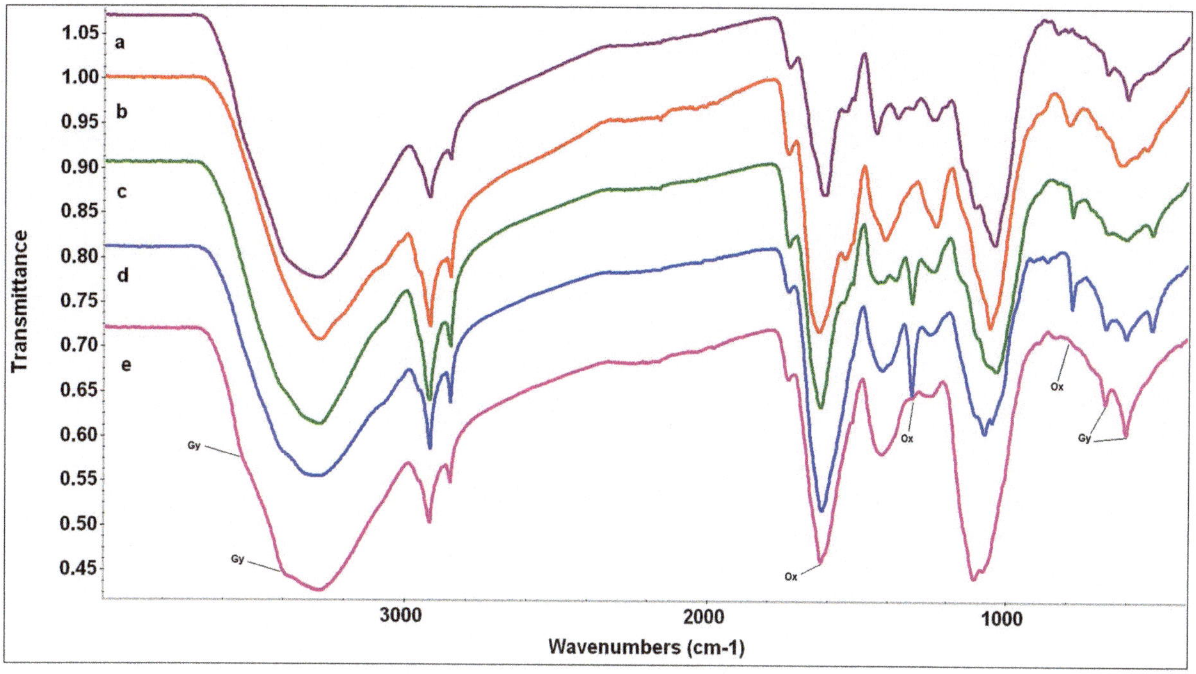

Figure 3. FTIR spectra of wide gypsophiles. The scale of the Y axis is offset for illustration purposes. a) *Helianthemum squamatum* (purple), b) *Lepidium subulatum* (red), c) *Herniaria fructicosa* (green), d) *Gypsophyla struthium* subsp. *hispanica* (blue), e) *Ononis tridentata* (pink). Abbreviations of compound peaks: Gy: Gypsum, Ox: Oxalate.

wide gypsophile. There was, however, some crossover between the spectra of gypsovags and narrow gypsophiles, with five cases being assigned to the wrong group, although generally they could be accurately distinguished. The cases of mis-classification mostly appear to relate directly to the spectral features noted above, i.e. the strong similarity, including the presence of oxalate, between the spectra of *Teucrium polium* subsp. *capitatum* (a gypsovag misclassified as narrow gypsophile) and those of *Teucrium pumilum* and *Thymus lacaitae* (misclassified as gypsovags).

Visual examination of the input datasets averaged over each group of species (Fig. 4) does not allow much distinction to be drawn between the three types of spectra, although it can be seen that between the frequencies of 3200 and 3600 cm^{-1}, and between 1500 and 1900 cm^{-1} (Fig. 5), wide gypsophiles do demonstrate differences to narrow endemisms and gypsovags, which have more subtle differences to each other in these ranges. These ranges, as described above for the FTIR results (Table 3), correspond to bands relating to gypsum present in wide gypsophiles, but not detectable in the other two groups. Also there is variation in oxalate and ester concentrations, with generally lower oxalate in the gypsovags, but higher ester (Fig. 4, Table 3).

Examination of the averaged input datasets over smaller ranges however, shows that, while at some ranges narrow gypsum endemisms show similar trends to gypsovags (Figs. 5 and 6), with higher ester (1740 cm^{-1}) than wide gypsophiles and no gypsum (1620 cm^{-1}), at other ranges narrow gypsum endemisms input value trends are more similar to wide gypsophiles with noticeable (although lower) presence of oxalate (1315 cm^{-1} and 1606 cm^{-1}), absent in most gypsovags (Figs. 5, 6). Therefore, it appears that

Table 4. Results of PLS regression analyses between FTIR spectra and the C (%), N (mg/g), C/N, Na (mg/g), K (mg/g), Ca (%), Mg (mg/g), P (mg/g), S (%) and Ash (%) concentrations of the samples analysed.

Variable	R^2	RPD values (Std Dev/Std error)
C (%)	95.7	4.81
N (mg/g)	79.8	2.24
Na (mg/g)	86.2	2.23
K (mg/g)	65.1	1.69
Ca (%)	88.1	2.69
Mg (mg/g)	79.9	2.9
P (mg/g)	85.5	2.62
S (%)	91.9	3.51
Ash (%)	94.5	4.28

Table 5. Confusion matrix of plant types, based on neural network model prediction.

Predicted groups	Observed groups		
	Gypsovag	Narrow endemism	Gypsophile
Gypsovag	15	2	0
Narrow endemism	3	9	0
Gypsophile	0	0	15
Model fit			
MAE	0.211	0.196	0.068
RMSE	0.305	0.287	0.098

Values give the number of each type given in the left-hand column that are categorised as each type given in the top row. Bottom rows indicate mean absolute error (MAE) and RMSE (Root Mean Squared Error) values for each output node in the neural network models, in relation to the output errors (range [0, 1]).

narrow gypsum endemisms, in terms of input values as developed using this FTIR-based approach, show some features both of gypsovags and gypsophiles and can thus be successfully discriminated from them.

Discussion

The FTIR spectra of the three groups of species analyzed (wide gypsophiles, narrow gypsum endemisms and gypsovags) showed distinct features that enabled their discrimination through neural network (NN) models with relatively high accuracy. This indicates that, although as predicted, elemental and spectral data correlated reasonably well, the analysis of FTIR spectra with NN provided an increased discriminating ability.

As expected, wide gypsophiles stood out for the strong presence of inorganic compounds in their leaves, particularly gypsum. This is in agreement with previous studies on the elemental composition of gypsum plants [12,13,14,16], finding high total ash, Ca and S (mostly in the form of sulphate) concentrations. Some of these studies suggested most of such S could be accumulated in the form of calcium sulphate [13,14], however, the widespread presence of mineral gypsum (i.e. $CaSO_4{\bullet}2H_2O$) in gypsophiles has never been reported before. With the sole exception of *L. subulatum*, all the wide gypsophiles analyzed accumulated mineral gypsum in their

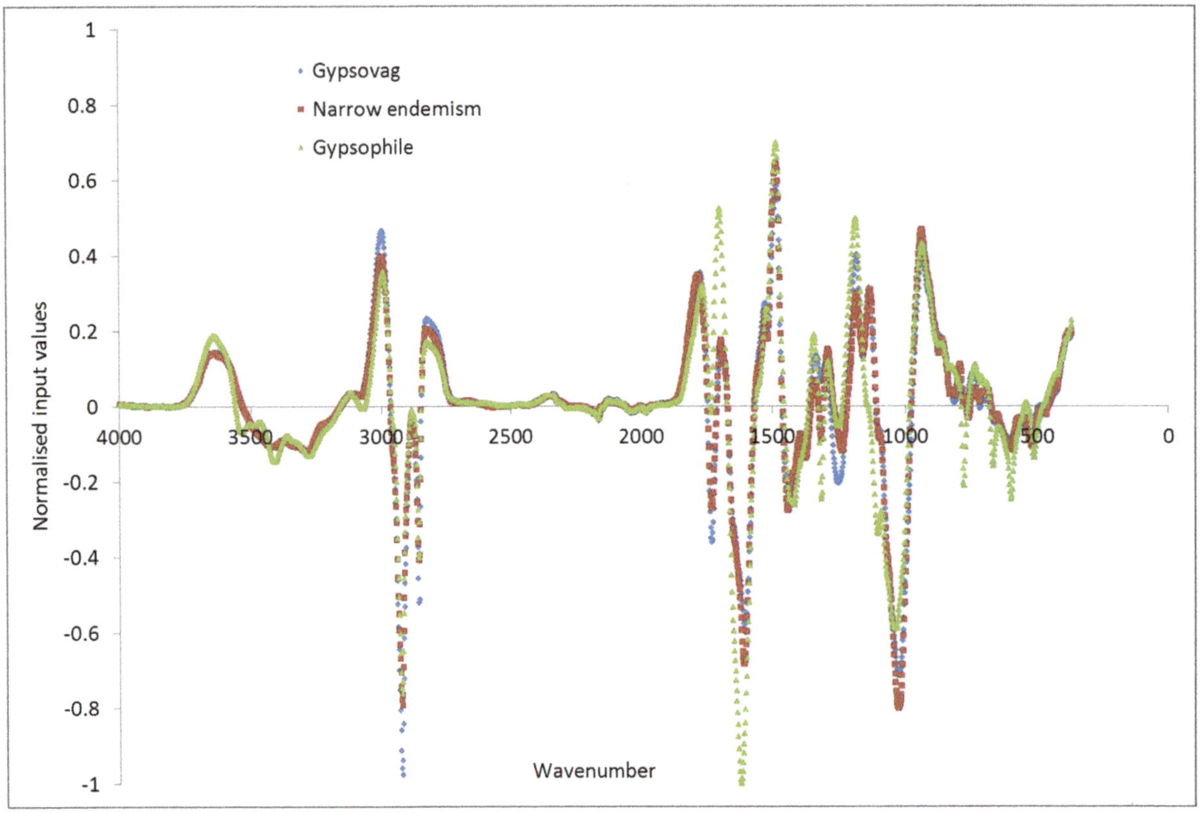

Figure 4. Input datasets averaged over each of the gypsum plant types (gypsovags in blue, narrow gypsum endemisms in red, wide gypsophiles in green), for the full wavenumber range.

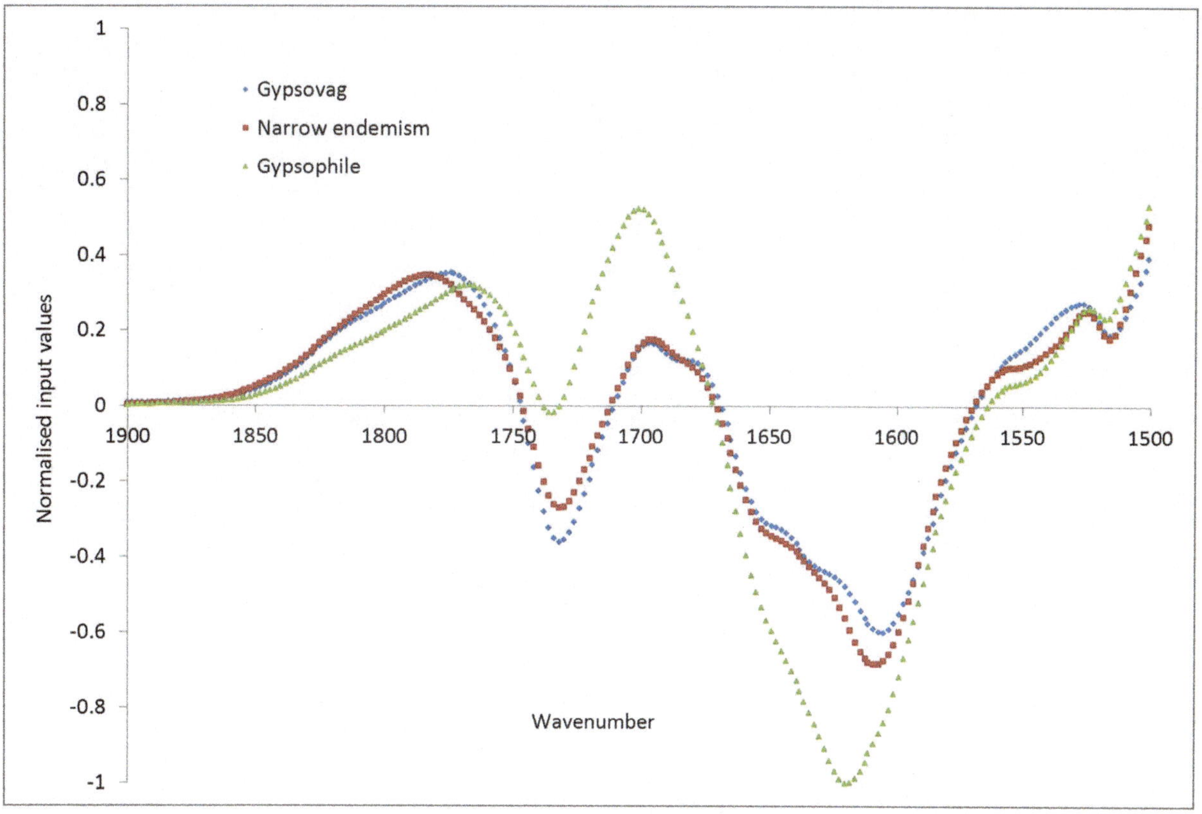

Figure 5. Input datasets averaged over each of the plant types (gypsovags in blue, narrow gypsum endemisms in red, wide gypsophiles in green), for the input wavenumber range 1500 to 1900 cm^{-1}.

leaves in the form of crystals. Although previous studies had reported the presence of crystals in some wide gypsophiles like *H. fruticosa* or *G. struthium* subsp. *hispanica* [40,41], this is the first evidence that most wide gypsophiles studied show such a distinct biochemical trait. Furthermore, the composition of crystals had only been analysed in gypsophiles growing on saline gypsum soils, like *Limonium* spp. or *Frankenia thymifolia* [6,42]. This is, therefore, the first empirical confirmation that some of the crystals observed in gypsophiles from non-saline soils, such as those considered in this study, are made of gypsum. Wide gypsophiles are commonly considered as "accumulators" of compounds such as S, Ca or Mg, being highly permeable to them and hence able to tolerate high concentrations (toxic for most plant species) in their leaves [12,13,14,43]. Soil gypsum contents above 20–30% are generally considered to be toxic for most plant species [44], but gypsum contents often surpass 50% in the soils where these plants were collected (Table 2). Similarly, normal average concentrations of S in plant leaves are ≈ 0.2% [45], but concentrations between 3 and 6% are common in wide gypsophiles [13,14,16,46]. The ability to "sequester" such excess S in the form of gypsum crystals that accumulate harmlessly in plant cells could be an adaptive mechanism of gypsum-specialist plants. Although gypsum crystals are rare among vascular plants [47], a recent study on the Australian endemic *Acacia robeorum* growing on high S soils, showed that this species was able to accumulate calcium sulphate crystals, presumably as a way to remove excess Ca and S [48]. The ability to accumulate excess gypsum in crystals may be widespread among different taxonomic groups, since the four species that showed gypsum crystals in their leaves in this study belong to three different taxonomic families (i.e. Cariophyllaceae, Fabaceae and Cistaceae). This ability is similar to the common mechanism of halophytes to excrete salts. Although it has been suggested that both mechanisms could be related [41], the question remains if lineages of halophytes and gypsophiles excreting salts are interconnected (i.e. share a common origin), or if both strategies have evolved independently. The general finding of gypsum in the leaves of wide gypsophiles opens up interesting questions regarding the regulatory mechanisms and the processes involved in the metabolism of gypsum in these plants, which is mostly unexplored.

Nevertheless, the formation of gypsum crystals is not the only strategy of wide gypsophiles to cope with high gypsum contents in the soil. *L. subulatum*, a gypsophile common to disturbed gypsum soils of the Iberian Peninsula [31,49], shows high S concentrations [13,16,46], but notably no gypsum. According to our results, this species seems to accumulate S in the form of other sulphates, possibly in combination with N. *L. subulatum* is known for its remarkably high N, amino acid and protein content [16,46,50], and high amide and protein bands were found for this species. It is hence possible that part of the S of this species is accumulated as ammonium sulphate (which presence could not be ruled out) or else in the form of S-rich proteins, peptides (glutathione), amino acids (methionine, cysteine) or other organic compounds rich in both N and S such as secondary metabolites (e.g. glucosinolates), which are common among the Brassicaceae. Indeed, the ability of *L. subulatum* to incorporate S into organic compounds has been suggested as a metabolic process explaining *L. subulatum* adaptation to gypsum soils [46].

The spectra of wide gypsophiles were also differentiated by their high calcium oxalate bands, a trait they shared with narrow gypsum endemisms. Calcium oxalate deposition is a common

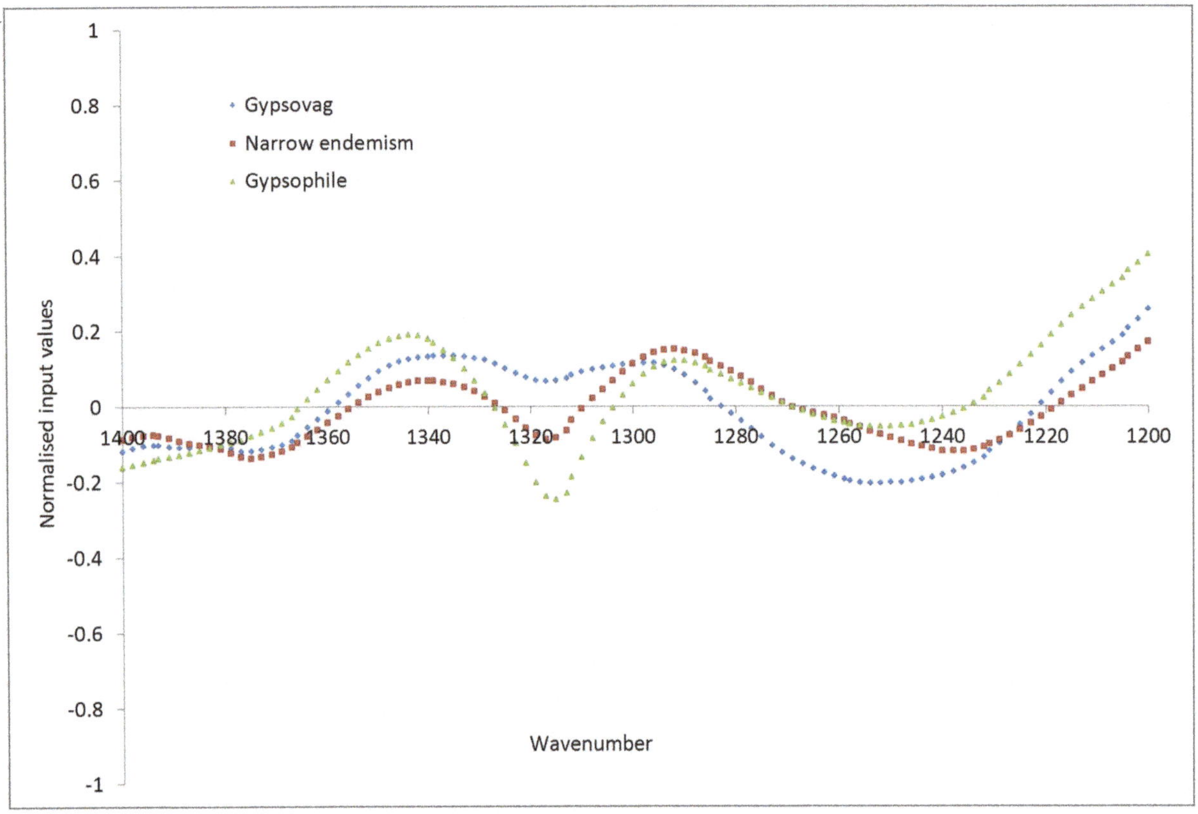

Figure 6. Input datasets averaged over each of the plant types (gypsovags in blue, narrow gypsum endemisms in red, wide gypsophiles in green), for the input wavenumber range 1200 to 1400 cm^{-1}.

phenomenon in plants, being present in more than 215 plant families [51]. The formation of calcium oxalate crystals in plant tissues is related to various important plant functions, among which the regulation of calcium concentrations seems primary [52]. Accordingly, several studies using a variety of plants have shown that the size and number of Ca oxalate crystals are responsive to changes in the soil concentrations of Ca ([52] and references therein). Previous studies indicated that gypsum specialists are Ca-accumulators [12,16,43], a trait that was interpreted as suggestive of the accumulation of mineral gypsum in this type of plants [14]. However, our results indicate that the accumulation of Ca in wide gypsophiles cannot solely be explained by the presence of mineral gypsum. Wide gypsophiles seem to show a further excess of Ca that is accumulated in the form of calcium oxalate. Interestingly, this is a common trait to most narrow gypsum endemisms, which have been suggested to be less specialized to live on gypsum than wide gypsophiles [16]. We therefore suggest that, given the widespread ability to form calcium oxalate crystals in different plants, the liberation of excess Ca in calcium oxalate could be a relatively easy to implement mechanism, as part of the process of adaptation to living on gypsum, of narrow gypsum endemisms.

Iberian gypsovags seem to show physiological mechanisms to block calcium and sulphate uptake by roots. This does not seem to be the case in all gypsovags from other regions of the world. For example, Boukhris and Lossaint [14,53] found that some species of Tunisian gypsovags were able to accumulate mineral S in their tissues when growing on gypsum, sometimes reaching similar (and even higher) values than gypsophiles. Similarly, Borer et al. [54] identified several strategies to accumulate and excrete excess Ca among gypsovags growing on gypsum soils in the Chihuahuan Desert, possibly involving the formation of calcium oxalate. Although the presence of gypsum was not evaluated by these authors, this discrepancy indicates results from Iberian gypsovags should be extrapolated with care. An increase in glutathione synthesis in leaves has been reported as a signalling factor to regulate sulphate uptake by roots in plants not adapted to gypsum soils [55]. Such a mechanism could also be operating in Iberian gypsovags and gypsum endemisms, but the sulphate metabolism of gypsum plants remains unexplored. The ability of Iberian gypsovags to grow in and out of gypsum soils could be related to their capacity to regulate the uptake of excess nutrients in the soil (particularly S and Ca). Nevertheless, such "blocking" of soil nutrients may also have negative consequences on their ability to uptake other nutrients, such as N and P, which are inherently poor in gypsum soils [18,19]. This suggestion agrees with the stress tolerant nature of Iberian gypsovags and their reported low nutrient concentrations [16].

Different to previous analyses based only on the elemental composition of gypsum plants [16] where narrow gypsum endemisms were found to be more similar to gypsovags than to wide gypsophiles, our results indicate that local gypsum endemisms share spectral features with the other two groups of gypsum plants that identify them as a separate group. These results have important implications for the understanding of plant adaptation to gypsum substrates as they suggest that narrow gypsum endemisms are not just stress-tolerant plants that find refuge from competition on gypsum without particular specialization to this special substrate. Although, like gypsovags, they do not seem to be able to cope with sulphate accumulation, they share the ability to eliminate excess Ca in calcium oxalate crystals with wide gypsophyles. The narrow gypsum endemisms studied might have recently evolved from stress-tolerant gypsovag species or ecotypes, and could be in the process of specializing to gypsum soils, developing adaptive mechanisms, such as the accumulation of oxalate, to survive on gypsum.

The analysis of FTIR spectra with NN provided an accurate tool to separate the three groups of gypsum plants analyzed. The discrimination ability was decreased for closely related species, such as *T. polium* subsp. *capitatum*, *T. pumilum* or *T. lacaitae*. However, the analysis worked generally well and only 5 out of 44 cases were misclassified. Therefore, this methodology arises as a promising tool for the analysis of plant adaptation to special substrates. Additional future work in this area could focus on determining the chemical composition of plant samples, rather than attempting to discriminate between different functional groups of species. This would provide useful information for nutrition studies and would allow us to determine relationships between soil and plant chemical composition. A NN model that could extract composition information (such as N or S) from FTIR spectra would be more useful than one that provides a categorisation, as it would provide more fundamental information about the samples. To achieve this, it would be necessary to develop calibrations for a wider range of samples and plants from different gypsum areas.

To conclude, FTIR spectroscopy linked to NN analysis is an efficient tool to discriminate plants with different specificity to gypsum substrates. Our results present evidence of the widespread presence of gypsum and calcium oxalate crystals in most gypsum specialist plants studied, although other mechanisms such as the accumulation of sulphates in organic molecules are also compatible with plant specialization to gypsum. While gypsovags seem to be stress tolerant plants that tightly regulate the uptake of S and Ca by their roots, narrow gypsum endemisms share the ability to accumulate excess Ca as oxalate with gypsophiles, possibly indicating their incipient specialization to live on gypsum. Further studies should focus on evaluating the generality of these conclusions and their extrapolation to gypsum plants from other regions of the world.

Supporting Information

Figure S1 Comparison of FTIR spectra of *Ononis tridentata* from different sites: a) red line, Chinchón (Madrid), b) blue line, Villamayor (Zaragoza).

Figure S2 Normalisation of FTIR spectra. Top – raw spectrum, bottom – after moving average preprocessing and normalisation to within range [-1, 1].

Table S1 Average chemical composition of study species.

Acknowledgments

The authors thank M.J. Albert, R. Milla, P. Fustero and A. Lamana for help with plant sampling and processing, to R. Hill for running the FTIR spectrometer, to E. Pérez-Fernández for assistance with FTIR figures and to J. F. Mota and an anonymous Referee for helpful comments on earlier versions of the manuscript.

Author Contributions

Conceived and designed the experiments: SP AE GM-M. Performed the experiments: SP AE GM-M MM. Analyzed the data: MA AHJR. Contributed reagents/materials/analysis tools: SP MA AE GM-M MM AHJR. Contributed to the writing of the manuscript: SP MA AE GM-M AHJR.

References

1. Mason H (1946) The edaphic factor in narrow endemism. The nature of environmental influences. Madroño 8: 209–226.
2. Raven PH (1964) Catastrophic selection and edaphic endemism. Evolution 18: 336–338.
3. Eswaran H, Gong ZT (1991) Properties, genesis, classification, and distribution of soils with gypsum. In: Nettleton WD, editor. Occurrence, characteristics, and genesis of carbonate, gypsum, and silica accumulations in soils.Madison: Soil Science Society of America. pp. 89–119.
4. Verheye WH, Boyadgiev TG (1997) Evaluating the land use potential of gypsiferous soils from field pedogenic characteristics. Soil Use and Management 13: 97–103.
5. Escudero A, Palacio S, Maestre FT, Luzuriaga AL (2014) Plant life on gypsum: an overview of its multiple facets. Biological Reviews (in press) DOI: 10.1111/brv.12092
6. Parsons RF (1976) Gypsophily in plants. A review. American Midland Naturalist 96: 1–20.
7. Guerrero Campo J, Alberto F, Hodgson J, García Ruiz JM, Montserrat Martí G (1999) Plant community patterns in a gypsum area of NE Spain. I- Interactions with topographic factors and soil erosion. Journal of Arid Environments 41: 401–410.
8. Guerrero Campo J, Alberto F, Maestro Martínez M, Hodgson J, Montserrat Martí G (1999) Plant community patterns in a gypsum area of NE Spain. II.-Effects of ion washing on topographic distribution of vegetation. Journal of Arid Environments 41: 411–419.
9. Meyer SE, García-Moya E (1989) Plant community patterns and soil moisture regime in gypsum grasslands of north central Mexico. Journal of Arid Environments 16: 147–155.
10. Meyer SE, García-Moya E, Lagunes-Espinoza LC (1992) Topographic and soil surface effects on gypsophile plant community patterns in central Mexico. Journal of Vegetation Science 3: 429–438.
11. Meyer SE (1986) The ecology of gypsophile endemism in the Eastern Mojave desert. Ecology 67: 1303–1313.
12. Duvigneaud P, Denaeyer-De Smet S (1968) Essai de classification chimique (éléments minéraux) des plantes gypsicoles du bassin de l'Ebre. Bulletin de la Société Royale de Botanique de Belgique 101: 279–291.
13. Duvigneaud P, Denaeyer-De Smet S (1966) Accumulation du soufre dans quelques espèces gypsophiles d'Espagne. Bulletin de la Société Royale de Botanique de Belgique 99: 263–269.
14. Boukhris M, Lossaint P (1970) Sur la teneur en soufre de quelques plantes gypsophiles de Tunisie. (Note préliminaire). Acta Oecologica 5: 345–354.
15. Romao RL, Escudero A (2005) Gypsum physical soil crusts and the existence of gypsophytes in semi-arid central Spain. Plant Ecology 181: 127–137.
16. Palacio S, Escudero A, Montserrat-Marti G, Maestro M, Milla R, et al. (2007) Plants living on gypsum: beyond the specialist model. Annals of Botany 99: 333–343.
17. Gankin R, Major J (1964) *Arctostaphylos myrtifolia*, its biology and relationship to the problem of endemism. Ecology 45: 792–808.
18. Delgado-Baquerizo M, Covelo F, Gallardo A (2011) Dissolved organic nitrogen in Mediterranean ecosystems. Pedosphere 21: 309–318.
19. Salmerón-Sánchez E, Martínez-Nieto MI, Martínez-Hernández F, Garrido-Becerra JA, Mendoza-Fernández AJ, et al. (2014) Ecology, genetic diversity and phylogeography of the Iberian endemic plant *Jurinea pinnata* (Lag.) DC. (Compositae) on two special edaphic substrates: Dolomite and gypsum. Plant and Soil 374: 233–250.
20. Berthomieu C, Hienerwadel R (2009) Fourier transform infrared (FTIR) spectroscopy. Photosynthesis Research 101: 157–170.
21. Smidt E, Böhm K, Schwanninger M (2011) The Application of FT-IR Spectroscopy in Waste Management. In: Nikolic G, editor. Fourier Transforms - New Analytical Approaches and FTIR Strategies. Rijeka, Croatia: InTech.pp. 405–430.
22. Euring D, Löfke C, Teichmann T, Polle A (2012) Nitrogen fertilization has differential effects on N allocation and lignin in two *Populus* species with contrasting ecology. Trees - Structure and Function 26: 1933–1942.
23. Jones R, Causton D, Emmett B, Mur L, Gwynn-Jones D (2012) Latent soil effects of grazing and ammonium deposition on *Deschampsia flexuosa* tillers inserted and grown in heather moorland soil. Environmental and Experimental Botany 81: 72–78.
24. Pitre FE, Lafarguette F, Boyle B, Pavy N, Caron S, et al. (2010) High nitrogen fertilization and stem leaning have overlapping effects on wood formation in poplar but invoke largely distinct molecular pathways. Tree Physiology 30: 1273–1289.

25. McNear Jr DH, Chaney RL, Sparks DL (2010) The hyperaccumulator *Alyssum murale* uses complexation with nitrogen and oxygen donor ligands for Ni transport and storage. Phytochemistry 71: 188–200.
26. Wei ZL, Dong L, Tian ZH (2009) Fourier transform infrared spectometry study on early stage of cadmium stress in clover leaves. Pakistan Journal of Botany 41: 1743–1750.
27. Xue S, Zhu F, Ye S, Wang J, Wu X (2011) Physiological response of *Mirabilis jalapa* Linn. to lead stress by FTIR spectroscopy. Shengtai Xuebao/Acta Ecologica Sinica 31: 6143–6148.
28. Dziuba B (2011) Identification of selected *Leuconostoc* species with the use of FTIR spectroscopy and artificial neural networks. Acta Scientiarum Polonorum, Technologia Alimentaria 10: 275–285.
29. Hong Q, Yu R, Zheng R, Wang H, Cheng C (2011) Identification between *Fimbristylis miliacea* (L.) Vahl and *Fimbristhlis stauntonii* Debeaux et Franch. by CWT-FTIR-RBFNN. Chinese Journal of Chemistry 29: 1024–1030.
30. Dharmaraj S, Gam LH, Sulaiman SF, Mansor SM, Ismail Z (2011) The application of pattern recognition techniques in metabolite fingerprinting of six different *Phyllanthus* spp. Spectroscopy 26: 69–78.
31. Braun-Blanquet J, Bolòs O (1957) Les groupements végétaux du Bassin Moyen de l'Ebre et leur dynamisme. Anales de la Estación Experimental de Aula Dei 5: 1–266.
32. Devesa JA, Ortega A (2004) Especies vegetales protegidas en España: plantas vasculares. Madrid: Ministerio de Medio ambiente.
33. Allen SE, editor (1989) Chemical analysis of ecological materials. Oxford: Blackwell Scientific Publ.
34. Becker (1961) Análisis y valoración de piensos y forrajes. Zaragoza, Spain: Ed. Acribia.
35. Aitkenhead MJ, Coull M, Towers W, Hudson G, Black HIJ (2013) Prediction of soil characteristics and colour using data from the National Soils Inventory of Scotland. Geoderma 200–201: 99–107.
36. Mouazen AM, Kuang B, De Baerdemaeker J, Ramon H (2010) Comparison among principal component, partial least squares and back propagation neural network analyses for accuracy of measurement of selected soil properties with visible and near infrared spectroscopy. Geoderma 158: 23–31.
37. Aitkenhead MJ, McDonald AJS, Dawson JJ, Couper G, Smart RP, et al. (2003) A novel method for training neural networks for time-series prediction in environmental systems. Ecological Modelling 162: 87–95.
38. Artz RRE, Chapman SJ, Jean Robertson AH, Potts JM, Laggoun-Défarge F, et al. (2008) FTIR spectroscopy can be used as a screening tool for organic matter quality in regenerating cutover peatlands. Soil Biology and Biochemistry 40: 515–527.
39. Farmer V (1974) The infrared spectra of minerals. London, UK: Mineralogical Society.
40. Grigore MN, Toma C, Zamfirache MM, Boscaiu M (2011) Anatomical considerations on Spanish gypsophytes. Where is their place within plant ecology? Scientific Annals of Alexandru Ioan Cuza University of Iasi New Series, Section 2 Plant Biology 57: 31–38.
41. Mota JF, Sánchez-Gómez P, Guirado JS, editors (2011) Diversidad vegetal de las yeseras ibéricas. El reto de los archipiélagos edáficos para la biología de la conservación. Almería: ADIF- Mediterráneo Asesores Consultores. 636 pp.
42. Denayer-De Smet S (1970) Note sur la composition chimique des sels recrétés par diverses espèces gypsohalophytes d'Espagne. Bulletin de la Société Royale de Botanique Belge 103: 273–278.
43. Merlo ME, Mota JF, Cabello J, Alemán MM (1998) La gipsofilia en plantas: un apasionante edafismo. Investigación y Gestión 3: 103–112.
44. FAO, editor (1990) Management of Gypsiferous Soils. Rome, Italy: FAO.
45. Tcherkez G, Tea I (2013) $^{32}S/^{34}S$ isotope fractionation in plant sulphur metabolism. New Phytologist 200: 44–53.
46. Ruiz JM, López-Cantarero I, Rivero RM, Romero L (2003) Sulphur phytoaccumulation in plant species characteristic of gypsipherous soils. International Journal of Phytoremediation 5: 203–210.
47. He H, Veneklaas EJ, Kuo J, Lambers H (2014) Physiological and ecological significance of biomineralization in plants. Trends in Plant Science 19: 166–174.
48. He HH, Bleby TM, Veneklaas EJ, Lambers H, Kuo J (2012) Morphologies and elemental compositions of calcium crystals in phyllodes and branchlets of *Acacia robeorum* (Leguminosae: Mimosoideae). Annals of Botany 109: 887–896.
49. Mota JF, Sola AJ, Dana ED (2003) Plant succession in abandoned gypsum quarries in SE Spain. Phytocoenologia 33: 13–28.
50. Alvarado JJ, Ruiz JM, López-Cantarero I, Molero J, Romero L (2000) Nitrogen metabolism in five plant species characteristic of gypsiferous soils. Journal of Plant Physiology 156: 612–616.
51. McNair JB (1932) The intersection between substances in plants: essential oils and resins, cyanogen and oxalate. American Journal of Botany 19: 255–271.
52. Nakata PA (2003) Advances in our understanding of calcium oxalate crystal formation and function in plants. Plant Science 164: 901–909.
53. Boukhris M, Lossaint P (1975) Aspects écologiques de la nutrition minérale des plantes gypsicoles de Tunisie. Revue d'Ecologie et de Biologie du Sol 12: 329–348.
54. Borer CH, Hamby MN, Hutchinson LH (2012) Plant tolerance of a high calcium environment via foliar partitioning and sequestration. Journal of Arid Environments 85: 128–131.
55. Herschbach C, Rennenberg H (1994) Influence of glutathione (GSH) on net uptake of sulphate and sulphate transport in tobacco plants. Journal of Experimental Botany 45: 1069–1076.
56. Jones F (2012) Infrared investigation of barite and gypsum crystallization: Evidence for an amorphous to crystalline transition. CrystEngComm 14: 8374–8381.
57. Ito K, Bernstein HJ (1956) The vibrational spectra of the formate, acetate and oxalate ions. Canadian Journal of Chemistry 34: 170–178.
58. Wilson MJ (1994) Clay mineralogy: spectroscopic and chemical determinative methods. London: Chapman &Hall.

14

Simulated Nitrogen Deposition Reduces CH_4 Uptake and Increases N_2O Emission from a Subtropical Plantation Forest Soil in Southern China

Yongsheng Wang[1,2], Shulan Cheng[2], Huajun Fang[1]*, Guirui Yu[1], Minjie Xu[2], Xusheng Dang[1,2], Linsen Li[1], Lei Wang[1,2]

1 Key Laboratory of Ecosystem Network Observation and Modeling, Institute of Geographical Sciences and Natural Resources Research, Chinese Academy of Sciences, Beijing, China, **2** University of Chinese Academy of Sciences, Beijing, China

Abstract

To date, few studies are conducted to quantify the effects of reduced ammonium (NH_4^+) and oxidized nitrate (NO_3^-) on soil CH_4 uptake and N_2O emission in the subtropical forests. In this study, NH_4Cl and $NaNO_3$ fertilizers were applied at three rates: 0, 40 and 120 kg N ha^{-1} yr^{-1}. Soil CH_4 and N_2O fluxes were determined twice a week using the static chamber technique and gas chromatography. Soil temperature and moisture were simultaneously measured. Soil dissolved N concentration in 0–20 cm depth was measured weekly to examine the regulation to soil CH_4 and N_2O fluxes. Our results showed that one year of N addition did not affect soil temperature, soil moisture, soil total dissolved N (TDN) and NH_4^+-N concentrations, but high levels of applied NH_4Cl and $NaNO_3$ fertilizers significantly increased soil NO_3^--N concentration by 124% and 157%, respectively. Nitrogen addition tended to inhibit soil CH_4 uptake, but significantly promoted soil N_2O emission by 403% to 762%. Furthermore, NH_4^+-N fertilizer application had a stronger inhibition to soil CH_4 uptake and a stronger promotion to soil N_2O emission than NO_3^--N application. Also, both soil CH_4 and N_2O fluxes were driven by soil temperature and moisture, but soil inorganic N availability was a key integrator of soil CH_4 uptake and N_2O emission. These results suggest that the subtropical plantation soil sensitively responses to atmospheric N deposition, and inorganic N rather than organic N is the regulator to soil CH_4 uptake and N_2O emission.

Editor: Dafeng Hui, Tennessee State University, United States of America

Funding: This project was supported by National Natural Science Foundation of China (No. 31290222, 31130009, 31290221, 31070435, and 41071166), Development Plan Program of State Key Basic Research (No. 2012CB41710, 32010CB833502, and 2010CB833501), Bingwei's Funds for Young Talents from Institute of Geographical Sciences and Natural Resources Research, CAS, (No. 2011RC202) and CAS Strategic Priority Program (No. XDA 05050600). The funders had no role in study design, data collection and analysis, decision to publish, or preparation of the manuscript.

Competing Interests: The authors have declared that no competing interests exist.

* E-mail: fanghj@igsnrr.ac.cn

Introduction

Humid tropical biome stores approximately 10% of global soil carbon (C) [1] and plays a vital role in the budget of ecosystem C and nitrogen (N) fluxes. The amount of nitrous oxide (N_2O) emission from the subtropical and tropical forest soils is estimated at 0.9–3.6 T g yr^{-1}, accounting for 14% to 23% of the global N_2O budget [2]. Simultaneously, well-aerated soils in the subtropical and tropical forests potentially function as a significant sink of atmospheric methane (CH_4) during the dry season [3–6]. The uptake of CH_4 from the subtropical and tropical forest soils is estimated to be 6.2 T g yr^{-1}, accounting for 28% of the global CH_4 sink [7]. Although the importance of the subtropical and tropical forest soils as atmospheric CH_4 sink and N_2O source is well understood, few observations can be available in this region [8–10]. Moreover, low-frequency measurement of gas fluxes in the few studies is unable to accurately estimate the annual amount of soil CH_4 uptake and N_2O emission, which leads to a high uncertainty in the budget of global soil CH_4 and N_2O fluxes.

Chronic N deposition input into terrestrial ecosystems alters plant physiology and soil microbial community, thereby changes the soil biogenic CH_4 and N_2O fluxes [11–13]. Based on a meta-analysis of N addition experimental data in globe, Liu and Greaver [14] concluded that N addition reduced CH_4 uptake by 38% and increased N_2O emission by 216%. In general, chronic N deposition will increase $NH_4^+$and NO_3^- availability in the forest ecosystems, thereby affects CH_4 uptake from the forest soils through changing the activity and composition of methanotrophic community [15–17]. Soil NH_4^+ accumulation can decrease, increase or have no effects on soil CH_4 uptake in the forest ecosystems, depending on forest types [5], duration of N application [18], and N fertilizer types and doses [19]. Three potential mechanisms have been proposed to clarify the inhibition of NH_4^+ accumulation to soil CH_4 uptake: (1) the competition of soil NH_3 to use CH_4 monooxygense with soil CH_4 [20], (2) the toxic inhibition of hydroxylamine and nitrite produced during soil NH_4^+ oxidation [21], and (3) the indirect effects of applied N and associated salt ions through osmotic stress [22]. On the contrary, elevated soil NH_4^+ availability can increase soil CH_4 uptake, which is related to the increase in the quantity of soil ammonia-oxidizing microorganisms [23]. Soil NO_3^- accumulation can also decrease or increase soil CH_4 uptake [18,24]. Osmotic stress caused by NO_3^--N fertilizer-associated salts [22] and anaerobic

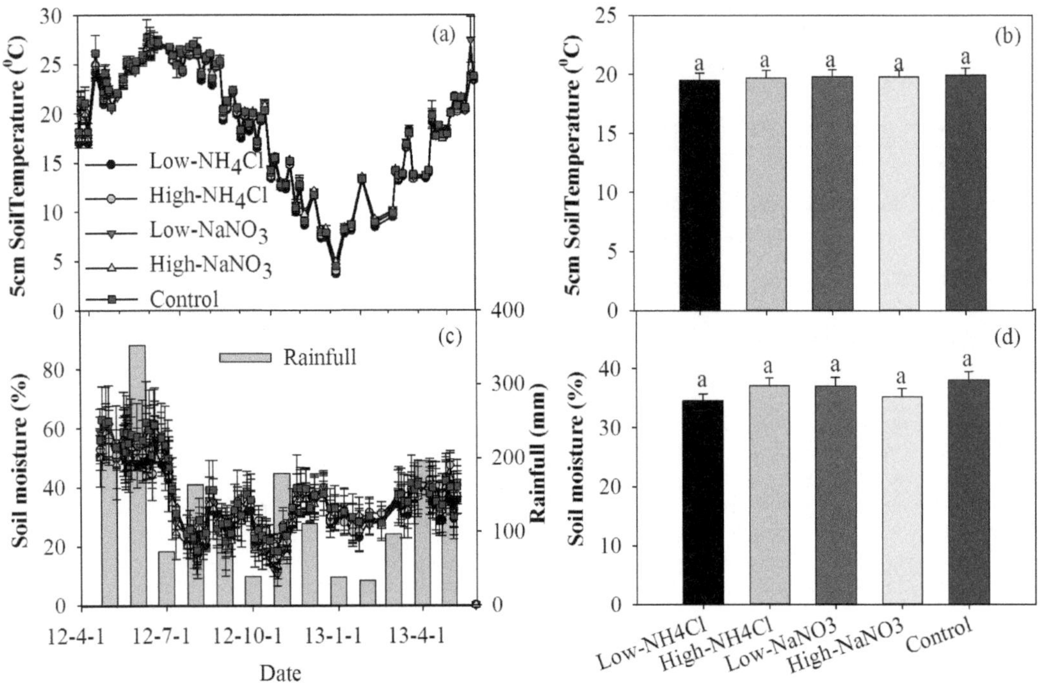

Figure 1. Temporal variations of 5 cm soil temperature and 10 cm soil moisture and their responses to N addition. Different letters above the columns mean significant differences between experimental treatments.

'microsites' produced by NO_3^- reduction [25] are toxic to CH_4-oxidizing bacteria. The mechanism responsible for the promotion of NO_3^- addition to soil CH_4 uptake is still unclear and need a number of experimental evidences to support [26]. A positive relationship between the amount of N addition and N_2O fluxes from the subtropical forest soils is mainly attributed to the promotion of soil nitrification or/and denitrification rates caused by increased N availability [27,28]. Some studies reported that denitrification was the main source of N_2O emission from the subtropical forest soils [29–31], whereas other studies claimed that nitrification dominated soil N_2O fluxes [32]. To date, single N fertilizer (i.e., NH_4NO_3) is widely used to simulate the effects of N deposition in all N manipulative experiments of subtropical forests in China [8,33,34]. The above studies have not evaluated the relative contributions of the deposited N ions (NH_4^+ and NO_3^-) to soil CH_4 uptake and N_2O emission. Moreover, most of soil CH_4 and N_2O fluxes are measured by low-frequency sampling over the short term, which is difficult to accurately assess the budget of soil CH_4 and N_2O fluxes and leads to great uncertainty.

In China, the plantations cover an area of 6.2×10^7 ha, accounting for 31.8% of China's forest area and ranking first in the world [35]. Approximately 63% of plantations concentrate in the subtropical region of southern China [36]. Meanwhile, the southern China is the most economically developed regions with high population density, and plantations, cities and farmlands are interspersed. Because a number of reactive N originated from fossil fuel combustion and fertilizer use is emitted to atmosphere, the forests in this region are receiving a high level of anthropogenic N deposition, mostly as ammonium [37]. Atmospheric N deposition rate via precipitation in southern China has been reported and ranges from 30 to 73 kg N ha^{-1} yr^{-1} [8]. So far, few studies are conducted to examine the effects of N deposition on CH_4 uptake and N_2O emission from the plantation of this region [33,34].

Humid subtropical forest soils are generally characterized by high N availability and high N turnover [38]. Therefore, we hypothesize that increased NH_4^+ and NO_3^- availability via experimental N deposition will inhibit soil CH_4 uptake and promote N_2O emissions from the subtropical plantation. Furthermore, NH_4^+-N fertilizer addition will decrease CH_4 uptake and increase N_2O emission due to soil NH_4^+-N accumulation. In contrast, the effects of NO_3^--N fertilizer addition on soil CH_4 uptake and N_2O emission depend on the concentration of soil NO_3^--N as well as associated salt ions. Our objectives were (1) to quantify the effects of NH_4^+-N and NO_3^--N fertilizer application on soil CH_4 and N_2O fluxes and soil variables in the subtropical plantation; (2) to examine the relationships between soil CH_4 and N_2O fluxes and the relevant soil properties.

Materials and Methods

Site description

This study was conducted in a subtropical slash pine plantation at the Qianyanzhou Ecological Station (QYZ, 26°44′39″N, 115°03′33″E) in southern China. The station belongs to the Institute of Geographic Sciences and Natural Resources Research, Chinese Academy of Sciences. All necessary permits were obtained for this field study. The field study did not involve endangered or protected species. According to local climate records from 1989 to 2008, mean temperature of QYZ site varies between 17 and 19°C. Mean annual precipitation ranges from 945 to 2145 mm, of which 24%, 41%, 23% and 12% occurs in four quarters in turn. The rainfall scarcity and high temperature in late summer often result in seasonal drought [39]. The exotic slash pine plantation was established in 1985. Mean tree height, diameter at breast height, stand basal area, and leaf area index were 12.0 m, 15.8 cm, 35 m^2 ha^{-1}, and 4.5 m^2 m^{-2}, respectively [40]. The main understory and midstory species are *Woodwardia japonica* (L.f.) Sm., *Dicranopteris dichotoma* (Thunb) Bernh, *Loropetalum*

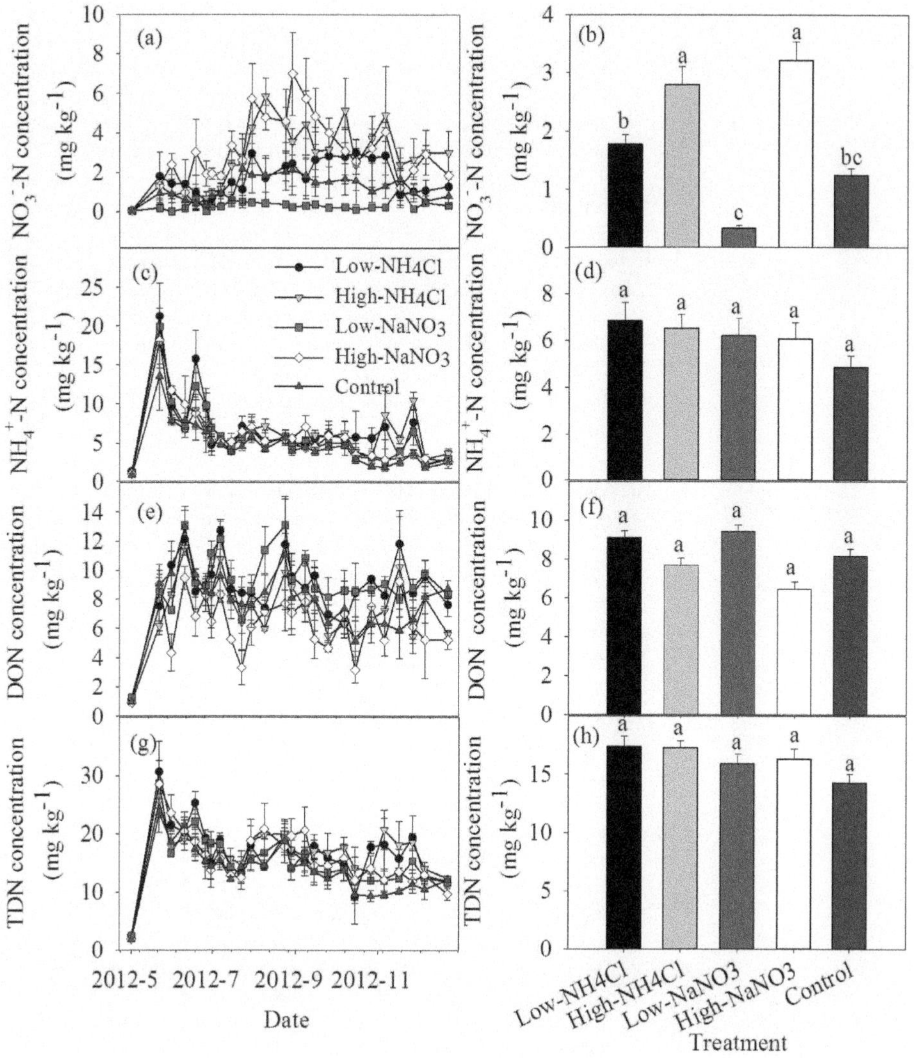

Figure 2. Temporal variations of soil NO_3^--N, NH_4^+-N, DON, and TDN concentrations and their responses to N addition. Different letters above the columns mean significant differences between experimental treatments.

chinense (R.Br.) Oliv, and *Quercus fabric* Hance. The red soil is weathered from red sand rock, and soil texture is divided into 2.0-0.05 mm (17%), 0.05-0.002 mm (68%), and <0.002 mm (15%) [39].

Experimental design

The N addition experiment is a randomized block design. In May 1, 2012, two N fertilizers (NH_4Cl and $NaNO_3$) were used to simulate the effects of deposited NH_4^+ and NO_3^- on ecosystem processes and functions. According to the level of atmospheric N deposition at the QYZ site, two levels referred to as low N (40 kg N ha^{-1} yr^{-1}) and high N (120 kg N ha^{-1} yr^{-1}) were used to simulate a future increase in the atmospheric N deposition by 1-, and 3-fold. A control treatment was designed at each block to calculate the net effect of N addition. Each N treatment was replicated three times, and a total of 15 plots were included. Each plot with 20 m×20 m was divided into four subplots with 5 m×5 m, and the plots were separated by 10 m wide buffer strips. Three subplots were used to measure soil CH_4 and N_2O fluxes, and the other one was used to investigate aboveground biomass and diversity. N fertilizer solutions were sprayed on the plots once a month in 12 equal applications over the entire year, and the control plots received equivalent deionized water only.

Measurement of soil CH_4 and N_2O fluxes

Flux measurements of soil CH_4 and N_2O fluxes were performed by using a static opaque chamber and gas chromatography method [41]. The static chambers were made of stainless steel and consisted of two parts: a square base frame (length×width×height = 50 cm×50 cm×10 cm) and a removable top (length×width×height = 50 cm×50 cm×15 cm). The installed equipments on the static chambers were detailed by Fang et al. [42]. The frames were inserted directly into the soil to a depth of 10 cm and remained intact during the entire observation period. To assess the spatial heterogeneity of soil C and N fluxes, a pre-experiment was conducted to examine the difference of CH_4 and N_2O fluxes among the three subplots of each plot before N addition. No significant difference of CH_4 and N_2O fluxes among the three subplots was found during the observation, suggesting a negligible effect of soil heterogeneity. Considering the practical reasons such as high labor intensity, we collected gas samples through changing the subplots within a month. The soil CH_4 and

Table 1. Results of repeated-measures ANOVAs on the effects of experimental treatment, month and their interaction on soil dissolved N concentrations.

Source of variation	Soil NO_3^--N		Soil NH_4^+-N		Soil DON		Soil TDN	
	F	p	F	p	F	p	F	p
Month	11.49	0.016	38.30	<0.001	16.71	<0.001	37.97	0.002
Treatment	4.40	0.026	1.62	0.24	1.81	0.20	1.51	0.27
Month×Treatment	1.31	0.24	1.41	0.13	1.33	0.17	0.75	0.77

N_2O fluxes were measured twice a week and conducted between 9:00 and 11:00 am (China Standard Time, CST). Five gas samples were taken using 100 ml plastic syringes at intervals of 0, 10, 20, 30, and 40 min after closing the chambers. CH_4 and N_2O concentrations of gas samples were analyzed within 24 h with a gas chromatography (Agilent 7890A, USA) equipped with an electron capture detector (ECD) for N_2O analysis and a flame ionization detector (FID) for CH_4 analysis. The high purity N_2 and H_2 were used as carrier gas and fuel gas, respectively. The ECD and FID were heated to 350°C and 200°C, respectively, and the column oven was kept at 55°C. The soil fluxes were calculated based on their rate of concentration change within the chamber, which was estimated as the slope of linear or nonlinear regression between concentration and time [41]. All the coefficients of determination (r^2) of the regression were greater than 0.90 in our study.

Measurements of soil temperature and moisture

Simultaneously, soil temperature at 5 cm (T_s) and soil moisture at 10 cm below soil surface (SM) were monitored at each chamber. Soil temperature was measured using a portable temperature probes (JM624 digital thermometer, Living-Jinming Ltd., China). Volumetric soil moisture ($m^3\ m^{-3}$) was measured using a moisture probe meter (TDR100, Spectrum, USA).

Soil sampling and mineral N analysis

During the measurement of soil CH_4 and N_2O fluxes, soil samples were collected weekly nearby the static chambers from a depth of 0–20 cm using an auger (2.5 cm in diameter). Five soils were collected and were pooled to one composite sample for each soil layer at each plot. Soils were immediately passed through a 2 mm sieve to remove roots, gravel and stones. Soil samples were extracted in 1.0 M KCl solution (soil: water = 1:10) and shaken for 1 h. The soil suspension was subsequently filtered through Whatman No. 40 filter papers for NH_4^+-N, NO_3^--N, and total dissolved nitrogen (TDN) determination on a continuous-flow autoanalyzer (Seal AA3, Germany). Dissolved organic nitrogen (DON) concentration was calculated as the difference between TDN and total inorganic nitrogen (NH_4^+-N and NO_3^--N).

Statistical analyses

Repeated measures analysis of variance (AVOVA) with Duncan test was applied to examine the differences of soil temperature, soil moisture, soil dissolved N, and soil CH_4 and N_2O fluxes between control and N addition plots. Experimental treatments were set as factors of between-subjects and measurement date was selected as a variable of within-subjects. Linear and nonlinear regression analyses were used to examine the relationships between soil CH_4 and N_2O fluxes and the measured soil variables in monthly scale. All statistical analyses were conducted using the SPSS software package (version 16.0), and statistical significant differences were set with P values<0.05 unless otherwise stated. All figures were drawn using the Sigmaplot software package (version 10.0).

Results

Soil temperature, moisture and precipitation

During the whole observation period, soil temperature at 5 cm depth fluctuated greatly, which correlated with the weather condition. Soil temperature varied as a single-peak and single-sink curve, i.e. temperature was the highest in early July, gradually reached the lowest value in early January, and then increased (Fig. 1a). There was no significant difference in surface temperature among various treatments (Fig. 1b).

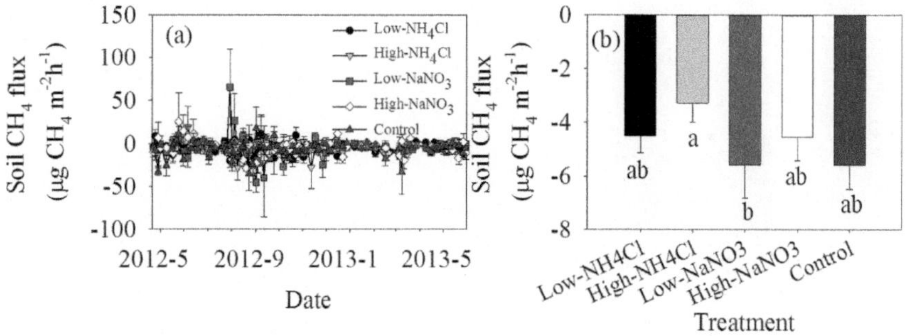

Figure 3. Temporal variations of soil CH_4 fluxes and their responses to N addition. Different letters below the columns mean significant differences between experimental treatments.

Soil moisture at 10 cm depth behaved as significant seasonal variation, with the maximum occurred in May and June and the minimum occurred in August and October (Fig. 1c). The seasonality of soil moisture was well consistent with that of precipitation (Fig. 1c). Similar to surface temperature, no significant difference in soil moisture was found among various treatments (Fig. 1d).

Soil dissolved N concentrations

Soil NO_3^--N concentration showed significant seasonal variation, with the minimum and maximum occurring in May and August (Fig. 2a, Table 1, $P=0.016$). In the control, the concentration of soil NO_3^--N ranged from 0.06 to 2.19 mg kg^{-1}, with an average of 1.25 mg kg^{-1} (Fig. 2a–b). N addition tended to alter soil NO_3^--N concentration, and the difference was significant among five experimental treatments (Table 1, $P=0.026$). Compared with the control, high level of $NaNO_3$ addition tended to increase soil NO_3^--N concentration, while an opposite pattern was found in the low level of $NaNO_3$ addition treatment (Fig. 2b). Furthermore, the promotion of high level of NH_4Cl addition to soil NO_3^--N concentration seemed to be stronger than that of low level of NH_4Cl addition (Fig. 2b).

Soil NH_4^+-N concentration peaked in the middle of May, and then continued to decrease (Fig. 2c). The seasonal variation of soil NH_4^+-N concentration was significant (Table 1, $P<0.001$). In the control, soil NH_4^+-N concentration ranged from 1.91 to 10.80 mg kg^{-1}, with an average of 4.84 mg kg^{-1} (Fig. 2d). Overall, although N addition treatments tended to increase soil NH_4^+-N concentration, the difference between N addition treatments and control was not significant (Fig. 2d, Table 1, $P=0.244$).

Soil DON concentration exhibited a significant seasonal variation (Fig. 2e, Table 1, $P<0.001$), and its seasonality was the same as that of soil NO_3^--N concentration (Fig. 2a and Fig. 2e). In the control, soil DON concentration ranged from 5.30 to 14.11 mg kg^{-1}, with an average of 8.18 mg kg^{-1} (Fig. 2e–f). Low level of N addition tended to increase the concentration of soil DON, while high level of N addition tended to reduce the concentration of soil DON (Fig. 2f). However, N addition did not change soil DON concentration at the level of 0.05 (Table 1, $P=0.203$).

The seasonal variation of soil TDN concentration was consistent with that of soil NH_4^+-N concentration, dramatically decreasing from May to December (Fig. 2c and Fig. 2g). The seasonal variation of soil TDN concentration was significant (Table 1, $P=0.002$). N addition tended to increase soil TDN concentration; moreover, the promotion of NH_4Cl application to soil TDN concentration was slightly higher than that of $NaNO_3$ addition (Fig. 2h). However, the difference of soil TDN concentration among the five experimental treatments was not significant (Table 1, $P=0.273$).

Soil CH_4 and N_2O fluxes

Soil CH_4 fluxes showed a significant seasonal pattern (Table 2, $P=0.008$). We observed both soil CH_4 uptake and emission in the control plots, ranging from −34.9 to 17.9 μg CH_4 m^{-2} h^{-1}, with an average of −5.56 μg CH_4 m^{-2} h^{-1} (Fig. 3a–b). A weak interaction between measurement date and treatment was found (Table 2, $P=0.079$). Significant differences in CH_4 fluxes between the control and N addition treatments were only found in July and September (Fig. 3a). For the same level of N addition, NH_4Cl fertilizer exhibited a higher inhibition to soil CH_4 uptake than $NaNO_3$ fertilizer. However, there was no significant difference in soil CH_4 fluxes between the control and N addition treatments (Fig. 3b).

Soil N_2O fluxes also showed a significant seasonality with the minimum occurring from early October to March next year (Fig. 4 a, Table 2, $P<0.001$). In the control, Soil N_2O flux ranged from −

Table 2. Results of repeated-measures ANOVAs on effects of experimental treatment, month and their interaction on soil CH_4 and N_2O fluxes.

Source of variation	CH_4 flux		N_2O flux	
	F	p	F	p
Month	2.34	0.008	23.83	<0.001
Treatment	0.66	0.064	2.47	0.011
Month×Treatment	1.37	0.079	2.35	<0.001

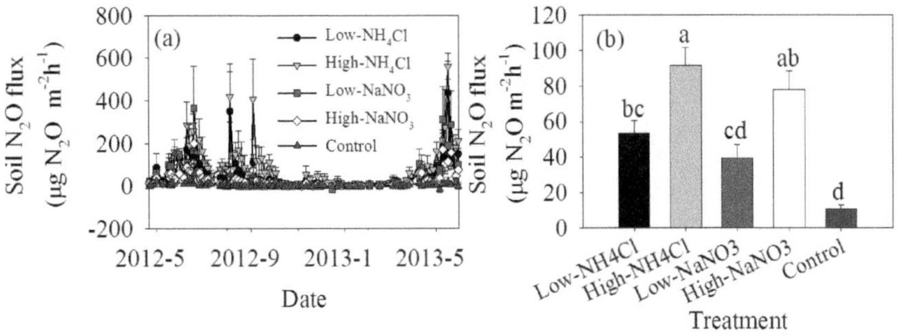

Figure 4. Temporal variations of soil N_2O fluxes and their responses to N addition. Different letters above the columns mean significant differences between experimental treatments.

15.26 to 559.30 μg N_2O $m^{-2}h^{-1}$, with an average of 10.60 μg N_2O $m^{-2}h^{-1}$ (Fig. 4b). Nitrogen addition produced obvious peaks of soil N_2O emission, which was detected within one week after N addition (Fig. 4a). Soil N_2O fluxes positively responded to N addition, and the promotion increased with the levels of N addition (Fig. 4b). In addition, there was a significant interaction between month and N treatment in the entire observation period (Table 2, $P<0.001$). For the same level of N addition, NH_4Cl fertilizer had a higher promotion to soil N_2O emission than $NaNO_3$ fertilizer (Fig. 4b).

Relationships between soil CH_4 and N_2O fluxes and soil properties

Both soil CH_4 and N_2O fluxes were positively correlated with soil temperature at 5 cm depth and soil moisture at 10 cm depth (Fig. 5, Table 3). The relationships between soil CH_4 fluxes and soil temperature and between soil CH_4 fluxes and soil moisture could be well fitted with quadratic and linear equations, respectively (Fig. 5a–b, Table 3). Similarly, soil N_2O fluxes were linearly correlated with soil temperature and soil moisture (Fig. 5c–d, Table 3).

Soil CH_4 fluxes were positively correlated with soil NO_3^--N concentrations, whereas no significant correlations between soil CH_4 fluxes and other dissolved N species were found (Fig. 6a–d, Table 3). Soil N_2O fluxes were linearly correlated with soil NO_3^--N and TDN concentrations (Fig. 6e, Fig. 6h, Table 3), and the relationship between soil N_2O fluxes and soil NH_4^+-N concentrations was well fitted with a logarithm equation (Fig. 6f, Table 3).

Discussion

Effects of N addition on soil CH_4 fluxes

The subtropical plantation soils can act as a sink of atmospheric CH_4. The mean annual soil CH_4 uptake in the control (0.49 kg CH_4 ha^{-1} yr^{-1}) was lower than those of subtropical plantation in Pingxiang (3.84 kg CH_4 ha^{-1} yr^{-1}) and Dinghushan of southern China (1.34 kg CH_4 ha^{-1} yr^{-1}) [5,6] as well as that of subtropical rainforest in Australia (3.13 kg CH_4 ha^{-1} yr^{-1}) [3]. Except low level of $NaNO_3$ treatment, the other three treatments decreased, on average, the rates of soil CH_4 uptake by 18.38% to 41.04% relative to control (Fig. 3). The decrease in soil CH_4 uptake caused by N addition in our site was higher than those of plantations in Dinghushan [5] and Heshan stations of southern China [8], despite the levels of N addition are similar in the three forest sites (120 vs. 150 kg ha^{-1} yr^{-1}). This result indicated that the response of soil CH_4 uptake to N addition was more sensitive in the northern subtropical plantations than in the southern subtropical

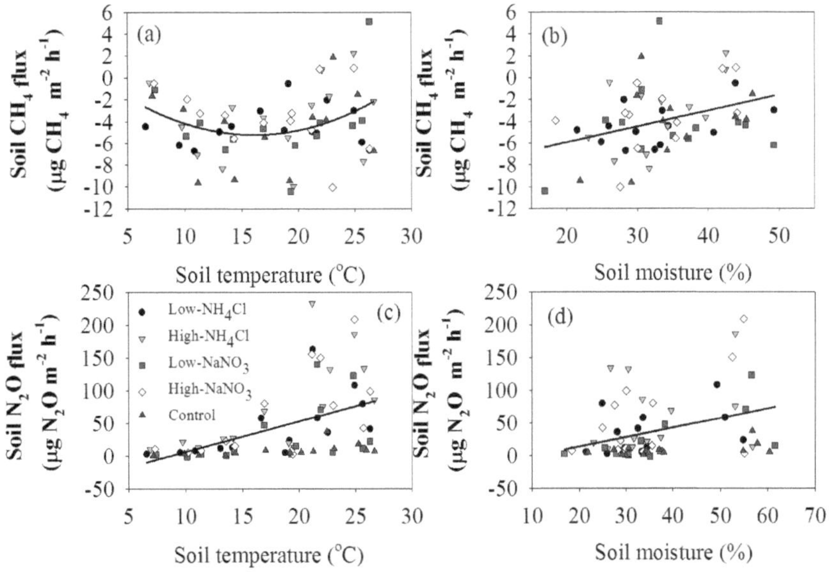

Figure 5. Relationships between soil CH_4 and N_2O fluxes, 5 cm soil temperature, and 10 cm soil moisture (n = 70).

Table 3. Regression models between soil CH_4 and N_2O fluxes and soil properties.

Flux	Soil variables[a]	Regression equation	R^2	*P* value
CH_4	T_s	$y = 0.028\ T_S^2 - 0.90\ T_S + 2.02$	0.094	0.044
	M_S	$y = 0.15\ M_S - 8.84$	0.123	0.006
	NO_3^-	$y = -1.28\ NO_3^- - 1.77$	0.21	0.004
	Combined	$y = 0.17\ M_S - 10.66$	0.21	0.003
N_2O	T_S	$y = 4.67\ T_S - 40.11$	0.25	<0.0001
	M_S	$y = 1.99\ M_S - 29.76$	0.10	0.001
	NO_3^-	$y = 12.29\ NO_3^- + 10.58$	0.22	0.005
	NH_4^+	$y = 81.52\ \ln(NH_4^+) - 85.34$	0.37	<0.0001
	TDN	$y = 9.26\ TDN - 93.50$	0.17	0.008
	Combined	$y = 0.01\ NH_4^+ + 0.013\ NO_3^- - 0.041$	0.50	<0.0001

[a]: T_S is soil temperature at 5 cm depth, M_S is soil moisture at 10 cm depth, NH_4^+, NO_3^-, and TDN are the concentrations of soil NH_4^+, NO_3^-, and TDN at 20 cm depth.

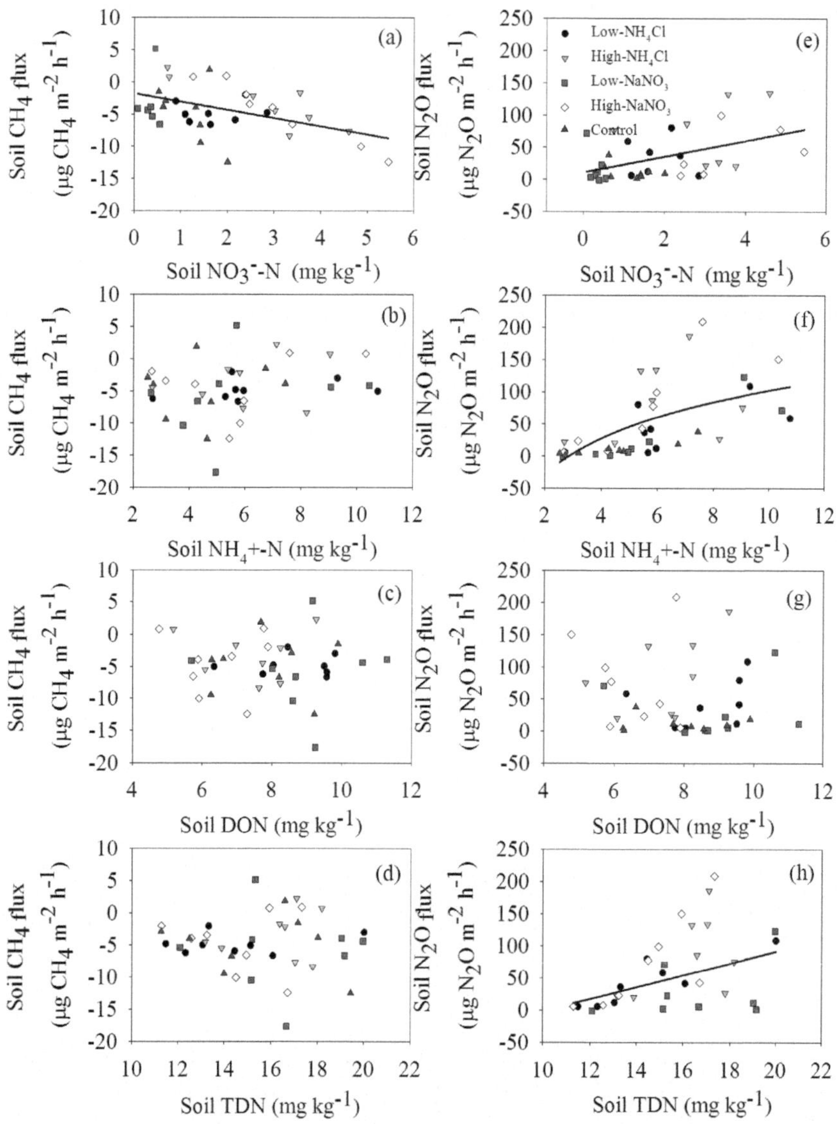

Figure 6. Relationships between soil CH_4 and N_2O fluxes and soil dissolved N concentrations (n = 40).

plantations. This could be attributed to the lower soil N availability, lower atmospheric deposition rate, and the shorter duration of N application in QYZ, compared with the southern subtropical plantations [5,8]. Furthermore, the subsurface mineral soils generally have higher capacity of oxidizing CH_4 than surface litter layer [22,24]. In our site, exogenous N input would directly affect the soil methanotrophic community as well as the amount of CH_4 oxidation due to the lacking of litter layer.

Generally, atmospheric N deposition increases NH_4^+ accumulation and thereby inhibits CH_4 uptake in the well-drained forest soils [8,12,43], despite contrasting results such as promotion and no effect have also been documented [44,45]. In this study, we found that various levels and forms of N addition did not significantly change soil CH_4 uptake over one year (Fig. 3). This could be related to the following third aspects. First, the short-term N fertilizers application did not significantly lead to soil NH_4^+-N accumulation (Fig. 2b), and no significant relationship between soil CH_4 fluxes and soil NH_4^+-N concentrations was found (Fig. 6b). Whalen and Reeburgh [46] also concluded that N inputs did not influence CH_4 uptake until they significantly increased soil NH_4^+ availability in the boreal forest soils. Although an inhibitory trend of soil CH_4 uptake under the NH_4^+-N addition treatments was found, the competition and toxic inhibition of accumulated NH_4^+ did not occur over the short term. Second, N addition enhances the availability of NH_4^+ to soil nitrifiers, which will accordingly decrease the extent to which CH_4 consumers are exposed to NH_4^+ [20]. A slight accumulation in soil NO_3^--N concentration under NH_4^+-N fertilizer treatments indirectly supported our deduction (Fig. 3a). Third, we also found that soil NO_3^--N accumulation could significantly promote soil CH_4 uptake (Fig. 6a), which had been documented in the subtropical plantations of southern China [6]. Especially, the low level of $NaNO_3$ treatment tented to reduce soil NO_3^--N concentration, and thereby it slightly stimulated soil CH_4 uptake (Fig. 2b, Fig. 3b). Moreover, stronger relationships were found between soil CH_4 fluxes and soil NO_3^--N concentrations than between soil CH_4 fluxes and other soil dissolved N concentrations (Fig. 6), suggesting that soil NO_3^- played a more important role in soil CH_4 uptake than other soil dissolved N species in the subtropical plantation.

Soil CH_4 flux is controlled by methanogens operating at anaerobic conditions and methanotrophs taking oxygen as a terminal electron acceptor [47]. Activities and population sizes of these microbial communities depend on a series of soil factors, including soil temperature, moisture, pH, substrate availability, and aeration of soil profile [19,48,49]. Soil CH_4 uptake is dominated by an optimal soil temperature [50]. In our study, the optimal soil temperature was about 15°C (Table 3), and the capacity of soil methanotrophs to oxidize CH_4 would decline when soil temperature was lower or higher than the threshold [51]. Also, soil moisture controls the mass flow of air and diffusion of atmospheric CH_4 into the soil by altering the water filled pore space (WFPS) of soils [52]. We also found that soil CH_4 fluxes under the N addition and control treatments were significantly related to soil moisture (Table 3). Based on the result of stepwise regression analysis, we found that the variation in soil CH_4 uptake was less affected by soil moisture (Table 3). Because N addition did not change soil moisture (Fig. 1), we reasonably deduced that the variation of CH_4 uptake elicited by N treatments was mainly attributed to the change in soil N availability.

Effects of N addition on soil N_2O fluxes

Our result showed that the subtropical slash pine plantation in QYZ exhibited a source of atmospheric N_2O under natural conditions. The average soil N_2O flux in the control (0.93 kg N_2O ha^{-1} yr^{-1}) was comparable with that of Heshengqiao station in Hubei province (0.71 kg N_2O ha^{-1} yr^{-1}) [53], but lower than that of Dinghushan station in South China (2.11 kg N_2O ha^{-1} yr^{-1}) [33]. In our study, NH_4Cl and $NaNO_3$ addition at rates of 40 and 120 kg N ha^{-1} yr^{-1} increased soil N_2O emission by 403% to 762%. On the contrary, in the pine, mixed and evergreen broadleaved forests of Dinghushan station, NH_4NO_3 addition at rates of 50, 100 and 150 kg N ha^{-1} yr^{-1} only increased soil N_2O fluxes by 38% to 58% [33]. These results indicated that the subtropical plantation had high turnover rates of soil N and sensitively responded to increased N deposition. The potential reasons include that the optimal hydrothermal conditions [54], low soil pH [55], and high clay content [39], which favor both soil nitrification and denitrification as well as soil N_2O emission.

Except soil DON concentration, soil N_2O fluxes were significantly correlated with concentrations of soil NH_4^+, NO_3^-, and TDN (Fig. 6), suggesting soil N_2O flux was dominated by both soil nitrification and denitrification processes. Furthermore, the promotion of $NaNO_3$ addition to N_2O emission was slightly lower than that of NH_4Cl addition (Fig. 4). Two potential mechanisms can be responsible for this phenomenon: (1) the high rates of NO_3^- immobilization and nitrification [38], and the low denitrification potential are found in the same type of subtropical plantations [56]; and (2) temperature regulates soil N_2O flux through influencing soil N_2O-producing microorganisms, such as nitrifers and denitrfiers [57]. Soil moisture effects on soil N_2O fluxes are a result of the limitation of O_2 diffusion into the soil and the expansion of soil anaerobic microbial community [58]. The relatively high temperature in wet season was benefit for soil nitrifers and denitrfiers activities, which partly explained the seasonal variation of soil N_2O fluxes with maximum occurring in between May and June (Fig. 4a). Because N addition did not change soil temperature and soil moisture (Fig. 1), the changes in soil N_2O emission under N addition treatments were unlikely to be caused by the changes in soil temperature and soil moisture. Therefore, soil NH_4^+-N and NO_3^--N concentrations were the dominant factors controlling soil N_2O emission in our study, and could explain 49.9% of the temporal variability of soil N_2O fluxes (Table 3).

Conclusions

This study emphasizes the contrasting effects of oxidized NO_3^- and reduced NH_4^+ inputs on the fluxes of CH_4 uptake and N_2O emission from a subtropical plantation soil based on high frequency observations. We found that N addition tended to inhibit soil CH_4 uptake, and dramatically promoted soil N_2O emission. Compared with NO_3^--N fertilizer application, NH_4^+-N fertilizer application had a stronger inhibition to soil CH_4 uptake and a stronger promotion to soil N_2O emission. Also, both soil CH_4 and N_2O fluxes were driven by soil moisture and temperature, but soil inorganic N availability was a key integrator of soil CH_4 uptake and N_2O emission. Overall, short-term N addition has already changed soil CH_4 and N_2O fluxes, which indicated that the subtropical plantation soil was sensitive to N deposition input. In the future, the long-term observation of soil fluxes and the measurement of key microbial functional groups are necessary to clarify the mechanisms responsible for the coupling between soil CH4 and N2O fluxes.

Author Contributions

Conceived and designed the experiments: YW HF SC GY. Performed the experiments: YW MX XD LL LW. Analyzed the data: YW HF SC. Wrote the paper: YW HF SC.

References

1. Post WM, Emanuel WR, Zinke PJ, Stangenberger AG (1982) Soil carbon pools and world life zones. Nature 298: 156–159.
2. IPCC (2001) Climate change 2001: the scientific basis. Contribution of working group I of intergovernmental panel on climate change. Cambridge University Press, Cambridge.
3. Rowlings DW, Grace PR, Kiese R, Weier KL (2012) Environmental factors controlling temporal and spatial variability in the soil-atmosphere exchange of CO_2, CH_4 and N_2O from an Australian subtropical rainforest. Global Change Biology 18: 726–738.
4. Tang XL, Liu SG, Zhou GY, Zhang DQ, Zhou CY (2006) Soil-atmospheric exchange of CO_2, CH_4, and N_2O in three subtropical forest ecosystems in southern China. Global Change Biology 12: 546–560.
5. Zhang W, Mo JM, Zhou GY, Gundersen P, Fang YT, et al. (2008) Methane uptake responses to nitrogen deposition in three tropical forests in southern China. Journal of Geophysical Research-Atmospheres 113, D11116. doi:10.1029/2007JD009195.
6. Wang H, Liu SR, Wang JX, Shi ZM, Lu LH, et al. (2013) Effects of tree species mixture on soil organic carbon stocks and greenhouse gas fluxes in subtropical plantations in China. Forest Ecology and Management 300: 4–13.
7. Dutaur L, Verchot LV (2007) A global inventory of the soil CH_4 sink. Global Biogeochemical Cycles 21: GB4013. doi: 10.1029/2006GB002734.
8. Zhang W, Zhu XM, Liu L, Fu SL, Chen H, et al. (2012) Large difference of inhibitive effect of nitrogen deposition on soil methane oxidation between plantations with N-fixing tree species and non-N-fixing tree species. Journal of Geophysical Research-Biogeosciences 117. doi:10.1029/2012JG002094.
9. Martinson GO, Corre MD, Veldkamp E (2013) Responses of nitrous oxide fluxes and soil nitrogen cycling to nutrient additions in montane forests along an elevation gradient in southern Ecuador. Biogeochemistry 112: 625–636.
10. Steudler PA, Garcia-Montiel DC, Piccolo MC, Neill C, Melillo JM, et al. (2002) Trace gas responses of tropical forest and pasture soils to N and P fertilization. Global Biogeochemical Cycles 16.
11. Butterbach-Bahl K, Gasche R, Huber C, Kreutzer K, Papen H (1998) Impact of N-input by wet deposition on N-trace gas fluxes and CH_4-oxidation in spruce forest ecosystems of the temperate zone in Europe. Atmospheric Environment 32: 559–564.
12. Kim YS, Imori M, Watanabe M, Hatano R, Yi MJ, et al. (2012) Simulated nitrogen inputs influence methane and nitrous oxide fluxes from a young larch plantation in northern Japan. Atmospheric Environment 46: 36–44.
13. Jassal RS, Black TA, Roy R, Ethier G (2011) Effect of nitrogen fertilization on soil CH_4 and N_2O fluxes, and soil and bole respiration. Geoderma 162: 182–186.
14. Liu LL, Greaver TL (2009) A review of nitrogen enrichment effects on three biogenic GHGs: the CO_2 sink may be largely offset by stimulated N_2O and CH_4 emission. Ecology Letters 12: 1103–1117.
15. Castro MS, Peterjohn WT, Melillo JM, Steudler PA, Gholz HL, et al. (1994) Effects of Nitrogen-Fertilization on the Fluxes of N_2O, CH_4, and CO_2 from Soils in a Florida Slash Pine Plantation. Canadian Journal of Forest Research-Revue Canadienne De Recherche Forestiere 24: 9–13.
16. Jang I, Lee S, Zoh KD, Kang H (2011) Methane concentrations and methanotrophic community structure influence the response of soil methane oxidation to nitrogen content in a temperate forest. Soil Biology & Biochemistry 43: 620–627.
17. Mohanty SR, Bodelier PLE, Floris V, Conrad R (2006) Differential effects of nitrogenous fertilizers on methane-consuming microbes in rice field and forest soils. Applied and Environmental Microbiology 72: 1346–1354.
18. Aronson EL, Helliker BR (2010) Methane flux in non-wetland soils in response to nitrogen addition: a meta-analysis. Ecology 91: 3242–3251.
19. Reay DS, Nedwell DB (2004) Methane oxidation in temperate soils: effects of inorganic N. Soil Biology & Biochemistry 36: 2059–2065.
20. Chan ASK, Steudler PA (2006) Carbon monoxide uptake kinetics in unamended and long-term nitrogen-amended temperate forest soils. Fems Microbiology Ecology 57: 343–354.
21. Nyerges G, Stein LY (2009) Ammonia cometabolism and product inhibition vary considerably among species of methanotrophic bacteria. Fems Microbiology Letters 297: 131–136.
22. Bodelier PLE, Laanbroek HJ (2004) Nitrogen as a regulatory factor of methane oxidation in soils and sediments. Fems Microbiology Ecology 47: 265–277.
23. King GM, Schnell S (1994) Effect of increasing atmospheric methane concentration on ammonium inhibition of soil methane consumption. Nature 370: 282–284.
24. Fang H, Cheng S, Yu G, Cooch J, Wang Y, et al. (2014) Low-level nitrogen deposition significantly inhibits methane uptake from an alpine meadow soil on the Qinghai–Tibetan Plateau. Geoderma 213: 444–452.
25. Xu XK, Inubushi K (2004) Effects of N sources and methane concentrations on methane uptake potential of a typical coniferous forest and its adjacent orchard soil. Biology and Fertility of Soils 40: 215–221.
26. Bodelier PLE (2011) Interactions between nitrogenous fertilizers and methane cycling in wetland and upland soils. Current Opinion in Environmental Sustainability 3: 379–388.
27. Matson P, Lohse KA, Hall SJ (2002) The globalization of nitrogen deposition: Consequences for terrestrial ecosystems. Ambio 31: 113–119.
28. Venterea RT, Groffman PM, Verchot LV, Magill AH, Aber JD, et al. (2003) Nitrogen oxide gas emissions from temperate forest soils receiving long-term nitrogen inputs. Global Change Biology 9: 346–357.
29. Chen GC, Tam NFY, Ye Y (2012) Spatial and seasonal variations of atmospheric N_2O and CO_2 fluxes from a subtropical mangrove swamp and their relationships with soil characteristics. Soil Biology & Biochemistry 48: 175–181.
30. Zhang JB, Cai ZC, Zhu TB (2011) N_2O production pathways in the subtropical acid forest soils in China. Environmental Research 111: 643–649.
31. Zhu J, Mulder J, Solheimslid SO, Dorsch P (2013) Functional traits of denitrification in a subtropical forest catchment in China with high atmogenic N deposition. Soil Biology & Biochemistry 57: 577–586.
32. Wang LF, Cai ZC (2008) Nitrous oxide production at different soil moisture contents in an arable soil in China. Soil Science and Plant Nutrition 54: 786–793.
33. Zhang W, Mo JM, Yu GR, Fang YT, Li DJ, et al. (2008) Emissions of nitrous oxide from three tropical forests in Southern China in response to simulated nitrogen deposition. Plant and Soil 306: 221–236.
34. Hu ZH, Zhang H, Cheng ST, Li T, Sheng SH (2011) Effects of simulated nitrogen deposition on N_2O and CH_4 fluxes of soils in forest belt. China Environmental Science 31: 892–897.
35. Department of Forest Resources Management S (2010) The 7th National forest inventory and status of forest resources. Forest Resource and Management 1: 3–10.
36. SFA (2007) China's Forestry 1999–2005. China Forestry Publishing House, Beijing.
37. Liu XJ, Duan L, Mo JM, Du EZ, Shen JL, et al. (2011) Nitrogen deposition and its ecological impact in China: An overview. Environmental Pollution 159: 2251–2264.
38. Zhang JB, Cai.Z.C,Zhu TB, Yang WY, Christoph M (2013) Mechanisms for retention of inorganic N in acid forest soils in southern China. Nature 3: 1–8.
39. Wen XF, Wang HM, Wang JL, Yu GR, Sun XM (2010) Ecosystem carbon exchanges of a subtropical evergreen coniferous plantation subjected to seasonal drought, 2003–2007. Biogeosciences 7: 357–369.
40. Wang YD, Wang ZL, Wang HM, Guo CC, Bao WK (2012) Rainfall pulse primarily drives litterfall respiration and its contribution to soil respiration in a young exotic pine plantation in subtropical China. Canadian Journal of Forest Research-Revue Canadienne De Recherche Forestiere 42: 657–666.
41. Wang YS, Wang YH (2003) Quick measurement of CH_4, CO_2 and N_2O emissions from a short-plant ecosystem. Advances in Atmospheric Sciences 20: 842–844.
42. Fang HJ, Cheng SL, Yu GR, Zheng JJ, Zhang PL, et al. (2012) Responses of CO_2 efflux from an alpine meadow soil on the Qinghai Tibetan Plateau to multi-form and low-level N addition. Plant and Soil 351: 177–190.
43. Le Mer J, Roger P (2001) Production, oxidation, emission and consumption of methane by soils: A review. European Journal of Soil Biology 37: 25–50.
44. Basiliko N, Khan A, Prescott CE, Roy R, Grayston SJ (2009) Soil greenhouse gas and nutrient dynamics in fertilized western Canadian plantation forests. Canadian Journal of Forest Research-Revue Canadienne De Recherche Forestiere 39: 1220–1235.
45. Borken W, Beese F, Brumme R, Lamersdorf N (2002) Long-term reduction in nitrogen and proton inputs did not affect atmospheric methane uptake and nitrous oxide emission from a German spruce forest soil. Soil Biology & Biochemistry 34: 1815–1819.
46. Whalen SC, Reeburgh WS (2000) Methane oxidation, production, and emission at contrasting sites in a boreal bog. Geomicrobiology Journal 17: 237–251.
47. Topp E, Pattey E (1997) Soils as sources and sinks for atmospheric methane. Canadian Journal of Soil Science 77: 167–178.
48. Werner C, Kiese R, Butterbach-Bahl K (2007) Soil-atmosphere exchange of N_2O, CH_4, and CO_2 and controlling environmental factors for tropical rain forest sites in western Kenya. Journal of Geophysical Research-Atmospheres 112.
49. Merino A, Perez-Batallon P, Macias F (2004) Responses of soil organic matter and greenhouse gas fluxes to soil management and land use changes in a humid temperate region of southern Europe. Soil Biology & Biochemistry 36: 917–925.
50. Fang HJ, Yu GR, Cheng SL, Zhu TH, Wang YS, et al. (2010) Effects of multiple environmental factors on CO_2 emission and CH_4 uptake from old-growth forest soils. Biogeosciences 7: 395–407.
51. Steinkamp R, Butterbach-Bahl K, Papen H (2001) Methane oxidation by soils of an N limited and N fertilized spruce forest in the Black Forest, Germany. Soil Biology & Biochemistry 33: 145–153.
52. Lin XW, Wang SP, Ma XZ, Xu GP, Luo CY, et al. (2009) Fluxes of CO_2, CH_4, and N_2O in an alpine meadow affected by yak excreta on the Qinghai-Tibetan plateau during summer grazing periods. Soil Biology & Biochemistry 41: 718–725.
53. Lin S, Iqbal J, Hu RG, Ruan LL, Wu JS, et al. (2012) Differences in nitrous oxide fluxes from red soil under different land uses in mid-subtropical China. Agriculture Ecosystems & Environment 146: 168–178.
54. Xu YB, Xu ZH, Cai ZC, Reverchon F (2013) Review of denitrification in tropical and subtropical soils of terrestrial ecosystems. Journal of Soils and Sediments 13: 699–710.

55. Xu YB, Cai ZC (2007) Denitrification characteristics of subtropical soils in China affected by soil parent material and land use. European Journal of Soil Science 58: 1293–1303.
56. Zhang JB, Cai ZC, Cheng Y, Zhu TB (2009) Denitrification and total nitrogen gas production from forest soils of Eastern China. Soil Biology & Biochemistry 41: 2551–2557.
57. Bijoor NS, Czimczik CI, Pataki DE, Billings SA (2008) Effects of temperature and fertilization on nitrogen cycling and community composition of an urban lawn. Global Change Biology 14: 2119–2131.
58. Luo GJ, Kiese R, Wolf B, Butterbach-Bahl K (2013) Effects of soil temperature and moisture on methane uptake and nitrous oxide emissions across three different ecosystem types. Biogeosciences 10: 3205–3219.

15

Bryophyte-Cyanobacteria Associations during Primary Succession in Recently Deglaciated Areas of Tierra del Fuego (Chile)

María Arróniz-Crespo[1,2]*, Sergio Pérez-Ortega[3], Asunción De los Ríos[3], T. G. Allan Green[1,4], Raúl Ochoa-Hueso[5], Miguel Ángel Casermeiro[6], María Teresa de la Cruz[6], Ana Pintado[1], David Palacios[7], Ricardo Rozzi[8,9], Niklas Tysklind[10], Leopoldo G. Sancho[1]

1 Dept. de Biología Vegetal II, Universidad Complutense de Madrid, Madrid, Spain, **2** School of Environment Natural Resources and Geography, Bangor University, Bangor, Wales, United Kingdom, **3** Dept. de Biología Ambiental, Museo Nacional de Ciencias Naturales, MNCN-CSIC, Madrid, Spain, **4** Biological Sciences, Waikato University, Hamilton, New Zealand, **5** Dept. de Biogeografía y Cambio Global, Museo Nacional de Ciencias Naturales, Madrid, Spain, **6** Departamento Edafología, Universidad Complutense de Madrid, Madrid, Spain, **7** Departamento Geografía Física, Universidad Complutense de Madrid, Madrid, Spain, **8** Sub-Antarctic Biocultural Conservation Program, University of North Texas, Denton, Texas, United States of America, **9** Institute of Ecology and Biodiversity, Universidad de Magallanes, Puerto Williams, Chile, **10** School of Biological Sciences, Bangor University, Bangor, Wales, United Kingdom

Abstract

Bryophyte establishment represents a positive feedback process that enhances soil development in newly exposed terrain. Further, biological nitrogen (N) fixation by cyanobacteria in association with mosses can be an important supply of N to terrestrial ecosystems, however the role of these associations during post-glacial primary succession is not yet fully understood. Here, we analyzed chronosequences in front of two receding glaciers with contrasting climatic conditions (wetter *vs* drier) at Cordillera Darwin (Tierra del Fuego) and found that most mosses had the capacity to support an epiphytic flora of cyanobacteria and exhibited high rates of N_2 fixation. Pioneer moss-cyanobacteria associations showed the highest N_2 fixation rates (4.60 and 4.96 μg N g^{-1} bryo. d^{-1}) very early after glacier retreat (4 and 7 years) which may help accelerate soil development under wetter conditions. In drier climate, N_2 fixation on bryophyte-cyanobacteria associations was also high (0.94 and 1.42 μg N g^{-1} bryo. d^{-1}) but peaked at intermediate-aged sites (26 and 66 years). N_2 fixation capacity on bryophytes was primarily driven by epiphytic cyanobacteria abundance rather than community composition. Most liverworts showed low colonization and N_2 fixation rates, and mosses did not exhibit consistent differences across life forms and habitat (saxicolous *vs* terricolous). We also found a clear relationship between cyanobacteria genera and the stages of ecological succession, but no relationship was found with host species identity. Glacier forelands in Tierra del Fuego show fast rates of soil transformation which imply large quantities of N inputs. Our results highlight the potential contribution of bryophyte-cyanobacteria associations to N accumulation during post-glacial primary succession and further describe the factors that drive N_2-fixation rates in post-glacial areas with very low N deposition.

Editor: Han Y.H. Chen, Lakehead University, Canada

Funding: This work was funded by the Ministerio de Ciencia e Innovación of Spain (Projects CTM2009-12838-C04-01 and CTM2009-12838-C04-03). SPO is supported by the grant CTM2012-38222-C02-02 from the Spanish Ministry of Economy and Competitiveness. The authors are grateful to the Universidad de Magallanes (UMAG) and the Institute of Ecology and Biodiversity (IEB-Chile, grants P05-002 ICM and CONICYT PFB-23) for logistic and scientific support. The funders had no role in study design, data collection and analysis, decision to publish, or preparation of the manuscript.

Competing Interests: The authors have declared that no competing interests exist.

* E-mail: maria.arroniz@gmail.com

Introduction

Cryptogamic covers account for nearly half of the biological nitrogen (N) fixation on land [1], representing an important supply of N to terrestrial ecosystems. In particular, dinitrogen (N_2) fixation by cyanobacteria-moss associations has been found to contribute significantly to the N economy in high-latitude ecosystems [2–6] where it can represent the main source of *new* N [7–10]. For example, moss-associated cyanobacteria can provide 2–58% of the N input to Arctic ecosystems and in Antarctica the percentage can reach up to 84% [11]. In boreal forests, up to 60% of N can come from N_2-fixation in mosses [12]. In early primary succession, however, the role of cyanobacteria-moss associations on N inputs has received considerably less attention, despite bryophytes being often the first plants to colonize newly exposed terrain [13]. The lack of studies is particularly marked in temperate and subpolar regions of the Southern Hemisphere, where bryophytes account for more than 50% of the plant species [14].

Rates of N_2 fixation in moss associated with cyanobacteria can be highly variable. Previous studies in the 1980s and early 1990s have shown that N_2 fixation varies greatly among bryophyte species [8,15–17] which may be due to differences in the microhabitat characteristics where the mosses grow [2]. Moderate increases in temperature leads to a general increase in N_2 fixation [18] whereas shading [19] and N fertilization (e.g. atmospheric N deposition or throughfall) results in reduced rates of N_2 fixation in these associations [4,12]. Moisture is also considered a major environmental factor influencing N_2-fixation on moss-cyanobacteria associations, with a positive relationship between water availability and N-fixation rates [3,20]. The diversity and density of cyanobacteria species associated with mosses have also been described as important factors explaining the variability in N_2 fixation rates [21,22]. However, little is still known about the existence of species-specific association between bryophyte and cyanobacteria species [23], although some studies have shown that host species identity could be a stronger driver of epiphytic cyanobacteria diversity and composition [22]. While these studies have already started to elucidate the factors that drive N_2-fixation rates, the influence of bryophyte-cyanobacteria associations on the biogeochemical cycling of N and, in particular, during early primary succession still needs further assessment.

In Tierra del Fuego, glacial retreat has been occurring since the end of the Little Ice Age (around 1750 to 1850) [24], exposing new terrain for ecosystem development by the process of primary succession. Glacier forelands are typically nutrient-poor [25]. However, our preliminary studies at the Cordillera Darwin (Tierra del Fuego) found rapid rates of vegetation growth and quick succession to stages dominated by *Nothofagus* tree species [26]. These tree-dominated stages were present on the southern side of the mountain range after 34 years of soil surface exposure [26], compared to 80 years on the northern side of the mountain range (Palacios *et al.*, unpublished). As inputs of N through atmospheric deposition are very low in such high latitude region, possibly < 0.7 kg N ha^{-1} yr^{-1} [27], biological fixation can greatly exceed bulk precipitation. For example, biological N fixation inputs from bryophytes during early succession, e.g. [28] reached 22 kg N ha^{-1} yr^{-1} in a New Zealand glacier foreland. Therefore, we considered that N_2 fixation and its interaction with other biotic and abiotic factors could be a key driver of early plant succession in recently deglaciated areas of Tierra del Fuego. The combination of three attributes: 1) the difference in the speed of ecological succession found between the two glaciers on southern and northern slopes; 2) the strong climatic gradients characteristic of this area: with high rates of precipitation on the southwestern side and dry conditions on the northern side of the mountain range [29]; and 3) the low rates of anthropogenic N deposition, make the Cordillera Darwin an unique site to further elucidate the potential drivers of N_2 fixation rates on bryophyte-cyanobacteria associations and its impact on C and N accumulation during post-glacial primary succession.

Specifically, we aimed to evaluate the role of bryophyte-cyanobacteria associations during post-glacial primary succession because (i) moss establishment is a positive feedback process that enhance N inputs in newly exposed terrain [13] and (ii) moss-associated cyanobacteria can be an important contributor to N inputs through the process of N_2 fixation [28,30]. Our aims were (1) to describe how soil properties and N_2 fixation related parameters change along two pro-glacial chronosequences at the Cordillera Darwin (Tierra del Fuego), (2) how species identity (both bryophytes and cyanobacteria), climatic conditions (wetter *vs.* drier), and the stage of ecological succession, influence N_2 fixation rates of bryophyte-cyanobacteria associations, and thereby ecosystem N accumulation, and (3) whether any of these associations are species specific.

Materials and Methods

Ethics Statement

The Chilean National Forestry Corporation (CONAF) issued the scientific collecting permit for each location considered in this study. This study did not involve endangered or protected species.

Site description and chronosequence establishment

The study area is located at the Cordillera Darwin in the southwestern part of Tierra del Fuego (Chile) (Fig. 1a). Around 80% of the mountain range is glaciated. One glacier was selected on each side of the mountain range for the present study; both glaciers have been receding steadily for several decades and present a well-defined sequence of moraine bands. The south-side glacier, SG (Pia Glacier, 54°46′S 69°40′W) flows into the east arm of Bahia Pia; plant succession goes from bare ground to full forest (stage of succession dominated by *Nothofagus* spp. >10 m high) in 34 years [26]. At the north-side glacier, NG (Parry Glacier, 54°41′S 69°23′W) the succession is slower and the equivalent stage of succession dominated by *Nothofagus* appears after 80 years of soil exposure (Palacios *et al.*, unpublished data). A characteristic feature of the area is the steep meteorological gradients across the Cordillera Darwin with heavy precipitation and solid cloud cover being typical over the southern and western fjords, whilst to the north and east conditions are much drier [29]. This results in a stark contrast in annual precipitation between both glaciers. Precipitation rates where SG flows, Bahia Pia (1600 mm) are twice as high as at where NG flows, Seno Almirantazgo fjord (800 mm) [29]. Mean annual air temperature, however, is relatively constant over the entire region, with only a slight drop from 5.9°C at Punta Arenas to 4.5°C at Bahia Pia [31].

The chronosequence in each glacier has been established by using several independent techniques to date surfaces: aerial photographs, dendrochronology, (sampling tree rings), and lichenometry (dating of rock surface from lichens growth) utilizing the growth rate and thallus size of the lichen *Rhizocarpon geographicum* (L.) DC [26].

Forest vegetation in the area consists of *Nothofagus antarctica* (G. Forst.) Oerst., and *N. betuloides* (Mirb.) as dominant tree species. After glacier retreat, the successional sequence at NG is characterized by pioneer lichens (*Placopsis perrugosa* (Nyl.) Nyl., *Rhizocarpon geographicum* (L.) DC., *Xanthoria candelaria* (L.) Th. Fr.) and mosses (*Dendroligotrichum squamosum* (Hook. f. & Wilson) Cardot and *Andreaea* spp.), followed by dwarf shrubs (*Empetrum rubrum* Wahl. Ex Willd., *Gaultheria mucronata* (L. f.) Hook. & Arn.), perennial herbs (*Gunnera magellanica* Lam) and moss (*Racomitrium* spp.) vegetation assemblage. At SG the successional sequence goes from pioneer lichen (same species as in NG plus *Placopsis stenophylla* (Hue) I.M. Lamb and *Placopsis pycnotheca* I.M. Lamb) and the moss *Ditrichum cylindricarpum* (Müll. Hal.) F. Muell, followed by perennial herbs vegetation assemblage (*Gunnera magellanica* Lam and *Uncinia tenuis* Poepp. ex Kunth) and the development of *Nothofagus* spp. forest (Fig. 1b).

Sample collection and transport

Field work took place in 2009 during the summer season which corresponds to the active period for N_2 fixation at high southern latitudes [32]. Sampling sites were established at increasing surface ages in front of both glaciers. There were six sites at SG with surface ages of 1, 4, 7, 10, 19 and 34 yrs and six sites at NG with surface ages of 5, 18, 20, 26, 66 and 80 yrs. At both glaciers, the younger site corresponds to bare ground close to the glacier front whereas the older site was covered within *Nothofagus* forests. Three transects were randomly established along the chronosequence in each glacier foreland and one independent sample was collected in each transect per site (surface age) so that there were three replicates per surface age for soil, each bryophyte species and cyanobacteria measurements.

Soil sampling. 200 g of soil samples were collected using the core method, samples were collected from the upper 0–4 cm of the soil. Samples were placed into, and sealed in, suitably sized plastic

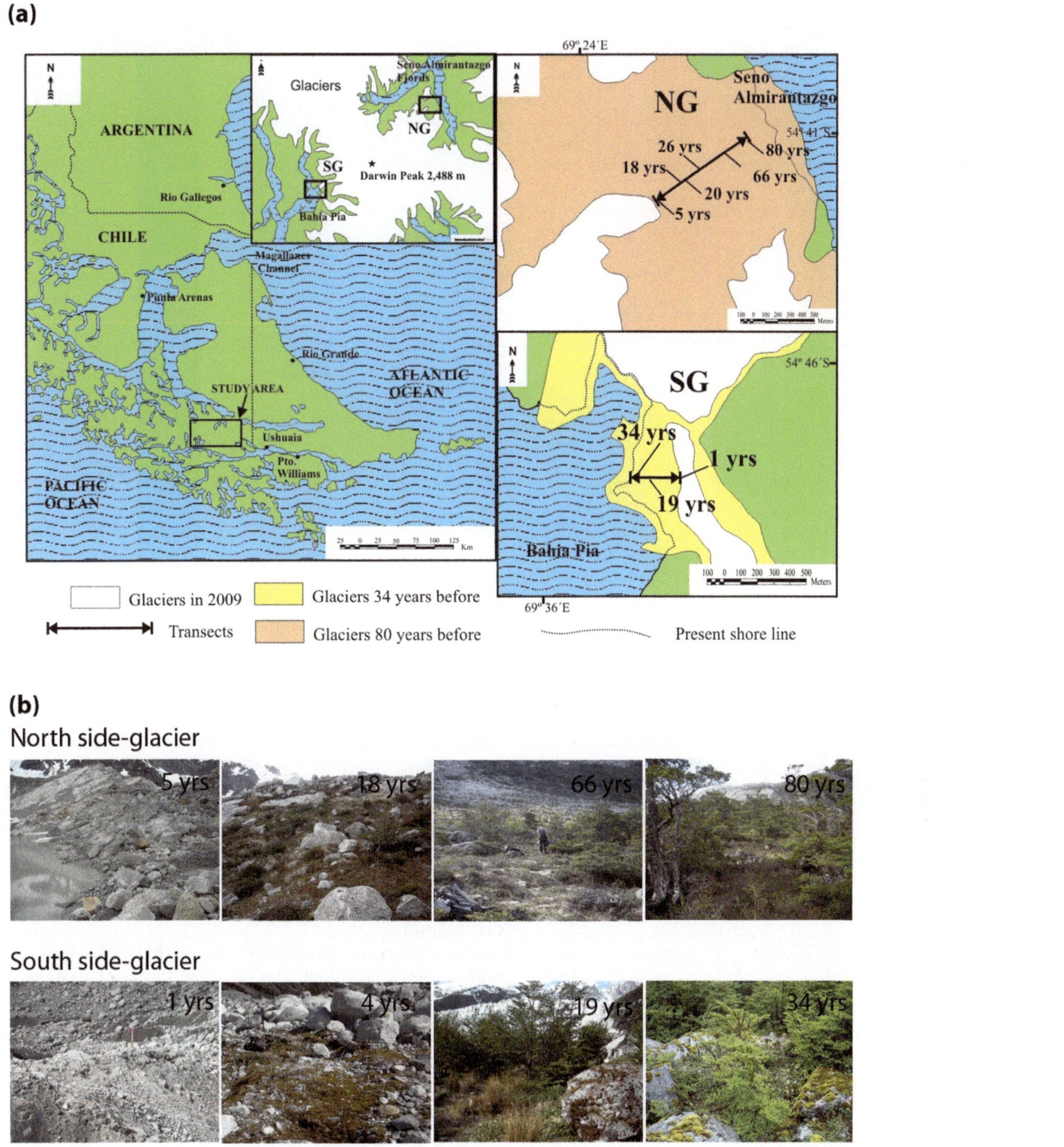

Figure 1. Study area and chronosequence description. (a) Map of the study area at the Cordillera Darwin in the southwestern part of Tierra del Fuego (Chile) showing the location of the south-side (SG) and north-side (NG) glaciers on each side of the mountain range. (b) Four representative sites along north-side glacier and south-side glacier chronosequences at Cordillera Darwin. At both glaciers, the younger site corresponds to bare ground close to the glacier front whereas the older site was covered within *Nothofagus* forests. Years since deglaciation per site are included in the picture.

bags. Collected samples were immediately frozen at −20°C and returned frozen to the Universidad Complutense de Madrid for soil characterization.

Bryophyte sampling. The dominant bryophyte species at each site (surface age) along the chronosequence were collected. In total 16 bryophyte species (including both mosses and liverworts) with varying morphologies, life forms and habitats (saxicolous and terricolous species) were selected. These included 10 mosses *Dendroligotrichum squamosum* (Hook. f. & Wilson) Cardot, *Andreaea laxifolia* (Hook. f. & Wilson), *A. alpina* (Hedw.), *Acroschisma wilsonii* (Hook. f.) A. Jaeger, *Racomitrium didymum* (Mont.) Lorentz, *R. laevigatum* (A. Jaeger), *R. lanuginosum* (Schimp. ex R.S. Williams), *R.*

subcrispipilum (Müll.Hal.) A. Jaeger, *Dicranoloma chilense* (De Not.) Ochyra & Matteri and *Ditrichum cylindricarpum* (Müll. Hal.) F. Muell; and 6 leafy liverworts *Cryptochila grandiflora* (Lindenb. & Gott.) Grolle, *Heteroscyphus magellanicus* (Steph.) Engel & Schust., *Blepharidophyllum densifolium* (Hook.) Ångström ex C. Massal., *Anastrophyllum involutifolium*, *Clasmatocolea humilis* (Hook. f. & Tayl.) Grolle and *Chiloscyphus leptanthus* (Hook. f. & Tayl.) Engel & Schust. Eight species were sampled at more than one site and the other eight were sampled at a single site on the chronosequence. A full description of bryophyte species sampled by site are shown in (Table S1). Monospecific patches of fresh bryophytes species were collected, air dried and sent to the Universidad Complutense de Madrid in paper bags for analysis that included: (1) rates of N_2 fixation, (2) quantification of the abundance of epiphytic cyanobacteria, and (3) abundance of stable isotopes (^{14}N and ^{15}N) and shoot N and C contents.

Cyanobacteria sampling. The diversity and distribution of cyanobacteria associated with bryophytes at each site (surface age) was determined by molecular analysis. A second set of bryophyte samples (around 10 shoots per species) were collected in the same bryophyte patches described above. Collectors used disposable laboratory latex gloves when taking samples. Samples were placed into, and sealed in suitably sized plastic bags. Collected samples were immediately frozen at −20°C and returned frozen to the Museo Nacional de Ciencias Naturales in Madrid (CSIC) where DNA from epiphytic cyanobacteria was extracted and analysed.

Soil chemical properties

Samples were stored at −20°C. Soils were air-dried, sieved (< 2 mm) homogenized before conducting the following analysis:

Soil pH. Soil pH was measured in water suspension (soil:-water, 1:2.5) [33].

Total soil N (TN) and total soil carbon (TC) were analysed in an autoanalyser system LECO CNS 2000I.

Soil extractable nitrogen. Soil nitrate (NO_3-N) and nitrite (NO_2-N) were extracted by shaking soil aliquots (30 g DW) in (1:5) water [33], filtered and analysed by ionic chromatography Metrohm 761 compact IC. Ammonium (NH_4-N) was extracted by shaking 10 g of soil in 50 ml of 0.5 M K_2SO_4 for 30 min, filtered and colorimetrically assessed [34].

Nitrogen-fixation measurements

Prior to N_2-fixation measurements, all bryophyte samples were reactivated for 48 h in a growth chamber, photosynthetic photon flux density (PPFD) of 100 µmol m^{-2} s^{-1} over the waveband 400–700 nm, 10°C, with a 18 : 6 h phototoperiod (light : dark) and maintained hydrated to ensure recovery (confirmed by chlorophyll fluorescence, data not shown). N_2-fixation rates of bryophyte-cyanobacteria associations were estimated under controlled conditions using the acetylene reduction assay (ARA) [35]. Revitalized shoot apices (around 10–15 shoots or 0.07–0.20 dry weight basis) were placed into 10 ml glass incubation vessels and sealed with a rubber septum. 1 ml of headspace was removed and replaced by 1 ml of acetylene (generated from CaC_2 and water), resulting in headspace enrichments of 10–20% (enrichment was determined individually for each species allowing for the different vial headspaces, see in (Appendix S1). Four samples were randomly chosen as control samples and were treated in the same manner except for the headspace substitution; another set of four samples with deionized water and acetylene was used as blank samples to assess the concentration of ethylene in the acetylene used. All incubation vessels were placed in a growth chamber (200 µmol m^{-2} s^{-1} PAR, 10°C) and incubation took 18 h. Before incubation, acetylene reduction was demonstrated to be linear over an 18-h period in 10 samples with known different cyanobacteria abundance (data not shown). Ethylene resulting from the reducing activity was measured by gas chromatography. N_2-fixation rates were calculated as nanomoles of acetylene reduced per gram bryophyte dry weight per hour (nmol C_2H_2 g^{-1} dw h^{-1}). Theoretical stoichiometric ratio of acetylene to nitrogen (3:1) is recommended to be corrected by parallel uptake of labelled $^{15}N_2$ for each organism and also, possibly, for location [36,37]. Thus, to estimate N_2 fixation rates ($^{15}N_2$ uptake), consecutive incubations of the same samples with $^{15}N_2$ after ARA was performed following [9]. 2 ml of headspace were replaced by 2 ml $^{15}N_2$ gas (98 atom % ^{15}N enriched), resulting in headspace enrichments of 20–30% and, after 18 h incubation at the same conditions as ARA, samples, were immediately oven dried (60°C for 48 h), ground in a ball mill and sent for ^{15}N enrichment and total N analysis (UCDavis Stable Isotope Facility, University of California, Davis, USA). Three control samples for each bryophyte-cyanobacteria association were used to determine the natural abundance ^{15}N. The amount of N fixed was calculated according to [6,38]. N_2-fixation rates were calculated as micrograms of ^{15}N incorporated per hour per gram of bryophyte on a dry weight basis and expressed as micrograms of fixed N per day (µg N g^{-1} dw bryophyte d^{-1}).

ARA was highly correlated with $^{15}N_2$ uptake ($N = 89$, $R^2 = 0.689$ $P<0.001$). ARA measurements were used to compare N_2 fixation rates between taxa and rates reported by other authors. $^{15}N_2$ uptake was used to evaluate the potential impact of N_2 fixation on ecosystem nitrogen (N) accumulation.

Stable isotope analysis and shoot N and C content

All isotopic analyses were conducted at the UCDavis Stable Isotope Facility, Department of Plant Sciences, University of California, Davis, USA. δ15N analyses were performed by elemental analyser/continuous flow isotope ratio mass spectrometry using a ANCA-GSL elemental analyser interfaced coupled with a PDZ Europa 20-20 isotope ratio mass spectrometer. Atom% was calculated from $\delta^{15}N$ values. Total tissue C and N was also analysed.

Cyanobacteria abundance, composition and phylogenetic analysis

Cyanobacteria abundance by HPLC analysis of echinenone. Bryophyte samples (~3–10 cm^2 of fresh tissue per sample depending on the species) were deep-frozen until pigment analyses. Pigments were extracted in 95% cool acetone in the presence of sodium ascorbate. Echinenone (a cyanobacterial pigment used as a proxy of cyanobacteria abundance) was separated by HPLC according to the method of [39] slightly modified. After passing through a 0.45 µm nylon filter, 25 µL of the extract was injected into a C18 column. The mobile phase rate was 1.2 mL min^{-1} and the elution time lasted 30 min. Solvents for HPLC analysis were degassed before use by bubbling helium. The HPLC system was equipped with a photodiode array detector. For peak identification and quantification, a pure commercial standard was used. The relative abundance of cyanobacteria was expressed as the amount of echinenone on a bryophyte surface area basis (nmol echinenone/cm^2 bryophyte). Digital pictures of each sample were taken before freezing and the surface area of each sample was calculated using a image analysis program.

DNA extraction and polymerase chain reaction (PCR), denaturing gradient gel electrophoresis (DGGE). For cyanobacteria DNA extraction, bryophyte samples (around 200 mg dry weight) were introduced into 15 ml falcon tubes with distilled water and vigorously shaken (maximum speed) for 10 minutes in a vortex using a falcon tube adapter to release the epiphytic cyanobacteria from bryophytes. The extracted cyanobacteria contained in the water were removed with a 5 ml pipette and cells were concentrated by centrifugation in an ultracentrifuge at 3000 rpm for 5 min. The concentrated cyanobacteria cells were used for DNA extraction.

Denaturing gradient gel electrophoresis (DGGE) of cyanobacterial PCR-amplified 16S rRNA gene fragments was used to survey the genetic diversity of isolated epiphytic cyanobacteria. Total genomic DNA for DGGE analysis was extracted in three replicates from the cyanobacteria isolates samples using CTAB extraction method described in [40]. A fragment of the cyanobacterial 16S rRNA gene suitable for DGGE analysis was amplified in each sample from total genomic DNA using the following primer pair: CYA359fGC and CY781r [41]. Each 25 µl volume of PCR mix [75 mM Tris pH 9.0/50 mM KCl/20 mM $(NH_4)_2SO_4$] contained 1 unit of Taq polymerase, 0.2 mM of each of the four dNTPs, 0.4 mM of each primer, 100 mg of bovine serum albumin, 1.5 mM of $MgCl_2$, 5 ml de 5xTaq Master PCR enhancer (Prime) and ca 10–50 ng genomic DNA. Annealing conditions were 60°C. The DGGE system was used for the analysis of cyanobacterial fragments of 16S rRNA gene according to the manufacturer's instructions). Acrylamide gels (6%) with a 30 to 60% urea-formamide denaturing gradient were prepared. Lanes were loaded with 22 µl of PCR product, run at a constant 200 V for 7 h at 60°C, and EtBr stained to visualize and photograph the resultant bands according to [42]. The most predominant bands were excised and incubated 1 h at 60°C in water before PCR amplification. The eluted DNA was reamplified as described above but using primers devoiding the GC clamp. The PCR products were cleaned on Quick Spin columns. Both complementary strands were sequenced separately at the MACROGEN sequencing company.

For sequence analysis, we obtained closely related sequences with known taxonomy using Blast searches in Gen Bank. Later, sequences obtained from DGGE bands were reduced to OTUs in *mothur* v.1.28 [43] using a cutoff of 0.01% [44]. A reference sequence for each OTU was chosen for subsequent analysis by means of the command *get.outrep* in *mothur* v.1.28. 20 sequences corresponding to each of the 0.01% cut-off OTUs were aligned with a total of 29 sequences retrieved from the GenBank (Table S2). Alignment of all sequences were carried out in Muscle 3.8.31 [45] and visually checked in Bioedit 7.0.5.2 [46]. Nucleotide substitution model was statistically selected with the help of jModelTest [47], program available at http://darwin.uvigo.es). Model selection was made according to the Akaike's information criterion (AIC, [48]; the General Time Reversible substitution model [49], with estimation assuming a gamma distribution (GTR+G) had the lowest –lnL value. Phylogenetic relationships among OTUs and sequences with known taxonomy were studied by means of Maximum Likelihood (ML) and Bayesian phylogenetic inference. ML analysis were carried out in PHYML 3.0 [50] through the webserver http://www.atgc-montpellier.fr/phyml/ using 1,000 bootstrap repetitions in order to assess node support. Bayesian analyses were carried out in Mr. Bayes 3.1.2 [51] running eight different chains and 5,000K generations. 50% majority rule consensus tree is shown in (Fig. 2).

Sequences generated in this study were deposited in GenBank under accession numbers KJ576732-KJ576783.

Statistical and data analyses

We calculated the Shannon index ($H' = -\Sigma\ (ni/N)\ \ln\ (ni/N)$ [52], where *ni* is the number of times we found each *i*th OTU and N is the total number of unique OTUs ($N = 20$) found in the study area, in order to evaluate changes in diversity of epiphytic cyanobacteria along the chronosequence (H'_{site}) and between bryophytes species ($H'_{species}$).

To be able to compare parameters between glacier chronosequences with different sampling site ages we first tested if the different stages of ecological succession, identified by the soil development transformation trends observed along the chronosequence, were comparable between the two glacier chronosequences; we used TC, TN and soil pH as descriptors of soil development status at each site and extracted one general factor component of soil development using principal component analysis [53]. The resulting scores were tested with two-way ANOVA to detect differences on the stage of ecological succession along the chronosequence and between glaciers, with sampling sites (0 to 5) and glacier location (north *vs.* south) as fixed factors. These analyses indicated that the two chronosequences were comparable despite the different absolute soil ages where the samples were collected (Fig. S1). Differences between glaciers on soil chemical properties, N_2-fixation rates ($^{15}N_2$ uptake), abundance of epiphytic cyanobacteria, and C : N ratio in bryophyte tissue were then tested with two-way ANOVAs. All data were tested for homoscedasticity (Levene's test for equal variances) and normality (IBM SPSS statistics, Armonk NY, USA). When data did not meet these assumptions, they were rank or Log_{10} +1 transformed prior to statistical analyses. Analyses were carried out with IBM SPSS Statistics 19 (IBM SPSS statistics, Armonk NY, USA).

A fundamental assumption about chronosequences is that the communities and ecosystems of the younger sites are developing in the same temporal sequence as the older sites developed (termed a space-for-time substitution). At any location on the chronosequence the vegetation present represents what will eventually form on younger sites and which will, itself, change to that found on the older sites present. As such, the assumption of independence is violated. We used non-parametric Krukall-Wallis test to evaluate changes of parameters measured at the different stages of ecological succession (mean values per site) along each glacier chronosequence separately, using stage of ecological succession (sampling site) as the grouping factor. When significant differences were found, Mann–Whitney U-test post hoc tests were run to detect differences between sites. Bonferroni's correction was used to determine the level of significance of the post hoc tests.

At species level, we used conditional inference trees (CTREE) [54] to assess and visualise the relative importance of bryophyte species identity (species effect), climatic conditions (glacier location: north *sv.* south), and stage of ecological succession (sampling site effect) in determining N_2 fixation capacity and cyanobacteria abundance on bryophytes. CTREE work by first testing the null hypothesis of independence between the predictors (here bryophyte species, glacier and site) and response variable (here N_2 fixation and cyanobacteria abundance); if dependence is found, then determining the best split value for the predictor with the strongest effect on the response; and third, repeating the first two steps with each branch until independence cannot be rejected. Relationships are represented as dichotomous trees with nodes indicating split points for significant predictor variables, and terminal nodes for the final groups of responses. Being nonparametric, CTREE make no assumptions about the data, and are appropriate where nonlinear relationships between predictive and response variables and complex interactions between variables, such as covariance, may be expected [54]. CTREE were constructed in R 3.0.1 (R Development Core Team, 2013) with the package PARTY [55], with 10,000 resamples of Bonferroni-adjusted $\alpha = 0.05$ significance levels and with at least one

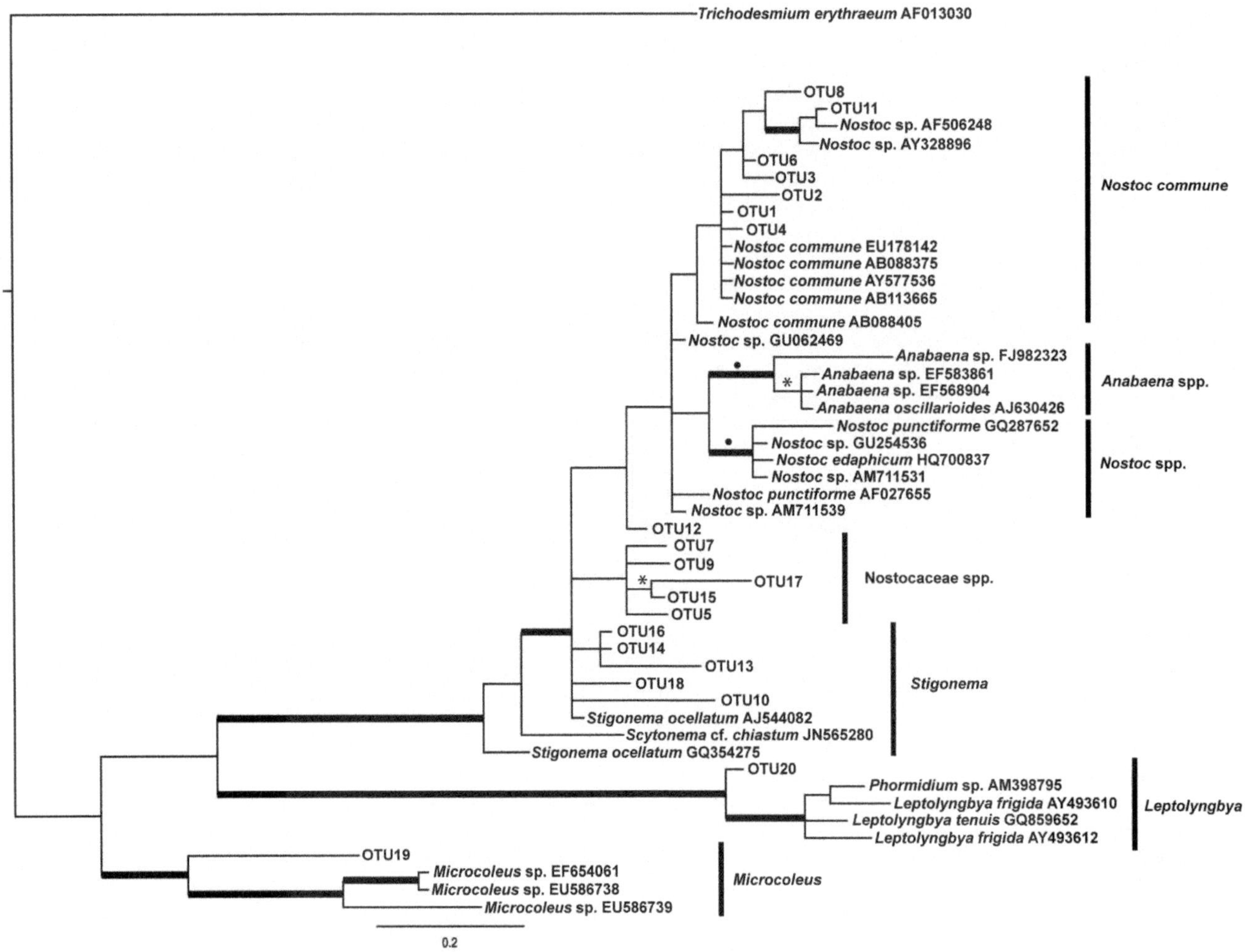

Figure 2. Maximum-likelihood phylogenetic tree. Maximum-likelihood phylogenetic tree (50% majority rule consensus tree) from alignments of 16S rRNA genes showing the phylogenetic relationships among all OTUs of epiphytic cyanobacteria found on bryophytes collected along the chronosequence in both glacier and sequences with known taxonomy, for Accession numbers (Table S1).

observation into each daughter node. N_2 fixation and cyanobacteria abundance data were log +1 transformed and site included as an ordered categorical variable.

Linear regression analysis was used to evaluate whether abundance of epiphytic cyanobacteria or its diversity on host bryophyte ($H'_{species}$) had a larger influence on the rate of N_2 fixation.

Correspondence analysis in each glacier separately were used to graphically examine the relationships between OTUs of cyanobacteria and the different stages of ecological succession and also to detect selectivity associations between cyanobacteria OTUs and bryophyte species (6 species with cyanobacteria colonies sampled at more than one site were used in this analyses). Analysis were carried out in R 3.0.1 (R Development Core Team, 2013) using the package *ca* [56].

Results

Genetic diversity of isolated epiphytic cyanobacteria

A total of 52 bands (18 from NG and 34 from SG) were recovered from DGGE analyses and subsequently sequenced, producing 28 unique sequences. After their collapse into OTUs (0.01% cutoff), 20 cyanobacteria putative taxa were obtained. The phylogenetic analysis of the relationships between these OTUs and sequences with known taxonomy obtained from the GenBank is depicted in Fig. 2. Seven 0.01% cutoff OTUs could be assigned to the species aggregate *Nostoc commune.* Six OTUs (5, 7, 9, 12, 15, 17) were assigned to unidentified Nostocaeae because no clear relationship was found in the phylogenetic tree with any certain genus. A similar case is presented for OTUs 10, 13, 14, 16, and 18, which did not show a clear and well supported clade with sequences of *Stigonema* included in the phylogenetic analysis. Nevertheless, BLAST searches of these OTUs recovered sequences identified as *Stigonema.* In both cases, the most likely explanation for such lack of phylogenetic support in our tree for these two OTU groups is the shortness of fragment of the 16S gene used in the analyses. Finally, OTU 20 is clearly and statistically supported related to *Leptolymbya* and OTU 19 is likewise related to *Microcoleus* (Fig. 2).

Soil properties of glacier forelands

Overall comparisons between both glaciers (two-way ANOVAs) showed no differences on pH, TC and TN between glaciers (NG *vs* SG) and higher concentration of inorganic N at SG (ANOVA $P = 0.037$). At NG, the concentration of inorganic N in young soils (younger than 26 years) was lower (from 0.16 to 8.23 mg kg^{-1})

compared to concentrations observed in soils at SG very early after glacier retreated (13.42 mg kg^{-1} after 4 years soil exposure and 34.83 mg kg^{-1} after 10 years, Fig. 3d). Within the *Nothofagus* forests inorganic N concentration in the soil was nearly 3-fold higher at SG (135.08 mg kg^{-1}) compared to NG (45.28 mg kg^{-1}). All soil parameters showed a significant overall site effect (ANOVA $P<0.001$) and no interaction between glacier location (north *vs.* south) and site (0 to 5). Within each chronosequence (non-parametric tests), soil pH decreases from 6.4 to 4.4 in NG ($P<0.01$) and from 6.81 to 4.55 in SG ($P<0.05$), with the higher values always found in the younger soils (Fig. 3a). TC increased from very low values (0.33% at NG and 0.24% at SG) to over 35% at NG ($P<0.05$) and 39% at SG ($P<0.05$) in the oldest soils (80 and 34 years of soil exposure respectively) inside the *Nothofagus* spp. forest (*Nothofagus*-dominated state, Fig. 3b); TN also increased along the chronosequence from 0.02% in NG and undetectable levels in SG to 0.99% in NG ($P<0.05$) and 1.51% in SG ($P<0.05$) inside the forest site (Fig. 3c). Soil extractable inorganic N ($NO_3^-+NO_2^-+NH_4^+$) followed a similar trend increasing with time since deglaciation at both glaciers ($P<0.05$ for NG and $P<0.01$ for SG, Fig. 3d).

Nitrogen-fixation, cyanobacteria abundance and shoot C : N contents

N_2 fixation rates ($^{15}N_2$ uptake) and epiphytic cyanobacteria relative abundance, when averaged across all bryophyte species per site, were higher at SG compare to NG (ANOVA $P<0.001$ and $P<0.01$ respectively). A significant overall interaction (ANOVA $P<0.001$) between glacier locations (north *vs.* south) and the stage of ecological succession (sampling site:1 to 5) suggested that variation of N_2 fixation rates and cyanobacteria abundance along each chronosequence differed between glaciers. At NG, N_2 fixation rate (Fig. 3e) showed an optimum increasing from the lowest rates at 18 and 20 year old sites (0.52 and 0.5 µg N g^{-1} DW bryo. d^{-1}, respectively) to rates that were 2-fold higher in intermediate-aged sites (0.94 and 1.42 µg N g^{-1} DW bryo. d^{-1}, 26 and 66 years old site respectively) and then declined in the 80 years old forest site (0.80 µg N g^{-1} DW bryo. d^{-1}). Epiphytic cyanobacteria abundance (Fig. 3f) followed a similar trend with the highest concentration at 66 years old site (42.09 nmol echin. cm^{-2}) and lowest concentration at 80 and 20 years old soil (4.5 and 7.94 nmol echin. cm^{-2}). In contrast, at SG the highest N_2 fixation rate and cyanobacteria abundance was found in young soils (4.60 and 4.96 µg N g^{-1} DW bryo. d^{-1} and 51.50 and 53.33 nmol echin. cm^{-2}, at 4 and 7 year old sites respectively) compared to older sites (ranging from 0.64 to 1.34 µg N g^{-1} DW bryo. d^{-1} and 6.46 to 23.96 nmol echin. cm^{-2}, at soil age 10 to 34 years old; Fig. 3e,f).

Shoot C : N ratios of bryophyte tissue, when averaged across all bryophyte species per site, did not differed between glacier (ANOVA $P=0.165$) but change with soil age (ANOVA $P<0.001$). Shoot C : N increased with time since deglaciation at NG ($P<0.001$) with the lower values always found in the younger soil (Fig. 3g). At SG shoot C : N ratios showed the same pattern but the effect of soil age was not significant (Fig. 3g).

Differences on bryophyte N_2 fixation capacity and cyanobacteria abundance

The conditional inference tree (Fig. 4a) revealed that both species identity and the stage of ecological succession (sampling site), were important variables in the N_2 fixation rates at species level (ARA, nmol C_2H_2 g^{-1} dm bryo. h^{-1}) Three species of mosses (*Ditrichum cylindricarpum*, *Racomitrium subcrispipilum* and *R. laevigatum*) showed the highest rates of N_2 fixation (Table 1 and node 3 in Fig. 4a); Intermediate N_2 fixation rates were found for a second group of species (node 4), namely *Racomitrium didymium, R. lanuginosum, Acroschisma wilsonii* and the liverwort *Cryptochila grandiflora.* An increment of N_2 fixation rates was detected between younger sites (node 5: sites 1–3) and older sites (node 6: sites 4 and 5) for node 4. However, a more detailed analysis by individual species, revealed that there was an effect of the stage of ecological succession on N_2 fixation rates only on the two most common species sampled in this study, *R. didymium* and *A. wilsonii* (statistical data not shown). The remainder of the species (node 7) had either low N_2 fixation rates as detected in *Dicranoloma chilense, Andreaea laxifolia, Andreaea alpina* and *Anastrophyllum involutifolium*, or zero N_2 fixation rates, as in the liverworts *Clasmatocolea humilis, Blepharidophyllum densifolium, Chiloscyphus leptanthus, Heteroscyphus magellanicus* and the endohydric moss *Dendroligotrichum squamosum* (Table 1 and Fig. 4a). With our dataset, no effect was detected for glacier location (north *vs.* south) at species level.

The conditional inference tree (Fig. 4b) also showed a significant effect of bryophytes species identity on cyanobacteria relative abundance (nmol echinenone/cm^2 bryophyte) with the highest colonization found in the mosses *Ditrichum cylindricarpum, Racomitrium subcrispipilum, R. laevigatum, R. lanuginosum* and the liverwort *Cryptochila grandiflora* (Table 1 and node 2 in Fig 4b). Variability of N_2 fixation rates between bryophyte species was best explained by changes in epiphytic cyanobacteria abundance (positive linear regression: $R^2=0.44$, $P<0.001$, Fig. 5a) rather than the Shannon diversity index of cyanobacteria on host bryophytes (positive linear regression: $R^2=0.19$, $P=0.012$, Fig. 5b).

Community structure of epiphytic cyanobacteria

The Shannon index of epiphytic cyanobacteria OTUs per site (H'_{site}) at NG rose from 0.29 to a maximum of 0.96 at 26 years followed by a steady decline to original level at the oldest site. At SG, H'_{site} increased sharply from 0.38 and 0.32 at the two youngest sites to 1.37 at the oldest site within *Nothofagus* forest (Fig. 3h). Correspondence analysis showed a clear relationship between cyanobacteria genera and the stage of ecological succession at both glaciers. At NG (Fig. 6a), sites 2 and 3 share similar cyanobacteria communities; *Nostoc commune* was present at site 2 (OTU 1 & 3) and site 3 (OTU 6) whereas unidentified Nostocaceae was confined to site 3 (OTU 5, 7 & 15) and site 4 (OTU 17); *Stigonema spp* was found just at site 4 (OTU 16) and *Microcoleus spp.* (OTU 19) at site 5. At SG (Fig. 6b), all sites differed in their cyanobacteria communities, *Nostoc commune* was present at sites 1 (OTU 2), site 2 (OTU 1), site 3 (OTU 4 & 8) and site 5 (OTU 11); unidentified Nostocaceae at site 4 (OTU 9) and site 5 (OTU 15 & 12) and *Stigonema spp.* at site 4 (OTU 18) and site 5 (OTU 16, 13, 10 & 14). *Leptolyngbya spp.* (OTU20) was found at site 1. No relationship was observed between the genus of cyanobacteria and bryophyte species (Table 1 and Fig. S2).

Discussion

Patterns of soil properties and N_2 fixation related parameters along chronosequences

The trend observed in soil development followed the classic sequence reported for glacier foreland chronosequences [57,58]. Soil pH fell and total C and N contents increased with surface age along both chronosequences. However, the time required differs significantly to rates reported in other proglacial areas. Soils developed from barren rocks to Histosol, an organic soil, with 39% C in SG and 35% C in NG after 34 and 80 years of soil exposure, respectively (Fig. 3b). This contrasts with 2% C reported in other

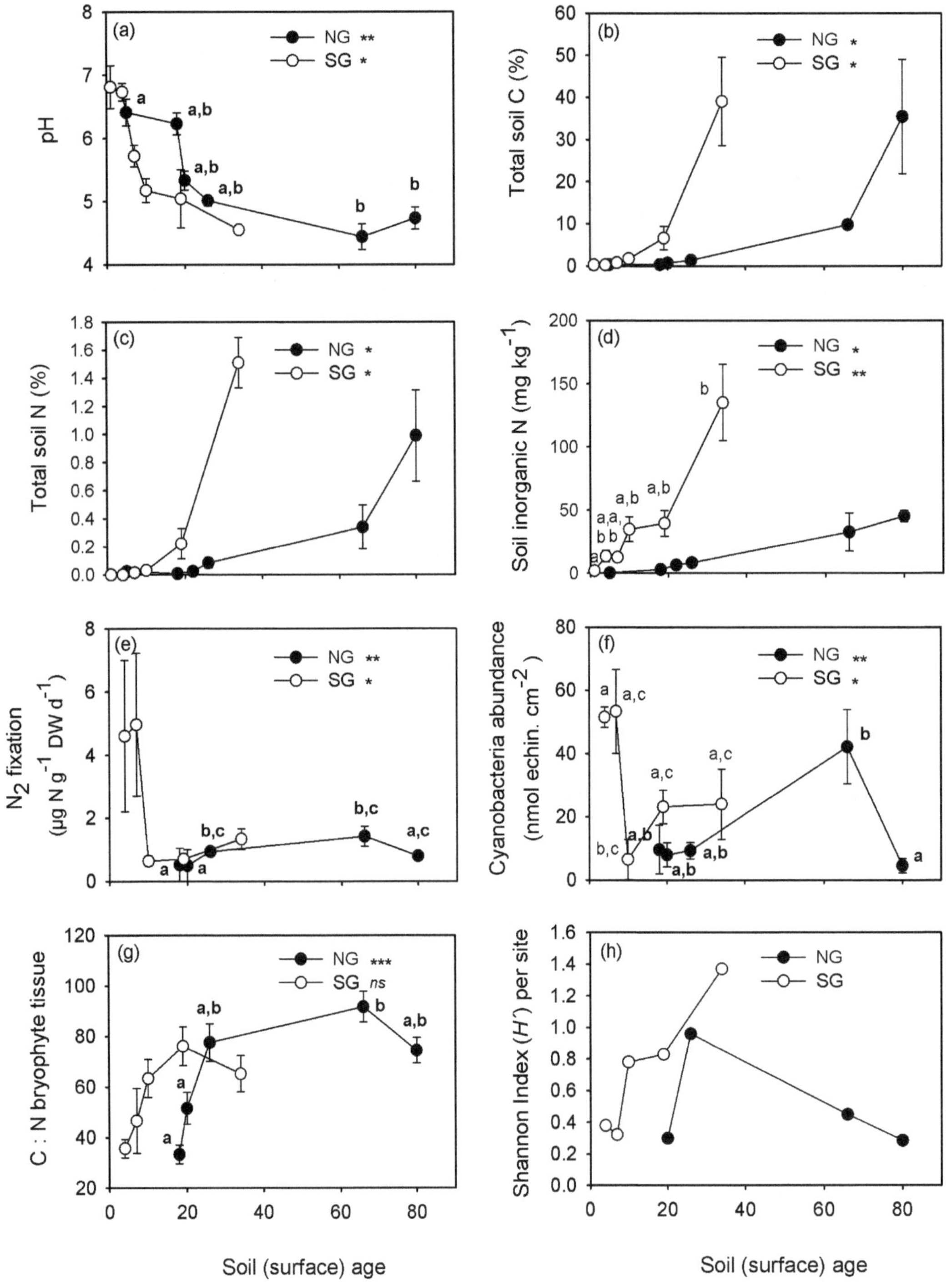

Figure 3. Patterns of soil properties and N_2 fixation related parameters along chronosequences. Site characteristics along the chronosequence in the north-side glacier (NG: black circles) with site soil (surface) age of 5, 18, 20, 26, 66 and 80 yrs and south-side glacier (SG: white circles) with site soil (surface) age of 1, 4, 7, 10, 19 and 34 yrs. At both glaciers, the younger site corresponds to bare ground close to the glacier front whereas the older site was covered within *Nothofagus* forests. Significant changes of parameters along each glacier chronosequence is indicated by *, $p<0.05$*, $p<0.01$**, $p<0.001$*** (Krukall-Wallis test). Different letters indicate significant differences between site ages after Mann–Whitney U-test post-hoc (Bonferroni corrected), bold letters are used for NG. Statistical differences on soil properties and N_2 fixation related parameters between the two glacier chronosequences are indicated in the result section.

studies reached 80 years after deglaciation [57]. Total soil N increased from very low or undetectable values at the youngest sites to 1.5% in SG and 0.99% in NG in the forest soil over this short period, again contrasting with concentration of 0.2% TN after 80 years exposure observed in the Central Alps [57]. These results suggest that conditions for vegetation succession are more favourable under the oceanic climatic conditions of Southern Hemisphere high latitude regions, such as Tierra del Fuego.

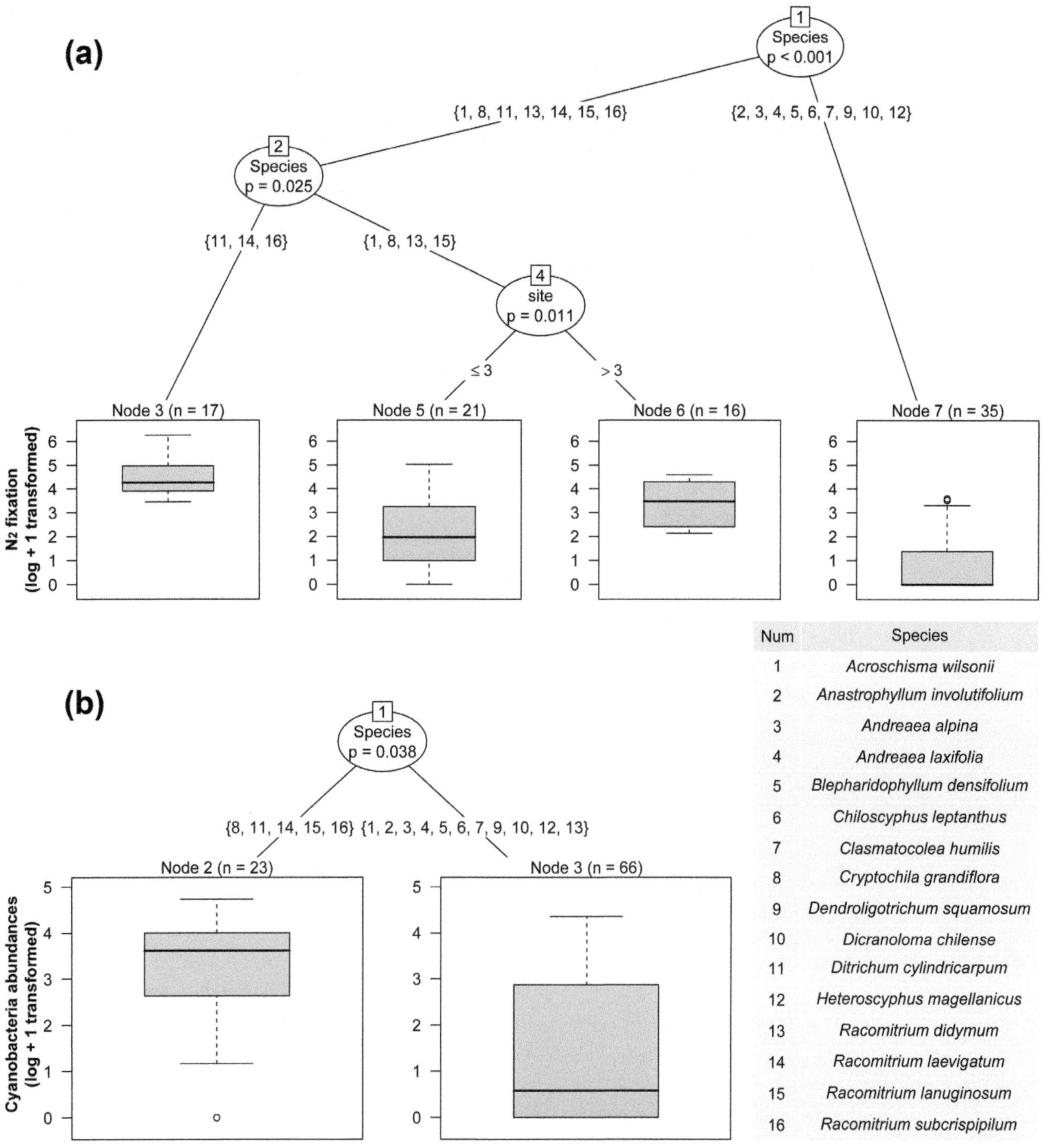

Num	Species
1	*Acroschisma wilsonii*
2	*Anastrophyllum involutifolium*
3	*Andreaea alpina*
4	*Andreaea laxifolia*
5	*Blepharidophyllum densifolium*
6	*Chiloscyphus leptanthus*
7	*Clasmatocolea humilis*
8	*Cryptochila grandiflora*
9	*Dendroligotrichum squamosum*
10	*Dicranoloma chilense*
11	*Ditrichum cylindricarpum*
12	*Heteroscyphus magellanicus*
13	*Racomitrium didymum*
14	*Racomitrium laevigatum*
15	*Racomitrium lanuginosum*
16	*Racomitrium subcrispipilum*

Figure 4. Conditional inference tree of: a) N_2 fixation capacity (nmol C_2H_2 g^{-1} dm bryo. h^{-1}) and b) cyanobacteria abundances (nmol echinenone/bryophyte cm^2) on bryophytes from both glaciers. Bubbles (nodes) indicate the variable with the strongest effect on the response, best split value, and associated probability (Bonferroni-adjusted $\alpha = 0.05$). Boxplots (terminal nodes) summarize the rate of N_2 fixation (log+1 transformed) or cyanobacteria abundances (log+1 transformed) in cases classified into each terminal node (n). Predictor variables included: bryophyte species identity (species effect), glacier locations (glacier effect) and the stage of ecological succession (sampling site effect). The bryophyte species name and their corresponded number are shown in the table within the graph.

Soil development along the two chronosequences, from bare ground to full forest, showed similarities in soil pH and total C and N concentration, indicating similar stages of soil development at both sites (Fig. 3a,b,c and Fig. S1). However, the time-required for C and N accumulation was much less at SG (Fig. 3d). In addition, after 4 and 7 years of soil exposure extractable inorganic N concentration was already much higher in the soil at SG (13.42 mg kg^{-1} and 12.51 mg kg^{-1} respectably, Fig. 3d) than at NG, where values remained low (0.16 to 8.23 mg kg^{-1}) even after 26 years of soil exposure. We also found the highest rates of N_2

Table 1. Mean ($N = 3$) N_2 fixation rates (nmol C_2H_2 g^{-1} dm bryo. h^{-1}) cyanobacteria abundance (nmol echinenone/bryophyte cm^2) and cyanobacteria genera associated with each individual bryophyte species along both glacier chronosequences.

			Mean reduction rate (nmol C_2H_2 g^{-1} dm bryo. h^{-1})		Cyanobacteria abundance (nmol echin. cm^{-2})	Cyanobacteria identification
Glacier	**Site**	**Bryophyte species**	**Mean**	**Range**		
SG	1	*Ditrichum cylindricarpum*	283.4	55–520	51.5 (3.2)	*N. commune. Leptolymbia sp*
SG	2	*Ditrichum cylindricarpum*	270.1	70–455	53.3 (8.2)	*N. commune*
SG	4	*Racomitrium subcrispipilum*	86.7	43–144	30.7 (12.1)	*N. commune*
SG	5	*Racomitrium subcrispipilum*	81.6	49–142	40.3 (36.2)	*N. commune.* Unidentified Nostocaceae *Stigonema*
NG	3	*Racomitrium laevigatum*	63.5	33–125	8.5 (6.9)	*N. commune*
NG	4	*Racomitrium laevigatum*	59.9	30–101	37.2 (9.4)	Unidentified Nostocaceae
NG	2	*Racomitrium didymum*	7.7	3–16	14.5 (14.5)	*N. commune*
NG	3	*Racomitrium didymum*	30.7	12–56	8.5 (5.3)	Unidentified Nostocaceae
NG	4	*Racomitrium didymum*	51.6	22–80	12.4 (7.1)	*Stigonema*
SG	3	*Racomitrium didymum*	0.0	0	0.3 (0.3)	*N. commune*
SG	4	*Racomitrium didymum*	20.7	7–34	21.5 (10.7)	Unidentified Nostocaceae
SG	5	*Racomitrium didymum*	66.6	53–75	39.4 (19.6)	*N. commune. Stigonema*
NG	1	*Andreaea laxifolia*	8.7	0–26	15.8 (15.5)	*Non found*
NG	3	*Andreaea laxifolia*	62.8	2–151	18.8 (9.6)	Unidentified Nostocaceae
NG	3	*Cryptochila grandiflora*	52.6	3–124	17.1 (4.5)	*N. commune*
NG	2	*Acroschisma wilsonii*	1.4	0–3	11.8 (6.0)	*Non found*
NG	3	*Acroschisma wilsonii*	6.1	1–14	0.0	Unidentified Nostocaceae
SG	3	*Acroschisma wilsonii*	25.7	25–40	12.5 (12.5)	*N. commune*
SG	4	*Acroschisma wilsonii*	18.4	9–28	19.5 (6.3)	Unidentified Nostocaceae
SG	5	*Acroschisma wilsonii*	39.8	8–72	9.9 (4.9)	Unidentified Nostocaceae
NG	4	*Racomitrium lanuginosum*	39.2	10–98	76.5 (22.0)	*Stigonema*
NG	3	*Dicranoloma chilense*	19.2	9–33	2.7 (2.7)	Unidentified Nostocaceae
NG	5	*Dicranoloma chilense*	2.9	0–9	5.9 (5.9)	*Non found*
NG	5	*Anastrophyllum involutifolium*	3.5	0–11	5.7 (5.1)	*Non found*
SG	4	*Clasmatocolea humilis*	1.5	0–3	15.2 (4.5)	*Stigonema*
NG	2	*Andreaea alpina*	0.9	0–3	2.7 (2.7)	*Non found*
NG	5	*Blepharidophyllum densifolium*	0.0	0	6.1 (6.1)	*Microcoleus*
SG	5	*Chiloscyphus leptanthus*	0.0	0	4.8 (4.8)	*Non found*
NG	1	*Dendroligotrichum squamosum*	0.0	0	3.2 (3.2)	*Non found*
NG	2	*Dendroligotrichum squamosum*	0.0	0	2.5 (2.5)	*Non found*
NG	5	*Heteroscyphus magellanicus*	0.0	0	0.1 (0.1)	*Non found*

In both glaciers the younger site (site 1) correspond to early succession site close to the glacier front whereas the older site (site 5) was settled within *Nothofagus* forests. Sites 3–4 represent intermediate stages of ecological succession. No species were sampled at site 0. SE for cyanobacteria abundances is shown in brackets. NG: north-side glacier; SG: south-side glacier of the mountain range.

fixation and cyanobacteria abundance in the pioneer terricolous moss *Ditrichum cylindricarpum* at 4 and 7 years soil exposure in SG (Fig 3e,f Table 1), where it forms a dense carpet in the soil (Fig. 1b). By contrast, in NG biological N_2 fixation and epiphytic cyanobacteria abundance on bryophytes was very low during the early stages of succession (Fig. 3e,f). It is likely that higher annual rainfall on SG is associated with higher rates of N_2 fixation on bryophyte-cyanobacteria associations in young soils, since fixation of N is strongly controlled by moisture availability [30]. Earlier studies have shown the importance of free living colonies of diazotrophs for C and N inputs during early stages of succession in glacial forelands [59–62] which could contribute to nutrient accumulation at our sites. Fixation of N by cyanobacteria associated with bryophytes that colonize soil surfaces very early after glacier retreat, as shown by our results, could also help accelerate soil development in recently deglaciated areas under wetter conditions.

As succession proceeds in SG, terricolous mosses are replaced by perennial herbs like *Gunnera magellanica,* which forms symbiotic associations with the cyanobacterial genus Nostoc [63], and becomes the dominant species together with *Uncinia tenuis,* which are likely to be a major factor promoting C and N accumulation on soil. In intermediate-aged sites, mosses are confined to rock outcrops (Fig. 1b) where, together with lichen species like *Placopsis*

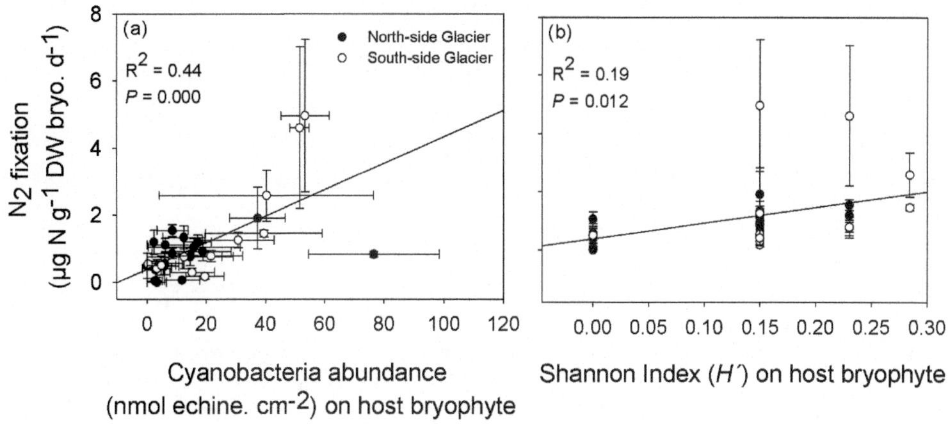

Figure 5. Relationship between N_2 fixation and the abundance and diversity of epiphytic cyanobacteria. Relationship between N_2 fixation (µg N g^{-1} DW bryo. d^{-1}) on each individual bryophyte species and the abundance (nmol echine. cm-2) and diversity (Shannon Index) of their host cyanobacteria. Bryophyte species from both glaciers are included. The results of the linear regression analysis are shown in the upper left side of the figure.

perrugosa [64], form a thick layer of organic material on rock surfaces (Fig. 1b). Despite rock outcrops possibly being a less favourable surface for N_2 fixation, we found high rates of fixation, abundance and diversity of epiphytic cyanobacteria on species like *Racomitrium subcrispipilum*, *Racomitrium didymium* and *Acroschisma wilsonii* which formed moss mats on rock surfaces at intermediate-aged sites (19 years old sites in SG) and even within *Nothofagus* forest (Table 1 and Fig. 3h). However, at the drier side of the mountain range where NG flows, intermediate-aged sites are dominated by the dwarf shrub *Empetrum rubrum* which allows the development of dense mats of terricolous bryophyte species such as *Racomitrium lanuginosum*, *R. laevigatum* and *Cryptochila grandiflora* (Fig. 1b). It is at this stage of succession that we found the highest rate of N_2 fixation and cyanobacteria colonization on bryophytes at NG (Fig. 3e,f and Table 1). In this drier area dominated by dwarf shrub vegetation, cyanobacteria in association with bryophytes are possibly one of the main sources of fixed N, since cyanobacteria in the soil decrease with increasing vegetation development, and the occurrence of N-fixing Nostoc host, *Gunnera magellanica*, is reduced under drier conditions.

It remains unclear, however, how the newly fixed N in the bryophyte-cyanobacteria system becomes available to vascular plants. We consider two pathways, described by [30], which may occur during primary succession in our study area. First, direct transfer of N from the moss system to vascular plants by mycorrhizal linkages: most vascular plant species in the study area form mycorrhizal association (e.g. ectomycorrhizal association with *Nothofagus* species or ericoid mycorrhizal with *Empetrum* spp.). Second, release of the newly fixed N retained in the bryophyte-cyanobacteria system after disturbance events, for example after membrane damage from desiccation following wet/dry cycles [65,66].

Overall, glacier forelands in Tierra del Fuego show fast rates of soil transformation, implying large quantities of N inputs to allow fast vegetation succession. Our results highlight that bryophyte-cyanobacteria associations have the potential to contribute to N accumulation and cycling in these areas with very low inputs of N from atmospheric deposition.

Differences on bryophyte N_2 fixation capacity and cyanobacteria abundance

Few studies have investigated interspecific or intraspecific variation in N_2 fixation capacity on bryophytes, particularly in the context of post-glacial primary succession. In our study, most bryophyte species have the ability to support an epiphytic flora of cyanobacteria. However, N_2 fixation rates varied between and within bryophyte species (Table 1). In line with results reported in previous studies [5], we detected differences in N_2 fixation rates between mosses and liverworts. Five of the six studied liverworts showed no or very low rate of N_2 fixation and cyanobacteria abundance (Table 1), and only one species, the terricolous liverwort *Cryptochila grandiflora*, showed high fixation rates (52.6 nmol C_2H_2 g^{-1} dm bryo. h^{-1}) and cyanobacteria colonization. Among moss species, *Dendroligotrichum squamosum* also showed no N_2 fixation or cyanobacteria colonization, possibly due to their rigid, water repellent leaves which may confer less favorable moisture conditions. However, morphology or habitat could not explain variations found among other moss species. Both terricolous (*Ditrichum cylindricarpum*, *Racomitrium laevigatum*, *R. lanuginosum*) and saxicolous species (*Racomitrium subcrispipilum*, *R. didymum*, *Acroschisma wilsonii*) showed high N_2 fixation rates (over 40 nmol C_2H_2 g^{-1} dm bryo. h^{-1}) compared to rates reported in Arctic [8] and sub-Antarctic [16] ecosystems. Regarding intraspecific variations, we observed that the rates of N_2 fixation of the saxicolous mosses *Racomitrium didymium* and *Acroschisma wilsonii* (the most frequently sampled species in the study area) were affected by the stages of ecological succession. The rate of N_2 fixation significantly increases on *R. didymium* and *Acroschisma wilsonii* populations that grow at sites that have been longer deglaciated, reflecting growth of the cyanobacterial population as vegetation develops when growing epiphytically on these mosses.

Community structure of epiphytic cyanobacteria

In the present study, we also aimed to evaluate the influence of cyanobacteria community structure on bryophyte N_2 fixation rates. We found that abundance of epiphytic cyanobacteria appears to be a better factor explaining differences in N_2 fixation rates (Fig. 5) than epiphytic cyanobacteria diversity on host bryophyte species (Fig. 5). It is also noteworthy that *Nostoc commune*-associated bryophyte species, both mosses and liverworts, showed the highest rate of N_2 fixation (Table 1), which is in agreement

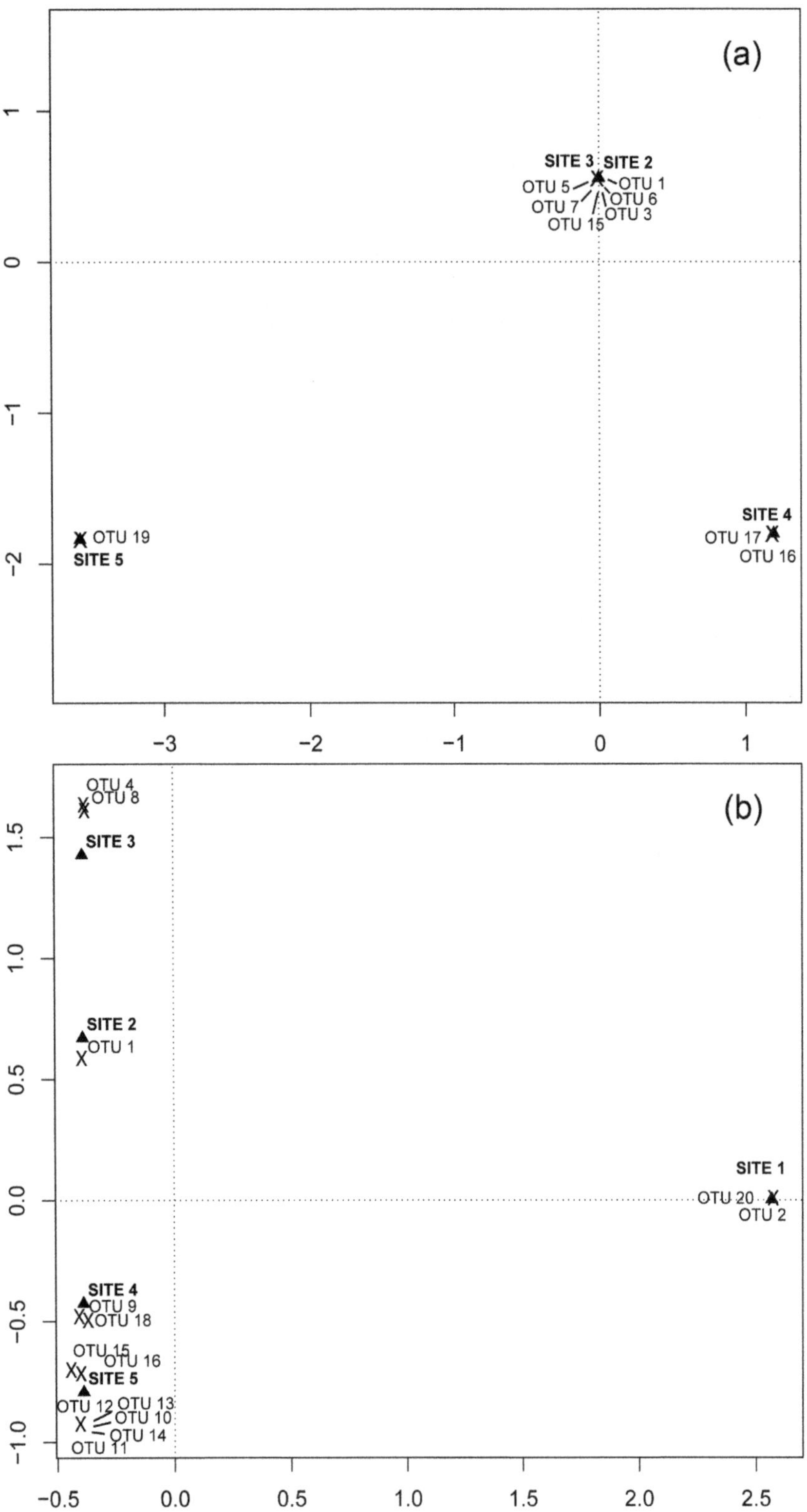

Figure 6. Relationships between cyanobacteria and the stages of ecological succession. Relationships between OTUs of epiphytic cyanobacteria obtained at a 0.01% cut-off and the different stages of ecological succession (triangles) -the younger site (site 1) correspond to bare ground close to the glacier front whereas the older site (site 5) correspond to *Nothofagus* forests- in the north-side glacier (a) and south-side glacier (b) after correspondence analysis. Site 1 is not included in (a) because cyanobacteria were not found epiphytically on bryophytes at that site. For OTUs identification please see Fig. 2.

with the suggestion that *Nostoc commune,* in particular, is an important contributor of soil N in both polar regions [32].

Specific affinities of individual cyanobacteria genera with particular bryophyte species were not apparent, as the same bryophyte species can support the three main cyanobacteria genera found here (*Nostoc,* unidentified Nostocaceae group and *Stigonema*) (Table 1 and Fig. S2). However, a relationship between cyanobacteria genera and the different stages of succession was apparent in both glaciers. With increasing age along the chronosequences, cyanobacteria shifted from: *Nostoc commune*-dominated communities at young sites; to unidentified Nostocaceae-dominated communities at intermediate sites; and finally to *Stigonema* spp-dominated communities at older sites (Fig. 6). Such procession correlated with the soil development trends found here and contrasts with results observed in mature boreal forests where moss species identity serves as an important determinant of cyanobacterial communities that inhabit mosses [22]. With our data set, our results suggest that the stage of succession acts as a strong driver in determining the composition of epiphytic cyanobacteria on bryophytes during primary succession. Despite this similarity between the two glaciers, diversity of epiphytic cyanobacteria at SG was higher than at NG and increased with time since deglaciation (Fig. 3h), reflecting more favourable conditions for cyanobacteria growth on the wetter southern side of Cordillera Darwin.

Conclusions

We have compared, from a multidisciplinary perspective, chronosequences in front of two receding glaciers with contrasting climatic conditions (wetter *vs* drier) and differing rates of ecological succession in Tierra del Fuego (Chile). In doing so, we have shown that bryophyte-cyanobacteria associations have the potential to contribute to N accumulation very early after glacier retreat, a process that could help promote soil development. At intermediate-aged sites, and under drier conditions, bryophytes can grow among dwarf shrub plants, forming dense mats that host cyanobacteria species and exhibit high N_2 fixation rates, contributing to N accumulation and cycling. We also found that differences in N_2 fixation capacity between bryophyte species were primarily driven by the abundance of epiphytic cyanobacteria rather than cyanobacterial community composition. Most liverworts showed low colonization, whilst in mosses rates of N_2 fixation did not exhibit consistent differences across life forms and habitat (saxicolous *vs* terricolous). Finally, we identified that the main factor structuring community composition of epiphytic cyanobacteria was the stage of ecological succession, with no relationship to host species identity.

Supporting Information

Figure S1 Scores variation of the general factor component of soil development along the glacier chronosequence in the north-side (black bars) and south-side (white bars) of Cordillera Darwin. Sites with higher loadings reflect a more developed soil. The general factor component of soil development was obtained by reduction of soil pH and total soil carbon and nitrogen by principal component analysis. Mean values (± SE) of three replicate transect are shown.

Figure S2 Correspondence analysis between OTUs of cyanobacteria obtained at a 0.01% cut-off and each individual bryophyte species, just the six most abundance bryophyte species with cyanbacteria colonies were used in this analysis, identification of OUTs can be see in Fig.2.

Table S1 Changes in bryophyte species along two proglacial chronosequence at the Cordillera Darwin (Tierra del Fuego, Chile).

Table S2 Gen Bank accession numbers corresponding to species and specimens used for tree inference.

Appendix S1 Calculation of different vial headspaces for ARA and $^{15}N_2$ uptake essays.

Acknowledgments

MA-C benefited from a postdoctoral Juan de la Cierva grant of the Ministerio de Ciencia e Innovación of Spain and a position as Associate Member at the University of Sheffield (UK). We are indebted to Gisela Oliván, Carolina Leon, William R. Buck, Juan ***Larraín***, John J. Engel, Barbara Murray and Esther Fuertes for their inestimable help with the identification of the bryophyte specimens and M[a] Jose Malo, Victoria Santonaya (CSIC) and José M. Blanquer, Carlos M Castilla and M. Pilar Villar Gil (UCM) for their technical assistant. We thank Dolores Saco for their support in the lab work and ***Víctor*** *Arroyo-Rodríguez* and Jesús Aboal for helping on statistical analysis. We are grateful to the Universidad de Magallanes (UMAG) and the Institute of Ecology and Biodiversity for logistic and scientific support. Francisca Massardo (UMAG and Fundación Omora) are particularly thanked for their excellent organization of the fieldwork in the complex area of the Darwin Mountains.

Author Contributions

Conceived and designed the experiments: MA-C SPO TGAG LGS. Performed the experiments: MA-C SPO. Analyzed the data: MA-C SPO ADLR NT MC MTDLC. Contributed reagents/materials/analysis tools: ROH DP. Wrote the paper: MA-C SPO. Carried out sampling field work: AP. Organised field logistics: AP RR LGS. Selected field sites: DP RR. Arranged required permissions and permits: RR. Discussed the results and commented on the manuscript: MA-C SPO TGAG LGS ADLR ROH RR NT MC MTDLC AP.

References

1. Elbert W, Weber B, Burrows S, Steinkamp J, Budel B, et al. (2012) Contribution of cryptogamic covers to the global cycles of carbon and nitrogen. Nature Geosci 5: 459–462.
2. Solheim B, Zielke M (2002) Associations Between Cyanobacteria and Mosses. In: Rai ANB, B.; Rasmussen, Ulla (Eds.), editor. Cyanobacteria in Symbiosis: Springer. pp. 137–152.
3. Zielke M, Solheim B, Spjelkavik S, Olsen RA (2005) Nitrogen fixation in the high arctic: Role of vegetation and environmental conditions. Arctic Antarctic and Alpine Research 37: 372–378.
4. Zackrisson O, DeLuca TH, Gentili F, Sellstedt A, Jaderlund A (2009) Nitrogen fixation in mixed Hylocomium splendens moss communities. Oecologia 160: 309–319.
5. Gavazov KS, Soudzilovskaia NA, van Logtestijn RSP, Braster M, Cornelissen JHC (2010) Isotopic analysis of cyanobacterial nitrogen fixation associated with subarctic lichen and bryophyte species. Plant and Soil 333: 507–517.
6. Stewart KJ, Coxson D, Grogan P (2011) Nitrogen Inputs by Associative Cyanobacteria across a Low Arctic Tundra Landscape. Arctic Antarctic and Alpine Research 43: 267–278.
7. Chapin DM, Bliss LC, Bledsoe LJ (1991) Environmental-regulation of nitrogen-fixation in a high arctic lowland ecosystem. Canadian Journal of Botany-Revue Canadienne De Botanique 69: 2744–2755.
8. Solheim B, Endal A, Vigstad H (1996) Nitrogen fixation in Arctic vegetation and soils from Svalbard, Norway. Polar Biology 16: 35–40.
9. DeLuca TH, Zackrisson O, Nilsson MC, Sellstedt A (2002) Quantifying nitrogen-fixation in feather moss carpets of boreal forests. Nature 419: 917–920.

10. Sorensen PL, Jonasson S, Michelsen A (2006) Nitrogen fixation, denitrification, and ecosystem nitrogen pools in relation to vegetation development in the subarctic. Arctic Antarctic and Alpine Research 38: 263–272.
11. Dodds WK, Gudder DA, Mollenhauer D (1995) The ecology of nostoc. Journal of Phycology 31: 2–18.
12. DeLuca TH, Zackrisson O, Gundale MJ, Nilsson MC (2008) Ecosystem feedbacks and nitrogen fixation in boreal forests. Science 320: 1181–1181.
13. Bowden RD (1991) Inputs, outputs, and accumulation of nitrogen in an early successional moss (polytrichum) ecosystem. Ecological Monographs 61: 207–223.
14. Rozzi R, Armesto JJ, Goffinet B, Buck W, Massardo F, et al. (2008) Changing lenses to assess biodiversity: patterns of species richness in sub-Antarctic plants and implications for global conservation. Frontiers in Ecology and the Environment 6: 131–137.
15. Basilier K (1979) Moss-associated nitrogen-fixation in some mire and coniferous forest environments around uppsala, sweden. Lindbergia 5: 84–88.
16. Smith VR, Russell S (1982) Acetylene-reduction by bryophyte-cyanobacteria associations on a sub-antarctic island. Polar Biology 1: 153–157.
17. Line MA (1992) Nitrogen-fixation in the sub-antarctic macquarie island. Polar Biology 11: 601–606.
18. Gundale MJ, Nilsson M, Bansal S, Jaderlund A (2012) The interactive effects of temperature and light on biological nitrogen fixation in boreal forests. New Phytologist 194: 453–463.
19. Sorensen PL, Lett S, Michelsen A (2012) Moss-specific changes in nitrogen fixation following two decades of warming, shading, and fertilizer addition. Plant Ecology 213: 695–706.
20. Gundale MJ, Gustafsson H, Nilsson MC (2009) The sensitivity of nitrogen fixation by a feathermoss-cyanobacteria association to litter and moisture variability in young and old boreal forests. Canadian Journal of Forest Research-Revue Canadienne De Recherche Forestiere 39: 2542–2549.
21. Pandey KD, Kashyap AK, Gupta RK (1992) Nitrogen fixation by cyanobacteria associated with moss communities in Schirmacher Oasis, Antarctica. Israel Journal of Botany 41: 187–198.
22. Ininbergs K, Bay G, Rasmussen U, Wardle DA, Nilsson MC (2011) Composition and diversity of nifH genes of nitrogen-fixing cyanobacteria associated with boreal forest feather mosses. New Phytologist 192: 507–517.
23. Bates JW (2009) Mineral nutrition, substratum ecology, and pollution. In: Shaw AJ, Goffinet B, editors. Bryophyte Biology, Cambridge: Cambridge University Press. pp. 2990–0356.
24. Masiokas MH, Villalba R, Luckman BH, Lascano ME, Delgado S, et al. (2008) 20th-century glacier recession and regional hydroclimatic changes in northwestern Patagonia. Global and Planetary Change 60: 85–100.
25. Kohls SJ, Baker DD, van Kessel C, Dawson JO (2003) An assessment of soil enrichment by actinorhizal N-2 fixation using delta N-15 values in a chronosequence of deglaciation at Glacier Bay, Alaska. Plant and Soil 254: 11–17.
26. Sancho LG, Palacios D, Green TGA, Vivas M, Pintado A (2011) Extreme high lichen growth rates detected in recently deglaciated areas in Tierra del Fuego. Polar Biology 34: 813–822.
27. Dentener F, Stevenson D, Ellingsen K, van Noije T, Schultz M, et al. (2006) The global atmospheric environment for the next generation. Environmental Science & Technology 40: 3586–3594.
28. Menge DNL, Hedin LO (2009) Nitrogen fixation in different biogeochemical niches along a 120 000-year chronosequence in New Zealand. Ecology 90: 2190–2201.
29. Koppes M, Hallet B, Anderson J (2009) Synchronous acceleration of ice loss and glacial erosion, Glaciar Marinelli, Chilean Tierra del Fuego. Journal of Glaciology 55: 207–220.
30. Lindo Z, Nilsson MC, Gundale MJ (2013) Bryophyte-cyanobacteria associations as regulators of the northern latitude carbon balance in response to global change. Global Change Biology 19: 2022–2035.
31. Santana A, Porter C, Butorovic N, Olave C (2006) First climatologic antecedents of automatic weather stations (AWS) in the Beagle Channel, Magallanes, Chile. An Inst Patagonia 34: 5–20.
32. Vincent WF (2000) Cyanobacterial dominance in the polar regions. In: Whitton BA, Potts M, editors. The ecology of cyanobacteria, Dordrecht: Kluwe. pp. 321–340.
33. ISRIC (2002) Procedures for soil analysis; Centre ISRal, editor.
34. America SSSo (1982) Methods of soil analysis. Part 2. Chemical and Microbiological Properties; Page AL, Miller RH, Keeny DR, editors. Winsconsin, USA: Madison.
35. Weber MG, Vancleve K (1984) Nitrogen transformations in feather moss and forest floor layers of interior alaska black spruce ecosystems. Canadian Journal of Forest Research-Revue Canadienne De Recherche Forestiere 14: 278–290.
36. Millbank JW (1981) The assessment of nitrogen-fixation and throughput by lichens. 1. the use of a controlled environment chamber to relate acetylene-reduction estimates to nitrogen-fixation. New Phytologist 89: 647–655.
37. Basilier K (1980) Fixation and uptake of nitrogen in sphagnum blue-green-algal associations. Oikos 34: 239–242.
38. Liengen T (1999) Conversion factor between acetylene reduction and nitrogen fixation in free-living cyanobacteria from high arctic habitats. Canadian Journal of Microbiology 45: 223–229.
39. Val J, Monge E, Baker NR (1994) An improved hplc method for rapid analysis of the xanthophyll cycle pigments. Journal of Chromatographic Science 32: 286–289.
40. Cubero OF, Crespo A, Fatehi J, Bridge PD (1999) DNA extraction and PCR amplification method suitable for fresh, herbarium-stored, lichenized, and other fungi. Plant Systematics and Evolution 216: 243–249.
41. Nubel U, GarciaPichel F, Muyzer G (1997) PCR primers to amplify 16S rRNA genes from cyanobacteria. Applied and Environmental Microbiology 63: 3327–3332.
42. de los Rios A, Valea S, Ascaso C, Davila A, Kastovsky J, et al. (2010) Comparative analysis of the microbial communities inhabiting halite evaporites of the Atacama Desert. International Microbiology 13: 79–89.
43. Schloss PD, Westcott SL, Ryabin T, Hall JR, Hartmann M, et al. (2009) Introducing mothur: Open-Source, Platform-Independent, Community-Supported Software for Describing and Comparing Microbial Communities. Applied and Environmental Microbiology 75: 7537–7541.
44. Stackebrand E, Ebers J (2006) Taxonomic parameters revisited: tarnished gold standards. Microbiology Today 33: 152–155.
45. Edgar RC (2004) MUSCLE: multiple sequence alignment with high accuracy and high throughput. Nucleic Acids Research 32: 1792–1797.
46. Hall TA (1999) BioEdit: a user-friendly biological sequence alignment editor and analysis. http://www.mbio.ncsu.edu/BioEdit/bioedit.html.
47. Posada D (2008) jModelTest: Phylogenetic model averaging. Molecular Biology and Evolution 25: 1253–1256.
48. Akaike H (1974) New look at statistical-model identification. Ieee Transactions on Automatic Control AC19: 716–723.
49. Tavare S (1986) Some probabilistic and statistical problems in the analysis of dna sequences. Miura, R M. pp. 57–86.
50. Guindon S, Dufayard JF, Lefort V, Anisimova M, Hordijk W, et al. (2010) New Algorithms and Methods to Estimate Maximum-Likelihood Phylogenies: Assessing the Performance of PhyML 3.0. Systematic Biology 59: 307–321.
51. Ronquist F, Huelsenbeck JP (2003) MrBayes 3: Bayesian phylogenetic inference under mixed models. Bioinformatics 19: 1572–1574.
52. Hill TCJ, Walsh KA, Harris JA, Moffett BF (2003) Using ecological diversity measures with bacterial communities. Fems Microbiology Ecology 43: 1–11.
53. Filzmoser P (1999) Robust principal component and factor analysis in the geostatistical treatment of environmental data. Environmetrics 10: 363–375.
54. Hothorn T, Hornik K, Strobl C, Zeileis A (2010) Party: A laboratory for recursive partytioning. Citeseer.
55. Hothorn T, Hornik K, Zeileis A (2006) Unbiased recursive partitioning: A conditional inference framework. Journal of Computational and Graphical Statistics 15: 651–674.
56. Nenadic O, Greenacre M (2007) Correspondence analysis in R, with two- and three-dimensional graphics: The ca package. Journal of Statistical Software 20.
57. Bernasconi SM, Bauder A, Bourdon B, Brunner I, Bunemann E, et al. (2011) Chemical and Biological Gradients along the Damma Glacier Soil Chronosequence, Switzerland. Vadose Zone Journal 10: 867–883.
58. Chapin FS, Walker LR, Fastie CL, Sharman LC (1994) Mechanisms of primary succession following deglaciation at glacier bay, alaska. Ecological Monographs 64: 149–175.
59. Schmidt SK, Reed SC, Nemergut DR, Grandy AS, Cleveland CC, et al. (2008) The earliest stages of ecosystem succession in high-elevation (5000 metres above sea level), recently deglaciated soils. Proceedings of the Royal Society B-Biological Sciences 275: 2793–2802.
60. Knelman JE, Legg TM, O'Neill SP, Washenberger CL, Gonzalez A, et al. (2012) Bacterial community structure and function change in association with colonizer plants during early primary succession in a glacier forefield. Soil Biology & Biochemistry 46: 172–180.
61. Zumsteg A, Luster J, Goransson H, Smittenberg RH, Brunner I, et al. (2012) Bacterial, archaeal and fungal succession in the forefield of a receding glacier. Microbial Ecology 63: 552–564.
62. Brankatschk R, Towe S, Kleineidam K, Schloter M, Zeyer J (2011) Abundances and potential activities of nitrogen cycling microbial communities along a chronosequence of a glacier forefield. Isme Journal 5: 1025–1037.
63. Fernandez-Martinez MA, de los Rios A, Sancho LG, Perez-Ortega S (2013) Diversity of Endosymbiotic Nostoc in Gunnera magellanica (L) from Tierra del Fuego, Chile. Microbial Ecology 66: 335–350.
64. Raggio J, Green TGA, Crittenden PD, Pintado A, Vivas M, et al. (2012) Comparative ecophysiology of three Placopsis species, pioneer lichens in recently exposed Chilean glacial forelands. Symbiosis 56: 55–66.
65. Wilson JA, Coxson DS (1999) Carbon flux in a subalpine spruce-fir forest: pulse release from Hylocomium splendens feather-moss mats. Canadian Journal of Botany-Revue Canadienne De Botanique 77: 564–569.
66. Wickland KP, Neff JC, Aiken GR (2007) Dissolved organic carbon in Alaskan boreal forest: Sources, chemical characteristics, and biodegradability. Ecosystems 10: 1323–1340.

16

Soil Bacterial Community Response to Differences in Agricultural Management along with Seasonal Changes in a Mediterranean Region

Annamaria Bevivino[1]*, Patrizia Paganin[1], Giovanni Bacci[2,3], Alessandro Florio[2], Maite Sampedro Pellicer[1], Maria Cristiana Papaleo[3], Alessio Mengoni[3], Luigi Ledda[4], Renato Fani[3], Anna Benedetti[2], Claudia Dalmastri[1]

1 ENEA (Italian National Agency for New Technologies, Energy and Sustainable Economic Development) Casaccia Research Center, Technical Unit for Sustainable Development and Innovation of Agro-Industrial System, Rome, Italy, **2** Consiglio per la Ricerca e la Sperimentazione in Agricoltura - Research Centre for the Soil-Plant System, Rome, Italy, **3** Laboratory of Microbial and Molecular Evolution, Department of Biology, University of Florence, Florence, Italy, **4** Dipartimento di Agraria, University of Sassari, Sassari, Italy

Abstract

Land-use change is considered likely to be one of main drivers of biodiversity changes in grassland ecosystems. To gain insight into the impact of land use on the underlying soil bacterial communities, we aimed at determining the effects of agricultural management, along with seasonal variations, on soil bacterial community in a Mediterranean ecosystem where different land-use and plant cover types led to the creation of a soil and vegetation gradient. A set of soils subjected to different anthropogenic impact in a typical Mediterranean landscape, dominated by *Quercus suber* L., was examined in spring and autumn: a natural cork-oak forest, a pasture, a managed meadow, and two vineyards (ploughed and grass covered). Land uses affected the chemical and structural composition of the most stabilised fractions of soil organic matter and reduced soil C stocks and labile organic matter at both sampling season. A significant effect of land uses on bacterial community structure as well as an interaction effect between land uses and season was revealed by the EP index. Cluster analysis of culture-dependent DGGE patterns showed a different seasonal distribution of soil bacterial populations with subgroups associated to different land uses, in agreement with culture-independent T-RFLP results. Soils subjected to low human inputs (cork-oak forest and pasture) showed a more stable bacterial community than those with high human input (vineyards and managed meadow). Phylogenetic analysis revealed the predominance of *Proteobacteria*, *Actinobacteria*, *Bacteroidetes*, and *Firmicutes* phyla with differences in class composition across the site, suggesting that the microbial composition changes in response to land uses. Taken altogether, our data suggest that soil bacterial communities were seasonally distinct and exhibited compositional shifts that tracked with changes in land use and soil management. These findings may contribute to future searches for bacterial bio-indicators of soil health and sustainable productivity.

Editor: Jack Anthony Gilbert, Argonne National Laboratory, United States of America

Funding: This research was funded by MIUR (Integrated Special Fund for Research - FISR) in the frame of the Italian National Project SOILSINK "Climate change and agro-forestry systems, impacts on soil carbon sink and microbial diversity", and partially supported by MIUR (Research Department of Italian Government) in the framework of the Agreement Program ENEA-CNR (Articolo 2, comma 44, Legge 23.12.2009 n. 191 - Legge Finanziaria 2010). The funders had no role in study design, data collection and analysis, decision to publish, or preparation of the manuscript.

Competing Interests: The authors have declared that no competing interests exist.

* Email: annamaria.bevivino@enea.it

Introduction

Soil microorganisms play an important role as regulators of major biogeochemical cycles and can significantly affect the ecosystem functioning [1], being involved in organic matter dynamics, nutrient cycling and decomposition processes [2]. The anthropogenic activities affect the diversity of natural habitats modifying the number of species occurring in the environment at the landscape scale. Soil management strongly influences soil biodiversity in agricultural ecosystems. Different practices can alter the below-ground ecosystem, often leading to depletion of soil carbon and loss of biodiversity, and thus affecting the structure of the resident microbial communities [3]. Therefore, characterizing genetic and functional diversity of soil bacterial communities in response to agricultural practices and/or climate is fundamental to better understand and manage the ecosystem processes.

The Mediterranean area is one of the most important biodiversity hotspots in the world and is increasingly threatened by intensive land use [4]. The high environmental diversity that characterizes the Mediterranean region is related to the integration of natural ecosystems and traditional human activities such as the agroforestry practices [5]. The collapse of the traditional agro-silvo-pastoral system that occurred during the past century has led to major changes in the extension of woodlands dominated by typical Mediterranean species, i.e. cork oak (*Quercus suber*) and/or holm oak (*Quercus ilex*) woodlands [6], [7]. Research on the influence of management practices on the biodiversity of these agro-silvo-pastoral systems is increasing but it has focused mostly on plants [8] and vertebrates [5], [9]. In the frame of the Italian Project SOILSINK (Climatic changes and agricultural and forest systems: impact on C reservoirs and on soil microbial diversity), a hilly basin in Gallura (Berchidda site, Sardinia, Italy) was selected as a reference Mediterranean site for studying the influence of land-use changes on diversity, function and seasonal variations of soil microbial communities [10], [11]. The site is within an area of about 1,450 ha and is characterised by extensive agro-silvo-pastoral systems, typical of north-eastern Sardinia (Italy) and similar areas of the Mediterranean basin [12]. The chosen site represents a sustainable balance between human activities and natural resources that have created a landscape of high heterogeneity and cultural value, whose importance has been recognized at the European level [13], [14]. Indeed, it is considered climatically (Mediterranean zone) and pedologically homogeneous with vegetation patterns similar to those called *dehesas* or *montados* of the south-western Iberian Peninsula [15], [16], [17]. In the past, this area was covered by cork-oak forests, which gradually were subjected to increasing under-storey grazing and usage for the extraction of cork. Today, there are different land-use and plant cover types that lead to a soil and vegetation gradient with an ecological progression: from a cork-oak forest undergoing minimum disturbance to managed vineyards with an intensive agricultural practice (grass covered and ploughed), passing through areas with temporary grassland, and pasture.

The different land uses altered soil potential, making possible to discriminate the role of human management on soil functioning. When forests are converted to grasslands, and grasslands turned into agricultural lands, a sharp switch from one type of soil microbial community to another one occurs. Since the ability of an ecosystem to withstand serious disturbances may partly depend on its microbial component(s), characterizing bacterial community composition and/or structure might help to better understand and manipulate ecosystem processes. The aim of the present study was to investigate the effects of soil characteristics and different agricultural managements on soil bacterial community in two seasons. Sampling was carried out in spring and autumn 2007, when the plant cover-growing season usually starts and ends. A combination of culture-based and molecular techniques along with statistical analysis of data obtained was applied to interrogate the diversity, function, and ecology of soil bacterial communities. The results obtained in the present study, along with the other ones obtained within the SOILSINK Project [11], [18], [19], [20], provide useful data on the impact of soil type, cover vegetation, and human activities on the distribution of the bacterial genetic resources in soil communities for this Mediterranean region. Our results confirm that the environments with low inputs (cork-oak forest and pasture) show a more stable soil microbial community than those subjected to increasing human input (vineyards and managed meadow) and suggest that soil bacterial communities are seasonally distinct with compositional shifts that track with changes in land use and soil management.

Materials and Methods

Ethics Statement

We carried out the study on the hilly basin in Gallura (Olbia-Tempio municipalities, Sardinia, Italy). Five soils uses were identified in private farms within Berchidda site (40°49′ 15″N, 9°17′ 32″ E) and were obtained. The soil sampling was carried out in the frame of a national research project (SOILSINK Project) and soils used in this study were collected under consent of the landowners. The responsible of the study site was Prof. P.P. Roggero (University of Sassari, Italy). We confirm that our study did not harm the environment and did not involve endangered or protected species. Specific geographic coordinates (referred to World Geodetic System, 1984) of our study area are reported in Table S1.

Sampling site

The study area (Olbia-Tempio) is representative of the climate, vegetation type and management of some of the most common agro-forestry systems in the Mediterranean basin [16].

The Berchidda site is made up of hydromorphic and granitic soil with a loamy sand texture. The altitude ranges from 275 m to 300 msl. This area is referred to as a meso-thermo Mediterranean, subhumid phytoclimatic belt with a mean annual rainfall of 862 mm and the mean annual temperature of 13.8°C [10]. In the

Figure 1. Long-term effects of different land-use with increasing level of intensification in spring and autumn. Both pasture and managed meadow included spotted cork oak trees, which are key components of the Dehesatype landscape typical of this area of Sardinia. The cork-oak formation, pasture, and managed meadow have been converted to the current use and maintained unchanged for more than 30 years, whereas the non-tilled cover cropped vineyard and the tilled one were planted in 1985 and 1994, respectively. From the left to right: cork-oak forest (CO), hayland pasture rotation (PA), managed meadow (MM), grass covered vineyard (CV), tilled vineyard (TV).

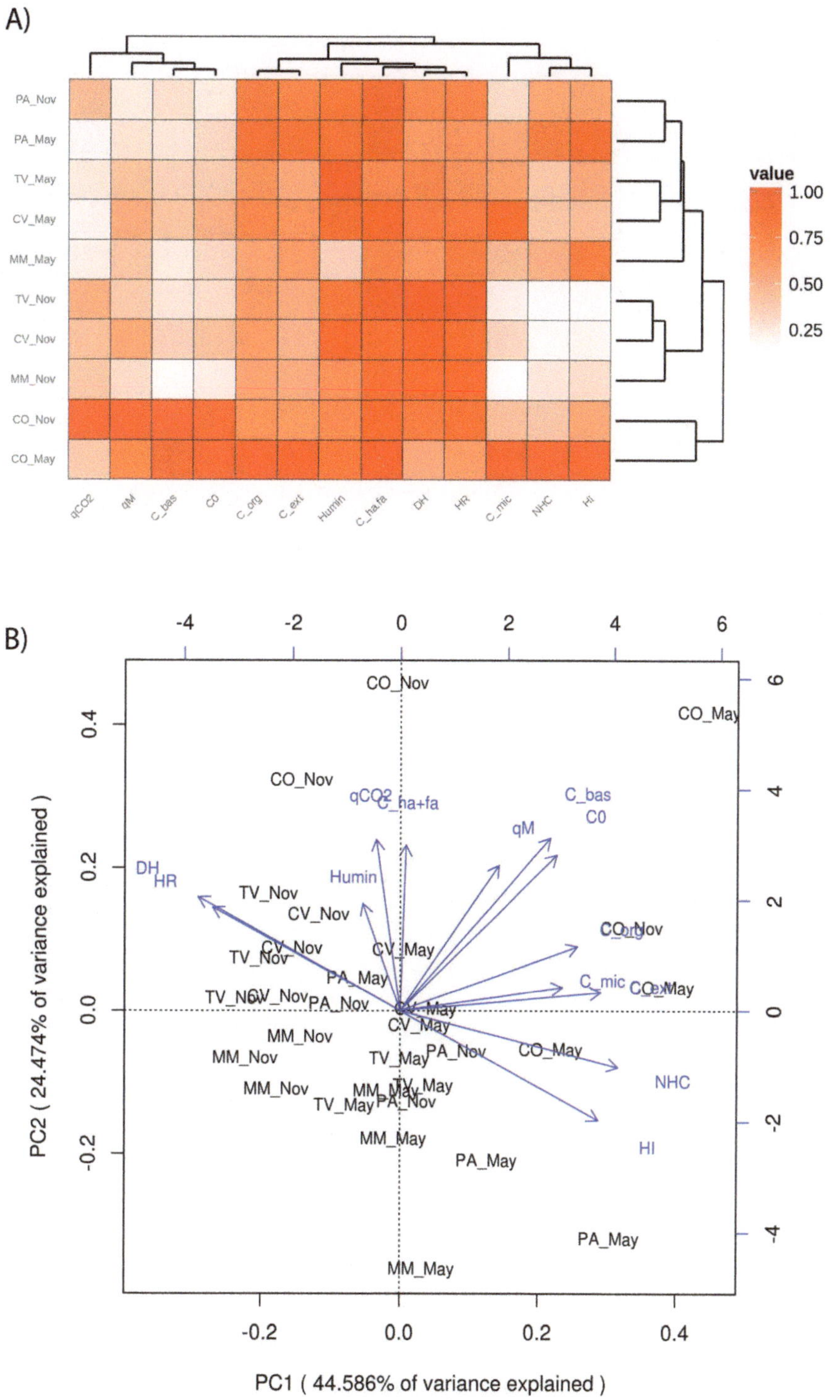

Figure 2. Effect of land-use and season on soil physical-chemical and biological parameters. A) Heat map with hierarchal clustering of physical-chemical and biological parameters across the five Sardinia soils with different land uses at the two different sampling time points (May and November). The heat map was constructed using a maximum-minimum normalization of the data in order to represent each value in a range between 0 and 1. Higher values are represented by darker colors whereas lower ones are represented by lighter colors. CO = cork-oak forest; PA = hayland-pasture rotation; MM = managed meadow; TV = tilled vineyard; CV = grass covered vineyard. B) PCA ordination of data (axes 1 and 2) generated from physical-chemical and biological properties of the different types of land use in May and November.

past, the Berchidda area was covered by cork-oak forests (dominated by *Quercus Suber* L.), which were subjected to intense usage for the extraction of cork. Today, there are five main different land-use units close together: cork-oak forest (CO), hayland-pasture rotation (PA), managed meadow (MM), and tilled (TV) and grass covered vineyard (CV). The cork-oak formation, pasture, and managed meadow have been converted to the current use and maintained unchanged for more than 30 years,

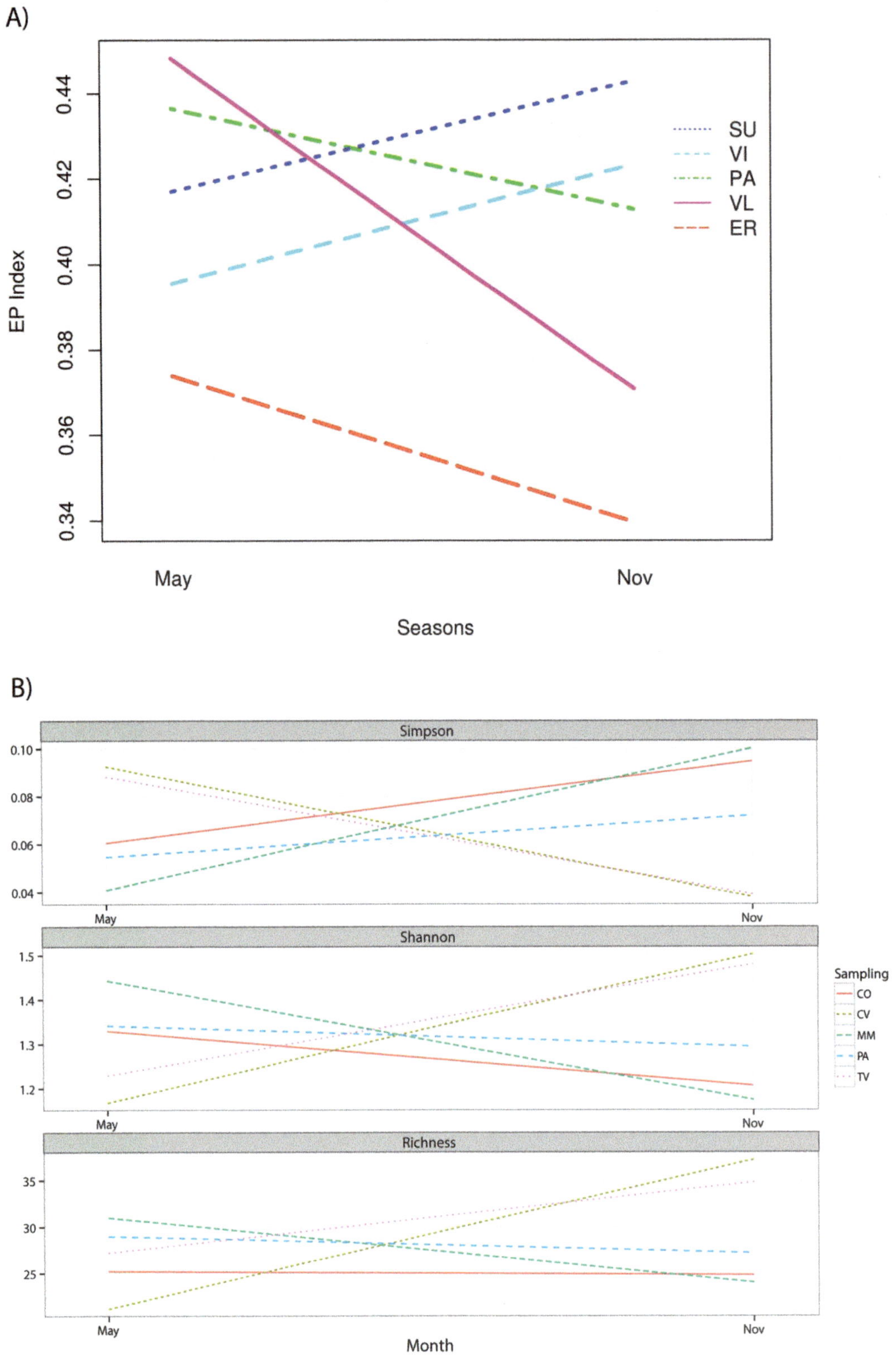

Figure 3. Effect of land-use and season on Eco-Physiological (EP) index of culturable bacteria (A) and diversity indices from CD-DGGE profiles (B).

whereas the non-tilled cover cropped vineyard and the tilled one were planted in 1985 and 1994, respectively. The five soils are located inside an area of 161.5 km^2. Detailed characteristic and management of the five soils have been previously described [12], [20], [21], [22], [23].

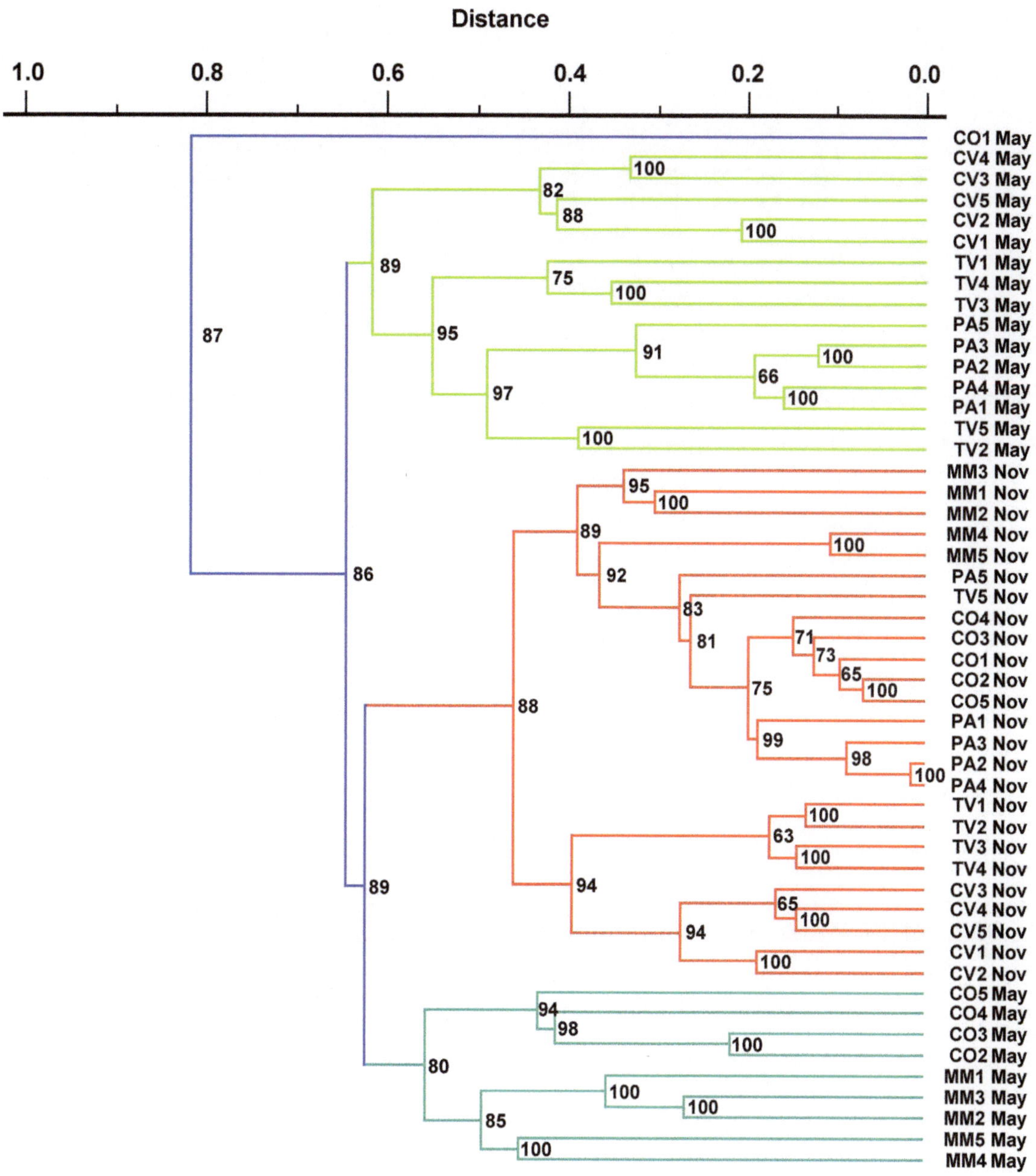

Figure 4. UPGMA dendrogram of DGGE profiles of amplified 16S rDNA of bacterial communities recovered in CO, PA, MM, CV and TV, in spring and autumn, generated using Phoretix ID advanced analysis package. Designation of samples is the same as in Fig. 1. The scale bar represents dissimilarity among samples. Consistency of each cluster was measured by the Cophenetic correlation coefficient shown at each node.

Pedological characterization of the study area

The pedogenic substrate of the study area consists of medium-grained granite, affected by localized presence of veins of quartz and porphyry. The morphology of the Berchidda area varies from flat to undulating. The processes of soil erosion by water channeled are evident only in the short-term forage crops made on soil with more than 15% of slope. All profiles have a horizons sequence of the type A-Bw-C or A-Bw-BC-C or more rarely, A-Bw-C-R (Table S1). The power of the profiles, limited to horizons A and Bw, varies from a minimum of 38 cm to a maximum of 100 cm. In three soil profiles in TV, the sequence between the horizon A and Bw and the substrate is gradually altered by the presence of a horizon BC, characterised by coarse texture, the power of which varies from about 30 cm to 90 cm. Direct contact with the unaltered rock, R horizon, was observed only in one profile in CV. The prevailing textural classes are sandy and sandy loam. The content of organic substance in the horizons A is never very high, with average value around to about 3%. The maximum values of 11.8% and 8.6% were observed on the A horizons in two soil profiles in CO. The exchange complex, in agreement with the reduced content in clays, is never high. The degree of base saturation is predominantly less than 60% (Dystric conditions of the USDA Soil Taxonomy). The profiles with the exchange complex with a degree of base saturation below 60% in all horizons between 25 and 75 cm (TV, CV, MM, and PA) were

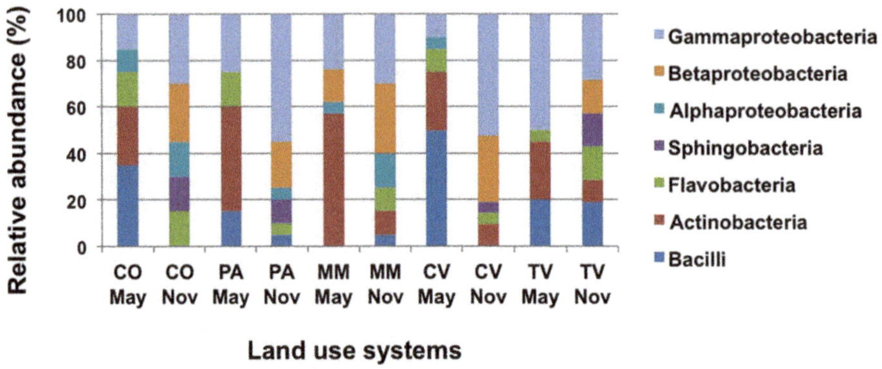

Figure 5. Relative abundances of major taxonomic groups across land use systems in spring and autumn. Detailed data of each class are listed in Additional file 1: Table S6. CO = cork-oak forest; PA = hayland-pasture rotation; MM = managed meadow; TV = tilled vineyard; CV = grass covered vineyard. Values presented are the mean percent.

classified as Typic Dystroxerepts [24]. The other one (CO) was classified as Lithic Xerorthents.

Soil sampling

Five soil replicates were collected from bulk soil of the five different managements (CO, PA, MM, CV, and TV) (Fig. 1). Soils were collected in May and November 2007. After removal of litter layer, soil core samples (50 to 100 g; diameter, 5 cm) were taken from each of the five locations, using a 5-on-dice sampling pattern with ca. 70 m distance between each sampling point. Sampling was performed at 20 cm depth, where most microbial activity is known to occur [25], [26]. In the vineyard soils, samples from along the rows and between the rows were pooled together to form a field replicate. The other soil samples (CO, PA, MM) were collected out of trees influence. At each season and for each soil type, five randomly field replicates were collected for a total of 25 soil samples (5 replicates × 5 land uses), each one being a composite sample of five soil cores. A total of 50 composite samples were taken for the two seasons. Soil samples were immediately sieved (<2 mm) to remove fine roots and large organic debris, air dried, and transported to the labs for microbiological analysis. The moisture content was adjusted to 60% of their water holding capacity (WHC) and soil samples were then left to equilibrate at room temperature in the dark for one day prior to analyses, in order to restore, within limits [27], the microbial activity of air- dried soils to that of soils in the field.

Chemical and biochemical analyses of soil samples

The chemical and biochemical analyses were performed on three replicates for each land use for a total of 15 soil samples (3 replicates × 5 land uses), each one being a composite sample of five soil cores. Total organic carbon (C_{org}) was estimated after oxidation with $K_2Cr_2O_7$ and subsequent titration of unreduced $Cr_2O_7^{2-}$ with $Fe(NH_4)_2(SO_4)_2$, as reported by Springer and Klee [28]. Soil Organic Matter (SOM) was determined by C_{org} multiplied by 1.724 van Belem coefficient. The C_{org} fractionation was set up as reported by Ciavatta and co-workers [29]. In particular, solid samples were extracted at 65°C for 24 h using 0.1 mol l^{-1} NaOH plus 0.1 mol l^{-1} $Na_4P_2O_7$ solution (1:50, solid:liquid ratio). The samples were then centrifuged at 5000×*g* and the supernatants were filtered through a 0.20-μm Millipore filter (Millipore, Billerica, MA) (total extractable C, C_{ext}). The humic-like acid (HA) fraction was separated from the fulvic-like acid (FA) and the non-humified carbon fractions (NHC) by precipitation after acidification of the alkaline solution (supernatant) to pH<2. Chromatography on a column of polyvinylpyrrolidone (PVP, Aldrich, Germany) was used to separate the NHC from the FA. The FA was then combined with the HA to obtain total humified fraction (HA+FA). Total extractable C (C_{ext} %), humic and fluvic acid C (C_{HA+FA} %) were determined by the dichromate oxidation method. The non-humified carbon (C_{NH}) was determined as the difference between C_{ext} and C_{HA+FA}. Humification indexes HI, DH, and HR were determined according to previous works [29], [30].

Microbial biomass C (C_{mic}) was determined by the fumigation-extraction method of Vance and co-workers [31] with some slight modifications. The measurements were performed on air-dried soils, pre-conditioned by a 10-d incubation in open glass jars, at −33 kPa water tension, and 30°C. The incubation was employed for restoring, within limits [27], the microbial activity of air-dried soils to that of soils in the field. Four replicates of each soil sample were used. Average values are given in mg C kg^{-1} of soil. For measuring microbial respiration, 20 g (oven-dry basis) of moist sample were placed in 1 L stoppered glass jars. The CO_2 evolved was trapped, after 1, 2, 4, 7, 10, 14, 21, 28 days of incubation, in 2 ml 1 M NaOH and determined by titration of the excess NaOH with 0.1 M HCl [32].

Non-linear least square regression analysis was used to calculate parameters affecting C mineralization from daily CO_2 evolution data (Stat Win 4.0 for Windows). The best fit was obtained with the exponential model of CO_2-C accumulation according to the negative exponential decay model:

$$C_m = C_0\left(1 - e^{-kt}\right),$$

where C_m is the cumulative value of mineralized C during *t* days, *k* is the rate constant, and C_o is the potentially mineralizable C [33]. The CO_2 emitted in 28 days of incubation was used as cumulative respiration (C_{cum}). The CO_2 evolved during the 28th day of incubation was used as the basal respiration value (C_{bas}). Microbial indices were calculated as follows [34], [35], [36]:

$$qCO_2 = \left[\left(\mu g\, CO_2 - C_{bas} \times \mu g^{-1} C_{mic}\right) h^{-1}\right] 10^3,$$
$$qM_{cum} = \left(\mu g\, CO_2 - C_0 \times \mu g^{-1} C_{org}\right),$$
$$qM_{bas} = \left(\mu g\, CO_2 - C_{bas} \times \mu g^{-1} C_{org}\right)$$

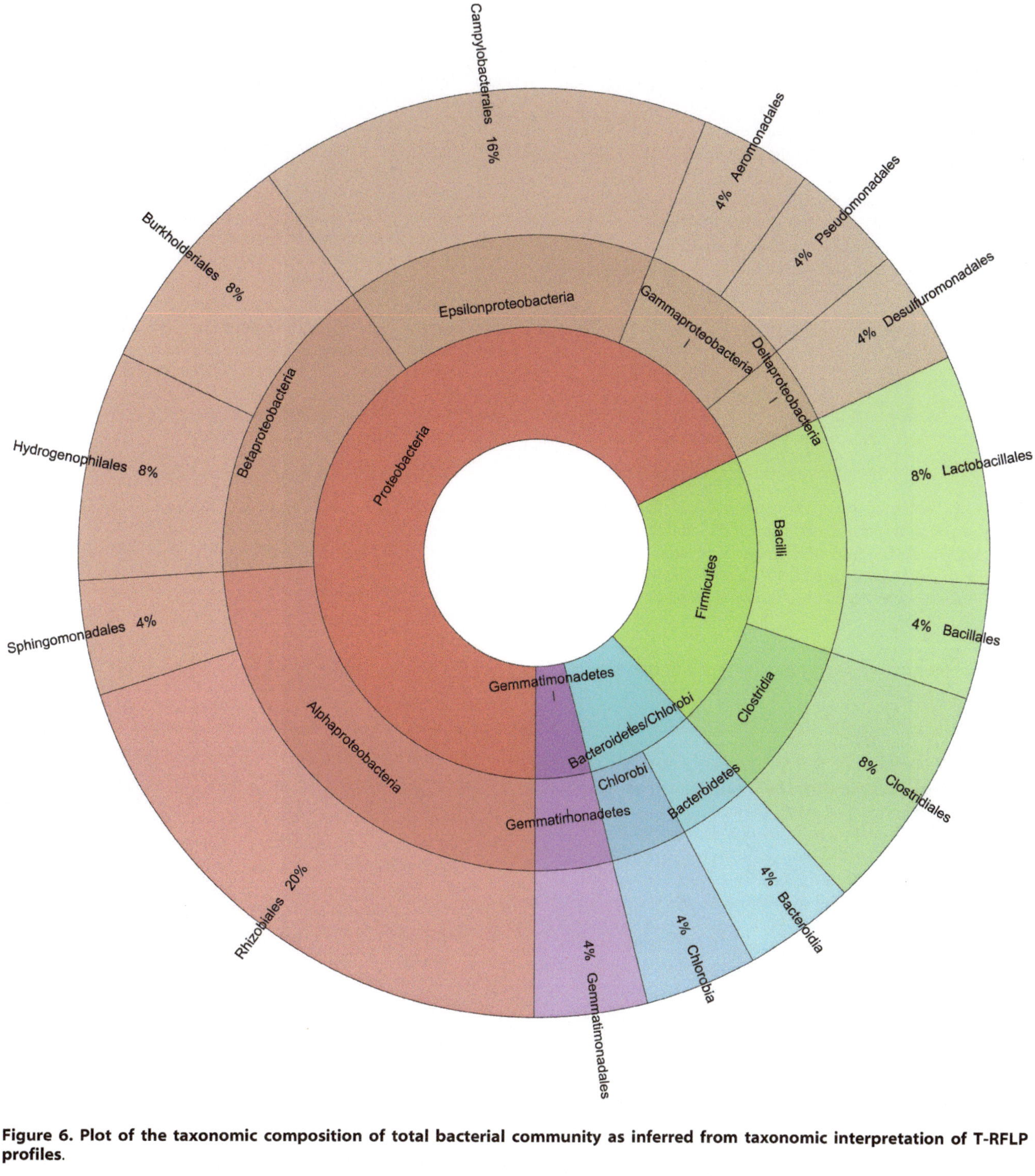

Figure 6. Plot of the taxonomic composition of total bacterial community as inferred from taxonomic interpretation of T-RFLP profiles.

Recovery of cultured bacterial cells from soil samples

Soil bacterial community analysis was performed on five soil replicates for each land use, as described above. Bacterial cell extraction was performed according to the recommendations of Smalla et al. [37] with minor modifications. Briefly, 1 g of soil was placed in a sterile 15 ml plastic tube containing 10 ml of phosphate buffered saline (PBS, pH 7.0). This mixture was homogenized for 30″ at low speed by using the Ultra-Turrax Thyristor Regle 50 (Janke & Kunkel IKA-Labortechnik). After homogenization, suspension was placed into a Erlenmeyer flask (100 ml) containing 10 g of glass beads (0,2 mm) and shaken for 1 h at 180 r.p.m. and 28°C to disperse bacteria. The flasks and glass beads were autoclaved for 20 min at 121°C before use. The soil suspension was transferred into a sterile 15 ml plastic tube and serially diluted with sterile saline solution (9 g l^{-1}NaCl) from 10^{-1} up to 10^{-7}. The, 100 µl aliquots of serially diluted soil suspensions were plated in triplicate on 0.1 Triptic Soy Broth (TSB; Difco) amended with 15 g l^{-1} agar (0.1 TSA) and 100 g μl^{-1} of cycloheximide (Sigma) to inhibit fungal growth. Plates were incubated at 28°C for 6 days.

Growth strategy and total bacterial populations

To determine the changes in the structure of culturable fraction of soil bacteria, the r/K-strategy concept proposed by De Leij and co-workers [38] was used. Bacterial colonies appearing within 48 h were designated as r-strategists, and the remaining as K-strategists. Colonies were enumerated at 1, 2 and 6 days of growth on 0.1 TSA; in this way, three counts (or classes) were generated per sample. Plates containing between 30 and 300 colonies were then selected for enumeration. Total bacterial counts obtained were expressed as colony forming units (CFU) per gram of soil. Distribution of bacteria in each class as a percentage of the total counts gave insight into the distribution of r- and K-strategists in each sample.

To evaluate the changes in the biodiversity of bacterial populations in soils, the eco-physiological (EP) index [38] was used. The EP index of each soils tested was calculated using the equation:

$$H' = -\sum(P_i)(log_e P_i),$$

where P_i represents the CFU on each day (1, 2 and 6 days of incubation) as a proportion of the total CFU in that sample after 6 days incubation, i.e., the proportion of colonies appearing on counting day i ($i = 1, 2, 6$) with $EP_{min} = 0$. Higher values of EP index imply a more even distribution of proportions of bacteria developing on different days (i.e., different classes of bacteria).

Terminal-Restriction Fragment Length Polymorphism (T-RFLP)

DNA was extracted from soil samples by using the FastDNA SPIN Kit for Soil (QBiogene). Terminal-Restriction Fragment Length Polymorphism (T-RFLP) was performed on 16SrRNA genes amplified from extracted DNA with primer pairs P0 and P6 as previously reported [39], [40]. Purified amplification products were digested separately with restriction enzymes *Rsa*I and *Msp*I and digestions were resolved by capillary electrophoresis on an ABI310 Genetic Analyzer (Applied Biosystems, Foster City, CA, USA) using LIZ 500 (Applied Biosystems) as size standard. T-RFLP analysis was performed as previously reported [41]. Diversity indices were calculated with PAST software [42] as previously reported [19], taking into account peak intensities of T-RFLP fragments. Taxonomic interpretation of T-RFLP profiles was performed by querying the Ribosomal Database Project Database by using MiCA3 web tool (http://mica.ibest.uidaho.edu/), as previously described [40].

Culture-dependent DGGE (CD DGGE) analysis

Culture-dependent DGGE (CD DGGE) fingerprinting of 16S rRNA gene was used to characterize mixed bacterial communities recovered on agar plates.

Collection of cultured bacterial communities for DGGE analysis. Cultured bacterial communities were collected following procedure proposed by Duineveld and co-workers [43] with minor modifications. Briefly, after one week of incubation at 28°C, colonies were removed from plates containing between 100 and 1000 colonies by adding 3.0 ml sterile physiological solution (0.9% NaCl) on each plate and scraping off all grown colonies with a sterile Drigalski spatula. The cell suspensions thus obtained were aliquoted into 1.5 ml Eppendorf tubes and centrifuged at 8,000 r.p.m. for 10 minutes. The pellets were stored at −80°C for subsequent DNA extraction and PCR-DGGE analysis.

DNA extraction and PCR amplification of 16S rRNA genes. Genomic DNA was extracted with sodium dodecyl sulfate-proteinase K lysis buffer, followed by a treatment with cetyl-trimethylammonium bromide (CTAB) as described in *Current Protocols for Molecular Biology* [44]. Briefly, the pellet was resuspended in 567 µl of TE [10 mM Tris-HCl - 1 mM EDTA (pH 8.0)] buffer, and glass beads 0.3 mm in diameter (250 mg) were added, followed by bead beating for 20 s. Then, 30 µl of 10% sodium dodecyl sulfate and 3 µl proteinase K 20 mg/ml (Sigma) were added and samples were incubated at 37°C for 1 h. Glass beads were removed by centrifugation 2 min at 2,800 rpm and samples were then incubated at 65°C for 10′ with100 µl of 5 M NaCl prepared with sterile water and 80 µl of CTAB/NaCl (10% CTAB in 0.7 M NaCl). Following incubation, extracts were purified by using phenol/phenol-chloroform/isoamyl alcohol (49.5:49.5:1) extraction and DNA was recovered by isopropanol precipitation at 4°C o/n. Pelleted DNA was washed twice with cold 70% ethanol, allowed to air dry, and re-suspended in 50 µl of sterile water. Quantity and purity of DNA were checked by NanoDrop (NanoDrop Technologies, USA) and gel electrophoresis. The DNA samples were stored at −20°C until required for use.

The 16S rRNA gene was amplified using 20 ng of lysate suspension and the universal bacterial primers P0 and P6 [45]. Dilutions 1:100 (2 µl) of the1450 bp PCR products were then used as template for the second PCR amplification with the forward primer 63F (5′- AGGCCTAACACATGCAAGTC -3′), with a GC clamp (5′-CGCCCGCCGCGCGCGGCGGGCGGGGCG-GGGGCACGGGGGG -3′) incorporated at the 5′ end, and the reverse primer 518R (5′-ATTACCGCGGCTGCTGG-3′), to produce 495 bp fragments suitable for DGGE analysis [46]. Both PCR reactions were performed in Qiagen Taq buffer (10X) containing 1.5 mM $MgCl_2$, with 150 ng of each primer, 250 µM (each) deoxynucleoside triphosphates, and 0.5 U of *Taq* DNA polymerase (Qiagen, Hilden, Germany) in a 25 µl reaction volume. Cycle parameters for PCR with the primer pairs P0–P6 and 63F-GC and 518R were previously described by Di Cello and co-workers [45] and El Fantroussi and co-workers [46], respectively.

Denaturing Gradient Gel Electrophoresis. 16S rRNA gene amplicons were separated by double gradient denaturing gradient gel electrophoresis (DG-DGGE) as described by Cremonesi et al. [47], in a DCode universal mutation detection system (Bio-Rad, CA, USA). Separation of purified PCR products (700 ng) was achieved in6%–12% polyacrylamide (acrylamide: N,N-methylenebisacrylamide, 37.5:1) gels containing an increasing linear gradient of denaturants ranging from 30% to 60% (100% denaturant corresponds to 7 M urea and40% deionized formamide). Each gel also included marker lanes represented by DGGE profiles containing a large number of discrete bands spanning the entire gradient, suitable for within- and between-gel alignment. Electrophoresis were carried out for 16 h at 75 mV in 1X TAE buffer at 60°C, stained with 50 µg/ml ethidium bromide for 30 min, destained in water and photographed with the UVIpro Platinum Gel Documentation System (GAS7500/7510; Eppendorf, Cambridge,UK).

Cluster analysis and diversity indices

Quantity One software package (Bio-Rad) and Phoretix 1D PRO software (Phoretix International, Newcastle upon Tyne, United Kingdom) were used for CD-DGGE profile analysis. The cluster analysis and dendrogram generation were carried out by using the Phoretix 1D Pro software according to the manufacturer's instructions (Phoretix International, Newcastle upon Tyne,

United Kingdom). Bands of CD-DGGE patterns were aligned and normalized using reference lanes. Background noise was subtracted by rolling ball algorithm with a radius of 50 pixels; the automatic band detection was performed with a minimum slope of 200 and a noise reduction of 10, and peaks smaller than 2% of the maximum peak were discarded. Bands were manually corrected and matched to create an absent/present binary matrix. The similarity between the band patterns was calculated using the Dice coefficient and the clustering analysis was performed with the unweighted pair group method with arithmetic averages (UPGMA) to generate a dendrogram by using mathematic averages algorithm programs integral to the Phoretix 1D Pro software. Coefficient of cophenetic correlation was used to measure the consistency of clusters.

DGGE banding data were used to estimate three diversity indices by treating each band as an individual operational taxonomic unit (OTU). The number of DGGE bands present in each sample was used to measure the Richness index (*R*). The Shannon-Weaver index of general diversity (H') [48] and the Simpson index of dominance (*D*) [49] were calculated from the number of bands present and the relative intensities of each band (P_i) in each lane. Relative signal intensities of detected bands, in each gel track, were determined by using the Quantity One software package (Bio-Rad) and calculated from the peak area of the densitometric curves.

The Shannon-Weaver diversity (H') was calculated using the following equation:

$$H' = -\sum (P_i)(\log_e P_i),$$

where P_i (the proportion of abundances of the ith band) is measured as:

$$P_i = n_i / N,$$

where n_i is the peak height of a band, and *N* is the sum of all peak heights in the densitometry profiles.

The Simpson index (D) was calculated with the formula:

$$D = \sum (P_i x P_i);$$

it measures the strength of dominance because it weights towards the abundance of the OTUs and varies inversely with species diversity [50].

Bacterial isolation and identification

A total of 100 bacterial colonies were randomly picked up for each soil sample from 0.1 TSA plates (containing approximately 50 to 500 colonies), previously used for the determination of the CFU counts and EPI-index, and repeatedly streaked onto 0.1 TSA fresh plates to obtain pure cultures. Isolated colonies were then grown overnight (o/n) in TSB medium at 28°C and 200 r.p.m., and stored at −80°C in 30% glycerol until further analysis. From all five soil samples, 500 colonies were isolated in each season, for a total of 1000 bacterial colonies. A total of 203 colonies with different morphologies (about 20 colonies per each sample) were taken up to investigate their taxonomic affiliation.

Genomic DNA and PCR amplification of the 16S rRNA gene were performed as described above. Sequencing reactions were prepared from PCR products using an Applied Biosystem Big Dye Terminator sequencing kit version 3.1, according to the manufacturer's instructions and analysed using a 3730 DNA Analyzer Applied Biosystem apparatus. The sequences were compared with those in the GenBank databases by using the BLAST program and Seqmatch tool of the RDP (http://www.ncbi.nlm.nih.gov/BLAST/and http://rdp.cme.msu.edu/, respectively) and aligned with the closest relatives with the Clustal W function of the BioEdit package [51]. Bacterial identification by 16S rRNA gene sequences assignment was performed using the RDP Classification Algorithm (http://rdp.cme.msu.edu/classifier/classifier.jsp).

Statistical analysis

Bacterial population data (CFU/g of soil) were log transformed and subsequently analysed by one-way ANOVA (STATISTICA, Release 3.0b, Copyright StatSoft Inc., CA, USA). Percentage data of EP index value were *logit*-transformed, as follows:

$$Logit\,(p) = log[p/(1-p)]$$

for the proportion *p*, and compared using one-way ANOVA (STATISTICA, Release3.0b, Copyright StatSoft Inc., CA, USA).

Analyses of variance (ANOVA) on biodiversity indexes (Shannon-Weaver, Richness and Simpson), principal component analysis (PCA) and clustering analysis on biochemical data were performed using R packages "stats" and "vegan" (http://cran.r-project.org/and http://cran.r-project.org/web/packages/vegan/index.html). All data clustering were performed using the "UPGMA" algorithm implemented in the "hclust" function of the R "stats" package. Distances among samples were calculated using "Bray-Curtis" distance implemented in "vegan" package as:

$$i = sample \text{ “}i\text{”};\ j = sample \text{ “}j\text{”},\ BC_{ij} = 2C_{ij}/(S_i + S_j),$$

where C_{ij} = sum of the smaller value for species in common between samples i and j, and S_i and S_j = total number of species in samples i and j, respectively [52]. Variation of biodiversity indexes of cultured bacteria was inspected using ANOVA analysis. Biodiversity indexes were first divided into groups depending on sampling season and different managements of soils and then the analysis was performed. Clustering analysis (UPGMA) on biochemical parameters was performed. Each parameter was first divided in groups, in the same way of previous ANOVA analysis, and averaged. Then, each result obtained was normalized using the maximum-minimum normalization technique in order to make the data comparable. PCA analysis using each biochemical data was performed.

Results and Discussion

Effect of land use on soil chemical and biochemical properties

Soil organic matter (SOM) represents a dynamic system influenced by several factors, including climate, clay content, mineralogy and soil management, which all affect the processes of organic matter transformation and evolution in soil [53], [54]. Both soil fertility and stability are related to the organic matter content of soil. Many functions of SOM are due to its more stabilised fractions, humified materials and balance between the labile and the stabilised fractions [55]. Changes in SOM content are related to changes in microbial biomass turnover, because they reflect the balance between rates of microbial organic matter accumulation and rates of organic matter degradation. The extent of organic matter's organization not only impacts the amount of carbon mineralized but also the type of carbon that is consumed by the microorganisms.

In this study two categories of soil quality indicators were used: organic matter quality indicators and microbial biomass activity indicators (Tables S2 and S3). Our results (Fig. S1) revealed that both land use and sampling season affect the chemical and structural composition of the most stabilised fractions of SOM. A higher content of C_{org} occurred in CO soil in May, and C_{ext} and C_{HA+FA} showed the same pattern, whereas the average values of these parameters were slightly higher in PA soil in November. These results were reflected in the humification parameters, where the humification rate (HR%) can provide quantitative information about the humic substances content normalised with respect to total SOM, while the degree of humification (DH%) provides the amount of the humified carbon in the extracted organic fraction and the humification index (HI) can be considered as an index of soil humification activity as well as of availability of non humified labile fractions [29]. Overall, land use change reduced soil C stocks and labile organic matter at both sampling times except for PA in November. Pasture has a great potential soil organic C stock and, in the long term, grass management systems have nearly equivalent potential to store soil organic C as forest [56]. Potentially mineralizable C (C_0), which indicates the amount of C in the labile fraction of soil organic matter, decreased over sampling time in all soils, and C_{mic} similarly declined. Cultivated soils are characterised by low microbial activity, mainly due to the disappearance of easily decomposable organic compounds through tillage and soil disturbance. As previously found for chemical properties, the metabolic activity responds to the different land uses; in both May and November samplings, qCO_2 was higher in CO soil when compared to the others, resulting in increases stress. Hence, unfavourable conditions result in a decrease in the size of the microbial biomass and the efficiency of C substrates degradation, conducting to an increase in respiration rate per unit of microbial biomass [57].

As beneficial and negative effects of soil management practices are strongly linked to microbial activities and regulate soil quality and functioning [58] the relationship existing between soil management practices and variation in chemical and biochemical parameters was evaluated, by performing a clustering analysis (UPGMA) as reported in Tables S2 and S3. The dendrogram obtained was linked to a heat-map representing all the biochemical parameters variation (Fig. 2A). Data analysis revealed that chemical and biochemical parameters clustered in two different groups each of which corresponding to one of the two parameters analyzed (chemical and biochemical). In both sampling times, TV and CV clustered together, suggesting that seasonal change rather than management regime was a major driving force contributing to vineyard soil fertility. Furthermore, according to the UPGMA clustering, CO formed a separate cluster from that of the other soil uses (Fig. 2A). Interestingly, the chemical and biochemical parameters clustered together in both CO and PA, regardless of the season, in contrast to MM, CV and TV, where they clustered separately in relation to the sampling season. This finding suggests that soils subjected to low human inputs (pasture and cork-oak forest) showed a more stable chemical and biochemical soil composition than those with high human input (managed meadow and vineyard).

The different positions of the variables in the plane of the first two principal components, as revealed by PCA analysis (Fig. 2B), indicated that chemical and biochemical parameters were differentially affected by the various land use types. The first and the second principal components (PC1 and PC2) accounted for 44.59% and 24.47% of the total variance in the data, respectively (Fig. 2B). Both chemical and biochemical parameters were positively affected by PC1 except for HR and DH, whereas PC2 was able to discriminate between the two variable sets, with the most of chemical and biochemical parameters being positively and negatively affected, respectively. The above analyses revealed an important effect of land uses on both chemical and biochemical; on the contrary, only biochemical parameters were affected by seasons.

Influence of land use on soil bacterial community

Community structure analysis by EP-index and r-K strategy. Environmental conditions select organisms that either grow rapidly in uncrowded, nutrient-rich conditions (r-strategists), or can efficiently exploit resources in crowded conditions (K-strategists). In this work, we applied the method developed by De Leij et al. [38], who used the r/K-strategy concept for the characterization of soil bacterial communities. First, we compared the microbial community structures found in the different land uses during each of the two seasons. Since sampling was performed at 20 cm depth, where most microbial activity is known to occur [26], the main changes are expected through conversion from one soil management system to another one [59]. Microbial community structure found in the different land uses during each of the two seasons was investigated by means of EP index that is a measure of both richness (i.e. total number of species in the community) and evenness (i.e. how evenly individuals in the community are distributed over the different species) of groups of microorganisms with similar developmental characteristics.

The cultured bacteria belonging to fast-growing organisms, especially the r-strategists detected on day 2, dominated in all land uses and in both seasons (Table S4). The MM soil exhibited a lower EP index and a lower percentage of r-strategists in both seasons, possibly due to amensalism from a dominant bacterial group in the community. Variation of EP index of cultured bacteria was inspected using an ANOVA analysis of variance (Fig. 3A). Interestingly, statistical analysis revealed a significant effect of land uses on bacterial community structure ($P<0.001$) as well as an interaction effect between land uses and season change ($P<0.001$) on EP index. Variation of EP index due to seasonal changes was significant in the soil with higher human impact (TV) when compared with the other soils. Total bacterial concentrations varied significantly in respect of season ($P<0.001$) being higher in spring than in autumn (Table S2) in all but CO soils, with significant differences between CO and PA in spring, and between CO and MM, and MM and PA in autumn.

Overall results suggest that land uses affected cultured bacterial communities. Soils with low human impact (cork-oak forest) have a more stable bacterial density and show a less variation of bacterial community structure across seasons than soils subjected to high human impact such as tilled vineyard.

Bacterial community profiling by CD-DGGE and T-RFLP analyses. Culture dependent DGGE (CD-DGGE) fingerprinting of the 16S rRNA was used to characterize mixed bacterial communities recovered on agar plates. CD-DGGE represents a useful technique to follow the dynamics of distinct culturable fractions of the soil bacterial community in relation to physical, chemical and biological changes in the soil environment [60]. Since culture-dependent and culture-independent methods likely profile distinct fractions of the soil bacterial community with unique ecological roles [61], culturable bacteria may provide an ecologically relevant complement to culture-independent community characterization [62].

By pooling the bacterial cells growing on individual agar plates, we obtained a culture-dependent bacterial community. The analysis of DGGE profiles revealed clear banding patterns for each land-use and plant-cover types of sufficient complexity to investigate differences in soil microbial communities and identified stable communities with highly reproducible profiles (Fig. S2). A different community composition among land-use types was found as evidenced by the presence of different dominant signals; this finding indicated a compositional shift among soils examined in spring and autumn and subjected to different anthropogenic impact. Bacterial diversity was investigated through richness (*R*), Shannon-Weaver (*H'*), and Simpson (*D*) indices. Analysis of variance (ANOVA) confirmed the differences in the distribution of bacterial species due to land uses and sampling seasons ($P<0.001$). Both *R* and *H'* indices decreased from May to November in MM samples and increased in TV and CV samples, and the complementary opposite trend was observed for *D* index (Fig. 3B). Otherwise, cultured bacterial community in CO and PA soils did not vary significantly over seasons. Most likely, these results reflect the impact of both land-use and vegetation type and coverage on soil microbial communities. In particular, the low shift of biodiversity indices observed in CO and PA seems to be correlated to a higher stability of bacterial populations in natural habitats with low human impact, whereas populations inhabiting more anthropogenic areas tend to be more variable. Land-use type and, in particular, differences in vegetation dynamics may have a large role in modulating the temporal variability in soil bacterial communities. As observed by Lauber and co-workers [3], soils from the different land-use types did not exhibit identical temporal dynamics even though all the soils were located in close proximity and exposed to the same climatic conditions. Diversity indices obtained from T-RFLP analysis performed on total bacterial DNA (Fig. S3) partially confirmed the trends of diversity shown by CD-DGGE, though observed differences were not statistically significant (Table S5).

The unweighted pair-group method using arithmetic averages (UPGMA) of bacterial community profiling by CD-DGGE revealed a high diversity of the bacterial communities in each land-use and plant-cover soils. The similarity between the DGGE patterns of the soil bacterial communities revealed three distinct clusters (Fig. 4). Samples collected in autumn grouped into a separate cluster with about 52% similarity, whereas samples collected in spring grouped into two separate clusters, each of which composed by samples sharing about 40% similarity. The highest similarity values were shown by clusters based on soil land uses, suggesting a low bacterial variation within each soil type. This finding is consistent with previous reports showing that land-use type was the most important factor in determining the composition of soil microbial community [3].

The clustering results based on CD-DGGE were in agreement with culture-independent T-RFLP analysis that confirmed a different seasonal distribution of soil bacterial populations with subgroups associated to different land uses. Most of samples retrieved from the same soil (CO, PA, and MM) clustered together, whereas samples retrieved from TV and CV were often intermixed in the UPGMA dendrogram (Fig. S4). As culture-dependent and culture-independent profiles can separately resolve unique, diverse, and equally complex fractions of the soil bacterial community [60], our combined results from both CD-DGGE and T-RFLP methods, applied on culturable and total fractions, respectively, revealed an interaction effect between land uses and season change in affecting soil bacterial communities.

Taxa responses to land use determined by phylogenetic affiliation of soil bacterial isolates. Taxonomic affiliation was investigated on a total of 203 bacterial isolates. In detail, 20–21 colonies recovered from each soil sample at each season and showing different morphologies (for a total of 101 bacterial isolates in spring and 102 in autumn) were subjected to DNA extraction and PCR amplification of 16S rRNA gene. An amplicon of the expected size (about 1500 bp) was obtained from each isolate, and its nucleotide sequence was determined and submitted to GenBank (Table S6). Although it is generally accepted that not all bacteria, including types of soil bacteria, are culturable, the isolation of bacteria by agar plate cultivation and subsequent phylogenetic analysis permit to isolate and identify previously uncultured representatives or even new members of certain bacterial species for further analysis of their metabolic function. Therefore, even if the metagenome sequencing is becoming the most powerful tool to investigate microbial communities, the ability to isolate indigenous strains actually remains the unique way to further characterize and select soil bacteria showing interesting properties. Additionally, standard cultivation techniques have been shown to be able to capture members of the soil rare biosphere which could not be detected by metagenome sequencing [63].

The 203 sequences obtained were compared with those present in the GenBank databases by using the BLAST [64] program. Results showed that most of 16S rRNA gene sequences matched NCBI database sequences at 99–100% of similarity at the genus level, with *Arthrobacter*, *Bacillus*, *Stenotrophomonas*, *Pseudomonas* and *Burkholderia* as the most representative genera. So, identification at the genus level was achieved in 103 isolates and at the species level in 72 isolates, while 12 isolates were only affiliated to taxa level higher than genus and 16 remained unidentified (Table S6). Further comparison with GenBank databases by using the Seqmatch tool of the RDP indicated that 16S rRNA gene sequences were affiliated to four phyla: *Proteobacteria* (classes α, β and γ), *Bacteroidetes* (classes *Flavobacteria* and *Sphingobacteria*), *Actinobacteria* and *Firmicutes*. Even though the RDP analysis does not permit to affiliate a bacterial isolate to a given species, it allowed to assess the genus of almost all isolates, including those not identified through the BLAST search. In fact, a boostrap value of at least 80% is satisfactory for RDP requirement. Almost all our sequences gave rise to 100% boostrap, 12 ranged from 89% to 99%, and only two of them only resulted in low values (24 and 54% respectively, referred to two spring isolates).

Data obtained from both approaches allowed to assess that Berchidda soil is colonized by bacteria included in the classes of γ-*Proteobacteria* (with a prevalence of *Pseudomonas* and *Stenotrophomonas* genera), *Actinobacteria* (in particular, the genus *Arthrobacter*), β-*Proteobacteria* (with a prevalence of *Burkholderia* genus), *Bacilli* (with the genus *Bacillus* as the dominant one), Flavobacteria and Sphingobacteria. The relative abundances of the different classes identified by RDP across the different samples related to soil uses and season are represented in Figure 5. A putative taxonomic description of total bacterial community, derived from the interpretation of T-RFLP data (Fig. S3) is reported in Figure 6, in which the largest fraction of the community is represented by members of *Proteobacteria*, with α-*Proteobacteria* as the most abundant class.

Differences in class composition across the site were observed suggesting that the microbial composition changes in response to land uses. In fact, all the seven classes (α, β and γ *Proteobacteria*, *Sphingobacteria Flavobacteria*, *Actinobacteria* and *Bacilli*) were present in CV, PA, and CO soils, while all classes but α-*Proteobacteria* in TV and all classes but *Sphingobacteria* in MM were found. The observed large diffusion of *Proteobacteria* in all soil-uses is in agreement with previously reported data concerning soil bacterial communities in different land-use systems [65]. Within *Proteobacteria*, the majority of the isolates fell into the gamma subgroup, which showed higher relative abundance compared to that of any other taxa at the class level while a few were alfa proteobacteria, especially in vineyards and pasture. In all soil-uses, were also found *Actinobacteria*, in agreement with previous work [65], and β-*Proteobacteria*, like *Burkholderia* sp., previously detected by Pastorelli and co-workers [19] who analysed the denitrifying bacterial communities present in the same Berchidda soil samples. Among Bacteroidetes, it has to be noted the relative abundance of the genera *Chryseobacterium* sp. (in all soil uses) and *Flavobacterium* sp. (in all soil uses, but MM), which include isolates already observed in Korea soils [66], and bacteria with plant growth promoting properties recovered in Iran soil [67], respectively. As already pointed out by Fierer and co-workers [68], the β-*Proteobacteria* and *Bacteroidetes* follow copiotrophic lifestyles and their relative abundance were highest in soils with high C availability. In general, copiotrophic bacteria should have higher growth rates and traditional culturing methods are likely to select for microorganisms that can grow rapidly in high resource environments.

Strong differences in class composition were observed in each soil when sampled in the two seasons: for instance, *Bacilli* and *Actinobacteria* dominate in spring (particularly, *Bacilli* in CV and CO, while *Actinobacteria* in MM and PA), whilst β-*Proteobacteria* tend to dominate in autumn. Six out of the seven identified classes (β and γ-*Proteobacteria*, *Flavobacteria*, *Actinobacteria* and *Bacilli*) were recovered in both seasons whilst *Sphingobacteria* were recovered only in autumn. In all soils, *Actinobacteria* were prevalent in spring, β *Proteobacteria* predominated in autumn, while β and γ-*Proteobacteria* in both spring and autumn. When genera composition was used to infer the diversity indices of cultured bacteria (Table S7) an increase of diversity (estimated as Shannon H and Evenness) from May to November was present for all soils, but CO, where Shannon H was slightly higher in May than in November. In particular, MM soil showed the highest increases for most indices, especially for alpha diversity (*i.e.* the species richness and evenness within a sample) that has often been correlated with ecosystem stability and functionality [69]. It must be noted that cultivated bacterial populations did not fluctuate with seasonal changes in soils with low human input (cork-oak forest). The bacterial communities of the cork-oak forest soil were similar in richness and composition, furthering the point that in a community with moderate disturbance, new individuals and groups could be introduced in a manner that promotes competition and diversity of the community, thus establishing a more stable community [65].

These data suggest that shifts from forest to managed meadow and vineyard result in changes of bacterial communities composition. Previous studies have also shown that shifts from forest to grassland soil [70], from cultivated system to pasture [71] and from grazed pine forest to cultivated crop and grazed pasture [65] resulted in significant changes in bacterial community composition. Our results suggest that the use of culture-dependent 16S rRNA gene sequencing along with traditional analysis of soil physiochemical properties may provide insight into the ecological relevance of soil bacterial taxa.

Conclusion

Overall, data obtained in this work revealed an important effect of land uses on both chemical and biochemical soil parameters. Soil bacterial communities were seasonally distinct and exhibited compositional shifts that tracked with changes in land use and soil management. This study, combining the pedological and biochemical data with microbiological and molecular analysis, furnishes a good methodological approach to describe the influence of different soil managements on soil microbial community structure. In fact, the results demonstrate that, in the same pedological conditions, long-term soil management influence the community structure; i.e., soils subjected to low human inputs (cork-oak forest and pasture) showed a more stable chemical and biochemical soil composition as well as bacterial community than those with high human input (vineyards and managed meadow). Further research is required to determine whether the observed shifts in bacterial community composition produce parallel changes in the functional attributes of these communities across soil types under different long-term management regimes. The use of culture-independent approaches, like metabarcoding and metagenome sequencing, will make it possible to identify the specific drivers of land-use dynamics exhibited by soil bacterial communities and to give a complete picture of the bacterial communities in a typical Mediterranean agro-silvo-pastoral system.

Supporting Information

Figure S1 Box-plot analysis showing the frequency distribution of physical-chemical and biological properties of the five Sardinia soils.

Figure S2 Examples of CD-DGGE profiles of the soil bacterial communities associated to the different land uses in May and November. A) From the left to right: hayland pasture rotation (PA), tilled vineyard (TV), grass covered vineyard (CV) in May; B) managed meadow (MM), cork-oak forest (CO), hayland pasture rotation (PA) in May; C) grass covered vineyard (CV) in May, grass covered vineyard (CV) in November, tilled vineyard (TV) in May; D) cork-oak forest (CO), hayland pasture rotation (PA), managed meadow (MM) in November.

Figure S3 Examples of T-RFLP profiles obtained after digestion with *Msp*I (A) and *Rsa*I (B) restriction enzymes of amplified of 16S rRNA gene sequences.

Figure S4 Cluster analysis of T-RFLP patterns generated by *Msp*I and *Rsa*I digestion of 16S rRNA gene sequences. The UPGMA cluster analysis based on Jaccard similarity matrix was calculated for each set of samples using the "hclust" function of the R "stats" package. The scale bar represents the percent of dissimilarity.

Table S1 Pedological profiles and classification of the soils investigated.

Table S2 Determination of total organic carbon soil (C_{org}), extractable carbon (C_{ext}), humified carbon (C_{HA+FA}), non humified carbon (C_{NH}) and humification parameters of the five Sardinian soils.

Table S3 Biochemical parameters measured in the five Sardinian soils.

Table S4 r/k bacterial strategists, total culturable bacteria and EPI index measured in soils under different long-term management practices.

Table S5 Diversity indices of total bacterial communities as inferred from T-RFLP profiles.

Table S6 Phylogenetic affiliations of 203 randomly selected soil bacterial isolates based on comparative analysis of their 16S rRNA gene sequences.

Table S7 The ratio of diversity indices related to the abundance of the different genera detected in cultured isolates between November and May.

Acknowledgments

We are grateful to P.P. Roggero (University of Sassari) and his team (G. Seddaiu, G. Urracci and L. Doro) for their contribution to collect the Sardinian soils and for providing data on experimental area. The experimental site was chosen on the basis of vegetation and soil surveys made in collaboration with S. Madrau, S. Bagella, R. Filigheddu, M.C. Caria, and E. Farris (University of Sassari). We thank R. Pastorelli for her help in soil sampling, S. Tabacchioni and L. Chiarini for bacterial isolation, C. Cantale and M. Sperandei for figure preparation, M.T. Rubino for chemical-biochemical analysis, and S. Cesarini and L. Pirone for their valuable suggestions. The research was carried out in the context of the FISR SOILSINK research project coordinated by R. Francaviglia (CRA-RPS, Rome) (?http://soilsink.entecra.it).

Author Contributions

Conceived and designed the experiments: A. Bevivino RF CD A. Benedetti. Performed the experiments: A. Bevivino PP MSP MCP CD LL. Analyzed the data: A. Bevivino PP GB AF AM CD LL. Contributed reagents/materials/analysis tools: A. Bevivino PP GB AF MSP AM LL MCP RF CD A. Benedetti. Contributed to the writing of the manuscript: A. Bevivino RF GB AF CD LL A. Benedetti.

References

1. Tiedje JM, Asuming-Brempong S, Nusslein K, Marsh TL, Flynn SJ (1999) Opening the black box of soil microbial diversity. Appl Soil Ecol 13: 109–122. doi:10.1016/S0929-1393(99)00026-8.
2. Nannipieri P, Ascher J, Ceccherini MT, Landi L, Pietramellara G, et al. (2003) Microbial diversity and soil functions. Eur J Soil Sci 54: 655–670. doi:10.1046/j.1365-2389.2003.00556.x.
3. Lauber CL, Ramirez KS, Aanderud Z, Lennon J, Fierer N (2013) Temporal variability in soil microbial communities across land-use types. ISME J 7: 1641–1650. doi:10.1038/ismej.2013.50.
4. Myers N, Mittermeier RA, Fonseca GAB, Fonseca GAB, Kent J (2000) Biodiversity hotspots for conservation priorities. Nature 403: 853–858. doi:10.1038/35002501.
5. Puddu G, Falcucci A, Maiorano L (2011) Forest changes over a century in Sardinia: implications for conservation in a Mediterranean hotspot. Agrofor Syst 85: 319–330. doi:10.1007/s10457-011-9443-y.
6. Médail F, Quézel P (1999) Biodiversity Hotspots in the Mediterranean Basin: Setting Global Conservation Priorities. Conserv Biol 13: 1510–1513. doi:10.1046/j.1523-1739.1999.98467.x.
7. Blondel J, Aronson J (1999) Biology and wildlife of the Mediterranean region. Oxford University Press, Oxford.
8. Salis L, Marrosu M, Bagella S, Sitzia M RP (2010) Grassland management, forage production and plant biodiversity in a Mediterranean grazing system. In: Porqueddu C, Ríos S, editors. The contributions of grasslands to the conservation of Mediterranean biodiversity. Zaragoza: CIHEAM/CIBIO/FAO/SEEP, Vol. 185. 181–185.
9. Gonçalves P, Alcobia S, Simões L, Santos-Reis M (2012) Effects of management options on mammal richness in a Mediterranean agro-silvo-pastoral system. Agrofor Syst 85: 383–395. doi:10.1007/s10457-011-9439-7.
10. Bacchetta G, Bagella S, Biondi E, Farris E, Filigheddu R, et al. (2004) A contribution to the knowledge of the order Quercetalia ilicis Br. -Bl. ex Molinier 1934 of Sardinia. Fitosociologia 41: 29–51.
11. Orgiazzi A, Lumini E, Nilsson RH, Girlanda M, Vizzini A, et al. (2012) Unravelling soil fungal communities from different Mediterranean land-use backgrounds. PLoS One 7: e34847. doi:10.1371/journal.pone.0034847.
12. Francaviglia R, Benedetti A, Doro L, Madrau S, Ledda L (2014) Influence of land use on soil quality and stratification ratios under agro-silvo-pastoral Mediterranean management systems. Agric Ecosyst Environ 183: 86–92.
13. Council of the European Communities (1992) Council Directive 92/43/EEC of 21 May 1992 on the conservation of natural habitats and of wild fauna and flora. Off J Eur Communities 35: 7–50.
14. Council of the European Communities (2001) Commission Regulation (EC) No 1808/2001 of 30 August 2001 laying down detailed rules concerning the implementation of Council Regulation (EC) No 338/97 on the protection of species of wild fauna and flora by regulating trade therein. Off J Eur Communities L250: 1–43.
15. Aru A, Baldaccini P, Delogu G, Dessena M, Madrau S, et al. (1990) Carta dei Suoli della Sardegna, scala 1/250.000. Assessorato alla programmazione e all'assestamento del territorio, Centro Regionale Programmazione, Dip. Sc. della Terra, Università di Cagliari, Italy.
16. Bacchetta G, Bagella S, Biondi E, Farris E, Filigheddu R, et al. (2009) Forest vegetation and serial vegetation of Sardinia (with map at the scale 1:350,000). Fitosociologia 46: 3–82.
17. Bagella S, Caria MC (2011) Vegetation series: a tool for the assessment of grassland ecosystem services in Mediterranean large-scale grazing systems. Fitosociologia 48: 47–54.
18. Lumini E, Orgiazzi A, Borriello R, Bonfante P, Bianciotto V (2010) Disclosing arbuscular mycorrhizal fungal biodiversity in soil through a land-use gradient using a pyrosequencing approach. Environ Microbiol 12: 2165–2179. doi:10.1111/j.1462-2920.2009.02099.x.
19. Pastorelli R, Landi S, Trabelsi D, Piccolo R, Mengoni A, et al. (2011) Effects of soil management on structure and activity of denitrifying bacterial communities. Appl Soil Ecol 49: 46–58. doi:10.1016/j.apsoil.2011.07.002.
20. Lagomarsino A, Benedetti A, Marinari S, Pompili L, Moscatelli MC, et al. (2011) Soil organic C variability and microbial functions in a Mediterranean agro-forest ecosystem. Biol Fertil Soils 47: 283–291. doi:10.1007/s00374-010-0530-4.
21. Francaviglia R, Coleman K, Whitmore AP, Doro L, Urracci G, et al. (2012) Changes in soil organic carbon and climate change - Application of the RothC model in agro-silvo-pastoral Mediterranean systems. Agric Syst 112: 48–54.
22. Seddaiu G, Porcu G, Ledda L, Roggero PP, Agnelli A, et al. (2013) Soil organic matter content and composition as influenced by soil management in a semi-arid Mediterranean agro-silvo-pastoral system. Agric Ecosyst Environ 167: 1–11. doi:http://dx.doi.org/10.1016/j.agee.2013.01.002.
23. Lai R, Lagomarsino A, Ledda L, Roggero PP (2014) Variation in soil C and microbial functions across tree canopy projection and open grassland microenvironments. Turkish J Agric For 38: 62–69. doi:10.3906/tar-1303-82.
24. Soil Survey Staff (2006) Keys to soil taxonomy. In: United States Department of Agriculture NRCS, editor. Natural Resoiurces Concervation Service.
25. Doran JW, Elliott ET, Paustian K (1998) Soil microbial activity, nitrogen cycling, and long-term changes in organic carbon pools as related to fallow tillage management. Soil Tillage Res 49: 3–18.
26. O'Brien HE, Parrent JL, Jackson JA, Moncalvo J-M, Vilgalys R (2005) Fungal community analysis by large-scale sequencing of environmental samples. Appl Environ Microbiol 71: 5544–5550.
27. Stotzky G, Goos RD, Timonin MI (1962) Microbial changes occurring in soil as a result of storage. Plant Soil 16: 1–18. doi:10.1007/BF01378154.
28. Springer U KJ (n.d.) Prüfung der Leistungsfähigkeit von einigen wichtigeren Verfahren zur Bestimmung des Kohlenstoffs mittels Chromschwefelsäure sowie Vorschlag einer neuen Schnellmethode. J Plant Nutr Soil Sci 64: 1–26 (in German).
29. Ciavatta C, Govi M, Vittori Antisari L, Sequi P (1990) Characterization of humified compounds by extraction and fractionation on solid polyvinylpyrrolidone. J Chromatogr 509: 141–146.
30. Sequi P, De Nobili M, Leita L, Cercignani G (1986) A new index of humification. Agrochimica 30: 175–179.
31. Vance ED, Brookes PC, Jenkinson DS (1987) An extraction method for measuring soil microbial biomass C. Soil Biol Biochem 19: 703–707. doi:10.1016/0038-0717(87)90052-6.
32. Badalucco L, Grego S, Dell'Orco S, Nannipieri P (1992) Effect of liming on some chemical, biochemical, and microbiological properties of acid soils under spruce (Picea abies L.). Biol Fertil Soils 14: 76–83. doi:10.1007/BF00336254.
33. Riffaldi R, Saviozzi A, Levi-Minzi R (1996) Carbon mineralization kinetics as influenced by soil properties. Biol Fertil Soils 22: 293–298. doi:10.1007/BF00334572.
34. Dilly O, Munch J-C (1998) Ratios between estimates of microbial biomass content and microbial activity in soils. Biol Fertil Soils 27: 374–379. doi:10.1007/s003740050446.
35. Anderson T-H, Domsch KH (1989) Ratios of microbial biomass carbon to total organic carbon in arable soils. Soil Biol Biochem 21: 471–479. doi:10.1016/0038-0717(89)90117-X.
36. Pinzari F, Trinchera A, Benedetti A, Sequi P (1999) Use of biochemical indices in the mediterranean environment: Comparison among soils under different forest vegetation. J Microbiol Methods 36: 21–28.
37. Smalla K, Wieland G, Buchner A, Zock A, Parzy J, et al. (2001) Bulk and Rhizosphere Soil Bacterial Communities Studied by Denaturing Gradient Gel Electrophoresis: Plant-Dependent Enrichment and Seasonal Shifts Revealed. 67: 4742–4751. doi:10.1128/AEM.67.10.4742.

38. De Leij FAAM, Whipps JM, Lynch JM (1994) The use of colony development for the characterization of bacterial communities in soil and on roots. Microb Ecol 27: 81–97.
39. Grifoni A, Bazzicalupo M, Di Serio C, Fancelli S, Fani R (1995) Identification of Azospirillum strains by restriction fragment length polymorphism of the 16S rDNA and of the histidine operon. FEMS Microbiol Lett 127: 85–91. doi:10.1016/0378-1097(95)00042-4.
40. Trabelsi D, Mengoni A, Elarbi Aouani M, Mhamdi R, Bazzicalupo M (2009) Genetic diversity and salt tolerance of bacterial communities from two Tunisian soils. Ann Microbiol 59: 25–32. doi:10.1007/BF03175594.
41. Mengoni A, Tatti E, Decorosi F, Viti C, Bazzicalupo M, et al. (2005) Comparison of 16S rRNA and 16S rDNA T-RFLP approaches to study bacterial communities in soil microcosms treated with chromate as perturbing agent. Microb Ecol 50: 375–384. doi:10.1007/s00248-004-0222-4.
42. Hammer Ø, Harper DAT, Ryan PD (2001) Past: Paleontological Statistics Software Package for Education and Data Analysi. Palaeontol Electron 4: 1–9. doi:10.1016/j.bcp.2008.05.025.
43. Duineveld B, Kowalchuk G, Keijzer A, van Elsas J, van Veen J (2001) Analysis of bacterial communities in the rhizosphere of chrysanthemum via denaturing gradient gel electrophoresis of PCR-amplified 16S rRNA as well as DNA fragments coding for 16S rRNA. Appl Environ Microbiol 67: 172–178.
44. Ausubel FM, Brent R, Kingston RE, Moore DD, Seidman JG, et al. (1987) Unit 2.4 Preparation of genomic DNA from bacteria. Curr Protoc Mol Biol: 2.4.1–2.4.2.
45. Di Cello F, Pepi M, Baldi F, Fani R (1997) Molecular characterization of an n-alkane-degrading bacterial community and identification of a new species, Acinetobacter venetianus. Res Microbiol 148: 237–249.
46. El Fantroussi S, Verschuere L, Verstraete W, Top EM (1999) Effect of phenylurea herbicides on soil microbial communities estimated by analysis of 16S rRNA gene fingerprints and community-level physiological profiles. Appl Environ Microbiol 65: 982–988.
47. Cremonesi L, Firpo S, Ferrari M, Righetti PG, Gelfi C (1997) Double-gradient DGGE for optimized detection of DNA point mutations. Biotechniques 22: 326–330.
48. Shannon C, Weaver W (1963) The mathematical theory of communication. University of Illinois Press, Urbana. doi:10.1145/584091.584093.
49. Simpson E (1949) Measurement of diversity. Nature 163: 688–688. doi:10.1038/163688a0.
50. Whittaker RH (1972) Evolution and measurement of species diversity. Taxon 21: 213. doi:10.2307/1218190.
51. Hall TA (1999) BioEdit: a user-friendly biological sequence alignment editor and analysis program for Windows 95/98/NT. Nucleic Acids Symp 41: 95–98. doi:citeulike-article-id:691774.
52. Bray J, Curtis J (1957) An ordination of the upland forest communities of southern Wisconsin. Ecol Monogr 27: 325–349.
53. Haider K (1992) Problems related to the humification processes in soils of temperate climates. In: Stotzky G, Bollag J-M, editors. Soil Biochemistry. Marcel Dekker, New York. 55–94.
54. Oades JM (1995) An overview of processes affecting the cycling of organic carbon in soils. In: Zepp R, Sonntag C, editors. Role of non living organic matter in the earth's carbon cycle. John Wiley, New York. 293–303.
55. Stevenson FJ, Cole MA (1999) Cycles of soil: carbon, nitrogen, phosphorus, sulfur, micronutrients. 2nd ed. John Wiley, New York.
56. Franzluebbers AJ, Stuedemann JA, Schomberg HH, Wilkinson SR (2000) Soil organic C and N pools under long-term pasture management in the Southern Piedmont USA. Soil Biol Biochem 32: 469–478. doi:10.1016/S0038-0717(99)00176-5.
57. Anderson T, Domsch KH (1993) The metabolic quotient for CO2 (qCO2) as a specific activity parameter to assess the effects of environmental conditions, such as pH, on the microbial biomass of forest soils. Soil Biol Biochem 25: 393–395.
58. Benedetti A, Brookes P, Lynch J (2005) Conclusive remarks. Part I: Approaches to defining, monitoring, evaluating and managing soil quality. In: Bloem J, Hopkins D, Benedetti A, editors. Soil Quality. CABI; Wallingford. 63–70.
59. Conant RT, Paustian K, Elliott ET (2001) Grassland management and conversion into grassland: effects on soil carbon. Ecol Appl 11: 343–355. doi:10.1890/1051-0761(2001)011[0343:GMACIG]2.0.CO;2.
60. Edenborn SL, Sexstone AJ (2007) DGGE fingerprinting of culturable soil bacterial communities complements culture-independent analyses. Soil Biol Biochem 39: 1570–1579.
61. Garland JL (1999) Potential and limitations of BIOLOG for microbial community analysis. In: Bell C, Brylinsky M, Johnson-Green P, editors. Microbial Biosystems: new frontiers. Proceedings of the 8th international symposium on microbial ecology. Atlantic Canada Society for Microbial Ecology. Halifax, Canada. 1–7.
62. Ellis RJ, Morgan P, Weightman AJ, Fry JC (2003) Cultivation-dependent and -independent approaches for determining bacterial diversity in heavy-metal-contaminated soil. Appl Environ Microbiol 69: 3223–3230. doi:10.1128/AEM.69.6.3223.
63. Shade A, Hogan CS, Klimowicz AK, Linske M, McManus PS, et al. (2012) Culturing captures members of the soil rare biosphere. Environ Microbiol 14: 2247–2252. doi:10.1111/j.1462-2920.2012.02817.x.
64. Altschul SF, Madden TL, Schäffer AA, Zhang J, Zhang Z, et al. (1997) Gapped BLAST and PSI-BLAST: a new generation of protein database search programs. Nucleic Acids Res 25: 3389–3402.
65. Shange RS, Ankumah RO, Ibekwe AM, Zabawa R, Dowd SE (2012) Distinct soil bacterial communities revealed under a diversely managed agroecosystem. PLoS One 7: e40338. doi:10.1371/journal.pone.0040338.
66. Weon H-Y, Kim B-Y, Yoo S-H, Kwon S-W, Stackebrandt E, et al. (2008) Chryseobacterium soli sp. nov. and Chryseobacterium jejuense sp. nov., isolated from soil samples from Jeju, Korea. Int J Syst Evol Microbiol 58: 470–473. doi:10.1099/ijs.0.65295-0.
67. Soltani A, Khavazi K, Asadi-Rahmani H, Omidvari M, Abaszadeh Dahaji P (2010) Plant growth promoting characteristics in some Flavobacterium spp. isolated from soils of Iran. J Agri Sci 2: 106–115. Available: http://www.ccsenet.org/journal/index.php/jas/article/view/8404/6221.
68. Fierer N, Bradford MA, Jackson RB (2007) Toward an ecological classification of soil bacteria. Ecology 88: 1354–1364. doi:10.1890/05-1839.
69. Girvan MS, Campbell CD, Killham K, Prosser JI, Glover LA (2005) Bacterial diversity promotes community stability and functional resilience after perturbation. Environ Microbiol 7: 301–313. doi:10.1111/j.1462-2920.2005.00695.x.
70. Nacke H, Thürmer A, Wollherr A, Will C, Hodac L, et al. (2011) Pyrosequencing-based assessment of bacterial community structure along different management types in German forest and grassland soils. PLoS One 6: e17000. doi:10.1371/journal.pone.0017000.
71. Acosta-Martínez V, Dowd S, Sun Y, Allen V (2008) Tag-encoded pyrosequencing analysis of bacterial diversity in a single soil type as affected by management and land use. Soil Biol Biochem 40: 2762–2770. doi:10.1016/j.soilbio.2008.07.022.

17

Nutrient Addition Dramatically Accelerates Microbial Community Succession

Joseph E. Knelman[1,2], Steven K. Schmidt[2], Ryan C. Lynch[2], John L. Darcy[2], Sarah C. Castle[3], Cory C. Cleveland[3], Diana R. Nemergut[1,4]*

1 Institute of Arctic and Alpine Research, University of Colorado, Boulder, Colorado, United States of America, **2** Department of Ecology and Evolutionary Biology, University of Colorado, Boulder, Colorado, United States of America, **3** Department of Ecosystem and Conservation Sciences, University of Montana, Missoula, Montana, United States of America, **4** Environmental Studies Program, University of Colorado, Boulder, Colorado, United States of America

Abstract

The ecological mechanisms driving community succession are widely debated, particularly for microorganisms. While successional soil microbial communities are known to undergo predictable changes in structure concomitant with shifts in a variety of edaphic properties, the causal mechanisms underlying these patterns are poorly understood. Thus, to specifically isolate how nutrients – important drivers of plant succession – affect soil microbial succession, we established a full factorial nitrogen (N) and phosphorus (P) fertilization plot experiment in recently deglaciated (~3 years since exposure), unvegetated soils of the Puca Glacier forefield in Southeastern Peru. We evaluated soil properties and examined bacterial community composition in plots before and one year after fertilization. Fertilized soils were then compared to samples from three reference successional transects representing advancing stages of soil development ranging from 5 years to 85 years since exposure. We found that a single application of +NP fertilizer caused the soil bacterial community structure of the three-year old soils to most resemble the 85-year old soils after one year. Despite differences in a variety of soil edaphic properties between fertilizer plots and late successional soils, bacterial community composition of +NP plots converged with late successional communities. Thus, our work suggests a mechanism for microbial succession whereby changes in resource availability drive shifts in community composition, supporting a role for nutrient colimitation in primary succession. These results suggest that nutrients alone, independent of other edaphic factors that change with succession, act as an important control over soil microbial community development, greatly accelerating the rate of succession.

Editor: Jack Anthony Gilbert, Argonne National Laboratory, United States of America

Funding: This work was supported by the National Science Foundation of the USA (http://nsf.gov/) through grants for studying microbial succession following glacial retreat (DEB-0922267) to DRN, SKS, and CCC; the Alpine Microbial Observatory (MCB-0455606) to SKS; and a graduate research fellowship to JEK. The funders had no role in study design, data collection and analysis, decision to publish, or preparation of the manuscript.

Competing Interests: The authors have declared that no competing interests exist.

* Email: nemergut@colorado.edu

Introduction

Deglaciated forefields have been valuable model systems for developing and testing theories of succession and have greatly enhanced our understanding of the relationship between community structure and function during ecosystem development [1–3]. Shifts in soil nutrient pools, including increases in available nitrogen (N) and phosphorus (P), have been well documented along early primary successional chronosequences [4–6] and have been shown to correlate with changes in plant community succession [1,7,8]. Recently, studies in such systems have revealed that – like plants – microbial communities also progress through successional stages [9–11]. However, the forces that control microbial succession are not well understood.

Some evidence suggests that shifts in nutrient availability may also, in part, drive microbial community succession. For example, in primary successional ecosystems, research has corroborated relationships between natural gradients in soil nutrients and microbial community composition [12,13]. Such correlations can be difficult to interpret, however, as changes in microbial community composition could be both a cause and consequence of shifts in soil fertility. Furthermore, the mechanisms underlying correlations between standing nutrient pools and microbial communities may be temporally disconnected, in that current soil biogeochemical status may not accurately reflect the historical nutrient conditions that structured the microbial community. Thus, manipulation experiments are essential in evaluating the direct impact of nutrients and their limitations on microbial communities. Indeed, fertilizer treatments are known to elicit changes in soil microbial community structure and function in more developed ecosystems [14,15] suggesting that nutrient availability may also be important in controlling successional changes in microbial community composition.

Yet, it would be surprising if nutrients alone drove microbial community succession for several reasons. First, other edaphic properties also undergo concomitant shifts with microbial community structure and function during succession, some of which are known to more strongly correlate with microbial community structure than nutrient pools in developed soils. For example, organic carbon (C) pools and pH, which typically show dramatic changes across primary successional chronosequences [2], are key determinants of soil microbial community composition at regional to global scales [16–19]. Second, soil microbial community structure can correlate with plant community composition [20,21], which can show strong spatial gradients in early succession [3]. Third, stochastic processes can be key in shaping early successional communities where the importance of dispersal events may be accentuated, [22–24] and arrival order may influence assembly through priority effects [25]. Given the large functional and phylogenetic diversity of microbial communities, it is possible that succession is influenced by a diverse combination of such factors [26].

Thus, the extent to which nutrients themselves influence microbial community assembly outside of the myriad of factors that change over succession is unknown. To specifically isolate the effects of nutrients, we performed a full factorial N×P fertilization experiment in soils that had been exposed for ~3 years in the forefield of the Puca Glacier in Southeastern Peru. We analyzed soil bacterial communities before and one year following nutrient additions and compared them with soils sampled from three different locations over an 85-year section of the Puca Glacier chronosequence. The Puca Glacier soils constitute an autotrophic successional sequence [27], and both photosynthesis and respiration respond strongly to P additions in microcosms [28,29]. Nitrogen appears to be limiting in this system as well and N-fixation rates in 4 year old unvegetated soils are comparable to rates measured in developed soil crusts [30]. Thus, given work that demonstrates relationships between nutrients and microbial community composition, we hypothesized that fertilizer additions to early successional soils would drive communities to be compositionally different than unfertilized (control) soils. However, due to the potential influence of other edaphic (e.g. pH, organic C, soil moisture) and stochastic factors on microbial succession, we hypothesized that fertilized communities would be unique from communities found along the natural chronosequence.

Materials and Methods

Study site description, fertilization, and sampling

The study site is located in the forelands of the Puca Glacier in the Cordillera Vilcanota of Peru (13°46′24″S, 71°04′17″W, ~5,000 m.a.s.l.). No specific permits were required for our field studies and our work did not involve endangered or protected species. Mean annual precipitation is roughly 100 cm and mean annual temperature is ~5°C. Moraine rocks at this site have high quartz and calcite mineral content. Further details of this site can be found in previous work [9,30] and soil characteristics are presented in Table S2.

We established permanent plots (1 m^2) near the terminus of the glacier, in soils that had been deglaciated for approximately 3 years at the time of initial sampling. Corners were marked with long nails (approximately 15 cm shank length) to guide resampling. Sampling occurred in August 2010 (pre-treatment) and August 2011 (post-treatment). All of the plots were unvegetated and no mosses and lichens were present at the time of establishment. Each of the 16 plots was randomly chosen to receive one of three nutrient amendments (nitrogen addition (+N), phosphorus addition (+P), the combination of the two (+NP)) or to serve as controls, resulting in a total of four plots per treatment and four control plots.

Pre-weighed amounts of fertilizer were dissolved in glacier-melt stream water and fertilizer solutions were applied with handheld sprayers. Each sprayer was designated for a particular treatment to avoid cross contamination. For the +N plots, nitrogen was added in the form of ammonium nitrate (NH_4NO_3) resulting in 15 g of NH_4NO_3 and 5.25 g of N/m^2. The +P plots received 0.5 g of phosphorus in the form of 2.2 g of potassium dihydrogen phosphate (KH_2PO_4). +NP plots received 15 g of NH_4NO_3 and 2.2 g of KH_2PO_4. For controls, stream water from the same source was sprayed onto the plots. These levels of nutrient addition were designed to result in a pulse of nutrients that would greatly overcome any possible natural limitations.

Plots were sampled prior to the application of fertilization treatment. In each plot surface soil was collected (0–5 cm) from 2 locations, and samples were composited to generate one sample per plot. Samples were obtained in the same manner one year following the fertilization treatment. Ethanol and paper towels were used to sterilize the tools before sampling each individual plot.

Samples were collected in a similar manner along three transects of varying age across the glacial forefield both years; molecular analyses were done on the samples collected in 2011. These reference soils represented advancing stages of succession: soils that had been exposed for approximately 5 years, soils with biological soil crust formation (approximately 20 years after exposure), and soils with 25–50% vegetation cover (approximately 85 years after exposure). At the field site, samples were kept in a cooler on ice for transport to Boulder, CO. Soils were sieved (to 2 mm), and then stored at 4°C for soil characterization. A subsample was immediately archived in a −80°C freezer for molecular analysis and later used for KCl extractions.

Soil Analysis

Gravimetric soil moisture and pH (using a ratio of 2 g soil to 4 mL DI H_2O) were assayed based on standard methods [9]. For total organic C analysis, carbonate (inorganic C) removal was first performed on dried, ground soils [9]. 50 mg of these processed soils were packed into tin capsules; %C and %N were determined using a Thermo Finnigan EA 1112 Series Flash Elemental Analyzer (Thermo Fisher Scientific, Inc., Waltham, Massachusetts, USA) [31]. Bio-available P concentrations were measured on air-dried and sieved soil (2 mm×2 mm) by extracting 3–5 g of soil with 0.5 M sodium bicarbonate for 30 minutes [32]. Extracts were filtered and analyzed colorimetrically using the ammonium molybdate-malachite green method [33] adapted for microplate analysis. NH_4^+ and NO_3^-/NO_2^- extractable N were analyzed from soils using 2M KCl with 1 hour shaking and a 22 hour extraction period [34]. This analysis was performed on soils that were frozen at −80°C. Although not fresh samples, these soils typically withstand extreme fluctuations in temperature [35] and the data presented here are intended for within study comparison only. NH_4^+ and NO_3^-/NO_2^- were measured on a Lachat QuikChem 8500 Flow Injection Analyzer (Lachat Instruments, Hach Company, Loveland, CO) and BioTek Synergy 2 Multi-detection Microplate Reader (BioTek, Winooski, VT) respectively.

DNA Extractions for 454 pyrosequencing

MO BIO PowerSoil™ DNA Isolation kits were used as per the manufacturer's instructions for DNA extractions of soil samples (Mo Bio Laboratories, Inc., Carlsbad, CA). PCR-amplified bacterial 16S rRNA genes from the genomic DNA of the soil samples were generated using a universal bacterial 27F and 338R primer set as described by Hamady et al. [36], and reaction conditions followed those described by Fierer et al. [37], though modified to 25 PCR cycles. Primers included a 2 bp linker, the 454 Roche Titanium A/B primer, and a unique, 12 base pair error-correcting Golay barcode for pyrosequencing as detailed by Knelman et al. [21]. 454 Life Sciences GS FLX Titanium pyrosequencing of the 16S rRNA gene amplicons was completed by the Duke Institute for Genome Sciences & Policy (Duke University, North Carolina).

Pyrosequence and statistical analysis

Using QIIME, sequences were limited to those of a sequence length of 200 to 400 base pairs, a maximum of 5 homopolymers, a minimum quality score of 25, and a maximum of ambiguous bases/primer mismatches of 0; reverse primers were removed, and all samples were then denoised using flowgram clustering in QIIME [38]. Chloroplast sequences were removed. OTUs were selected at a 97% identity level by clustering based on representative sequences via UCLUST [39]. The Ribosomal Database Classifier [40], a naïve Bayesian classifier, was employed to assign taxonomic identification to OTUs. After sequence alignments based on the NAST algorithm [41], a phylogeny was constructed with the FastTree algorithm [42]. OTU tables were rarified to the lowest number of sequences in a sample: 407 for community dissimilarity analyses of fertilization plots. Reference transects of advancing age included 6, 5, and 3 sequenced replicate samples, respectively, and were rarefied to 71 to include all of these samples. For comparison of reference samples and fertilization plots this workflow was repeated. In order to examine differences among bacterial communities, pairwise distance matrices based on weighted UniFrac, a phylogenetic distance metric, were generated for entire communities and the cyanobacterial subset of communities in fertilization plots [43,44]. The Principal Coordinate Analysis (PCoA) ordinations were constructed based on OTU tables and weighted UniFrac distance matrices for overall communities. The QIIME-generated OTU tables were used to evaluate the relative abundance of all taxa.

Primer v6 software [45] was used to perform permutational ANOVAs (PERMANOVA) to compare phylogenetic distances among bacterial communities. PERMANOVA tests were used on both UniFrac beta diversity matrices of the entire communities and cyanobacterial portions of communities. PERMANOVA analysis was also employed to assess differences among treatment-affected communities and successional reference communities. For all comparisons with reference communities, data were rarefied to the lowest sampling depth among both fertilization plot and reference plot samples.

R software [46] was used for further statistical analysis. The PERMDISP2 procedure (with permutational P-values) from the R vegan package to test homogeneity of group dispersions (variances) was also employed via QIIME in order to test for differences in community phylogenetic dispersion (UniFrac) in fertilized samples and reference successional communities [47,48]. As well, the pgirmess package in R was used to evaluate comparisons among reference chronosequence soil relative abundance data via the Kruskal Wallis test. To assess treatment vs. temporal effects underlying shifts in overall phylogenetic community composition, a Tukey's HSD post-hoc test was used to compare UniFrac distances of paired pre- and post-treatment plots with paired control plots from both years. Additionally, to assess the relative abundances of bacterial taxa, we compared the differences in paired pre- to post-treatment taxon relative abundances for each treatment with that of paired control plots via Tukey's HSD post-hoc tests. To examine the relationship between treatment-related community shifts from our fertilization experiment and reference communities across advancing stages of soil development, we examined the relationship between weighted UniFrac phylogenetic dissimilarity and time between +NP communities and reference communities via a Spearman correlation Mantel test. The Mantel test tests the null hypothesis that there is no correlation between +NP and reference community dissimilarity and chronosequence age rank.

All relative abundance data and environmental variables were evaluated for normality. Taxon relative abundances and fertilizer plot NO_3^-/NO_2^- were square root transformed to achieve a normal distribution prior to statistical analysis. All other edaphic factors were natural log transformed. ANOVAs, Tukey's HSD, and Kruskal Wallis post-hoc tests were used to assess differences in pH, %C, P, N pools and soil moisture in fertilization plots and reference chronosequence soils. Percent N was below the detection limit in a majority of samples and thus removed from statistical evaluations.

Sequences and metadata have been deposited in FigShare and are available with the DOIs: 10.6084/m9.figshare.1050042 (metadata) and 10.6084/m9.figshare.1048992 (sequences).

Results and Discussion

Together, our analyses demonstrate that a single +NP application caused the bacterial community structure of the 3-year-old barren soils to converge with the structure of 85-year-old vegetated soils after only one year. First, paired pre- and post-treatment plot community differences (weighted UniFrac distance) were assessed among all plot categories using an ANOVA. The + NP plots showed a significant community shift in response to the treatment (Tukey's HSD; $P = 0.037$); no other significant differences in community structure were detected between treatments and controls (Tukey's HSD; $P > 0.05$). A PCoA ordination (Fig. 1) revealed a successional trend in community composition across the reference chronosequence, with post-treatment +NP communities clustering with the oldest reference communities. A PERMANOVA analysis demonstrated that there were no significant differences among pre-treatment communities (Table 1). However, communities in post-treatment +NP plots were significantly different from both pre- and post-treatment controls, including the paired pre-treatment +NP plots (PERMANOVA, $P < 0.05$, Table 1). When +NP communities were compared to reference communities across the natural chronosequence, a Mantel test of pairwise average UniFrac [43,44] distances between +NP plots and reference samples revealed significant patterns of decreasing dissimilarity: +NP communities were most similar to the 85 year old successional soils (Fig. 2, $\rho_M = -0.35$ $P = 0.01$). The PERMANOVA analysis also showed that +NP communities were significantly different than communities of all successional stages except those of the oldest transect (85 years old) (Table 1). These results suggest that fertilization drives community composition away from early successional stages and results in convergence with communities of older soils. Likewise, the phylogenetic dispersion [47,48] of +NP communities was significantly different from all reference communities except those in the 85 year old soils (Table 2). We note that our PERMANOVA analysis was not corrected for multiple comparisons due to the low statistical power of our study, but the general results of this analysis were nonetheless corroborated by our other statistical analyses of treatment effect (ANOVA/Tukey's HSD of pre- and post-treatment community shifts) and convergence of the +NP plots to the oldest successional soils (Mantel test of +NP community distance compared to successional reference samples over time).

Our results suggest that nutrient colimitation is an important control on microbial primary succession in this system. Because of low statistical power, it is difficult to discern whether this colimitation is simultaneous, meaning that both nutrients need to be present for a community response, or independent, meaning that each nutrient in isolation may elicit some response [49].

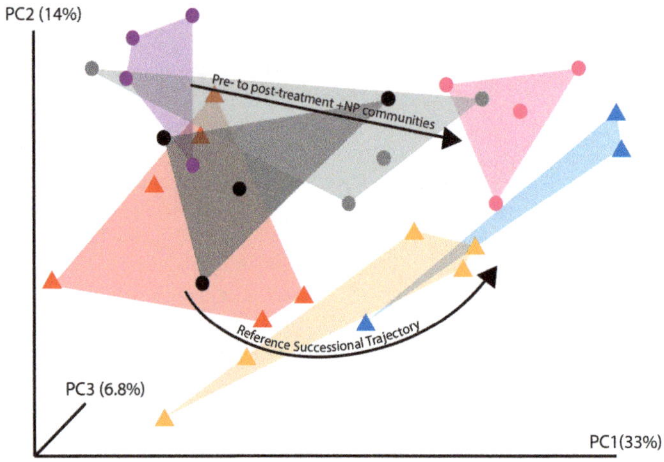

Figure 1. Principal Coordinates Analysis (PCoA) ordination plot of bacterial communities from the field fertilization experiment and bacterial communities from the successional chronosequence. Only the +NP treatment communities are shown because the +N and +P treatments did not result in significant community shifts. PCoA visually represents differences among community composition as the distance between points. Triangles represent communities from the natural chronosequence: red = 5 years old; orange = 20 years old; blue = 85 years old. Circles represent communities from the fertilization experiment: black = pre-treatment control; grey = post-treatment control; purple = pretreatment +NP; Pink = post-treatment +NP. Our analysis revealed significant community shifts over the reference chronosequence (triangles) as well as a significant response to +NP fertilization (circles). As well, the PCoA analysis demonstrates that the +NP communities (pink circles) group with the oldest soils from the chronosequence (blue triangles).

However, there is some evidence that single nutrient additions may cause a smaller response than when both nutrients are abundant. For example, our results show that post-treatment +P communities are not significantly different from post treatment + NP communities (Table 1). As well, both +N and +P plots show patterns of convergence similar to +NP plots in comparison with ongoing natural succession; by contrast, control plots do not display convergence (Table 1). Thus, +N and +P communities may represent intermediate states between control and +NP plots, but we were not able to statistically demonstrate an underlying treatment effect.

While our study is unique as we established and resampled nutrient addition plots in a remote glacial forefield, the rapidly changing nature of the Puca Glacier landscape and criteria for setting up plots on a stable and relatively homogenous surface limited replication and necessitated rarefaction of sequencing depth to 71 to include all available samples. As such, we acknowledge the need to be circumspect in drawing conclusions as such factors curbed the statistical power of our study and potentially our ability to detect smaller magnitude treatment effects in the +N and +P additions, for example. However, we note that the patterns shown here are robust to even lower rarefaction depths (55–70); thus, it is likely that observed patterns are real. Nonetheless, our research shows the greatest, and only statistically significant treatment effect on microbial communities under +NP additions, suggesting the effect of both nutrients in tandem is important in succession.

Interestingly, standing nutrient pool analysis lends some insight into particular dynamics that may underlie nutrient colimitation in this autotrophic chronosequence. For example, +P and +NP soils both show significant increases in ammonium pools in comparison with control plot soils (Table S1), which is consistent with a body of research that demonstrates P limitation is a strong control of N-fixation [50,51], and may be particularly strong in this autotrophic

Table 1. Post-treatment +NP phylogenetic community structure was significantly different from controls and from all communities from the reference chronosequence with the exception of communities in the oldest soils ($P<0.05$).

Permutational MANOVA (PERMANOVA) contrast P-values

Sample vs. Sample	post-treatment control	post-treatment +N	post-treatment +P	post-treatment +NP
pre-treatment control	0.415	0.066	**0.031**	**0.026**
post-treatment control	---	0.422	0.072	**0.036**
pre-treatment +NP-paired	0.114	0.085	**0.024**	**0.032**
post-treatment +NP plots	**0.036**	**0.023**	0.18	---
succession timepoint 1	0.124	**0.003**	**0.005**	**0.006**
succession timepoint 2	0.105	**0.009**	**0.018**	**0.022**
succession timepoint 3	0.152	0.055	0.179	0.162
Significant P-values (P<0.05) bolded.				

Controls showed no differences from any contrasts ($P>0.05$). Significant P-values ($P<0.05$) are bolded.

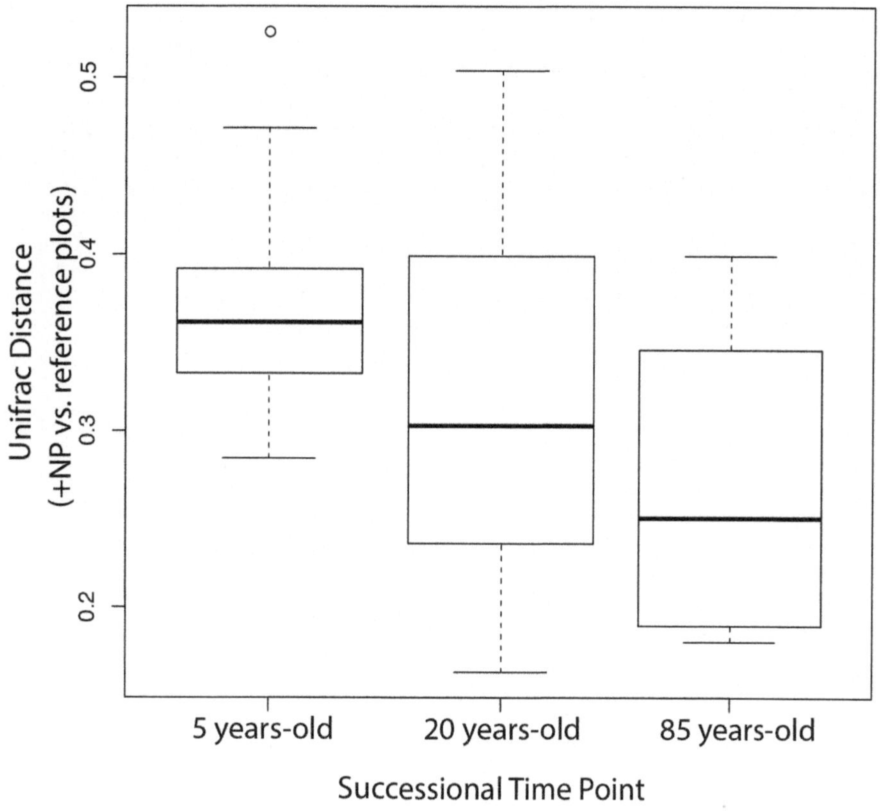

Figure 2. Relationship between +NP treatment-affected communities and reference communities. A box plot shows the average weighted UniFrac [43,44] distance between +NP-treated communities and reference communities with increasing successional time. A Mantel test demonstrates that +NP communities show decreasing dissimiliarty as compared to the reference communities over advancing stages of succession ($\rho_M = -0.35$ P = 0.01).

chronosequence that features cyanobacterial N-fixers [9]. Likewise, +N plots show a significant increase in bioavailable-P relative to control plots (Table S1), a pattern supported by research that shows N availability may limit the production of phosphatase enzymes [51–53]. Thus, these particular biochemical pathways lead to a coupling of nutrient cycles, which appears to be reflected in a colimitation to successional processes.

Despite the multitude of well documented changes across successional gradients including shifts in pH, C pools, plant cover and biotic historical factors, nutrient addition alone not only caused changes in early successional community structure, but induced convergence with late successional soil communities of the natural chronosequence (Fig. 1 and 2 and Tables 1 and 2). For example, strong changes in %C, another known filter on microbial communities, were observed across the natural chronosequence but not in +NP plots (Tables S1 and S2). In other ecosystems, the effects of fertilization on microbial community structure have been attributed to changes in plant productivity or community structure [14]. However, it is important to note that while the +NP fertilization caused sparse vegetation (<15 cm tall) to colonize after one year at our site, soils were collected at least 75 cm from these small plants. Altogether, our results suggest that the effects of the +NP fertilization on microbial community succession were direct and not mediated through changes in other aspects of the

Table 2. Post-treatment +NP communities showed differences from all reference succession communities with the exception of the oldest transect (P<0.05).

Homogeneity of Dispersion (PERMDISP) P-values		
Sample vs. Sample	**post-treatment +NP**	**post-treatment control**
succession timepoint 1	**0.022**	0.508
succession timepoint 2	**0.042**	0.588
succession timepoint 3	0.555	0.997
Significant P-values (P<0.05) bolded.		

Controls showed no difference in community dispersion from communities of any of the reference succession transects (P>0.05). Significant P-values (P<0.05) are bolded.

abiotic environment or through the effects of plants on soil communities.

Our field-based fertilization experiment helps to extend existing ecological theory regarding the role of nutrient limitations in succession [4,5,54] to microbial communities present in the earliest primary successional soils, which are important for biogeochemical cycling, physical soil development, and plant colonization [9,21,30]. While it is widely acknowledged that microbes can alter soil fertility and nutrient cycling processes, and that changes in soil nutrient pools and microbial communities occur over primary succession [9,12,13,30], to what extent nutrients directly structure soil microbial communities is not clear. Our fertilization experiment allowed us to decouple the effects of changes in microbial communities on nutrient cycles and to directly demonstrate the influence of nutrient pools on microbial succession. Correlative studies are less powerful because they cannot isolate the impact of individual factors amidst the multiplicity of soil properties that change with succession, and because measured soil properties may be decoupled from microbial community composition in time.

Despite the high fertilization rate we used, the nutrient addition treatment did not push communities to an alternative or novel state, but simply accelerated succession, rapidly producing a community that was structurally most similar to the community in the 85 year old soils in the chronosequence (Fig. 1 and 2 and Tables 1 and 2). Thus, our data highlight the stability of soil microbial communities [55]. Few studies have explicitly evaluated nutrients in the context of longer-term successional reference plant communities to understand how nutrients may either drive succession or shape alternative stable states in communities. However, in a study of salt marsh vegetation, Van Wijnen and Bakker [56] observed that fertilization of young marsh communities resulted in plant communities that resembled those of older, unfertilized marshes. These results further suggest that nutrient-related mechanisms for succession may be generalizable between plant and microbial communities.

The relative abundance of cyanobacteria significantly increased in the +NP plots and the phylogenetic structure of the cyanobacterial communities in post-treatment +NP plots was significantly different from the paired pre-treatment +NP and pre-/post-treatment control plots (PERMANOVA, $P<0.05$, Table S3). Although not significant, cyanobacterial relative abundance nearly doubled between the oldest and youngest stages of the reference chronosequence and past work at this site has documented similar successional changes in cyanobacterial community structure (Table S2) [9,30]. Consistent with these results, a laboratory experiment evaluating microbial autotrophs from this site demonstrated that P additions resulted in significant increases in the growth rate of photoautotrophic crusts [28]. Both N fixation rates and the relative abundance of N-fixing cyanobacteria show successional trends at this site as well [30], suggesting that N availability may also limit microbial growth and activity. The current study adds to this work and demonstrates that both N and P together are important colimiting controls over community successional processes in this system (Tables 1 and 2).

The increase in the relative abundance of cyanobacteria in the + NP plots may reflect their ecological advantage in this low C environment. In a laboratory study, Drakare [57] observed that P additions enhanced cyanobacterial populations, but only in an environment where low C concentrations constrained heterotrophic growth. Incubation studies of early successional soils that found increases in heterotrophic activity in response to both N and C (but not to N alone) are also consistent with this interpretation [58,59]. These results indicate that C often limits the response of the heterotrophic community to nutrient additions, whereas cyanobacteria can readily take advantage of such nutrients to fuel photosynthesis. By extension, we argue that the observed effects of N and P additions on microbial community succession are likely to apply only to autotrophic successional sequences, and that heterotrophic succession (sensu Fierer et. al [27]) may be controlled by a different suite of resources, including C availability.

Microbes are fundamental to soil physical and chemical development and underlie ecosystem function, thus understanding the factors that drive soil microbial community succession is key to predicting and managing ecosystem development. Particularly in low nutrient environments, microbial activity has major effects on soil, plant community, and ecosystem development [9,30,60,61]. Likewise, low nutrient environments may feature more prominent nutrient colimitations [49]. As such, the results of this study have important implications for understanding nutrient controls on ecosystem development and relevant models for microbial succession. Furthermore, while early successional microbial communities may vary strongly in both composition and in terms of the specifics of resource availability (e.g., heterotrophic vs. autotrophic), our study provides evidence that nutrient colimitation may provide a generalizable mechanism for microbial community succession in autotrophic successional sequences. Our data also support recent evidence for the stability of soil microbial communities, as fertilization simply accelerated succession and did not push communities into a novel state. Overall, the details of microbial nutrient limitations presented herein are essential to understanding the factors that structure early successional microbial communities, the profound contributions they make to soil development, and the ecosystem processes they mediate.

Supporting Information

Table S1 Mean of Edaphic Properties and Tukey's HSD Comparisons for Post-Treatment Plots.

Table S2 Mean of Edaphic Properties and Cyanobacterial Relative Abundance for Reference Chronosequence.

Table S3 Mean Relative Abundances of Major Taxa at Puca Glacier Site.

Acknowledgments

We thank S.P. O'Neill, A.J. King, P. Sowell and K. Yager for assistance in the field and laboratory. We also thank A.R. Townsend, E.B. Graham, and J.S. Prevéy for contributing to discussions regarding this research. We are grateful for the constructive feedback provided via the review process.

Author Contributions

Conceived and designed the experiments: JEK SKS RCL CCC DRN. Performed the experiments: JEK SKS RCL JLD SCC CCC DRN. Analyzed the data: JEK JLD DRN. Contributed to the writing of the manuscript: JEK SKS RCL JLD SCC CCC DRN.

References

1. Chapin FS, Walker LR, Fastie CL, Sharman LC (1994) Mechanisms of primary succession following deglaciation at Glacier Bay, Alaska. Ecological Monographs 64: 149–175.
2. Walker LR, Moral R del (2003) Primary succession and ecosystem rehabilitation. Cambridge: Cambridge University Press. 348 p.
3. Matthews JA (1992) The ecology of recently-deglaciated terrain: a geoecological approach to glacier forelands and primary succession. Cambridge: Cambridge University Press. 409 p.
4. Walker TW, Syers JK (1976) The fate of phosphorus during pedogenesis. Geoderma 15: 1–19.
5. Crews TE, Kitayama K, Fownes JH, Riley RH, Herbert DA, et al. (1995) Changes in soil phosphorus fractions and ecosystem dynamics across a long chronosequence in Hawaii. Ecology 76: 1407–1424.
6. Ugolini FC (1968) Soil development and alder invasion in a recently deglaciated area of Glacier Bay, Alaska. In: Trappe JM, Franklin JF, Tarrant RF, Hansen GM, editors. Biology of Alder pp. 115–148 Portland: Pacific Northwest Forest and Range Experiment Station, Portland, OR. pp. 115–148.
7. Vitousek PM, Walker LR, Whiteaker LD, Matson PA (1993) Nutrient limitations to plant growth during primary succession in Hawaii Volcanoes National Park. Biogeochemistry 23: 197–215.
8. Richardson SJ, Peltzer DA, Allen RB, McGlone MS, Parfitt RL (2004) Rapid development of phosphorus limitation in temperate rainforest along the Franz Josef soil chronosequence. Oecologia 139: 267–276.
9. Nemergut DR, Anderson SP, Cleveland CC, Martin AP, Miller AE, et al. (2007) Microbial community succession in an unvegetated, recently deglaciated soil. Microbial Ecology 53: 110–122.
10. Brown SP, Jumpponen A (2013) Contrasting primary successional trajectories of fungi and bacteria in retreating glacier soils. Molecular Ecology 23: 481–497.
11. Schütte UME, Abdo Z, Bent SJ, Williams CJ, Schneider GM, et al. (2009) Bacterial succession in a glacier foreland of the High Arctic. ISME Journal 3: 1258–1268.
12. Zumsteg A, Luster J, Göransson H, Smittenberg RH, Brunner I, et al. (2012) Bacterial, archaeal and fungal succession in the forefield of a receding glacier. Microbial Ecology 63: 552–564.
13. Edwards IP, Bürgmann H, Miniaci C, Zeyer J (2006) Variation in microbial community composition and culturability in the rhizosphere of Leucanthemopsis alpina (L.) Heywood and adjacent bare soil along an alpine chronosequence. Microbial Ecology 52: 679–692.
14. Ramirez KS, Lauber CL, Knight R, Bradford MA, Fierer N (2010) Consistent effects of nitrogen fertilization on soil bacterial communities in contrasting systems. Ecology 91: 3463–3470.
15. Nemergut DR, Townsend AR, Sattin SR, Freeman KR, Fierer N, et al. (2008) The effects of chronic nitrogen fertilization on alpine tundra soil microbial communities: implications for carbon and nitrogen cycling. Environmental Microbiology 10: 3093–3105.
16. Fierer N, Jackson RB (2006) The diversity and biogeography of soil bacterial communities. Proceedings of the National Academy of Sciences 103: 626–631.
17. Deiglmayr K, Philippot L, Tscherko D, Kandeler E (2006) Microbial succession of nitrate-reducing bacteria in the rhizosphere of Poa alpina across a glacier foreland in the Central Alps. Environmental Microbiology 8: 1600–1612.
18. Noll M, Wellinger M (2008) Changes of the soil ecosystem along a receding glacier: Testing the correlation between environmental factors and bacterial community structure. Soil Biology and Biochemistry 40: 2611–2619.
19. Fierer N, Bradford MA, Jackson RB (2007) Toward an ecological classification of soil bacteria. Ecology 88: 1354–1364.
20. Jangid K, Whitman W, Condron L, Turner B, Williams M (2013) Soil bacterial community succession during long-term ecosystem development. Molecular Ecology 22: 3415–3424.
21. Knelman JE, Legg TM, O'Neill SP, Washenberger CL, González A, et al. (2012) Bacterial community structure and function change in association with colonizer plants during early primary succession in a glacier forefield. Soil Biology and Biochemistry 46: 172–180.
22. Ferrenberg S, O'Neill SP, Knelman JE, Todd B, Duggan S, et al. (2013) Changes in assembly processes in soil bacterial communities following a wildfire disturbance. ISME Journal 7: 1102–1111.
23. Cline LC, Zak DR (2013) Dispersal limitation structures fungal community assembly in a long-term glacial chronosequence. Environmental Microbiology 16:..1538–1548
24. Meola M, Lazzaro A, Zeyer J (2014) Diversity, resistance and resilience of the bacterial communities at two alpine glacier forefields after a reciprocal soil transplantation. Environmental Microbiology 16: 1918–1934.
25. Fukami T, Morin PJ (2003) Productivity–biodiversity relationships depend on the history of community assembly. Nature 424: 423–426.
26. Nemergut DR, Schmidt SK, Fukami T, O'Neill SP, Bilinski TM, et al. (2013) Patterns and processes of microbial community assembly. Microbiology and Molecular Biology Reviews 77: 342–356.
27. Fierer N, Nemergut D, Knight R, Craine JM (2010) Changes through time: integrating microorganisms into the study of succession. Research in Microbiology 161: 635–642.
28. Schmidt SK, Nemergut DR, Todd BT, Lynch RC, Darcy JL, et al. (2012) A simple method for determining limiting nutrients for photosynthetic crusts. Plant Ecology and Diversity 5: 513–519.
29. Schmidt SK, Cleveland CC, Nemergut DR, Reed SC, King AJ, et al. (2011) Estimating phosphorus availability for microbial growth in an emerging landscape. Geoderma 163: 135–140.
30. Schmidt S, Reed SC, Nemergut DR, Stuart Grandy A, Cleveland CC, et al. (2008) The earliest stages of ecosystem succession in high-elevation (5000 metres above sea level), recently deglaciated soils. Proceedings of the Royal Society B: Biological Sciences 275: 2793–2802.
31. Matejovic I (1997) Determination of carbon and nitrogen in samples of various soils by the dry combustion. Communications in Soil Science and Plant Analysis 28: 1499–1511.
32. Jeannotte R, Sommerville DW, Hamel C, Whalen JK (2004) A microplate assay to measure soil microbial biomass phosphorus. Biology and Fertility of Soils 40: 201–205.
33. Lajtha K, Driscoll C, Jarrell W, Elliot ET (1999) Standard soil methods for long-term ecological research. New York: Oxford University Press. 486 p.
34. Weaver RW, Angler S, Bottomly P, Bezdicek D, Smith S, et al. (1994) Methods of soil analysis, part 2. Microbiological and biochemical properties. Soil Society of America, Inc.
35. Schmidt SK, Nemergut DR, Miller AE, Freeman KR, King AJ, et al. (2009) Microbial activity and diversity during extreme freeze–thaw cycles in periglacial soils, 5400 m elevation, Cordillera Vilcanota, Perú. Extremophiles 13: 807–816.
36. Hamady M, Walker JJ, Harris JK, Gold NJ, Knight R (2008) Error-correcting barcoded primers for pyrosequencing hundreds of samples in multiplex. Nature Methods 5: 235–237.
37. Fierer N, Hamady M, Lauber CL, Knight R (2008) The influence of sex, handedness, and washing on the diversity of hand surface bacteria. Proceedings of the National Academy of Sciences 105: 17994–17999.
38. Reeder J, Knight R (2010) Rapid denoising of pyrosequencing amplicon data: exploiting the rank-abundance distribution. Nature Methods 7: 668–669.
39. Edgar RC (2010) Search and clustering orders of magnitude faster than BLAST. Bioinformatics 26: 2460–2461.
40. Wang Q, Garrity GM, Tiedje JM, Cole JR (2007) Naïve Bayesian classifier for rapid assignment of rRNA sequences into the new bacterial taxonomy. Applied Environmental Microbiology 73: 5261–5267.
41. DeSantis TZ, Hugenholtz P, Keller K, Brodie EL, Larsen N, et al. (2006) NAST: a multiple sequence alignment server for comparative analysis of 16S rRNA genes. Nucleic Acids Research 34: 394–399.
42. Price MN, Dehal PS, Arkin AP (2009) FastTree: Computing large minimum evolution trees with profiles instead of a distance matrix. Molecular Biology and Evolution 26: 1641–1650.
43. Lozupone C, Hamady M, Knight R (2006) UniFrac - An online tool for comparing microbial community diversity in a phylogenetic context. BMC Bioinformatics 7: 371.
44. Lozupone CA, Hamady M, Kelley ST, Knight R (2007) Quantitative and qualitative β-diversity measures lead to different insights into factors that structure microbial communities. Applied Environmental Microbiology 73: 1576–1585.
45. Clarke, Gorley (2006) PRIMER v6: User Manual/Tutorial, PRIMER–E. Plymouth, UK.
46. R Development Core Team (2009) R: A language and environment for statistical computing. R Foundation for Statistical Computing Vienna, Austria. Available: http://www.R-project.org.
47. Oksanen J, Blanchet FG, Kindt R, Legendre P, Minchin P, et al. (2013) vegan: community ecology package. Available: http://CRAN.R-project.org/package = vegan.
48. Caporaso JG, Kuczynski J, Stombaugh J, Bittinger K, Bushman FD, et al. (2010) QIIME allows analysis of high-throughput community sequencing data. Nature Methods 7: 335–336.
49. Harpole WS, Ngai JT, Cleland EE, Seabloom EW, Borer ET, et al. (2011) Nutrient co-limitation of primary producer communities. Ecology Letters 14: 852–862.
50. Vitousek PM (1999) Nutrient limitation to nitrogen fixation in young volcanic sites. Ecosystems 2: 505–510.
51. Vitousek PM, Porder S, Houlton BZ, Chadwick OA (2010) Terrestrial phosphorus limitation: mechanisms, implications, and nitrogen–phosphorus interactions. Ecological Applications 20: 5–15.
52. Olander LP, Vitousek PM (2000) Regulation of soil phosphatase and chitinase activity by N and P availability. Biogeochemistry 49: 175–191.
53. Wang Y-P, Houlton BZ, Field CB (2007) A model of biogeochemical cycles of carbon, nitrogen, and phosphorus including symbiotic nitrogen fixation and phosphatase production. Global Biogeochemical Cycles 21: GB1018.
54. Vitousek PM, Farrington H (1997) Nutrient limitation and soil development: Experimental test of a biogeochemical theory. Biogeochemistry 37: 63–75.
55. Griffiths BS, Philippot L (2013) Insights into the resistance and resilience of the soil microbial community. FEMS Microbiology Reviews 37: 112–129.
56. Van Wijnen HJ, Bakker JP (1999) Nitrogen and phosphorus limitation in a coastal barrier salt marsh: the implications for vegetation succession. Journal of Ecology 87: 265–272.

57. Drakare S (2002) Competition between picoplanktonic Cyanobacteria and heterotrophic bacteria along crossed gradients of glucose and phosphate. Microbial Ecology 44: 327–335.
58. Yoshitake S, Uchida M, Koizumi H, Nakatsubo T (2007) Carbon and nitrogen limitation of soil microbial respiration in a High Arctic successional glacier foreland near Ny-Ålesund, Svalbard. Polar Research 26: 22–30.
59. Göransson H, Olde Venterink H, Bååth E (2011) Soil bacterial growth and nutrient limitation along a chronosequence from a glacier forefield. Soil Biology of Biochemistry 43: 1333–1340.
60. Borin S, Ventura S, Tambone F, Mapelli F, Schubotz F, et al. (2010) Rock weathering creates oases of life in a High Arctic desert. Environmental Microbiology 12: 293–303.
61. Van Der Heijden MG, Bardgett RD, Van Straalen NM (2008) The unseen majority: soil microbes as drivers of plant diversity and productivity in terrestrial ecosystems. Ecology Letter 11: 296–310.

18

Similar Processes but Different Environmental Filters for Soil Bacterial and Fungal Community Composition Turnover on a Broad Spatial Scale

Nicolas Chemidlin Prévost-Bouré[1]*, Samuel Dequiedt[2], Jean Thioulouse[3], Mélanie Lelièvre[2], Nicolas P. A. Saby[4], Claudy Jolivet[4], Dominique Arrouays[4], Pierre Plassart[2], Philippe Lemanceau[1], Lionel Ranjard[1,2]

1 Unité Mixte de Recherche 1347 Agroécologie, Institut National de la Recherche Agronomique-AgroSup Dijon-Université de Bourgogne, Dijon, France, **2** Unité Mixte de Recherche 1347 Agroécologie-Plateforme GenoSol, Institut National de la Recherche Agronomique-AgroSup Dijon-Université de Bourgogne, Dijon, France, **3** Unité Mixte de Recherche 555 Laboratoire de Biométrie et Biologie Evolutive, Université Lyon 1-Centre National de la Recherche Scientifique, Villeurbanne, France, **4** Unité de Services 1106 InfoSol, Institut National de la Recherche Agronomique, Orléans, France

Abstract

Spatial scaling of microorganisms has been demonstrated over the last decade. However, the processes and environmental filters shaping soil microbial community structure on a broad spatial scale still need to be refined and ranked. Here, we compared bacterial and fungal community composition turnovers through a biogeographical approach on the same soil sampling design at a broad spatial scale (area range: 13300 to 31000 km^2): i) to examine their spatial structuring; ii) to investigate the relative importance of environmental selection and spatial autocorrelation in determining their community composition turnover; and iii) to identify and rank the relevant environmental filters and scales involved in their spatial variations. Molecular fingerprinting of soil bacterial and fungal communities was performed on 413 soils from four French regions of contrasting environmental heterogeneity (Landes<Burgundy≤Brittany<<South-East) using the systematic grid of French Soil Quality Monitoring Network to evaluate the communities' composition turnovers. The relative importance of processes and filters was assessed by distance-based redundancy analysis. This study demonstrates significant community composition turnover rates for soil bacteria and fungi, which were dependent on the region. Bacterial and fungal community composition turnovers were mainly driven by environmental selection explaining from 10% to 20% of community composition variations, but spatial variables also explained 3% to 9% of total variance. These variables highlighted significant spatial autocorrelation of both communities unexplained by the environmental variables measured and could partly be explained by dispersal limitations. Although the identified filters and their hierarchy were dependent on the region and organism, selection was systematically based on a common group of environmental variables: pH, trophic resources, texture and land use. Spatial autocorrelation was also important at coarse (80 to 120 km radius) and/or medium (40 to 65 km radius) spatial scales, suggesting dispersal limitations at these scales.

Editor: Maarja Öpik, University of Tartu, Estonia

Funding: This study was granted by ADEME (Energy and Environment Management Agency) and by the French National Research Agency (ANR). RMQS soil sampling and physico-chemical analyses were supported by a French Scientific Group of Interest on soils: the "GIS Sol", involving the French Ministry for Ecology and Sustainable Development (MEDAD), the French Ministry of Agriculture (MAP), the French Institute for Environment (IFEN), the French Agency for Energy and Environment (ADEME), the French Institute for Research and Development (IRD) and the National Institute for Agronomic Research (INRA). This work, through the involvement of technical facilities of the GenoSol platform of the infrastructure ANAEE378 Services, received a grant from the French state via the National Agency for Research under the program "Investments for the Future" (reference ANR-11-INBS-0001), as well as a grant from the Regional Council of Burgundy. The funders had no role in study design, data collection and analysis, decision to publish, or preparation of the manuscript.

* Email: n.chemidlin@agrosupdijon.fr

Introduction

For over two centuries, biogeographical studies have been carried out on macroorganisms and have provided a better understanding of species distribution, extinction and interactions [1–2]. For microorganisms, the first biogeographic postulate was developed by Baas Becking in 1934 [3]: "Everything is everywhere, *but*, the environment selects" suggesting that microbial "species" may be everywhere due to huge dispersal potentials, but that their abundances are constrained by contemporary environmental context, which may be especially true at broad spatial scales (spatial scales larger than 100 km^2 are considered as broad in this study). The number of studies in microbial biogeography has increased exponentially over the past decade thanks to new molecular tools applicable in routine on wide scale sampling networks constituted of several hundreds of samples [4–5]. These studies revealed that soil microorganisms are not strictly cosmopolitan since their distributions are systematically heterogeneous and structured into biogeographical patterns [2], [6–9].

One way to discriminate the spatial processing of microbial diversity is to evaluate either the Taxa-Area Relationship (TAR), *i.e.* the accumulation of new taxa with increasing sampling area, or the Distance-Decay Relationship (DDR), *i.e.* the rate of change in compositional similarity with increasing distance. [10–11]. Although significant TAR and DDR have recently been demonstrated for both soil fungal [12] and bacterial [8], [11], [13–14] communities, the relative importance of the ecological processes shaping these communities is still under debate. Therefore, it needs to be more deeply considered at the community level. According to Vellend [15], four processes are involved in shaping microbial community composition: selection, dispersal, ecological drift and speciation. Speciation is difficult to consider at the community level because the molecular markers used to discriminate microbial taxa mainly target highly conserved regions (*e.g.* ribosomal genes) with low mutation rates. The stochastic demographic processes underlying ecological drift are also difficult to consider since it remains a challenge to fully characterize demographic evolutions within complex microbial communities in environmental samples. Consequently, most biogeographical studies have focused on environmental selection and dispersal limitations, the later leading to a spatial autocorrelation between sites independently of environmental factors. Numerous studies have identified environmental selection as relevant in shaping soil bacterial community composition [8–9], [16–24]. Conversely, dispersal limitation is still under debate regarding the high dispersal potentials of microorganisms and because some environmental variables always remain unmeasured. Nevertheless, recent publications also suggest that bacteria may be dispersal limited [9], [21–23] or that part of soil bacterial communities is endemic [17]. As regards soil fungi, the relevance of environmental selection and dispersal limitations has been demonstrated at the community level and for ectomycorrhizal groups [16], [19], [22], [24–29]. Nevertheless, most of these studies were performed on different sampling designs with different molecular techniques. Only few studies have investigated such processes for both soil fungal and bacterial communities simultaneously to compare their biodiversity turnover. Most of them were performed in particular ecosystems and lead to diverging conclusions: Pasternak et al [18] concluded that bacterial and fungal communities were primarily shaped by environmental selection rather than dispersal limitations at the scale of the Israeli desert. On the contrary, Talbot et al. [27] highlighted a strong endemism for fungi in pine forests and Hovatter [22] suggested that the ecological processes shaping soil bacterial community could differ at a local scale due to the presence/absence of a particular plant. Altogether, this suggests that environmental heterogeneity may determine the relative importance of the ecological processes at work and therefore affect the distance-decay relationship for both soil bacteria and fungi [9–11], also suggested by other macrobial studies [30–31]. The comparison of different microbial communities along different levels of environmental heterogeneity may therefore help to reach a consensus.

Both selection and dispersal are based on the various ecological attributes of soil bacteria and fungi in terms of soil colonization, dispersal forms, trophic requirements, biological interactions and adaptation to environmental conditions, together with stochastic factors. Consequently, studies focusing particularly on soil bacteria or soil fungi have identified numerous environmental filters involved in shaping these particular communities but no consensus could be reached regarding microbial community as a whole on broad spatial scales. The filter most frequently identified for bacteria is soil pH [5], [17], [20], [23–24], [32–34] and it is commonly assumed that this is an important driver for fungal communities [18], [24]. Soil texture and carbon content have also been identified as important filters for bacteria [6], [22–23], [33–34]. Similarly, the quality of soil organic matter, represented by the C:N ratio, and the amount of N were shown to have a significant effect on the abundance and composition of bacterial and fungal communities [24], [34–35]. These edaphic factors are often considered as the main determinants of bacterial diversity since the importance of climate may vary across biomes at a continental scale [5], [12], [16]. Land-use, agricultural practices, and plant community composition are also important filters for both bacteria and fungi on a wide scale [18], [19], [22–24], [36–38]. Therefore, this suggests that common filters determine the composition of both bacterial and fungal communities, but they still need to be ranked according to their relative importance to reach a consensus. This may be achieved by comparing bacterial and fungal communities over regions contrasted in terms of habitat heterogeneity but also with wide ranges of variations for the identified filters.

The objectives of this study were: i) to examine the spatial structuring of bacterial and fungal communities on a broad spatial scale; ii) to investigate the relative importance of environmental selection and spatial autocorrelation in determining the community composition turnover of these communities; and iii) to identify and rank the relevant environmental filters and scales involved in their spatial variations. To attain these objectives, four regions in the RMQS data set ("Réseau de Mesures de la Qualité des Sols" = French Monitoring Network for Soil Quality, recovering 2,200 soils over the whole of France) were selected along a gradient of environmental heterogeneity, representing a total of 413 soils. This gradient was chosen in order to confront the community composition turnover rates of bacterial and fungal communities to soil habitat heterogeneity [9]. Bacterial and fungal communities were characterized by Automated RISA fingerprinting of soil DNA. Community composition turnover (z) was estimated by means of a similarity DDR using an exponential model as suggested by Harte et al [39–40] for microorganisms. Together with this measure of community composition turnover over broad spatial scales, the initial similarity of communities was evaluated [30]. It represents the variability of community composition at finer spatial scales. High initial similarity corresponds to low local variability. The relative influence of environmental selection and spatial autocorrelation was investigated through a variance partitioning approach involving pedoclimatic characteristics and land-use and spatial variables (geographic coordinates and Principal Coordinates of Neighbour Matrices; PCNM), respectively.

Methods

Soil samples

Soil samples were provided by the Soil Genetic Resource Conservatory (platform GenoSol, http://www.dijon.inra.fr/plateforme_genosol, [41]) and obtained from the soil storage facility of the RMQS ("Réseau de Mesures de la Qualité des Sols" = French Monitoring Network for Soil Quality). The RMQS database consists of observations of soil properties on a 16-km regular grid across the 550000 km^2 French metropolitan territory and was designed to monitor soil properties [42]. The baseline survey consisting of 2,200 sites (each corresponding to a composite soil sample constituted of 25 soil cores) was completed in 2009. The sites were selected at the centre of each 16×16-km cell. In this study, we focused on a subset of 413 sites from the RMQS data set. The samples were organized into four regions: Brittany (131 sites), Burgundy (109 sites), Landes (52 sites) and South-East (121 sites, Fig. 1A) which are contrasted in terms of soil type, land-use (coarse level of the CORINE Land Cover classification; IFEN, http://

www.ifen.fr; 7 classes: forest, crop systems, grasslands, particular natural ecosystems, vineyards/orchards, parkland and wild land), climate and geomorphology (Table S1). Within a region, sites were separated by 16 km at least. For each soil, the pedo-climatic characteristics considered were particle-size distribution, pH in water (pH_{water}), organic carbon content (C_{org}), N content, C:N ratio, soluble P contents, $CaCO_3$ and exchangeable cations (Ca, K, Mg), sum of annual temperature (°C) and annual rainfall (mm). Physical and chemical analyses were performed by the Soil Analysis Laboratory of INRA (Arras, France) which is accredited for such analyses by the French Ministry of Agriculture.

Bacterial and fungal community fingerprinting

Soil DNA extraction. For each soil sample, the equivalent of 1.5 g of dry soil was used for DNA extraction, following the procedure optimized by platform GenoSol [35]. Briefly, extraction buffer (100 mM Tris pH 8.0, 100 mM EDTA pH 8.0, 100 mM NaCl and 2% (w/v) SDS) was added to the sample in the proportion 3:1 (v/w), with two grams of glass beads (106 μm diameter) and eight glass beads (2 mm diameter) in a bead-beater tube. All beads were acid washed and sterilized. The samples were homogenized for 30 s at 1600 rpm in a mini bead-beater cell disruptor (Mikro-dismembrator, S. B. Braun Biotech International), incubated for 30 min at 70°C in a water bath and centrifuged for 5 min at 7000 g and room temperature. The supernatant was collected, incubated on ice with 1/10 volume of 3 M potassium acetate (pH 5.5) and centrifuged for 5 min at 14000 g. DNA was precipitated with one volume of ice-cold isopropanol and centrifuged for 30 min at 13000 rpm. The DNA pellet was washed with ice-cold 70% ethanol and dissolved in 100 μl of ultra pure water. For purification, aliquots (100 μL) of crude DNA extracts were loaded onto PVPP (polyvinyl polypyrrolidone) minicolumns (BIORAD, Marne la Coquette, France) and centrifuged for 4 min at 1000 g and 10°C. This step was repeated if the eluate was opaque. The eluate was then collected and purified for residual impurities using the Geneclean Turbo kit as recommended by the manufacturer (Q Biogene, France).

PCR conditions. The bacterial ribosomal IGS was amplified using the PCR protocol described in Ranjard et al [43]. 12.5 ng of DNA was used as the template for PCR volumes of 25 μl. The fungal ribosomal ITS was amplified using the primer set ITS1F/ITS4-IRD800 (5′- CTTGGTCATTTAGAGGAAGTAA -3′/5′-IRD800-TCCTCCGCTTATTGATATGC -3′). 20 ng of DNA was used as the template for PCR volumes of 25 μl with the following PCR conditions: denaturation at 95°C for 3 min, 35 cycles of 30 s at 95°C, 45 s at 55°C and 1 min at 72°C, and a final elongation of 7 min at 72°C. The primer Tm was the same for bacterial IGS and fungal ITS. Every PCR products were purified using the MinElute Kit (QIAGEN, Courtaboeuf, France) and quantified using a calf thymus DNA standard curve.

ARISA fingerprinting conditions. 2 μL of the PCR product was added to deionized formamide and denatured at 90°C for 2 min. Bacterial and Fungal ARISA fragments were resolved on 3.7% polyacrylamide gels under denaturing conditions as

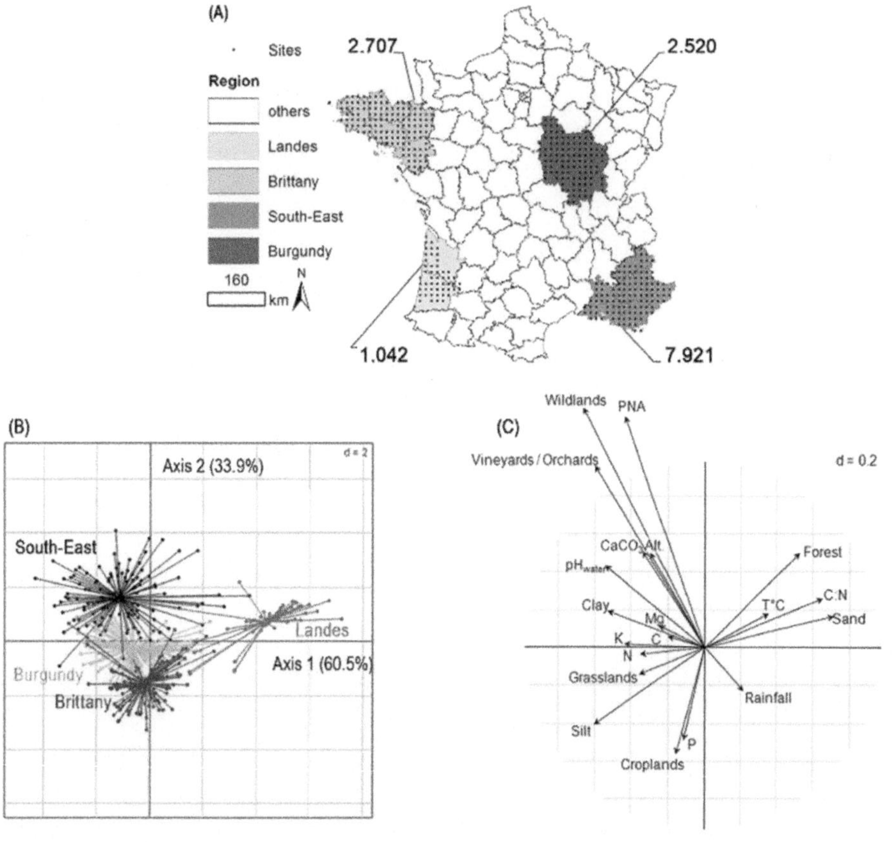

Figure 1. Comparison of the regions considered on the basis of their soil habitats. A. Soil average dissimilarity of soil habitat for the different regions (number linked to the corresponding region) and position of sites; B. Between group analysis of soil habitats according to the region; C. Correlation circle of the variables defining soil habitat in the between group analysis. The length of the arrow corresponds to the Pearson's correlation coefficient for quantitative variables and to the correlation ratio for qualitative variables. Symbols: Alt.: Elevation; T°C: Sum of annual temperatures; P_{ass}: Assimilable P; C:N: Carbon to Nitrogen ratio; C_{org}: Organic Carbon content.

described in Ranjard et al [9] on a LiCor DNA sequencer (ScienceTec).

Image analysis. The data were analyzed using the 1D-Scan software (ScienceTec), converting fluorescence data into electrophoregrams, where peaks represented PCR fragments (100 to 110 peaks retained per sample, the resolution limit to avoid considering background noise). The height of the peaks was calculated in conjunction with the median filter option and the Gaussian integration in 1D-Scan, and represented the relative proportion of the fragments in the total products. Lengths (in base pairs) were calculated by using a size standard with bands ranging from 200 to 1659 bp. The data were then converted into a contingency table with prepRISA package in R.

Statistical analyses

Characterization of habitat variability and average dissimilarity across regions. Habitats were compared between regions in a Hill & Smith multivariate analysis [44] using the ade4 package in R [45–46]. The analysis was applied to pedoclimatic characteristics, land-use and geomorphology, by centering and scaling the quantitative variables, and converting the qualitative ones into weighted binary variables (weight equal to 1/n; n is the number of classes for the qualitative variables). Differences between these regions were examined by between group analysis and tested by applying a Monte-Carlo permutation test (1000 permutations). The average dissimilarity between soil habitats was determined by transposing a method based on the dissimilarity matrix for communities [47] to soil habitat. The dissimilarity matrix for soil habitat was derived from the site coordinates in the Hill & Smith analysis, following equation 1 [9]:

$$D_{i,j} = \frac{ED_{i,j}}{ED\max} \qquad (1)$$

Where $D_{i,j}$ and $ED_{i,j}$ are the dissimilarity and the Euclidean distance between sites i and j, respectively. ED_{max} is the maximum Euclidean distance observed between sites. The average dissimilarity between soil habitats was then calculated as follows [47]:

$$\overline{D}_{habitat} = \frac{1}{n}\sum_{h=1}^{n-1}\sum_{i=h+1}^{n} D_{h,i}^{2} \qquad (2)$$

$\overline{D}_{habitat}$ is the average dissimilarity between soil habitats of soil habitat and n the number of sites in the region.

Evaluation of the similarity distance-decay relationship and initial similarity of Bacterial and Fungal community composition. The *similarity distance-decay* relationship was estimated as proposed by Harte et al [39] for organisms with large populations per taxa. From this relationship, the community composition turnover rates (z) for bacterial and fungal communities composition were derived as described in Ranjard *et al.* [9] following the method and the exponential model (equation 3) proposed by Harte et al [39–40] for microorganisms.

$$\log_{10}(\chi_d) = (-2 * z) * \log_{10}(d) + b \qquad (3)$$

Where χ_d is the observed Sørensen's similarity between two soil samples that are d meters apart from each other; b is the intercept of the linear relationship and z the turnover rate of the community composition. The z estimate and its 95% confidence interval were derived from the slope (–2*z) of the relationship between similarity and distance by weighted linear regression. The overlap of the 95% confidence intervals was used to test for significant differences in community composition turnover rates between regions or between bacteria and fungi. The initial similarity of community composition was taken as the average similarity between sites 16 km apart and the 95% confidence intervals of the mean were determined [30].

Variance partitioning of community composition variations according to environmental filters and space. The relative importance of spatial variables, pedoclimatic characteristics and land-use in determining community composition turnover was tested by db-RDA [47–48]. Quantitative data were centered and scaled. Spatial variables were constructed from site coordinates (x, y, elevation) to reveal potential spatial trends at scales larger than the region, and of Principle Coordinates of Neighbour Matrices [49] in each region. The PCNM approach creates independent spatial descriptors that can be introduced in canonical analysis models to consider the spatial autocorrelation between sites in the model [49]. PCNMs with a significant Moran index ($P<0.001$) were selected. Land-use corresponded to the Corinne Land Cover classes recoded into dummy variables. Pedo-climatic characteristics consisted of climate and all the physico-chemical variables except sand. The most parsimonious model was obtained by forward selection from null to full model in two steps: a first step for selecting environmental variables and a second step for selecting the relevant PCNMs. Then, the pure effects of each set of filters or each individual filter were tested with an anova-like permutation test for canonical analyses (anova.cca function in vegan package, [50]). The PCNMs approach does not provide directly the range of the spatial descriptor. Nevertheless, Bellier et al [51] demonstrated that kriging approach could be applied to PCNMs to estimate their spatial range. Consequently, when PCNMs were selected in the most parsimonious model, their ranges were determined by standard kriging techniques (ordinary kriging with a Gaussian model). The hierarchy of these filters must nevertheless be considered with caution due to the small amounts of variance explained by each one. Land-use was not included in the filter-ranking since it corresponded to a set of categories and a global "land-use" category was already taken into account in the processes section. Maps of soil fungal community structure variations are provided as (Fig. S1, mapping methodology is described in the legend).

Results

Heterogeneity of soil habitat

The four regions were selected for their contrasting environmental heterogeneity as demonstrated by the between group analysis (Fig. 1B) and comparison of the calculated average dissimilarity between soil habitats ($\overline{D}_{habitat}$) (Fig. 1A and C). Multivariate analysis revealed a clear discrimination of the four regions on the first and second axes (Monte-Carlo permutation test, $P<0.001$): Landes was significantly discriminated from Brittany, Burgundy and South-east on the first axis and these three regions were discriminated from each other on the second axis. In addition, the environmental variability strongly differed between the four regions as demonstrated by the dispersal of sites in the factorial map. Sites from Landes were less dispersed on the factorial map than the sites from Brittany or Burgundy, which were less dispersed than sites from South-East. The calculated $\overline{D}_{habitat}$ ranged from 1.042 in the Landes to 7.921 in the South-East, with intermediate values for Burgundy and Brittany (2.520 and 2.707, respectively; Fig. 1A). It provided the same discrimination between regions. Fig. 1C shows that the four regions could mainly be distinguished according to land-use (*e.g.*: 86% of the

Landes sites are forest sites), a restricted set of soil physico-chemical characteristics (sand and silt contents, pH_{water} and $CaCO_3$ content, assimilable P content and organic matter quality as measured by C:N ratio) and by differences in elevation. Climatic conditions did not play a significant role in regional discrimination.

Distance-Decay Relationship for bacterial and fungal soil communities

Bacterial and fungal community similarity decreased with increasing distance in each region (Fig. 2). Community similarity was systematically higher for bacteria than for fungi in all regions at small or large distances. For bacteria, the linear regression model was highly significant in each region ($P<0.001$, Fig. 2) except in Landes where it was just below the significance threshold of 5% ($P<0.02$). For fungi, the linear regression model was highly significant in each region ($P<0.001$) except Landes (Fig. 2).

Community composition turnover rates were derived from the parameters of the linear regression. The community composition turnover rates for the bacterial and fungal communities ranged from 0.006 to 0.013 (Table 1). No significant differences were highlighted between these organisms when the community composition turnover rates were compared within each region. When the community composition turnover of bacterial or fungal communities was compared between regions, a significant difference was only found between Brittany and the South-East ($P<0.05$) for bacteria.

The initial similarity was always higher for bacterial communities than for fungal communities within each region, ranging respectively from 54.4% to 61.4% and from 39.7% to 43.1%. The initial similarity of the bacterial community in Landes was significantly lower than in the other regions, which did not differ from each other. For fungi, the initial similarities in Landes and Brittany were similar and significantly lower than those in Burgundy and the South-East.

These results were confirmed by a covariance analysis comparing the models between organisms within a region and between regions for a given organism (data not shown).

Variance partitioning of community composition variations

The relative importance of the sets of spatial variables, land-use and pedo-climatic characteristics on variations in bacterial and fungal community composition was tested by db-RDA using the Sørensen index (Fig. 3). The amount of variance in bacterial and fungal community composition explained by the three sets of filters

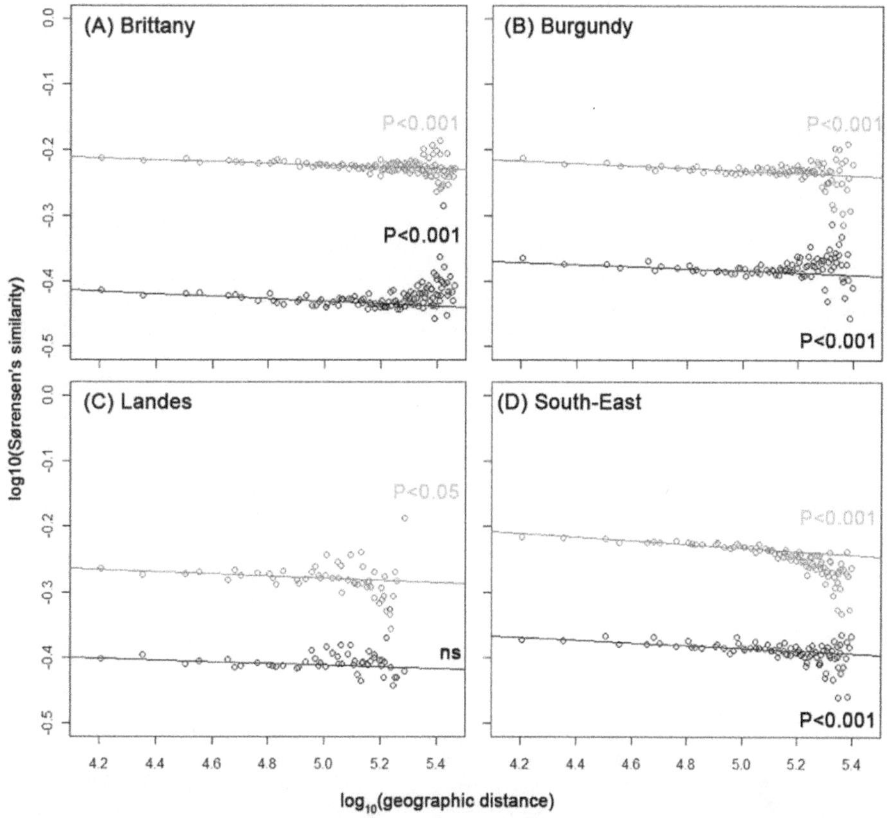

Figure 2. Distance-Decay Relationships for bacteria and fungi. Each panel correspond to a region: Brittany (A), Burgundy (B), Landes (C) and South-East (D) Points the average Sørensen's similarity between sites for each distance class. Lines represent the regression model based on the whole set of paired comparisons; for bacteria (grey) and fungi (black). The equations for the regression models were as follows: **(A)** Brittany: Bacteria***: log10(Sørensen's similarity) = −0.014×log10(geographic distance)−0.156; Fungi***: log10(Sørensen's similarity) = −0.017×log10(geographic distance)−0.350; **(B)** Burgundy: Bacteria*** log10(Sørensen's similarity) = −0.018×log10(geographic distance)−0.144; Fungi***: log10(Sørensen's similarity) = −0.015×log10(geographic distance)−0.316; **(C)** Landes: Bacteria*: log10(Sørensen's similarity) = −0.017×log10(geographic distance)−0.198; Fungi ^{ns}: log10(Sørensen's similarity) = −0.012×log10(geographic distance)−0.357; **(D)** South-East: Bacteria***: log10(Sørensen's similarity) = −0.027×log10(geographic distance)−0.101; Fungi***: log10(Sørensen's similarity) = −0.019×log10(geographic distance)−0.298. A graph with points representing all paired-comparisions between sites as points can be found in Figure S2. Significance of the model is indicated as an exponent for each organism: ns: not significant; P<0.05: *; P<0.01: **, P<0.001: ***.

Table 1. Regression parameters of the Distance-Decay Relationships for Bacteria and Fungi.

Region	Parameter	Organism	Estimate	95% Confidence interval
Brittany (131)	Z	Bacteria	0.007	[0.005; 0.009]
		Fungi	0.009	[0.006; 0.011]
	Initial similarity	Bacteria	61.4%	[60.4%; 62.3%]
		Fungi	38.5%	[37.7%; 39.4%]
Burgundy (109)	z	Bacteria	0.009	[0.006; 0.012]
		Fungi	0.008	[0.003; 0.012]
	Initial similarity	Bacteria	61.4%	[60.4%; 62.5%]
		Fungi	43.1%	[42.0%; 44.3%]
Landes (52)	z	Bacteria	0.009	[0.001; 0.016]
		Fungi	0.006	[−0.001; 0.013]
	Initial similarity	Bacteria	54.4%	[52.9%; 56.0%]
		Fungi	39.7%	[38.4%; 40.9%]
South-East (121)	z	Bacteria	0.013	[0.011; 0.016]
		Fungi	0.009	[0.006; 0.013]
	Initial similarity	Bacteria	60.9%	[59.9%; 61.9%]
		Fungi	42.4%	[41.5%; 43.3%]

The number of observations per region is provided in brackets beside the name of the region. The community composition turnover rate (z) and the initial similarity are derived from the slope of the regression (−2z) and the mean of similarity at 16 km; respectively. The statistical comparison between region and organism was performed by examining the overlap of the 95% confidence intervals of turnover rates or initial similarities.

ranged from 17% to 32%. The significance of the interactions between the three sets of filters could not be tested but always explained a small amount of the total variance (from 1.2% to 8.4%).

Among the sets of filters, soil pedo-climatic characteristics were the main contributor to variations in bacterial and fungal community composition. Spatial variables systematically explained a lower amount of variance than pedo-climatic characteristics, but a higher amount than land-use for both bacteria and fungi except for bacteria in Landes region where neither spatial variables nor land-use were significant.

The amount of variance explained by pedo-climatic characteristics for bacteria or fungi was always significant, ranging from 5% to 15%, and was similar between regions. Nevertheless, within each region, the amount of variance explained by soil pedo-climatic characteristics was always higher for bacteria than for fungi (Fig. 3).

Spatial variables significantly explained part of the community composition variations in all regions except for bacteria in Landes region. When significant, spatial variables represented from 3% to 9% of the total variance and were of the same order of magnitude both between bacteria and fungi and between regions (Fig. 3).

Similarly, land-use explained a significant amount of community composition variations in all regions except for bacteria in Landes region. The amount of explained variance ranged from 2.6% to 6.5% of the total variance. Within each region, land-use explained similar amounts of bacterial and fungal community variations when significant. Between regions, amounts of variance explained by land-use were similar for both bacteria and fungi (Fig. 3).

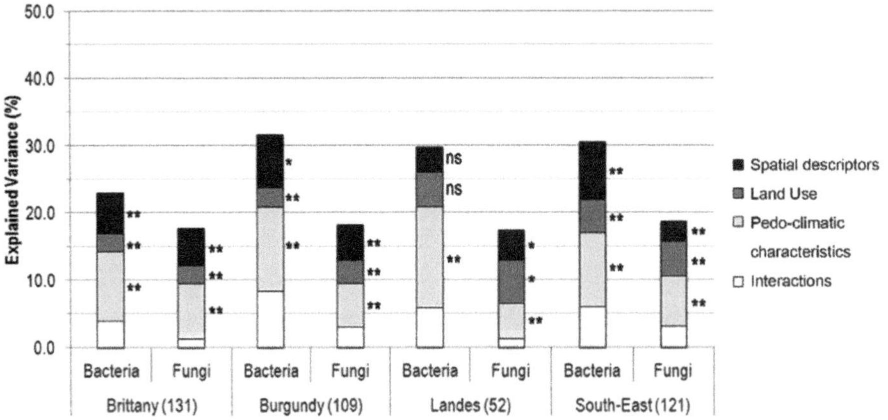

Figure 3. Variance partitioning of bacterial and fungal community composition. The number indicated in brackets corresponds to the number of samples for the region. Significance levels: ns: not significant; P<0.05: *; P<0.01: **, P<0.001: ***.

Hierarchy of environmental filters

Within the sets of soil pedo-climatic characteristics and spatial variables, the pure effects of individual filters on bacterial and fungal community structure are presented in Figure 4. These pure effects account for relatively small proportions of the total variance (from 0.7% to 6.5%) because of the large number of filters explaining the total variance of bacterial or fungal community compositions. Regarding the pedo-climatic characteristics, the significant filters for soil bacterial communities in Brittany were first pH and secondly N content. For fungi, these were the quality of organic matter resource as indicated by the selection of N content, C_{org} content and C:N ratio. The least significant filters corresponded to soil texture and other soil nutrients for bacteria (clay and silt contents, Mg and $CaCO_3$ concentrations) and fungi (clay content, K and P concentrations, Figure 4). In Burgundy, as in Brittany, the soil bacterial and fungal communities were principally affected by pH. Beside pH, bacterial community composition was shaped by the quality of organic matter resource (N content, C_{org} content, C:N ratio) followed by clay content and annual rainfall. The fungal community composition was shaped by C:N ratio only. The only filter that had a significant effect in the Landes region, on both soil bacteria and soil fungi, was the C:N ratio. The pH was the most important filter, followed by clay content and K concentration, for bacteria in the South-East. In this region, N content and K concentration were more important filters for fungi than pH and clay content.

The spatial variables corresponded to the sites coordinates and 16 significant PCNM eigenfunctions, each representing a different spatial scale of analysis (coarse, medium and fine scales, Figure 4). Longitude, latitude or altitude coordinates did not influence community composition except latitude for fungi in Burgundy region (Figure 4). PCNMs representing spatial structures of 80–120 km radius explained significant amounts of variance in the composition of bacterial and fungal communities in Brittany and South-East. This highlighted that these communities were spatially structured at a coarse spatial scale. Similarly, PCNMs representing spatial structures of 40–65 km radius explained significant amounts of variation in the composition of bacterial community in Brittany and of fungal community in Burgundy, South-East and Landes. This showed the spatial structuration of these communities at medium spatial scales. PCNMs accounting for fine scale variables were neither significant for bacteria nor fungi in any region.

Discussion

The four regions were selected to challenge the hypothesis that different levels of environmental heterogeneity, *i.e.* different habitat diversity and fragmentation for soil bacteria and fungi, results in different community composition turnovers [9], [14] and to compare their determinism. The multivariate analysis and the calculated $\overline{D}_{habitat}$ both highlighted a significant gradient in environmental heterogeneity following the sequence: Landes< Burgundy≤Brittany<<South-East. The four regions were mainly discriminated by environmental variables already demonstrated to influence soil microbial community abundance and diversity, such as land-use [52] and soil characteristics (texture, pH_{water}, P content and C:N ratio; [6], [14], [33], [35]). Among these four regions, Landes and South-East represent two extremes of environmental heterogeneity. Landes was distinct in having sites with conifer forest on acidic soils and low altitudinal variations. This specificity may turn Landes into an outlier for its environmental variability. On the other hand, South-East is characterized by a strong altitudinal gradient and a mosaic of land-use. In comparison, Burgundy and Brittany were also characterized by a mosaic of land-use but were more marked by croplands than South-East and less by forests. In addition, these two regions presented intermediate soil types regarding Landes and South-East: silty-clay calcic soils and silty acidic soils, respectively; associated to intermediate levels of organic Carbon content. Altogether, these observations highlight that these four regions allow the consideration of a large range of environmental conditions associated to different levels of environmental heterogeneity. Therefore, community similarity turnover rates can be confronted to environmental heterogeneity and their comparison may lead to a consensus regarding the environmental filters shaping soil microbial communities.

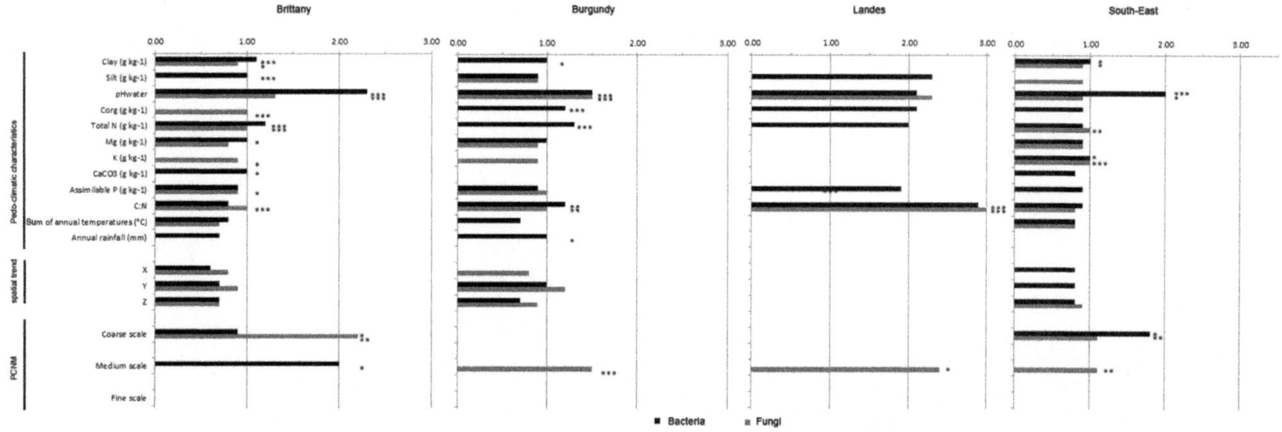

Figure 4. Variations of microbial communities partitionned according to edaphic variables and space. For each organism and region, only variables retained in the most parsimonious model are presented and their pure effect is tested by a permutation test. Significance levels are: P< 0.05: *; P<0.01: **, P<0.001: ***. Missing values or variables indicate that the variable was not retained in the model. Sand was removed prior to model evaluation since it was represented by the opposite of the sum of silt and clay content. Rainfall: Sum of annual rainfall (mm). Temperature: Sum of annual temperature (°C). Spatial components were summarized according to the scale considered: trend (x, y and z coordinates), coarse, medium or fine. The interval in brackets indicates the numbers of PCNMs retained in the model for each scale. The proportion of variance for each scale was determined as the sum of the pure effects of each PCNM when these were significant. Coarse, medium and fine scales correspond to PCNM with a spatial range of 80 to 120 km, 40 to 65 km and less than 40 km; respectively.

In all regions, except Landes, the soil bacterial and fungal communities were spatially structured as indicated by the significant DDR. This supports that the concept of DDR for soil microorganisms may be generalized, as suggested by several other studies [1], [9], [12], [24], [53]. The estimated bacterial and fungal community composition turnover rates ranged from 0.006 to 0.013. This is in agreement with the recent community composition turnover rates observed for soil microorganisms [10], [14], [24]. However, the turnover rates estimates were low. This could be due to technical limitations; particularly the low taxonomic resolution of DNA fingerprinting. Species variations were aggregated by the DNA fingerprints into few dominant bands, which precluded the accumulation of new minor species with increasing distance [54–55]. This would be supported by the higher community composition turnovers observed in Zinger [24] with the high resolution level provided be the pyrosequencing approach. The sampling design could also have led to low estimates of the community composition turnover rate. Indeed, lower composition turnover rates are commonly observed in regions composed of contiguous habitats [13] with gradual variations of habitat characteristics across sites and higher rates of community composition turnover have been observed at finer taxonomic levels and spatial scales [56].

Beside these technical points, community composition turnover rates were very close within each region for bacteria and fungi. This observation was in agreement with maps of bacterial and fungal communities' structures which revealed large patches of *ca.* 100–140 km radius (see Fig. 2 in [7] for bacteria and Fig. S1 for fungi). Altogether, this indicated a low-level of aggregation of both communities on a broad spatial scale [57]. This observation would suggest that differences in terms of biological and ecological features (habitat characteristics, colonization modes, trophic requirements, biotic interaction, dormancy; [6], [58–59]) would not lead to different rates of community composition turnover on a broad spatial scale. This would be supported by strong biotic interactions between bacterial and fungal communities highlithed by a significant correlation between bacterial and fungal communities' compositions. Nevertheless, this correlation was much lower than that between each community composition and environmental variables (data not shown). This strong dependency to environmental conditions resulted in a trend towards higher turnover rates in regions with higher $\overline{D}_{habitat}$. This would corroborate our first hypothesis that the microbial DDR is positively related to environmental heterogeneity. This observation was more marked and significant for bacterial communities but only a tendency for fungal communities. Although this conclusion is based on analyses of only 4 regions, it is in agreement with Ranjard *et al.* [9] who demonstrated a correlation between habitat heterogeneity and turnover rate for soil bacterial communities on a broad spatial scale, and with other microbial and macrobial studies [24], [30–31], [60].

The differences between the sensitivities of soil bacteria and fungi to environmental heterogeneity could be due to the ecology of soil fungi [58] at the spatial scale of this study. Nevertheless, according to Peay et al [25–26], ectomycorhizal fungi are spatially structured at very fine spatial scales. This would suggest that these organisms are strongly dependent on environmental conditions but that the grain size of this study (252 km^2) did not allow perceiving this dependence. This hypothesis would be supported by the estimated similarity of the microbial communities at a local spatial scale (initial similarity, [30]). Indeed, the initial similarity was systematically higher for bacterial than for fungal communities. According to Morlon et al [57], this would indicate that fungal communities are more aggregated than bacterial communities, *i.e.* more variable than bacterial communities, at a local spatial scale. This would lead to the conclusion that fungi are somehow more dependent on environmental conditions than bacteria at a local scale, in agreement with Peay et al [25–26] and the numerous biological interactions they are involved in.

The observed patterns and significant turnover rates result from different ecological processes [8] which may shape differently bacterial and fungal community composition on a broad spatial scale according to their relative importance. Recent studies have demonstrated that the main processes involved in increasing or decreasing community composition turnover rate are environmental selection and dispersal, respectively [8–9]. Dispersal is commonly supposed, but not empirically demonstrated to be high for microbes [61]. The maintenance of significant turnover rates, as recorded in our study, thus requires environmental selection to be high and dispersal not infinite. This hypothesis was tested by partitioning the β-diversity variations of microbial communities within regions. This partition was made according to filters involved in environmental selection: soil pedo-climatic characteristics and land-use; and to spatial variables characterizing spatial autocorrelation unexplained by the environmental variables. The residual spatial autocorrelation could result from dispersal limitations, but also from unmeasured spatially autocorrelated environmental variables despite the extensive number of environmental variables considered in this study. The amount of explained variance ranged from 23% to 32% for bacteria and from 17.5% to 18.8% for fungi. These values are within the range reported in the literature for the whole communities of bacteria [8], [14] and fungi [25–26], [62].

Soil characteristics accounted for higher amounts of variance, for both bacteria and fungi, than spatial variables. This supports the idea that soil microbial communities are primarily affected by environmental selection and secondly by other processes leading to spatial structuration independently of the environment [2], [7–8], [38]. More precisely, environmental selection was mainly driven by pedo-climatic characteristics, which accounted for larger amounts of variance than land-use. This is in agreement with several studies conducted on a broad spatial scale evidencing the higher dependency of bacterial and fungal communities on soil physico-chemical characteristics [18], [38], [63] than on land-use. Nevertheless, the amount of variance explained by pedo-climatic characteristics was higher for bacteria than for fungi, highlighting the greater effect of soil habitat on shaping the bacterial community. This observation is supported by the higher diversity and reactivity of the soil bacterial community to changes in surrounding conditions which, in turn, leads to a community structure that is better fitted to the habitat characteristics [13], [58]. On the contrary, land-use explained a higher amount of fungal community variance than of bacterial community variance. This latter could be due to plant-soil microbe interactions resulting from the type of vegetative cover [19], [38], [52], [64] as well as from human activities, especially agricultural or industrial practices [35]; [37] potentially affecting fungi more strongly than bacteria. This difference would be in agreement with a highly patchy distribution of soil fungi at a local scale as suggested above by the low level of initial similarity. Altogether, these results support the postulate of Baas-Becking "the environment selects" in summarizing microbial biogeography.

Nevertheless, variance partitioning also highlighted that bacterial and fungal communities are spatially structured independently of environmental characteristics since spatial variables accounted for a significant amount of community variance. A hypothesis in explaining this spatial autocorrelation is that soil bacterial and fungal communities might be dispersal limited, even if this result may also be related to unmeasured, spatially structured, environmental variables. This hypothesis would be supported by the study of Ranjard et al [9] demonstrating significant turnover rates of bacterial diversity in fully homogenous regions and by the consideration of several soil physico-chemical characteristics in the variance partitioning approach of the present study. Nevertheless, this must be confirmed by more in-depth studies since few evidence remains to date on the shaping of bacterial and fungal communities by limited dispersal [8], [17], [26–27], [65]. Comparisons have indicated that fungi tend to be less dependent on spatial variables than bacteria. This result is surprising since fungi are demonstrated to be dispersal limited [27] and bacteria are generally expected to disperse over larger distances, even wider than our regional scale [14], [66]. It was estimated in the literature that 10^{18} viable bacteria were transported annually in the atmosphere between continents [61]. Moreover, particular fungal populations have been demonstrated to disperse over short distances, *e.g.* ectomycorrhiza at the "plant island scale" [25–26], or to be endemic [27]. On the other hand, variance explained by spatial variables was similar in each region for fungi, whereas it followed the environmental heterogeneity gradient: Landes< Brittany≤Burgundy<South-East for bacteria. This observation supports the hypothesis that bacteria are more sensitive to geomorphology and that the presence of natural barriers (*e.g.* mountains), or a higher fragmentation of landscape in the South-East, due to a mosaic of agricultural and natural plots [9] may lead to significant spatial variations.

Our last objective was to identify and rank the environmental filters and the spatial variables involved in environmental selection process and spatial autocorrelation of bacteria and fungi. To do so, the environmental and spatial filters selected in the variance partitioning approach were ranked according to their pure effect on bacterial and fungal community composition turnover. First, environmental selection was mainly based on soil characteristics but not on climatic conditions. This is in accordance with previous studies highlighting the weak influence of climate on soil microbial diversity on a broad spatial scale [6–7], [35], [37], [67], [68]. Nevertheless, recent studies also demonstrated strong differences between bacterial communities across different biomes, especially cold deserts *versus* temperate biomes and hot deserts, suggesting that climate may play a role [32] at very large spatial scales. Overall, the main soil characteristics identified as important filters for soil bacterial and fungal community composition were pH, trophic resources (N, C, K, P contents; C:N ratio) and texture (clay, silt content). These findings are consistent with the literature where i) pH is regularly identified as the main filter for both soil bacteria [6], [16–17] and fungi [25], [62], ii) soil organic C and N contents and the C:N ratio constitute important components of microbial niches [18], [32], [35], [58], and iii) texture determines the size and stability of soil micro-habitats [18], [69], [70]. The hierarchy of these filters depended on the type of organisms and the regions. The above-described hierarchical sequence (pH> trophic resources≥texture) was observed for three of the four regions, but not for Landes. In the Landes region, the filter variability is low and environmental selection is basically driven by organic matter quality (C:N ratio). This is related to the large number of conifer forests sites leading to soil organic matter with a high C:N ratio and strong recalcitrance to microbial decomposition. A strong selection of particular populations with enzymatic ability to transform this organic matter is occurring in such soils [58], which is deeply influencing the corresponding community structure. Comparison of the overall filter hierarchies for bacteria and fungi in a given region did not reveal any discrepancy for the primary drivers. The only differences were observed for secondary filters, including C:N ratio and mineral nutrients (such as K and P) for fungi, and clay and N content for bacteria. This is consistent with the well-known dependency of the soil fungal community on soil P content [34] and soil organic matter quality [58]. As regards bacteria, clay content is positively correlated with the biotic capacity of soil as well as its indigenous bacterial diversity by enhancing the level of protection of the soil habitat and the retention of nutrients [69–70]. Second, regarding spatial autocorrelation, different scales were derived from the hierarchy of spatial variables and their range. The main scale identified was the coarse scale (80 to 120 km radius), for both bacteria and fungi in Brittany and South-east regions. This spatial scale is smaller than the one at which soil habitat changed on the RMQS Network (150 to 470 km, [9]), whereas it is in agreement with the large patches obtained by mapping the bacterial and fungal community structure over these regions [7] and Fig. S1). In addition, finer spatial scales (medium scales; 40 to 65 km radius) were also identified as significant in Burgundy, Landes and South-East for fungi and in Brittany for bacteria. Altogether, this scale dependency would support a hypothesis for the dispersal limitation of bacterial and fungal communities [33]. This scale dependency is in agreement with Martiny et al [14] for bacteria and is supported by the observations of Peay et al [25–26] on soil fungi. As in other studies at continental scales [14], this highlights the importance of considering multiple scales to better understand microbial ecology.

Altogether, our study demonstrated the spatial structuring of soil bacterial and fungal communities on local to coarse scales, which was based on environmental selection and on an unexplained spatial autocorrelation that could be related to limited dispersal. Selection and spatial autocorrelation were shown to have a similar influence on soil bacteria and soil fungi but the filters involved could differ depending on the environmental heterogeneity. Nevertheless, the comparison of bacterial and fungal communities helped to propose a primary consensus regarding the environmental filters shaping soil microbial community composition as a whole: Land-Use and pH are the primary filters, followed by trophic resources quantity (organic carbon content and nitrogen content) and then quality (C:N ratio). The results of this study increase our knowledge on the effects of soil habitat and provide insights in the scale at which dispersal may occur according to the ecological attributes of bacteria and fungi. However further investigations, based on up-scaling approaches, are now required to: i) provide a direct measurement of bacterial cells dispersal which is now crucial to demonstrate limited dispersal of bacteria; ii) identify the filters operating at each spatial scale. Especially, these upscaling approaches could be associated to high-throughput sequencing to achieve a finer resolution on the communities. This would improve our ability to sustainably manage soil biodiversity.

Supporting Information

Figure S1 Maps of interpolated MULTISPATI scores for the first three MULTISPATI axes (columns) and for the four geographical regions (rows). Each map was generated as described in Dequiedt et al (2009) and corresponds to the spatial synthesis of the F-ARISA genetic structure of indigenous fungal communities from the corresponding soils sampled in the four regions of France. Colours on the map are proportional to the score of each soil sample on each MULTISPATI axis following the scale provided at the bottom of the figure. Below each column, the empirical variogram is provided for each MULTISPATI axis.

Figure S2 Distance-Decay Relationship for bacteria and fungi. Each panel correspond to: **(A–D): Bacteria** in Brittany,

Burgundy, Landes and South-East; **(E–H): Fungi** in Brittany, Burgundy, Landes and South-East. Points represent paired-comparisons between sites and line the linear model. The equations for the regression models were as follows: **(A)** log10(Sørensen's similarity) = −0.014×log10(geographic distance)−0.156; **(B)** log10(Sørensen's similarity) = −0.018×log10(geographic distance)−0.144; **(C)** log10(Sørensen's similarity) = −0.017×log10(geographic distance)−0.198; **(D)** (Sørensen's similarity) = −0.027×log10(geographic distance)−0.101; **(E)** log10(Sørensen's similarity) = −0.017×log10(geographic distance)−0.350; **(F)** log10(Sørensen's similarity) = −0.015×log10(geographic distance)−0.316; **(G)** log10(Sørensen's similarity) = −0.012×log10(geographic distance)−0.357; **(H)** log10(Sørensen's similarity) = −0.019×log10(geographic distance)−0.298. Significance of the model is indicated as an exponent for each organism: ns: not significant; P<0.05: *; P<0.01: **, P<0.001: ***.

Table S1 Summary statistics of regions characteristics. PNE: Particular Natural Ecosystems, SE: standard error of the mean.

Author Contributions

Conceived and designed the experiments: NCPB SD NPAS CJ DA LR PL. Performed the experiments: NCPB ML. Analyzed the data: NCPB SD JT PP. Contributed reagents/materials/analysis tools: NCPB SD JT ML. Wrote the paper: NCPB LR.

References

1. Horner-Devine MC, Lage M, Hughes JB, Bohannan BJM (2004) A taxa-area relationship for bacteria. Nature 432: 750–753.
2. Martiny JBH, Bohannan BJM, Brown JH, Colwell RK, Fuhrman JA, et al. (2006) Microbial biogeography: putting microorganisms on the map. Nat Rev Microbiol 4: 102–112.
3. Baas Becking LGM (1934) *Geobiologie of inleiding tot de milieukunde*, The Hague, The Netherlands.
4. Maron P-A, Mougel C, Ranjard L (2011) Soil microbial diversity: Methodological strategy, spatial overview and functional interest. C R Biol 334: 403–411.
5. Gilbert J, Meyer F, Jansson J, Gordon J, Pace N, et al. (2010) *The Earth Microbiome Project: Meeting report of the "1st EMP meeting on sample selection and acquisition"* at Argonne National Laboratory October 6th 2010.
6. Fierer N, Jackson RB (2006) The diversity and biogeography of soil bacterial communities. Proc Natl Acad Sci U 103: 626–631.
7. Dequiedt S, Thioulouse J, Jolivet C, Saby NPA, Lelievre M, et al. (2009) Biogeographical patterns of soil bacterial communities. Environ Microbiol Rep 1: 251–255.
8. Hanson CA, Fuhrman JA, Horner-Devine MC, Martiny JBH (2012) Beyond biogeographic patterns: processes shaping the microbial landscape. Nat Rev Microbiol 10: 497–506.
9. Ranjard L, Dequiedt S, Chemidlin Prévost-Bouré N, Thioulouse J, Saby NPA, et al. (2013) Turnover of soil bacterial diversity driven by wide-scale environmental heterogeneity. Nat Commun 4: 1434.
10. Gleason AH (1922) On the relationship between species and area. Ecology (N. Y.), 3: 158–162.
11. Zinger L, Boetius A, Ramette A (2014) Bacterial taxa–area and distance–decay relationships in marine environments. Mol Ecol 23: 954–964.
12. Green JL, Holmes AJ, Westoby M, Oliver I, Briscoe D, et al. (2004) Spatial scaling of microbial eukaryote diversity. Nature 432: 747–750.
13. Prosser JI, Bohannan BJM, Curtis TP, Ellis RJ, Firestone MK, et al. (2007) Essay - The role of ecological theory in microbial ecology. Nat Rev Microbiol 5: 384–392.
14. Martiny JBH, Eisen JA, Penn K, Allison SD, Horner-Devine MC (2011) Drivers of bacterial β-diversity depend on spatial scale. P Natl Acad Sci USA 108: 7850–7854.
15. Vellend M (2010) Conceptual synthesis in community ecology. Q Rev Biol 85: 183–206.
16. Green JL, Bohannan BJM (2006) Spatial scaling of microbial biodiversity. Trends Ecol Evol 21: 501–507.
17. Chu H, Fierer N, Lauber CL, Caporaso JG, Knight R, et al. (2010) Soil bacterial diversity in the Arctic is not fundamentally different from that found in other biomes. Environ Microbiol DOI: 0.1111/j.1462–2920.2010.02277.x.
18. Pasternak Z, Al-Ashhab A, Gatica J, Gafny R, Avraham S, et al. (2013) Spatial and Temporal Biogeography of Soil Microbial Communities in Arid and Semiarid Regions. PLoS ONE 8: e69705. doi:10.1371/journal.pone.0069705.
19. Brodie E, Edwards S, Clipson N (2002) Bacterial Community Dynamics across a Floristic Gradient in a Temperate Upland Grassland Ecosystem. Microb Ecol 44: 260–270.
20. Rousk J, Baath E, Brookes PC, Lauber CL, Lozupone C, et al. (2010) Soil bacterial and fungal communities across a pH gradient in an arable soil. ISME J 4: 1340–1351.
21. Langenheder S, Szekely AJ (2011) Species sorting and neutral processes are both important during the initial assembly of bacterial communities. ISME J 5: 1086–1094.
22. Hovatter SR, Dejelo C, Case AL, Blackwood CB (2010) Metacommunity organization of soil microorganisms depends on habitat defined by presence of Lobelia siphilitica plants. Ecology 92: 57–65.
23. Yergeau E, Bezemer TM, Hedlund K, Mortimer SR, Kowalchuk GA, et al. (2010) Influences of space, soil, nematodes and plants on microbial community composition of chalk grassland soils. Environ Microbiol 12: 2096–2106.
24. Zinger L, Lejon DPH, Baptist F, Bouasria A, Aubert S, et al. (2011) Contrasting Diversity Patterns of Crenarchaeal, Bacterial and Fungal Soil Communities in an Alpine Landscape. PLoS ONE 6: e19950.
25. Peay KG, Bruns TD, Kennedy PG, Bergemann SE, Garbelotto M (2007) A strong species–area relationship for eukaryotic soil microbes: island size matters for ectomycorrhizal fungi. Ecol Lett 10: 470–480.
26. Peay KG, Schubert MG, Nguyen NH, Bruns TD (2012) Measuring ectomycorrhizal fungal dispersal: macroecological patterns driven by microscopic propagules. Mol Ecol 21: 4122–4136.
27. Talbot JM, Bruns TD, Taylor JW, Smith DP, Branco S, et al. (2014) Endemism and functional convergence across the North American soil mycobiome. P Natl Acad Sci USA 111: 6341–6346.
28. Taylor DL, Hollingsworth TN, McFarland JW, Lennon NJ, Nusbaum C, et al. (2013) A first comprehensive census of fungi in soil reveals both hyperdiversity and fine-scale niche partitioning. Ecol Monogr 84: 3–20.
29. Taylor DL, Herriott IC, Stone KE, McFarland JW, Booth MG, et al. (2010) Structure and resilience of fungal communities in Alaskan boreal forest soilsThis article is one of a selection of papers from The Dynamics of Change in Alaska's Boreal Forests: Resilience and Vulnerability in Response to Climate Warming. Can J For Res 40: 1288–1301.
30. Soininen J, McDonald R, Hillebrand H (2007) The distance decay of similarity in ecological communities. Ecography 30: 3–12.
31. Drakare S, Lennon JJ, Hillebrand H (2006) The imprint of the geographical, evolutionary and ecological context on species–area relationships. Ecol Lett 9: 215–227.
32. Fierer N, Leff JW, Adams BJ, Nielsen UN, Bates ST, et al. (2012) Cross-biome metagenomic analyses of soil microbial communities and their functional attributes. Proc Natl Acad Sci USA 109: 21390–21395.
33. Ramette A, Tiedje JM (2007) Multiscale responses of microbial life to spatial distance and environmental heterogeneity in a patchy ecosystem. P Natl Acad Sci USA 104: 2761–2766.
34. Lauber CL, Strickland MS, Bradford MA, Fierer N (2008) The influence of soil properties on the structure of bacterial and fungal communities across land-use types. Soil Biol Biochem 40: 2407–2415.
35. Dequiedt S, Saby NPA, Lelievre M, Jolivet C, Thioulouse J, et al. (2011) Biogeographical patterns of soil molecular microbial biomass as influenced by soil characteristics and management. Global Ecol Biogeog 20: 641–652.
36. Brodie E, Edwards S, Clipson N (2003) Soil fungal community structure in a temperate upland grassland soil. FEMS Microbiol Ecol 45: 105–114.
37. Lienhard P, Tivet F, Chabanne A, Dequiedt S, Lelièvre M, et al. (2013) No-till and cover crops shift soil microbial abundance and diversity in Laos tropical grasslands. Agron Sustain Dev 33: 1–10.
38. Griffiths RI, Thomson BC, James P, Bell T, Bailey M, et al. (2011) The bacterial biogeography of British soils. Environ Microbiol 13: 1642–1654.
39. Harte J, Kinzig A, Green J (1999) Self-Similarity in the Distribution and Abundance of Species. Science 284: 334–336.
40. Harte J, Smith AB, Storch D (2009) Biodiversity scales from plots to biomes with a universal species–area curve. Ecol Lett 12: 789–797.
41. Ranjard L, Dequiedt S, Lelievre M, Maron PA, Mougel C, et al. (2009) Platform GenoSol: a new tool for conserving and exploring soil microbial diversity. Environ Microbiol Rep 1: 97–99.
42. Arrouays D, Jolivet C, Boulonne L, Bodineau G, Saby NPA, et al. (2002) A new projection in France: a multi-institutional soil quality monitoring networkUne initiative nouvelle en France : la mise en place d'un réseau multi-institutionnel de mesure de la qualité des sols (RMQS). Comptes Rendus de l'Académie d'Agriculture de France 88: 93–105.
43. Ranjard L, Poly F, Nazaret S (2000) Monitoring complex bacterial communities using culture-independent molecular techniques: application to soil environment. Res Microbiol 151: 167–177.
44. Hill MO, Smith AEJ (1976) Principal Component Analysis of Taxonomic Data with Multi-State Discrete Characters. Taxon 25: 249–255.

45. Thioulouse J, Dray S (2007) Interactive multivariate data analysisin R with the ade4 and ade4TkGUI packages. J stat softw 22: 1–14.
46. R Development Core Team (2011) *R: A Language and Environment for Statistical Computing*, Vienna, Austria.
47. Legendre P, Borcard D, Peres-Neto PR (2005) Analyzing beta diversity: Partitioning the spatial variation of community composition data. Ecol Monogr 75: 435–450.
48. Legendre P, Fortin MJ (2010) Comparison of the Mantel test and alternative approaches for detecting complex multivariate relationships in the spatial analysis of genetic data. Mol Ecol Resour 10: 831–844.
49. Dray SP, Legendre P, Peres-Neto PR (2006) Spatial modelling: a comprehensive framework for principal coordinate analysis of neighbour matrices (PCNM) Ecol Model 196: 483–493.
50. Oksanen J, Blanchet FG, Kindt R, Legendre P, Minchin PR, et al. (2011) vegan: Community Ecology Package.
51. Bellier E, Monestiez P, Durbec J-P, Candau J-N (2007) Identifying spatial relationships at multiple scales: principal coordinates of neighbour matrices (PCNM) and geostatistical approaches. Ecography 30: 385–399.
52. Drenovsky RE, Steenwerth KL, Jackson LE, Scow KM (2010) Land use and climatic factors structure regional patterns in soil microbial communities. Global Ecol Biogeog 19: 27–39.
53. Bell T, Ager D, Song J-I, Newman JA, Thompson IP, et al. (2005) Larger Islands House More Bacterial Taxa. Science 308: 1884.
54. Woodcock S, Curtis TP, Head IM, Lunn M, Sloan WT (2006) Taxa–area relationships for microbes: the unsampled and the unseen. Ecol Lett 9: 805–812.
55. MacArthur RH, Wilson EO (1967) *The theory of island biogeography/Robert H. MacArthur and Edward O. Wilson*. Princeton University Press, Princeton, N.J.
56. Noguez AM, Arita HT, Escalante AE, Forney LJ, Garcia-Oliva F, et al. (2005) Microbial macroecology: highly structured prokaryotic soil assemblages in a tropical deciduous forest. Global Ecol Biogeog 14: 241–248.
57. Morlon H, Chuyong G, Condit R, Hubbell S, Kenfack D, et al. (2008) A general framework for the distance–decay of similarity in ecological communities. Ecol Lett 11: 904–917.
58. de Boer W, Folman LB, Summerbell RC, Boddy L (2005) Living in a fungal world: impact of fungi on soil bacterial niche development. FEMS Microbiol Rev 29: 795–811.
59. Lennon JT, Jones SE (2011) Microbial seed banks: the ecological and evolutionary implications of dormancy. Nat Rev Micro 9: 119–130.
60. Jobe RT (2008) Estimating ladscape-scale species richness: reconciling frequency- and turnover-based approaches. Ecology 89: 174–182.
61. Fenchel T (2003) Biogeography for Bacteria. Science 301: 925–926.
62. Dumbrell AJ, Nelson M, Helgason T, Dytham C, Fitter AH (2010) Relative roles of niche and neutral processes in structuring a soil microbial community. ISME J 4: 337–345.
63. Hossain Z, Sugiyama S-I (2011) Geographical structure of soil microbial communities in northern Japan: Effects of distance, land use type and soil properties. Eur J Soil Biol 47: 88–94.
64. Bryant JA, Lamanna C, Morlon H, Kerkhoff AJ, Enquist BJ, et al. (2008) Microbes on mountainsides: Contrasting elevational patterns of bacterial and plant diversity. P Natl Acad Sci USA 105: 11505–11511.
65. Monroy F, van der Putten WH, Yergeau E, Mortimer SR, Duyts H, et al. (2012) Community patterns of soil bacteria and nematodes in relation to geographic distance. Soil Biol Biochem 45: 1–7.
66. Cho JC, Tiedje JM (2000) Biogeography and Degree of Endemicity of Fluorescent Pseudomonas Strains in Soil. Appl Environ Microb 66: 5448–5456.
67. Lozupone CA, Knight R (2007) Global patterns in bacterial diversity. Proceedings of the National Academy of Sciences 104: 11436–11440.
68. Auguet J-C, Barberan A, Casamayor EO (2009) Global ecological patterns in uncultured Archaea. ISME J 4: 182–190.
69. Ranjard L, Richaume AS (2001) Quantitative and qualitative microscale distribution of bacteria in soil. Res Microbiol 152: 707–716.
70. Kong AYY, Scow KM, Cordova-Kreylos AL, Holmes WE, Six J (2011) Microbial community composition and carbon cycling within soil microenvironments of conventional, low-input, and organic cropping systems. Soil Biol Biochem 43: 20–30.

19

Comparison of the Rhizosphere Bacterial Communities of Zigongdongdou Soybean and a High-Methionine Transgenic Line of This Cultivar

Jingang Liang[1]◔, Shi Sun[2]◔, Jun Ji[1], Haiying Wu[3], Fang Meng[1], Mingrong Zhang[3], Xiaobo Zheng[1], Cunxiang Wu[2]*, Zhengguang Zhang[1]*

1 Department of Plant Pathology, College of Plant Protection, Nanjing Agricultural University, and Key Laboratory of Integrated Management of Crop Diseases and Pests, Ministry of Education, Nanjing, China, **2** The National Key Facility for Crop Gene Resources and Genetic Improvement (NFCRI), MOA Key Laboratory of Soybean Biology (Beijing), Institute of Crop Science, The Chinese Academy of Agricultural Sciences, Beijing, China, **3** Nanchong Academy of Agricultural Science, Nanchong, China

Abstract

Previous studies have shown that methionine from root exudates affects the rhizosphere bacterial population involved in soil nitrogen fixation. A transgenic line of Zigongdongdou soybean cultivar (ZD91) that expresses *Arabidopsis* cystathionine γ-synthase resulting in an increased methionine production was examined for its influence to the rhizosphere bacterial population. Using 16S rRNA gene-based pyrosequencing analysis of the V4 region and DNA extracted from bacterial consortia collected from the rhizosphere of soybean plants grown in an agricultural field at the pod-setting stage, we characterized the populational structure of the bacterial community involved. In total, 87,267 sequences (approximately 10,908 per sample) were analyzed. We found that Acidobacteria, Proteobacteria, Bacteroidetes, Actinobacteria, Chloroflexi, Planctomycetes, Gemmatimonadetes, Firmicutes, and Verrucomicrobia constitute the dominant taxonomic groups in either the ZD91 transgenic line or parental cultivar ZD, and that there was no statistically significant difference in the rhizosphere bacterial community structure between the two cultivars.

Editor: Raffaella Balestrini, Institute for Sustainable Plant Protection, C.N.R., Italy

Funding: This research was supported by Genetically Modified Organisms Breeding Major Projects of China (Grant No. 2014ZX08011-003) and Jiangsu Provincial Coordination Panel for Recruitment and Selection of Specially-Appointed Professors. The funders had no role in study design, data collection and analysis, decision to publish, or preparation of the manuscript.

* Email: zhgzhang@njau.edu.cn (ZZ); wucx@mail.caas.net.cn (CW)

◔ These authors contributed equally to this work.

Introduction

The global commercial cultivation of transgenic crops has increased from 1.7 million hectares in 1996 to 170.3 million hectares in 2012 [1]. Modern agricultural biotechnology and genetic engineering have allowed the development of crops with improved properties, such as the transgenic high-methionine soybean line Zigongdongdou 91 (ZD91) [2]. Most studies to date have suggested that the release of transgenic plants results in only minor changes to the microbial community structure and that these minor changes are often transient and short lived [3–15].

Soil microorganisms are affected by soil characteristics, environmental conditions, and crop management strategies [16–19]. Organic compounds in root exudates and rhizosphere microbial communities perform fundamental processes that contribute to nutrient cycling and plant health and root growth, which are governed by complex interactions driven by various factors such as soil types and plant species [3,20–23]. The type and amount of root exudation may be an inherent property of a plant, and the rhizosphere microbial community is also expected to be unique [3,24–26]. Thus, an altered exudate composition may result in a corresponding shift in the rhizosphere microorganism community [27–30]. According to Faragova et al. [31], genetically modified plants release root exudates different from those of non-transgenic counterparts, therefore affecting the microbial community of the rhizosphere. This occurs even if the genetically modified plant is of different cultivars of the same plant species [32–34]. It was reported that methionine has the effect of inhibiting nitrification in soil, and transgenic alfalfa that produces high levels of methionine has a profound effect on bacteria population involved in the nitrogen cycle of the soil [31,35]. In a recent comparative study that employed Polymerase Chain Reaction-Denaturing Gradient Gel Electrophoresis (PCR-DGGE) technology to examine the bacterial communities in the rhizosphere of a transgenic high-methionine soybean, we found that the growth stage was the main factor affecting the community structure of the soil bacteria [36]. This finding is consistent with that of van Overbeek and van Elsas [25], Jin et al. [37] and Inceoglu et al. [38], in which bacterial community structures vary per growth stage changes. In addition, a study of rape (*Brassica napus*) indicated that the rhizosphere bacterial composition of a transgenic cultivar could be distinguished from that of its non-transgenic cultivar [39]. Conversely, in a study of potatoes, the microbial community structure in the rhizosphere was not significantly affected by genetic engineering [40,41].

The biological nitrogen cycle is one of the most important nutrient cycles in the terrestrial ecosystem. It includes four major processes: nitrogen fixation, mineralization (decay), nitrification and denitrification. Studies have shown that there are three functional genes: *nifH*, *amoA*, and *nosZ*, which encode the key enzymes involved in nitrogen fixation, ammonia oxidization and complete denitrification, respectively. Microorganisms capable of nitrogen fixation are called diazotrophs whose composition is determined by associated plant species. In addition, agricultural practices can also influence soil diazotrophs [42–45]. There are two types of diazotrophs: free-living and symbiotic [46,47]. The Rhizobia, *Frankia*, and Cyanobacteria are three important symbiotic diazotrophs. Rhizobia make up a paraphyletic group that falls into two classes in the phylum Proteobacteria: Alphaproteobacteria and Betaproteobacteria. Nitrification, which converts ammonium to nitrate, includes two steps: ammonia oxidation to nitrite, and nitrite oxidation to nitrate. The oxidation of ammonia to nitrite is performed by two groups of organisms, ammonia-oxidizing bacteria (AOB) and ammonia-oxidizing archaea (AOA) [48]. AOB is found among the Betaproteobacteria and Gammaproteobacteria [49,50]. Denitrification is a microbially-facilitated process of nitrate reduction. So far, more than 60 genera of denitrifying microorganisms have been identified [51].

High-throughput TAG-encoded FLX amplicon pyrosequencing (using 16S rRNA genes) is a great technical advance for microbial ecology studies [52]. This technique can produce ~400,000–1,000,000 reads in a single 4 h run, with an accuracy of approaching 99%, and the tag sequence contains adequate information for taxonomic assignment [53]. Several recent studies have used this technique with the 16S rRNA gene to determine the microbial diversity in various environmental samples [3,52–60]. Given the wealth of information available due to the deep sequencing of environmental DNA, it is possible to determine whether these lineages are ecologically coherent [3]. It also has been proposed that some taxa are extremely coherent ecologically. Such taxa can be said to follow an r- or K-type life strategy [61–63].

In the present study, we used 16S rRNA gene-based pyrosequencing to compare the community dynamics of bacteria inhabiting the rhizosphere of field-grown transgenic cultivar ZD91 during the pod-setting stage to those of the corresponding non-transgenic ZD plants. Our study showed a wide variety of rhizosphere bacterial population associated with these cultivars but there was no significant difference in the bacterial population between transgenic cultivar ZD91 and parental cultivar ZD.

Results

Quantification of *nifH*, *amoA*, and *nosZ* genes, and carbon, nitrogen concentrations in the rhizospheres

To examine the effect of transgenic soybean on the abundance of diazotrophs, ammonia-oxidizing bacteria, and denitrifiers, we used quantitative real-time PCR (qPCR) assays to quantify *nifH*, *amoA*, and *nosZ* genes to inform nitrogen-cycling related bacteria collected from the rhizosphere soil. The abundance of *nifH*, *amoA*, and *nosZ* genes between ZD and ZD91 showed no significant differences (Table 1).

We also estimated whether there are any differences in carbon and nitrogen levels in the rhizosphere of ZD and ZD91 and found no significant differences (Table 1).

Amino acids in root exudates

Amino acids included aspartic acid, threonine, serine, glutamic acid, glycine, alanine, valine, methionine, isoleucine, leucine, tyrosine, phenylalanine, lysine and histidine in root exudates were determined. However, there were no differences in amino acids contents between ZD and ZD91 (Table 2).

Pyrosequencing analysis of the bacterial communities of the rhizospheres

We performed 16S rRNA gene-based pyrosequencing analysis of the V4 region to characterize the bacterial community composition in the rhizospheres surrounding ZD and ZD91 cultivars. A total of 87,267 high-quality sequences was obtained, with an average of 10,908 sequences per sample. The results were summarized in Table 3 with detailed sample characteristics shown in Table S1. We used the grouping of cultivars based on the number of replicates (ZD, four replicates: ZD_1, ZD_2, ZD_3, and ZD_4; ZD91, four replicates: ZD91_1, ZD91_2, ZD91_3, and ZD91_4).

The rarefaction curves tended towards the saturation plateau while Good's coverage estimations revealed that 78–84% of the species were from eight samples, indicating that the sequence coverage was sufficient to capture the diversity of the bacterial communities in the samples (Fig. 1). The estimators of community richness (Chao and ACE) and diversity (Shannon and Simpson) were shown in Table 3, and there were no significant differences in these indices between ZD and ZD91. The null hypothesis was tested through AMOVA analysis (Table 4) and it did show that there were no significant differences between ZD and ZD91.

Taxonomic composition

Although probably present, this study did not attempt to characterize any Archaea. All sequences were classified from phylum to genus according to the RDP classifier using the default settings. 22 different phyla, including 1 archaeal phylum, were identified from the samples. The overall microbiota structure for ZD and ZD91 at the phylum level was shown in Figure 2 and Table S2. Acidobacteria, Proteobacteria, Bacteroidetes, Actinobacteria, Chloroflexi, Planctomycetes, Gemmatimonadetes, Firmicutes, and Verrucomicrobia accounted for >90% of the reads. Again, there were no observable differences among these categories between ZD and ZD91 at this taxonomic level (Fig. 3).

A total of 13,299 operational taxonomic units was found in the complete data set (Table S3). The ten most abundant OTUs within each sample were identified (Table S4). There were 12 dominant OTUs in total, including Gp4, Gp6, *Terrimonas, Sphingosinicella, Sphingomonas, Flavisolibacter, Sphingomonadaceae, Lysobacter, Chryseobacterium, Acinetobacter, Levilinea and Opitutus.* A total of 2047 common OTUs was determined in rhizosphere soil and there was no significant difference between ZD and ZD91 ($P<0.01$) (Fig. 4). The ten most abundant genera from each of the eight samples were classified as a percentage of the total sequences per sample (Table S5). This analysis showed that there were 17 dominant genera in total, including four members of the Acidobacteria, five of the Bacteroidetes, three Alphaproteobacteria, two Gammaproteobacteria, and one each of the Actinobacteria, Gemmatimonadetes and Verrucomicrobia.

To analyze whether the distribution of low-abundance genera differs between samples, sequences with abundances below 0.1% were analyzed. A total of 481 such low-abundance genera was found. Of the 481 genera, 214 (70.63%), 203 (68.35%), 200 (61.54%), and 219 (64.60%) were observed in ZD_1, ZD_2,

Table 1. Changes in abundance of *nifH*, *amoA*, and *nosZ* genes (Log_{10} copies/g soil), and carbon (C), nitrogen (N) contents in rhizosphere soil ($P<0.01$).

	ZD	ZD91
nifH	9.05±0.02 A	9.06±0.12 A
amoA	8.03±0.02 A	8.13±0.01 A
nosZ	7.70±0.08 A	7.90±0.02 A
Total C (%)	1.68±0.02 A	1.65±0.03 A
Total N (%)	0.06±0.01 A	0.07±0.01 A

ZD_3, and ZD_4, respectively. In comparison, 208 (67.97%), 205 (61.75%), 219 (68.65%), and 206 (65.19%) were detected in ZD91_1, ZD91_2, ZD91_3, and ZD91_4, respectively. There were no genera unique to cultivar ZD or cultivar ZD91. Rare genera (abundances below 0.1%) accounted for over 60% in all samples, and there was no significant difference between cultivar ZD (66.28±4.02%) and cultivar ZD91 (65.89±3.14%).

Root-colonizing pseudomonads are known to be involved in plant growth and health. These bacteria increase plant growth either directly through the production of phytohormones or other stimulants and increasing the bioavailability of nutrients in the soil, or indirectly by the suppression of plant diseases and the induction of systemic resistance in plants [10]. A total of 6 pseudomonad species was found in our samples and there was no significant difference between ZD (0.07±0.03%) and ZD91 (0.05±0.04%).

Nitrogen-fixing Bacteria

In the present study, *Frankia* was not found in any samples. Sequences related to Cyanobacteria, Alphaproteobacteria, and Betaproteobacteria were not significantly different between cultivar ZD (0.22±0.07%, 14.04±5.07%, and 1.23±0.30%) and cultivar ZD91 (0.14±0.08%, 13.56±4.88%, and 1.11±0.25%) samples.

Nitrifying bacteria

In the present study, there were 35 Gammaproteobacteria genera detected. There was no significant difference between the sequences for cultivar ZD (3.47±0.17%) and ZD91 (4.49±1.38%). Nitrifying bacteria, which oxidize nitrite, include *Nitrobacter*, *Nitrospina*, *Nitrococcus*, and *Nitrospira* [65], but only *Nitrospira* was detected in this study. In addition, there was no significant difference between the sequences for cultivar ZD (0.36±0.16%) and cultivar ZD91 (0.44±0.16%).

Denitrifying bacteria

In the present study, 36 such bacterial genera were found. The sequences related to denitrifying bacteria were not significantly different between cultivar ZD (3.13±1.31%) and cultivar ZD91 (2.80±0.28%).

Shared OTUs in ZD and ZD91 libraries

The bacterial OTUs in the cultivar ZD and cultivar ZD91 libraries were further investigated for shared OTUs. As shown in Table S6, ZD_1, ZD_2, ZD_3, and ZD_4 had 700 OTUs in common. A statistical analysis revealed that the OTUs common to the four libraries comprised 58.96%, 59.62%, 52.19%, and 46.82% of the reads in the ZD_1, ZD_2, ZD_3, and ZD_4 libraries, respectively. Acidobacteria, Proteobacteria, and Bacteroidetes comprised 531 of the shared OTUs (75.86% in propor-

Table 2. Amino acids secreted by soybean roots in the root box at pod-setting stage (μmol/ml) ($P<0.01$).

	ZD	ZD91
Asp	3.44±1.56 A	1.13±0.48 A
Thr	2.85±1.28 A	0.99±0.41 A
Ser	12.13±6.05 A	4.11±1.82 A
Glu	2.31±0.59 A	1.30±0.23 A
Gly	4.83±2.43 A	1.60±0.86 A
Ala	3.63±1.72 A	1.32±0.51 A
Val	2.14±0.83 A	1.01±0.25 A
Met	0.32±0.28 A	0.31±0.27 A
Ile	1.04±0.41 A	0.40±0.17 A
Leu	1.54±0.55 A	0.77±0.24 A
Tyr	2.03±0.58 A	1.33±0.20 A
Phe	1.22±0.47 A	0.65±0.17 A
Lys	1.03±0.57 A	0.33±0.20 A
His	2.12±1.36 A	0.68±0.46 A

Table 3. Pyrosequencing data summary.

Sample	Sequences	OTUs	ACE	Chao	Shannon	Simpson	Coverage
ZD	10534±1353A	3391±505A	8900±1019A	6381±780A	7.34±0.21A	0.002±0.001A	0.82±0.02A
ZD91	11283±1818A	3649±370A	9637±795A	6911±688A	7.47±0.17A	0.001±0.000A	0.82±0.03A

Values are the mean±standard deviation of four replicates. Within each vertical column, values followed by the same letter are not statistically different from each other ($P<0.01$). The number of OTUs, richness estimators Chao and ACE, diversity estimators Shannon and Simpson, and Good's coverage were calculated at 3% distance.

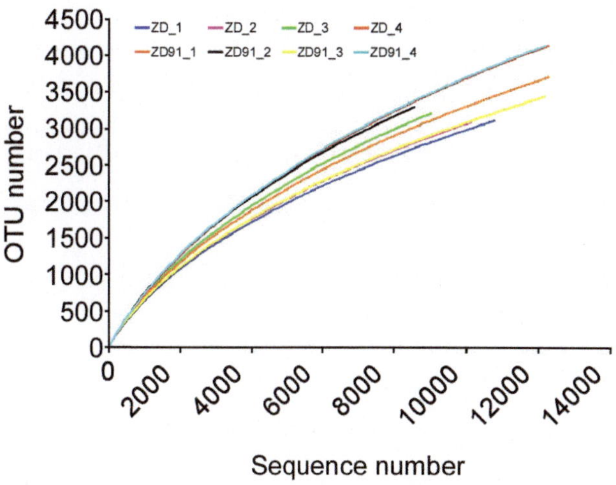

Figure 1. Rarefaction analysis. Rarefaction curves of OTUs clustered at 97% sequence identity across the samples. The sample labeled with ZD_1, ZD_2, ZD_3 and ZD_4 correspond to four replicates of cultivar ZD; ZD91_1, ZD91_2, ZD91_3 and ZD91_4 represent four replicates of transgenic cultivar ZD91.

tion). Further, the OTUs common to the four libraries of the transgenic cultivar comprised 54.04%, 49.01%, 52.17%, and 42.50% of the reads in the ZD91_1, ZD91_2, ZD91_3, and ZD91_4 libraries, respectively. Acidobacteria, Proteobacteria, and Bacteroidetes included 534 of the shared OTUs (76.29% in proportion).

Bacterial community dynamics

To examine the effect of cultivar on the total distribution of phyla and genera, we performed principal coordinate analysis (PCoA) of our data. The PCoA analysis of weighted UniFrac distances was used to calculate pairwise distances between the rhizosphere bacterial communities of the eight samples [64]. The first two axes of the PCoA represented 40.52% and 19.06% of the total variation, respectively (Fig. 5). AMOVA analysis (Table S7) was used to find out whether the separation of cultivar ZD and cultivar ZD91 in the PCoA was statistically significant. The result showed *p*-value was 0.859 indicating no significantly different between cultivar ZD and cultivar ZD91. Furthermore, we calculated ADONIS differences between cultivar ZD and cultivar ZD91 (Table 5), and found no significant difference between these two cultivars. Thus, the bacterial communities in the cultivars were not different from each other.

Discussion

In this study, we monitored the quantitative changes of the key genes involved in N-cycling as well as elements carbon, nitrogen concentrations in rhizosphere soil of ZD and ZD91, and analyzed the amino acids contents in the root exudates of different cultivars. We found that there were no significant differences between cultivar ZD and cultivar ZD91 in nitrogen-cycling genes and carbon, nitrogen concentrations. The results showed that the amino acids contents in the root exudates of ZD91 were lower than ZD, but they were not statistically significant.

We assessed the dynamics of the relative soil abundance of bacteria and the structure of soil bacterial communities as a function of cultivar type using the transgenic soybean cultivar ZD91 and its parental non-transgenic soybean cultivar ZD in a single experimental field. According to the pyrosequencing results,

Table 4. AMOVA analysis between ZD and ZD91.

	ZD91-ZD	Among	Within	Total
Based on OTU (Jclass distance coefficient)	SS	0.261484	1.603340	1.86483
	df	1.000000	6.000000	7.00000
	MS	0.261484	0.267224	
	Fs: 0.978522			
	p-value: 0.745			
Based on OTU (Thetayc distance coefficient)	SS	0.0321963	0.290100	0.322296
	df	1.0000000	6.000000	7.000000
	MS	0.0321963	0.048350	
	Fs: 0.665901			
	p-value: 0.565			
Based on weighted UniFrac distances	SS	0.0172751	0.1477470	0.165022
	df	1.0000000	6.0000000	7.000000
	MS	0.0172751	0.0246245	
	Fs: 0.701539			
	p-value: 0.861			

Experiment-wise error rate: 0.05.

genetic modification via the insertion and expression of *AtD-CGS* did not affect the composition of the bacterial community in the rhizosphere of soybean. However, the profiles of independent replicates differed in terms of the abundance of various phyla (Fig. 5 and Table S2), but this variation did not correlate with the experimental variables. All plots were on the same field, but due to the large size of the field, the replicates were located an average of 20 meters apart from each other. Thus, field heterogeneity was not considered when evaluating the impact of transgenic soybean on rhizosphere bacterial community. Furthermore, we analyzed the OTUs of high abundances and repeated well (occurred in more than three samples, defined as the common OTUs) in the soil.

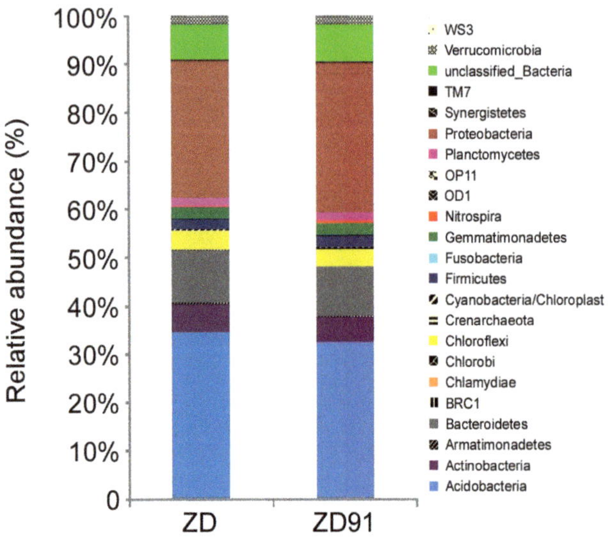

Figure 2. Bacterial composition at the phylum level. Relative read abundance of bacterial phyla within the communities. Sequences that could not be classified into any known group were assigned as unclassified_Bacteria.

Since the singletons and low abundance OTUs were detected randomly, and there was no significant difference within the treatment (among the 4 replicates) and between the treatments (Fig. 4).

Due to logistic and technical (e.g., low sequence numbers) constraints, studies of soil microbial diversity and function have focused mainly on the most dominant species [3]. Previous studies have demonstrated weaknesses in traditional molecular methods

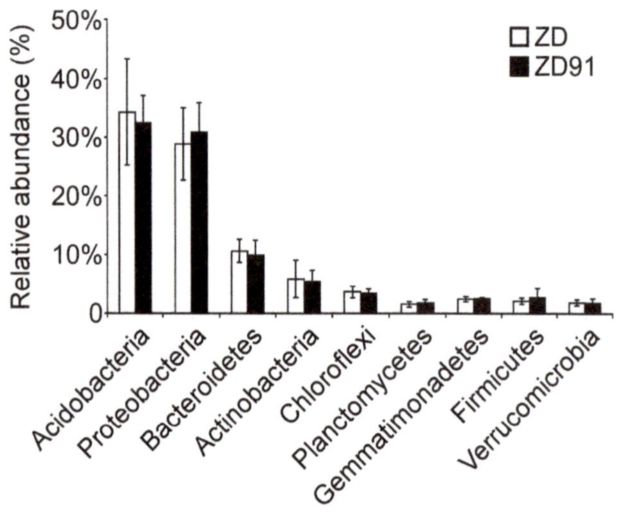

Value less than ten:

	Bacteroidetes	Actinobacteria	Chloroflexi	Planctomycetes	Gemmatimonadetes	Firmicutes	Verrucomicrobia
ZD		5.83%	3.63%	1.54%	2.43%	2.09%	1.83%
ZD91	9.92%	5.43%	3.35%	1.70%	2.51%	2.69%	1.73%

Figure 3. Relative abundance of main phyla in the rhizosphere of soybean cultivars. Error bars indicate standard errors. The *p*-values of the nine groups of bacteria, in order, were 0.734, 0.633, 0.662, 0.834, 0.687, 0.690, 0.751, 0.493, and 0.840 when testing for differences between cultivar ZD and transgenic cultivar ZD91.

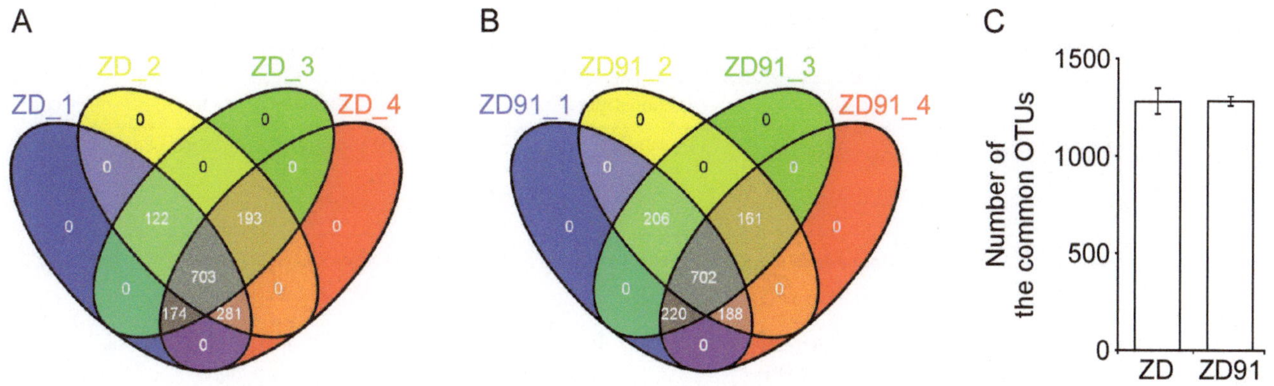

Figure 4. Common OTUs recovered from soil. (A) Venn diagram showing variable overlap between four replicates of ZD. (B) Venn diagram showing variable overlap between four replicates of ZD91. (C) Number of the common OTUs between ZD and ZD91.

(e.g., DGGE, RFLP, ARDRA, and SSCP) in terms of resolution; specifically, it is difficult to detect subtle changes in banding patterns or changes in closely related species, leading to the underestimation of bacterial diversity by one to two orders of magnitude [4,54,57,66]. Moreover, some rare species may have large effects on soil function, in spite of their low total biomass or rarity [67,68]. In this study, pyrosequencing provided a promising, fast, economical, and unique opportunity to access the less abundant bacterial taxa surrounding soybean plants, as well as the chance to compare their presence across cultivars in the pod-setting stage [54,55,57]. Similar to the results of Liu et al. [20] and Xu et al. [21,69], our sequencing results revealed that numerous bacterial phyla (e.g., Acidobacteria, Proteobacteria, Bacteroidetes, Actinobacteria, Chloroflexi, Planctomycetes, Gemmatimonadetes, Firmicutes, Nitrospirae, Verrucomicrobia, BRC1, Chlamydiae, Cyanobacteria, Fusobacteria, Chlorobi, OD1, OP11, Synergistetes, TM7, Armatimonadetes, and WS3) commonly inhabit the soybean rhizosphere.

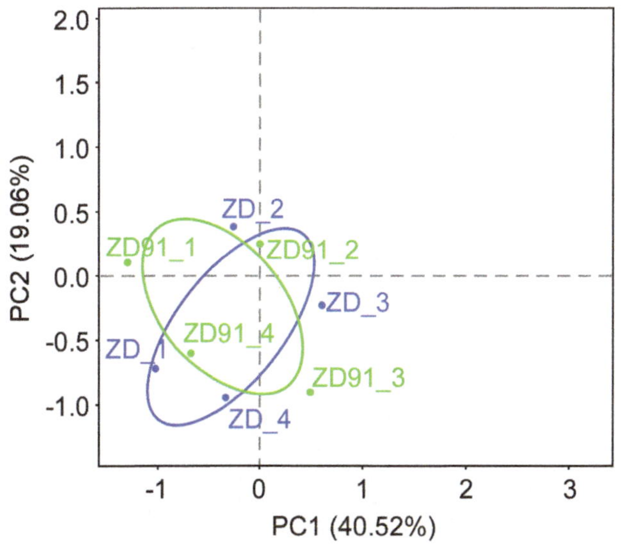

Figure 5. Principal coordinate analysis (PCoA) plot of samples using the weighted UniFrac distance metric. The variance explained by each principal coordinate axis is shown in parentheses. Datasets were subsampled to equal depth prior to the UniFrac distance computation. A circle was drawn with R around the samples of the same type (blue represents cultivar ZD, and green represents transgenic cultivar ZD91).

Moreover, dominant members of particular phyla or classes were used as indicators to reveal the prevailing ecological pattern (e.g., along the copiotroph/oligotroph scale) [3]. Members of the Betaproteobacteria and Bacteroidetes, as well as species belonging to the genus *Pseudomonas*, tend to be favored in r-selection/copiotrophic soils, which have higher carbon availability, while Acidobacteria are favored under K-selection/oligotrophic conditions [61–63]. Consequently, such groups may have higher abundances in soils with higher versus lower levels of easily available carbon, respectively. Furthermore, the ratio of Proteobacteria to Acidobacteria can be used as an indicator of the soil trophic level [70]. Although it is rather implausible that a whole phylum would have shared ecological characteristics [3], our indicators revealed variability (based on the ratio of Proteobacteria to Acidobacteria) occurs in the response of Proteobacteria versus Acidobacteria to shifting ecological conditions in the soil. The ratio is 0.16 in oligotrophic soil [71,72], 0.34 in low-input agricultural soil [73], 0.46 in low-nutrient systems [62], and 0.87 in high-input agricultural systems [74]. The values obtained in this study, 0.93 ± 0.46 in cultivar ZD, and 0.98 ± 0.31 in cultivar ZD91, indicating a high-nutrient system, but again there was no significant difference between the two cultivars.

The pyrosequencing protocol used in previous investigations and here carries certain limitations [3,52–60]. Any PCR primer set for the amplification of 16S rDNA genes may miss a considerable amount of the extant microbial diversity; thus, at present, no clear answer can be given with respect to the real extant community make-up [3]. Because PCR biases mostly affect the detection of rare sequences in a sample, it is important to view such data on the rare biosphere with proper cautions.

Of the ten most abundant bacterial OTUs in the samples, several were related to Acidobacteria. Of the ten most abundant bacterial genera, several were related to both Acidobacteria and Proteobacteria. The Acidobacteria are physiologically diverse and ubiquitous, especially in soil, but are underrepresented in culture [75–77]. Because most Acidobacteria have not been cultured, their ecology and metabolism are not well understood [76]. However, these bacteria may be important contributors to ecosystems because they are particularly abundant in soil [78]. The Proteobacteria are a major group of bacteria, and include many of the species responsible for nitrogen fixation.

We also monitored structural and quantitative changes in the bacterial communities of each soybean regarding nitrogen cycling.

Table 5. ADONIS differences between ZD and ZD91.

	Df	SumsOfSqs	MeanSqs	F. Model	R^2	Pr(>F)
qiime.data$map [[opts$category]]	1	0.017275	0.017275	0.70124	0.10464	0.8492
Residuals	6	0.147814	0.024636		0.89536	
Total	7	0.165089			1.00000	

ADONIS differences is a non-parametric method to measure statistical significance of sample grouping. It was calculated based on weighted UniFrac distance metric.

The abundance of nitrogen-transforming bacteria were not significantly different between cultivars ZD and ZD91.

In conclusion, we found that 454-pyrosequencing can be a useful tool for risk assessment studies to analyze the immediate impact of plants on the diversity of soil bacteria in rhizospheres. Of all DNA reads, 7.56% belonged to unclassified bacteria, containing a suite of diverse taxa. To understand the ecological impact of unclassified sequences, they must be identified and ideally an organism should be cultured. This once again stresses that culture-independent techniques should go hand-in-hand with culture-dependent ones [3]. For ecological risk identification, however, the observation of shifts during one stage of plant development and in a one-year period is insufficient. The positive identification of risk requires that changes be demonstrated at all stages of soybean development and over an extended time frame. The fact that no effect between transgenic and non-transgenic plants was detected suggests only minor physiological changes caused by the insertion of *AtD-CGS* into the soybean genome.

Materials and Methods

Soybean Cultivars

Transgenic soybean cultivar (ZD91) contains the *Arabidopsis* cystathionine γ-synthase (*AtD-CGS*) gene which has been introduced artificially into the soybean cultivar Zigongdongdou (ZD) using *Agrobacterium*-mediated transformation, and which exhibits a high content of methionine in the seeds.

Field tests setup and sampling

An experimental field - Nanchong (30°48′N, 106°04′E), Sichuan Province, China, in which a completely randomized block design was set out in 2011, was used. The basic properties of the soil were 46.98 g/kg organic matter, 0.51 g/kg total nitrogen, 11.73 mg/kg available phosphorus, 220.40 mg/kg available potassium, and 52.81 mg/kg alkali-hydrolyzed nitrogen. The studied region was 2.8 acres. For each soybean cultivar, four replicate plots, which were randomly distributed over the field, were used. The field was under standard agricultural practice. The samples were collected at the pod-setting stage (80 days after seedling emergence) of the two soybean cultivars. Rhizosphere soil samples were collected as described previously [38]. Briefly, five plants were carefully sampled using five-point sampling method and the plants with adhering soil were immediately taken to the laboratory. The loosely adhering soil on the roots was shaken off, and the resulting roots (containing rhizosphere soil) were pooled per plot. Using the pooled sample, soil tightly adhering to the roots was brushed off and collected (constituting rhizosphere soil).

Ethics statements

This study was approved by the Ministry of Agriculture of the People's Republic of China and the genetically modified organisms safety team of Nanjing Agricultural University, China. The field studies did not involve endangered or protected species. The land was not privately owned or protected in any way.

Soil DNA extraction, PCR amplification and pyrosequencing

Soil DNA was extracted from rhizosphere soil samples by employing the FastDNA SPIN Kit for Soil (MP Biomedicals, USA) as recommended by the manufacturer. DNA concentrations were quantified by using a NanoDrop 1000 Spectrophotometer (Thermo Scientific, USA) according to the manufacturer's protocol. The 16S rDNA gene (V4 region) was amplified using the barcode primers 515F [79] and 926R [80]. The primers were designed as 5′-adapter+barcode+gene specific primer-3′ (Table S8). The PCR reaction mixture (20 μL) contained 2 μL 10-fold buffer, 1.6 μL dNTP mix (2.5 mM each), 1 μL BSA (20 mg/ml), 0.25 μL rTaq polymerase (2.5 U/μL), 0.2 μL of each of the primers (10 μM), 0.5–5 μL (1–50 ng) of isolated DNA as template. The mixtures were placed in a TaKaRa PCR Thermal Cycler Dice (TaKaRa, Japan) and thermal cycling was performed as follows: initial denaturation consisting of 5 min at 95°C; followed by 30 cycles consisting of 30 sec at 94°C, 30 sec at 57°C and 30 sec at 72°C; and final extension for 10 min at 72°C. Analysis of PCR products on 1% agarose gel revealed bands of the corresponding size. These bands were cut and purified using the QIAquick Gel Extraction Kit as recommended by the manufacturer (Qiagen, Netherlands), and then purified by Agencourt AMPure XP (Beckman Coulter, USA). Quantification of the purified PCR products was performed using the Quant-iT PicoGreen dsDNA Assay Kit (Invitrogen, Netherlands) and a TBS-380 Fluorometer (Turner Biosystems, USA) as recommended by the manufacturer. The PCR products from different samples were then mixed in equal ratios (7.2 ng per sample) for pyrosequencing with a Roche Genome Sequencer GS-FLX Titanium platform [56].

Real-time PCR

The abundance of *nifH*, *amoA*, and *nosZ* genes in all samples were quantified using real-time PCR. Quantitative real-time PCR was performed according to the methods from previous studies: *nifH* (as a measure of N-fixing bacteria) used primers nifH-F (5′-AAA GGY GGW ATC GGY AAR TCC ACC AC-3′) and nifH-R (5′-TTG TTS GCS GCR TAC ATS GCC ATC AT-3′) [81]; *amoA* (as a measure of ammonia-oxidizing bacteria) used primers amoA-1F (5′-GGG GTT TCT ACT GGT GGT-3′) and amoA-2R (5′-CCC CTC KGS AAA GCC TTC TTC-3′) [82]; and *nosZ* (as a measure of denitrification bacteria) used primers nosZ-F (5′-CGY TGT TCM TCG ACA GCC AG-3′) and nosZ 1622R (5′-CGS ACC TTS TTG CCS TYG CG-3′) [83]. Purified PCR products from a common DNA mixture (equal amounts of DNA from all samples) were used to prepare sample-derived quantification standards as previously described [84]. In comparison to using a clone (plasmid) as standard, this method avoids the difference of PCR amplification efficiency between standards and samples caused by the different sequence composition in the PCR templates (single sequence in a plasmid for the standard vs. mixture of thousands of sequences in a soil sample) [83]. The copy number of gene in each sample was calculated by http://cels.uri.edu/gsc/cndna.html and the data were Log_{10} transformed [84]. The 20 μL reaction mixture contained 10 μL SYBR *Premix Ex Taq* (Tli RNaseH Plus) (2×) (TaKaRa, Japan), 0.4 μL of each primer (10 μM), 0.4 μL ROX Reference Dye II (50×) (TaKaRa, Japan), 2 μL template, and 6.8 μL dH_2O (sterile distilled water). Real-time amplification was performed in an Applied Biosystems 7500 Fast Real-Time PCR System using the following program: 95°C for 30 s; 40 cycles of 95°C for 5 s, (58°C for *nifH* gene; 57°C for *amoA* gene; 64°C for *nosZ* gene) for 31 s, 72°C for 34 s. A dissociation step was added at the end of the qPCR to assess amplification quality. The R^2 of all these standard curves were higher than 0.99. All quantitative PCR reactions were performed in triplicate.

Root box design and determination of amino acids secreted by roots

The root box used in this study was designed as reported previously [85]. There were three replicates of root box for each genotype. The lower chamber of the root box was filled with 500 mL of sterile ultrapure water, which was used to collect the amino acids exuded by the roots during a 72-h period at the pod-setting stage. Samples of root exudate were collected, filtered through a 0.45 μm membrane, freeze-dried, and redissolved in 1 mL of deionized water. Amino acids contents in the root exudates were analyzed by L-8900 Amino Acid Analyzer (Hitachi, Japan).

Chemical analysis of carbon and nitrogen contents in rhizosphere soil

The concentrations of carbon and nitrogen in rhizosphere soil were determined by Vario MICRO cube (Elementar, Germany). The detection was based on JY/T 017–1996 General rules for elemental analyzer which was approved by Ministry of Education of the People's Republic of China.

Bioinformatic analysis

All sequences generated in this study can be downloaded from NCBI Sequence Read Archive, accession number: SRP026381. On the basis of several previous reports describing sources of errors in 454 sequencing runs, the valid reads should comply with the following rules: the sequences with low quality (length< 200 bp, with ambiguous base 'N', and average base quality score< 20) were removed; the sequences that matched the primer and one of the used barcode sequences and had at least an 80% match to a previously determined 16S rRNA gene sequence were retained [54,56,83,86]. The sequences were used for chimera check using UCHIME method in Mothur [87], and chimeric sequences were removed.

The high-quality sequences were assigned to samples according to barcodes. Sequences were aligned in accordance with RDP (Ribosomal Database Project) alignment [88] and clustered into operational taxonomic units (OTUs). OTUs were defined using a threshold of 97% identity, within and between groups, which was a criterion for species level delineation in previous studies [56]. OTUs that reached 97% similarity level were used for diversity (Shannon and Simpson), richness (Chao and ACE), Good's coverage, and Rarefaction curve analysis by using Mothur [87]. The OTUs which occurred in more than three samples from either ZD or ZD91 were defined as the common OTUs [89]. Taxonomic assignments of OTUs exhibiting 97% similarity were performed by using Mothur in accordance with RDP Classifier at 80% confidence threshold. The null hypothesis was tested using OTU and weighted UniFrac distances. The study objective was to test whether the transgenic soybean had an effect on rhizosphere bacterial communities. AMOVA analysis was done by using Mothur. ADONIS differences were calculated by using R.

To compare bacterial community structures across all samples based on the relative abundance of bacterial phyla, Principal coordinate analysis (PCoA) was performed by using Qiime [90]. AMOVA analysis was used to find out whether the separation of ZD and ZD91 in the PCoA was statistically significant.

Statistical analysis

One-way analysis of variance and Duncan pair-wise comparisons ($P<0.01$) were used to determine the minimum significant difference between soybean cultivars (rhizosphere soils) by employing SPSS version 17.0 for Windows.

Supporting Information

Table S1 Pyrosequencing data and estimator index of each sample. The number of OTUs, richness estimator Chao and ACE, diversity estimator Shannon and Simpson, and Good's coverage were calculated at 3% distance.

Table S2 Bacterial composition of the communities at the phylum level. The number in the table (except the first line and column) represent pyrosequencing reads.

Table S3 Different OTUs for each sample.

Table S4 Classification of the 10 most abundant bacterial OTUs in the eight samples. Relative abundance (%, a percentage of the total sequences per sample) of each OTU is under the genus name/group number. OTUs were identified using 97% cutoffs. All Gp4 OTU classify in the same class 'Acidobacteria_Gp4' and all Gp6 OTU classify in the same class 'Acidobacteria_Gp6'.

Table S5 The ten most abundant genera found in the eight samples. Relative abundance (%, a percentage of the total sequences per sample) of each genus is included in parentheses. All Gp6 classify in the same class 'Acidobacteria_Gp6', all Gp4 classify in the same class 'Acidobacteria_Gp4' and all Gp3 classify in the same class 'Acidobacteria_Gp3'.

Table S6 Shared OTUs in the samples. (A) Shared OTUs in the ZD libraries. (B) Shared OTUs in the ZD91 libraries.

Table S7 AMOVA analysis between ZD and ZD91 in PCoA. AMOVA analysis in PCoA (classfily method: UPGMA). AMOVA analysis was used to find out whether the separation of ZD and ZD91 in the PCoA is statistically significant.

Table S8 Primers with tags and adapters used in pyrosequencing.

Acknowledgments

We are grateful to Ping Wang of LSUHSC (USA), Ruifu Zhang and Zhongli Cui of Nanjing Agricultural University (China) for their helpful suggestions.

Author Contributions

Conceived and designed the experiments: ZZ CW SS XZ. Performed the experiments: JL JJ FM. Analyzed the data: JL. Contributed reagents/materials/analysis tools: CW SS HW MZ. Wrote the paper: JL ZZ.

References

1. James C (2012) Global Status of commercialized biotech/GM crops: 2012. Ithaca: The International Service for the Acquisition of Agri-biotech Applications. 1 p.
2. Song S, Hou W, Godo I, Wu C, Yu Y, et al. (2013) Soybean seeds expressing feedback-insensitive cystathionine gamma-synthase exhibit a higher content of methionine. J Exp Bot 64: 1917–1926.
3. Inceoglu O, Al-Soud WA, Salles JF, Semenov AV, van Elsas JD (2011) Comparative analysis of bacterial communities in a potato field as determined by pyrosequencing. PLoS One 6: e23321.
4. Liu B, Zeng Q, Yan FM, Xu HG, Xu CR (2005) Effects of transgenic plants on soil microorganisms. Plant Soil 271: 1–13.
5. Milling A, Smalla K, Maidl FX, Schloter M, Munch JC (2004) Effects of transgenic potatoes with an altered starch composition on the diversity of soil and rhizosphere bacteria and fungi. Plant Soil 266: 23–39.
6. Li XG, Zhang TL, Wang XX, Hua K, Zhao L, et al. (2013) The composition of root exudates from two different resistant peanut cultivars and their effects on the growth of soil-borne pathogen. Inter J Biol Sci 9: 164–173.
7. Hur M, Kim Y, Song HR, Kim JM, Choi YI, et al. (2011) Effect of genetically modified poplars on soil microbial communities during the phytoremediation of waste mine tailings. Appl Environ Microbiol 77: 7611–7619.
8. Hannula SE, de Boer W, van Veen J (2012) A 3-year study reveals that plant growth stage, season and field site affect soil fungal communities while cultivar and GM-trait have minor effects. PLoS One 7: e33819.
9. Duc C, Nentwig W, Lindfeld A (2011) No adverse effect of genetically modified antifungal wheat on decomposition dynamics and the soil fauna community - A Field Study. PLoS One 6: e25014.
10. Meyer JB, Song-Wilson Y, Foetzki A, Luginbuhl C, Winzeler M, et al. (2013) Does wheat genetically modified for disease resistance affect root-colonizing Pseudomonads and Arbuscular Mycorrhizal Fungi? PLoS One 8: e53825.
11. Di Giovanni GD, Watrud LS, Seidler RJ, Widmer F (1999) Comparison of parental and transgenic alfalfa rhizosphere bacterial communities using Biolog GN metabolic fingerprinting and enterobacterial repetitive intergenic consensus sequence-PCR (ERIC-PCR). Microb Ecol 37: 129–139.
12. Rasche F, Hodl V, Poll C, Kandeler E, Gerzabek MH, et al. (2006) Rhizosphere bacteria affected by transgenic potatoes with antibacterial activities compared with the effects of soil, wild-type potatoes, vegetation stage and pathogen exposure. FEMS Microbiol Ecol 56: 219–235.
13. Heuer H, Kroppenstedt RM, Lottmann J, Berg G, Smalla K (2002) Effects of T4 lysozyme release from transgenic potato roots on bacterial rhizosphere communities are negligible relative to natural factors. Appl Environ Microbiol 68: 1325–1335.
14. Duke SO, Lydon J, Koskinen WC, Moorman TB, Chaney RL, et al. (2012) Glyphosate effects on plant mineral nutrition, crop rhizosphere microbiota, and plant disease in glyphosate-resistant crops. J Agric Food Chem 60: 10375–10397.
15. Demaneche S, Sanguin H, Pote J, Navarro E, Bernillon D, et al. (2008) Antibiotic-resistant soil bacteria in transgenic plant fields. Proc Natl Acad Sci U S A 105: 3957–3962.
16. Govaerts B, Mezzalama M, Unno Y, Sayre KD, Luna-Guido M, et al. (2007) Influence of tillage, residue management, and crop rotation on soil microbial biomass and catabolic diversity. Appl Soil Ecol 37: 18–30.
17. McLaughlin A, Mineau P (1995) The impact of agricultural practices on biodiversity. Agric, Ecosyst & Environt 55: 201–212.
18. Garbeva P, Postma J, Van Veen J, Van Elsas J (2006) Effect of above-ground plant species on soil microbial community structure and its impact on suppression of *Rhizoctonia solani* AG3. Environ Microbiol 8: 233–246.
19. Salles J, Van Elsas J, Van Veen J (2006) Effect of agricultural management regime on Burkholderia community structure in soil. Micro Ecol 52: 267–279.
20. Liu J, Wang G, Jin J, Liu J, Liu X (2011) Effects of different concentrations of phosphorus on microbial communities in soybean rhizosphere grown in two types of soils. Ann Microbiol 61: 525–534.
21. Xu Y, Wang G, Jin J, Liu J, Zhang Q, et al. (2009) Bacterial communities in soybean rhizosphere in response to soil type, soybean genotype, and their growth stage. Soil Biol Biochem 41: 919–925.
22. Wang G-H, Jin J, Xu M-N, Pan X-W, Tang C (2007) Inoculation with phosphate-solubilizing fungi diversifies the bacterial community in rhizospheres of maize and soybean. Pedosphere 17: 191–199.
23. Albareda M, Dardanelli MS, Sousa C, Megías M, Temprano F, et al. (2006) Factors affecting the attachment of rhizospheric bacteria to bean and soybean roots. FEMS Microbiol Lett 259: 67–73.
24. Garbeva P, van Veen JA, van Elsas JD (2004) Microbial diversity in soil: selection microbial populations by plant and soil type and implications for disease suppressiveness. Annu Rev Phytopathol 42: 243–270.
25. van Overbeek L, van Elsas JD (2008) Effects of plant genotype and growth stage on the structure of bacterial communities associated with potato (Solanum tuberosum L.). FEMS Microbiol Ecol 64: 283–296.
26. Germida JJ, Siciliano SD (2001) Taxonomic diversity of bacteria associated with the roots of modern, recent and ancient wheat cultivars. Biol Fertility Soils 33: 410–415.
27. Duineveld BM, Rosado AS, van Elsas JD, van Veen JA (1998) Analysis of the dynamics of bacterial communities in the rhizosphere of the chrysanthemum via denaturing gradient gel electrophoresis and substrate utilization patterns. Appl Environ Microbiol 64: 4950–4957.
28. Gelsomino A, Keijzer-Wolters AC, Cacco G, van Elsas JD (1999) Assessment of bacterial community structure in soil by polymerase chain reaction and denaturing gradient gel electrophoresis. J Microbiol Methods 38: 1–15.
29. Schmalenberger A, Tebbe CC (2002) Bacterial community composition in the rhizosphere of a transgenic, herbicide-resistant maize (Zea mays) and comparison to its non-transgenic cultivar Bosphore. FEMS Microbiol Ecol 40: 29–37.
30. Nannipieri P, Ascher J, Ceccherini M, Landi L, Pietramellara G, et al. (2008) Effects of root exudates in microbial diversity and activity in rhizosphere soils. Molecular mechanisms of plant and microbe coexistence: Springer. 339–365.
31. Faragova N, Faragó J, Drabekova J (2005) Evaluation of abundance of aerobic bacteria in the rhizosphere of transgenic and non-transgenic alfalfa lines. Folia Microbiol 50: 509–514.
32. da Silva KRA, Salles JF, Seldin L, van Elsas JD (2003) Application of a novel Paenibacillus-specific PCR-DGGE method and sequence analysis to assess the diversity of Paenibacillus spp. in the maize rhizosphere. J Microbiol Methods 54: 213–231.
33. Miller HJ, Henken G, Vanveen JA (1989) Variation and composition of bacterial-populations in the rhizospheres of maize, wheat, and grass cultivars. Can J Microbiol 35: 656–660.
34. Rengel Z, Ross G, Hirsch P (1998) Plant genotype and micronutrient status influence colonization of wheat roots by soil bacteria. J Plant Nutrition 21: 99–113.
35. Vasantharajan V, Bhat J (1968) Interrelations of micro-organisms and mulberry. Plant Soil 29: 156–169.
36. Pu CL, Liang JG, Gao JY, Wu CX, Zhang MR, et al. (2012) Effects of high producing methionine soy-bean transferred cystathionine γ-synthase gene on community structure of bacteria in soil. J Nanjing Agric Univ 35: 8–14.
37. Jin J, Wang GH, Liu XB, Liu JD, Chen XL, et al. (2009) Temporal and spatial dynamics of bacterial community in the rhizosphere of soybean genotypes grown in a black Soil. Pedosphere 19: 808–816.
38. Inceoglu O, Salles JF, van Overbeek L, van Elsas JD (2010) Effects of plant genotype and growth stage on the Betaproteobacterial communities associated with different potato cultivars in two fields. Applied and Environ Microbiol 76: 3675–3684.
39. Siciliano SD, Germida JJ (1999) Taxonomic diversity of bacteria associated with the roots of field-grown transgenic Brassica napus cv. Quest, compared to the non-transgenic B-napus cv. Excel and B. rapa cv. Parkland. FEMS Microbiol Ecol 29: 263–272.
40. Lukow T, Dunfield PF, Liesack W (2000) Use of the T-RFLP technique to assess spatial and temporal changes in the bacterial community structure within an agricultural soil planted with transgenic and non-transgenic potato plants. FEMS Microbiol Ecol 32: 241–247.
41. Lottmann J, Berg G (2001) Phenotypic and genotypic characterization of antagonistic bacteria associated with roots of transgenic and non-transgenic potato plants. Microbiol Res 156: 75–82.
42. Tan XY, Hurek T, Reinhold-Hurek B (2003) Effect of N-fertilization, plant genotype and environmental conditions on nifH gene pools in roots of rice. Environ Microbiol 5: 1009–1015.
43. Shen JP, Zhang LM, Zhu YG, Zhang JB, He JZ (2008) Abundance and composition of ammonia-oxidizing bacteria and ammonia-oxidizing archaea communities of an alkaline sandy loam. Environ Microbiol 10: 1601–1611.
44. Briones AM, Okabe S, Umemiya Y, Ramsing NB, Reichardt W, et al. (2002) Influence of different cultivars on populations of ammonia-oxidizing bacteria in the root environment of rice. Appl Environ Microbiol 68: 3067–3075.
45. Wang YA, Ke XB, Wu LQ, Lu YH (2009) Community composition of ammonia-oxidizing bacteria and archaea in rice field soil as affected by nitrogen fertilization. Syst Appl Microbiol 32: 27–36.
46. Cleveland CC, Townsend AR, Schimel DS, Fisher H, Howarth RW, et al. (1999) Global patterns of terrestrial biological nitrogen (N-2) fixation in natural ecosystems. Global Biogeochem Cycles 13: 623–645.
47. Hsu SF, Buckley DH (2009) Evidence for the functional significance of diazotroph community structure in soil. Isme J 3: 124–136.
48. Treusch AH, Leininger S, Kletzin A, Schuster SC, Klenk HP, et al. (2005) Novel genes for nitrite reductase and Amo-related proteins indicate a role of

uncultivated mesophilic crenarchaeota in nitrogen cycling. Environ Microbiol 7: 1985–1995.
49. Purkhold U, Pommerening-Roser A, Juretschko S, Schmid MC, Koops HP, et al. (2000) Phylogeny of all recognized species of ammonia oxidizers based on comparative 16S rRNA and amoA sequence analysis: Implications for molecular diversity surveys. Appl Environ Microbiol 66: 5368–5382.
50. Kowalchuk GA, Stephen JR (2001) Ammonia-oxidizing bacteria: A model for molecular microbial ecology. Annu Rev Microbiol 55: 485–529.
51. Philippot L, Hallin S, Schloter M (2007) Ecology of denitrifying prokaryotes in agricultural soil. Adv Agronomy 96: 249–305.
52. Zhou HW, Li DF, Tam NFY, Jiang XT, Zhang H, et al. (2011) BIPES, a cost-effective high-throughput method for assessing microbial diversity. Isme J 5: 741–749.
53. Qiu MH, Zhang RF, Xue C, Zhang SS, Li SQ, et al. (2012) Application of bio-organic fertilizer can control Fusarium wilt of cucumber plants by regulating microbial community of rhizosphere soil. Biol Fertility Soils 48: 807–816.
54. Nacke H, Thurmer A, Wollherr A, Will C, Hodac L, et al. (2011) Pyrosequencing-based assessment of bacterial community structure along different management types in German forest and grassland soils. PLoS One 6: e17000.
55. Hirsch J, Strohmeier S, Pfannkuchen M, Reineke A (2012) Assessment of bacterial endosymbiont diversity in Otiorhynchus spp. (Coleoptera: Curculionidae) larvae using a multitag 454 pyrosequencing approach. Bmc Microbiol 12: S6.
56. Zhang CH, Zhang MH, Wang SY, Han RJ, Cao YF, et al. (2010) Interactions between gut microbiota, host genetics and diet relevant to development of metabolic syndromes in mice (vol 4, pg 232, 2010). Isme J 4: 312–313.
57. Lee OO, Wang Y, Yang JK, Lafi FF, Al-Suwailem A, et al. (2011) Pyrosequencing reveals highly diverse and species-specific microbial communities in sponges from the Red Sea. Isme J 5: 650–664.
58. Yu K, Zhang T (2012) Metagenomic and metatranscriptomic analysis of microbial community dtructure and gene expression of activated sludge. PLoS One 7: e38183.
59. Zhang T, Zhang XX, Ye L (2011) Plasmid metagenome reveals high levels of antibiotic resistance genes and mobile genetic elements in activated sludge. PLoS One 6: e26041.
60. Chen WG, Liu FL, Ling ZX, Tong XJ, Xiang C (2012) Human intestinal lumen and mucosa-associated microbiota in patients with colorectal cancer. PLoS One 7: e39743.
61. Fierer N, Bradford MA, Jackson RB (2007) Toward an ecological classification of soil bacteria. Ecology 88: 1354–1364.
62. Smit E, Leeflang P, Gommans S, van den Broek J, van Mil S, et al. (2001) Diversity and seasonal fluctuations of the dominant members of the bacterial soil community in a wheat field as determined by cultivation and molecular methods. Appl Environ Microbiol 67: 2284–2291.
63. Castro HF, Classen AT, Austin EE, Norby RJ, Schadt CW (2010) Soil microbial community responses to multiple experimental climate change drivers. Appl Environ Microbiol 76: 999–1007.
64. Yashiro E, McManus PS (2012) Effect of streptomycin treatment on bacterial community structure in the apple phyllosphere. PLoS One 7: e37131.
65. Gerardi MH (2002) Nitrification and denitrification in the activated sludge process. New York: John Wiley and Sons, Inc. 43–55.
66. Sogin ML, Morrison HG, Huber JA, Welch DM, Huse SM, et al. (2006) Microbial diversity in the deep sea and the underexplored "rare biosphere". Proc Natl Acad Sci U S A 103: 12115–12120.
67. Power ME, Tilman D, Estes JA, Menge BA, Bond WJ, et al. (1996) Challenges in the quest for keystones. BioSci 46: 609–620.
68. Hooper D, Chapin Iii F, Ewel J, Hector A, Inchausti P, et al. (2005) Effects of biodiversity on ecosystem functioning: a consensus of current knowledge. Ecol Monogr 75: 3–35.
69. Xu YX, Wang GH, Jin J, Liu JD, Zhang QY, et al. (2007) Comparison of bacterial community in rhizosphere of soybean by anaylysis of 16S rDNA obtained directly from soil and culture plate. Soybean Sci 26: 907–913.
70. Hartman WH, Richardson CJ, Vilgalys R, Bruland GL (2008) Environmental and anthropogenic controls over bacterial communities in wetland soils. Proc Natl Acad Sci U S A 105: 17842–17847.
71. Dunbar J, Takala S, Barns SM, Davis JA, Kuske CR (1999) Levels of bacterial community diversity in four arid soils compared by cultivation and 16S rRNA gene cloning. Appl Environ Microbiol 65: 1662–1669.
72. Kuske CR, Barns SM, Busch JD (1997) Diverse uncultivated bacterial groups from soils of the arid southwestern United States that are present in many geographic regions. Appl Environ Microbiol 63: 3614–3621.
73. Borneman J, Skroch PW, O'Sullivan KM, Palus JA, Rumjanek NG, et al. (1996) Molecular microbial diversity of an agricultural soil in Wisconsin. Appl Environ Microbiol 62: 1935–1943.
74. McCaig AE, Glover LA, Prosser JI (1999) Molecular analysis of bacterial community structure and diversity in unimproved and improved upland grass pastures. Appl Environ Microbiol 65: 1721–1730.
75. Barns SM, Cain EC, Sommerville L, Kuske CR (2007) Acidobacteria phylum sequences in uranium-contaminated subsurface sediments greatly expand the known diversity within the phylum. Appl Environ Microbiol 73: 3113–3116.
76. Quaiser A, Ochsenreiter T, Lanz C, Schuster SC, Treusch AH, et al. (2003) Acidobacteria form a coherent but highly diverse group within the bacterial domain: evidence from environmental genomics. Mol Microbiol 50: 563–575.
77. Rappé MS, Giovannoni SJ (2003) The uncultured microbial majority. Annu Rev Microbiol 57: 369–394.
78. Eichorst SA, Breznak JA, Schmidt TM (2007) Isolation and characterization of soil bacteria that define Terriglobus gen. nov., in the phylum Acidobacteria. Appl Environ Microbiol 73: 2708–2717.
79. Turner S, Pryer KM, Miao VP, Palmer JD (1999) Investigating deep phylogenetic relationships among Cyanobacteria and plastids by small subunit rRNA sequence analysis1. J Eukaryot Microbiol 46: 327–338.
80. Liu Z, Lozupone C, Hamady M, Bushman FD, Knight R (2007) Short pyrosequencing reads suffice for accurate microbial community analysis. Nucleic Acids Res 35: e120.
81. Rösch C, Mergel A, Bothe H (2002) Biodiversity of denitrifying and dinitrogen-fixing bacteria in an acid forest soil. Appl Environ Microbiol 68: 3818–3829.
82. Rotthauwe J-H, Witzel K-P, Liesack W (1997) The ammonia monooxygenase structural gene amoA as a functional marker: molecular fine-scale analysis of natural ammonia-oxidizing populations. Appl Environ Microbiol 63: 4704–4712.
83. Mao Y, Yannarell AC, Mackie RI (2011) Changes in N-transforming archaea and bacteria in soil during the establishment of bioenergy crops. PLoS One 6: e24750.
84. Chen J, Yu Z, Michel FC, Wittum T, Morrison M (2007) Development and application of real-time PCR assays for quantification of erm genes conferring resistance to macrolides-lincosamides-streptogramin B in livestock manure and manure management systems. Appl Environ Microbiol 73: 4407–4416.
85. Yang T, Liu G, Li Y, Zhu S, Zou A, et al. (2012) Rhizosphere microbial communities and organic acids secreted by aluminum-tolerant and aluminum-sensitive soybean in acid soil. Biol Fertility Soils 48: 97–108.
86. Wu S, Wang G, Angert ER, Wang W, Li W, et al. (2012) Composition, diversity, and origin of the bacterial community in grass carp intestine. PLoS One 7: e30440.
87. Schloss PD, Westcott SL, Ryabin T, Hall JR, Hartmann M, et al. (2009) Introducing mothur: open-source, platform-independent, community-supported software for describing and comparing microbial communities. Appl Environ Microbiol 75: 7537–7541.
88. Wang Q, Garrity GM, Tiedje JM, Cole JR (2007) Naive Bayesian classifier for rapid assignment of rRNA sequences into the new bacterial taxonomy. Appl Environ Microbiol 73: 5261–5267.
89. Yang W, Zheng Y, Gao C, He X, Ding Q, et al. (2013) The arbuscular mycorrhizal fungal community response to warming and grazing differs between soil and roots on the Qinghai-Tibetan Plateau. PLoS One 8: e76447.
90. Caporaso JG, Kuczynski J, Stombaugh J, Bittinger K, Bushman FD, et al. (2010) QIIME allows analysis of high-throughput community sequencing data. Nat methods 7: 335–336.

20

Microbial Community Structure of Relict Niter-Beds Previously Used for Saltpeter Production

Takashi Narihiro[1]*, Hideyuki Tamaki[1], Aya Akiba[1], Kazuto Takasaki[2], Koichiro Nakano[2], Yoichi Kamagata[1,3], Satoshi Hanada[1], Taizo Maji[4]

1 Bioproduction Research Institute, National Institute of Advanced Industrial Science and Technology (AIST), Tsukuba, Ibaraki, Japan, **2** FASMAC Co., Ltd., Atsugi, Kanagawa, Japan, **3** Bioproduction Research Institute, National Institute of Advanced Industrial Science and Technology (AIST), Sapporo, Hokkaido, Japan, **4** Shubun University, Ichinomiya, Aichi, Japan

Abstract

From the 16th to the 18th centuries in Japan, saltpeter was produced using a biological niter-bed process and was formed under the floor of gassho-style houses in the historic villages of Shirakawa-go and Gokayama, which are classified as United Nations Educational, Scientific and Cultural Organization (UNESCO) World Heritage Sites. The relict niter-beds are now conserved in the underfloor space of gassho-style houses, where they are isolated from destabilizing environmental factors and retain the ability to produce nitrate. However, little is known about the nitrifying microbes in such relict niter-bed ecosystems. In this study, the microbial community structures within nine relict niter-bed soils were investigated using 454 pyrotag analysis targeting the 16S rRNA gene and the bacterial and archaeal ammonia monooxygenase gene (*amoA*). The 16S rRNA gene pyrotag analysis showed that members of the phyla *Proteobacteria*, *Actinobacteria*, *Bacteroidetes*, *Chloroflexi*, *Firmicutes*, *Gemmatimonadetes*, and *Planctomycetes* were major microbial constituents, and principal coordinate analysis showed that the NO_3^-, Cl^-, K^+, and Na^+ contents were potential determinants of the structures of entire microbial communities in relict niter-bed soils. The bacterial and archaeal *amoA* libraries indicated that members of the *Nitrosospira*-type ammonia-oxidizing bacteria (AOB) and "*Ca.* Nitrososphaera"-type ammonia-oxidizing archaea (AOA), respectively, predominated in relict niter-bed soils. In addition, soil pH and organic carbon content were important factors for the ecological niche of AOB and AOA in relict niter-bed soil ecosystems.

Editor: A. Mark Ibekwe, U. S. Salinity Lab, United States of America

Funding: This research was supported by a Grant-in-Aid for Young Scientists (B) (Grant Number 23760512) from JSPS KAKENHI. http://www.jsps.go.jp/english/e-grants/. The funders had no role in study design, data collection and analysis, decision to publish, or preparation of the manuscript.

Competing Interests: TN, HT, AA, YK, SH and TM have neither financial nor occupational affiliations with FASMAC Co., Ltd. at the time of the study. KT and KN were employees of FASMAC Co., Ltd. at the time of the study. There are no patents, products in development or marketed products to declare.

* Email: t.narihiro@aist.go.jp

Introduction

Until the early 20th century, saltpeter (i.e., niter or potassium nitrate [KNO_3]), which is known as the main source of gunpowder, was produced via an artificial bioprocess, the so-called "niter-bed" [1,2]. The basic concept of saltpeter production via niter-bed involves the nitrification of ammonia-containing wastes (e.g., plant residue, horse manure, and human urine) mixed with a certain amount of soil [1,2]. This nitrification reaction was thought to be performed by ammonia-oxidizing and nitrite-oxidizing microbes [1,2]. The first description of a niter-bed was found in the military technological manual "Bellifortis," which was written by Konrad Kyeser and published in 1405. This manual contains brief instructions describing how to construct a niter-bed [3,4], and this process was employed for saltpeter production in European countries [1,5,6], India [7,8], and Japan [9,10] until the discovery of a large natural deposit in Chile [11]. In Finland, the export of saltpeter produced by niter-bed reached a maximum level (433 tons) in 1816 during the last stages of the Napoleonic wars [6]. In India, a large amount of saltpeter was produced annually (20,000 tons/year) using the niter-bed methodology in the early 20th century [7]. Thus, the niter-bed process is a historically important industry in several nations.

From the 16th to the 18th centuries in Japan, saltpeter was produced for military purposes using the niter-bed formed under the floor of gassho-style houses [10], which is a specific architectural style used in the historic villages of Shirakawa-go and Gokayama, which are listed as UNESCO World Heritage Sites [12]. Because these villages were located in the mountainous area of the former Kaga domain in Japan (currently the Toyama and Ishikawa prefectures), there was limited land for the production of agricultural crops. Hence, people in this area faced economic difficulties. To overcome this issue, the people practiced sericulture to produce silk as an alternative income source in the loft spaces of gassho-style houses [10]. Moreover, according to the historical records, the people produced saltpeter in the underfloor space of gassho-style houses using ammonia-containing wastes such as plant residues (e.g., the stems of Japanese millet, buckwheat, and tobacco), silkworm feces from sericulture, and human urine as primary substrates [10]. These substrates were mixed with soil under the floor of the gassho-style house to create a niter-bed, and nitrate (NO_3) was produced via a microbial nitrification reaction. The resulting nitrate contained in the niter-bed soil was further extracted with water and purified by adding wood ash to remove unnecessary minerals (e.g., Ca, Mg, K, and Na). Purified saltpeter was used as the raw material for gunpowder production, and this gunpowder also became an important source of income for people who lived in this area [9,10]. Thus, saltpeter production via the niter-bed process was a cornerstone of the lifestyle in this area.

Although saltpeter production in the villages of Shirakawa-go and Gokayama ended more than 100 years ago, the relict niter-beds are now conserved in the underfloor space of gassho-style houses, where they are isolated from rainfall and other destabilizing factors [10]. Surprisingly, bulk soils collected from relict niter-beds in Shirakawa-go village retained nitrification activity [13]. This finding implies that nitrifying microbes have persisted in the relict niter-bed soils; however, there is no information available regarding the microbial community structure responsible for nitrification in the relict niter-bed ecosystem. In this study, we investigated the microbial community structure of nine soil samples obtained from the underfloor areas of three gassho-style houses by using 454 pyrotag analysis targeting the 16S rRNA gene and the ammonium monooxygenase subunit A gene (*amoA*) of ammonia-oxidizing bacteria (AOB) and ammonia-oxidizing archaea (AOA). In addition, a previously reported *amoA* pyrotag dataset [14] was used to elucidate the ecological niche of AOB and AOA in soil environments.

Materials and Methods

Soil sample

A total of nine soil samples were collected from three relict niter-beds under the floors of gassho-style houses located in Shirakawa Village, referred to as OVA, OVB, and OVC. We collected five, two, and two soil samples from the OVA, OVB, and OVC houses; these samples are referred to as OVA1–5, OVB2 and OVB3, and OVC1 and OVC2, respectively. Each 500–1,000 g of soil sample was collected from relict niter-beds (for OVA and OVC, a depth of ca. 10 cm from the surface; for OVB, a depth of ca. 20 cm from the surface) by permission from the education board of Shirakawa-Village, Gifu, Japan. Detailed physicochemical parameters, including pH, organic carbon content, and ion contents, are summarized in Table S1 [13,15–17]. Pebbles were removed from the samples using a 30-mesh filter, and the samples were subjected to DNA extraction.

DNA extraction, PCR, and pyrosequencing

DNA was extracted from the soil samples using the FastDNA SPIN Kit for Soil (MP Biomedicals, Carlsbad, CA, USA) according to the manufacturer's instructions. Approximately 50 ng of template DNA was used for PCR amplification in a reaction volume of 50 μL. The 16S rRNA, bacterial *amoA*, and archaeal *amoA* genes were amplified with the forward/reverse primer sets Univ519F/Univ926R [18], amoA-1F/amoA-2R [19], and Arch-amoAF/Arch-amoAR [20], respectively. The forward and reverse primers were barcoded with fusion A (FA; 5′-CGTATCGCCTCCCTCGCGCCATCAG-3′) and fusion B (FB; 5′-CTATGCGCCTTGCCAGCCCGCTCAG-3′) adaptor sequences, respectively. The primers also included Multiplex Identifier (MID) sequences. Detailed information regarding the primer sequences is provided in Table S2. These genes were amplified with PrimeSTAR HS DNA Polymerase (TaKaRa, Otsu, Japan) using the following thermocycling conditions: the appro-

Table 1. Pyrosequencing results of 16S rRNA, bacterial *amoA*, and archaeal *amoA* genes and observed diversity indexes.

Sample ID	16S rRNA						Bacterial *amoA*						Archaeal *amoA*					
	Total reads	OTU	Chao 1	Shannon	Simpson	Coverage (%)	Total reads	OTU	Chao 1	Shannon	Simpson	Coverage (%)	Total reads	OTU	Chao 1	Shannon	Simpson	Coverage (%)
OVA1	7,770	514	757.8	4.57	37.42	97.52	N.D.[a]						4,240	6	7.00	1.16	1.06	99.95
OVA2	17,480	641	1,042.8	4.63	47.06	98.59	N.D.[a]						3,455	9	10.00	1.70	1.34	99.91
OVA3	16,615	990	1,485.7	4.92	33.83	97.80	7,826	6	6.00	0.07	1.02	99.99	3,872	6	7.00	2.18	1.83	99.95
OVA4	19,266	1,496	2,248.9	5.60	84.90	96.99	10,197	5	5.00	0.15	1.07	99.99	2,705	9	10.50	2.37	1.87	99.89
OVA5	18,029	1,360	2,124.7	5.50	77.00	97.06	14,558	6	6.00	0.22	1.10	99.99	3,670	8	8.33	2.02	1.72	99.95
OVB2	19,304	1,020	1,411.7	5.06	44.17	98.23	4,606	6	6.00	0.32	1.19	99.98	2,957	4	4.00	2.02	1.96	100.00
OVB3	22,764	1,028	1,538.0	4.99	36.75	98.42	6,896	9	10.50	0.32	1.17	99.96	3,695	8	8.33	2.13	1.94	99.95
OVC1	23,172	1,144	1,816.1	4.56	17.29	98.09	8,016	3	3.00	0.40	1.23	100.00	3,550	10	10.00	2.13	1.55	99.97
OVC2	28,616	1,238	1,824.5	4.66	20.59	98.42	7,305	5	5.00	0.75	1.89	100.00	4,943	12	18.00	1.85	1.48	99.92
Total	173,016	3,146					59,404	13					33,087	16				

[a]N.D., not detected.

priate number of cycles (25 cycles for 16S rRNA, 40 cycles for bacterial *amoA*, and 35 cycles for archaeal *amoA* genes) of denaturation at 98°C for 10 sec, annealing at 55°C for 15 sec, and extension at 72°C for 1 min. PCR amplicons were purified using a Wizard SV Gel and PCR Clean-Up System (Promega, Fitchburg, WI, USA) and pooled for subsequent 454 pyrotag analysis. Pyrosequencing was performed using the GS-FLX Titanium platform (Roche/454 Life Sciences, Branford, CT, USA) at the FASMAC Co., Ltd. (Atugi, Japan). Sequencing was performed using 1/2-region gaskets. For the 16S rRNA gene, we used pyrotag reads obtained from both fusion sides (A and B) because the 454 platform used in this study generated full-length amplicons (approximately 370 bp). For bacterial and archaeal *amoA* genes, we used pyrotag reads obtained from only the fusion A side because the length of each sequenced read did not cover the entire region of the amplicon.

Operational taxonomic unit (OTU) analysis for 16S rRNA gene pyrotags

Raw 16S rRNA gene pyrotag sequences were qualitatively screened and trimmed with QIIME 1.7.0 [21], maintaining the sequence length at ≥300 nt. The sequences had an average quality score of ≥25. A denoising function was applied to the filtered sequence data [22]. The denoised sequence data were grouped into OTUs with the UCLUST algorithm [23]. The sequences exhibiting >97% identities were grouped into one OTU. The representative sequences of each OTU were aligned with PyNAST [24], and chimeric sequences were detected and removed using ChimeraSlayer [25]. The taxonomic assignment of each OTU was performed with the RDP classifier trained by a dataset from Greengenes at a minimum confidence level of 0.8 [26]. The sequence data were compared with the 16S rRNA database of RDP release 10 [27] using the BLAST+ program [28].

OTU analysis for *amoA* pyrotags

The procedures for quality screening, trimming, denoising, and OTU selection were same as those described above for 16S rRNA gene pyrotag analysis, except that sequence identity (>85%) was determined according to the threshold of Pester *et al.* [14]. The filtered sequences were manually checked after aligning with MEGA program [29]. Chimeric sequences were detected and removed using the UCHIME algorithm [30] on the FunGene platform [27]. Sequence data were compared with the GenBank database using the BLAST+ program [28].

Statistical and phylogenetic analysis

After taxonomy assignments, α-diversity indices (e.g., Chao1, Shannon, and Simpson) of the 16S rRNA and *amoA* genes pyrotag libraries were calculated by QIIME [21] and EstimateS (version 9.1.0) [31] software, respectively. The coverage values were calculated using equation [1− (n/N)], where n is the number of OTUs in a single read (singleton) and N is the total number of reads analyzed [32]. The weighted UniFrac distances were calculated and used for principal coordinate analysis (PCoA) [33]. Phylogenetic trees for the bacterial and archaeal *amoA* libraries for Fast UniFrac analysis were constructed with the

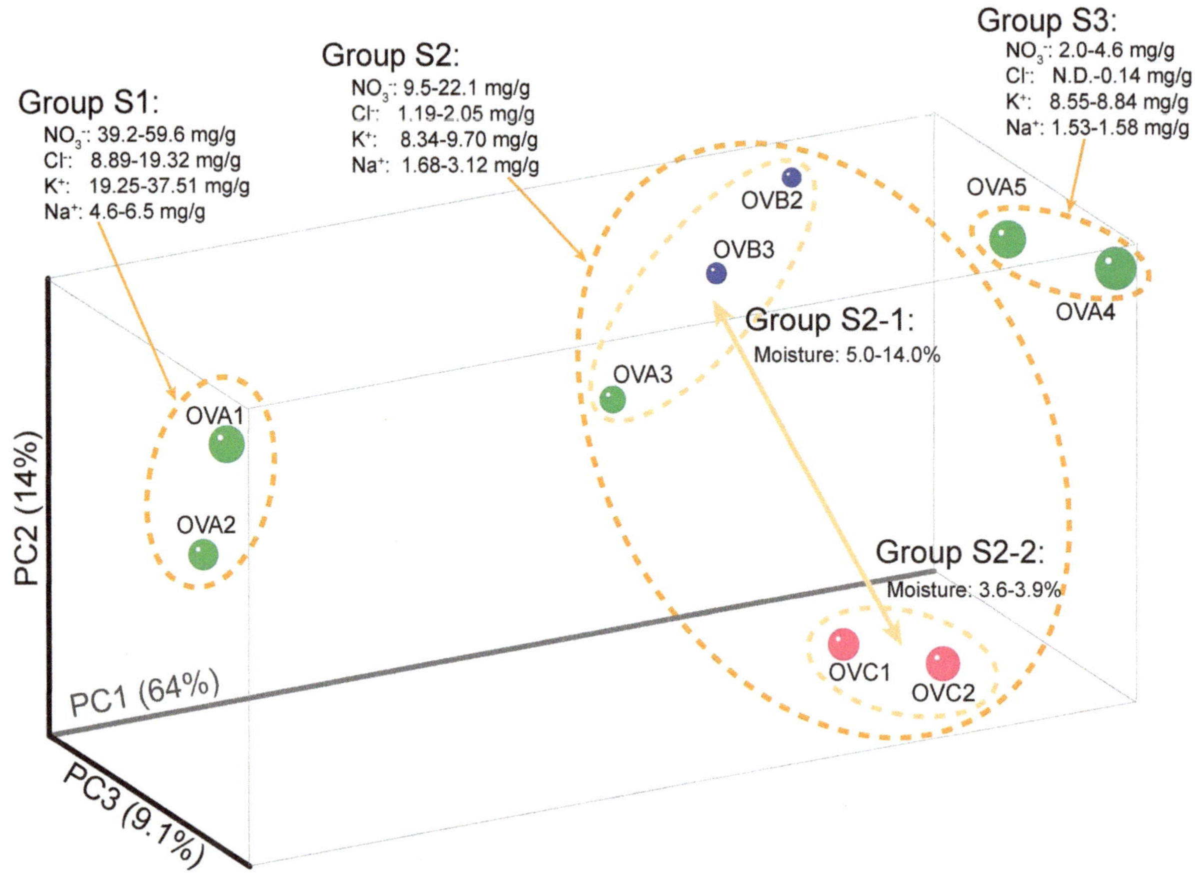

Figure 1. Principal coordinate analysis based on the abundances of 16S rRNA gene OTUs (weighted UniFrac) and separating soils according to their ion contents. For this analysis, observed 16S rRNA gene OTUs were normalized to 7,770 reads per soil.

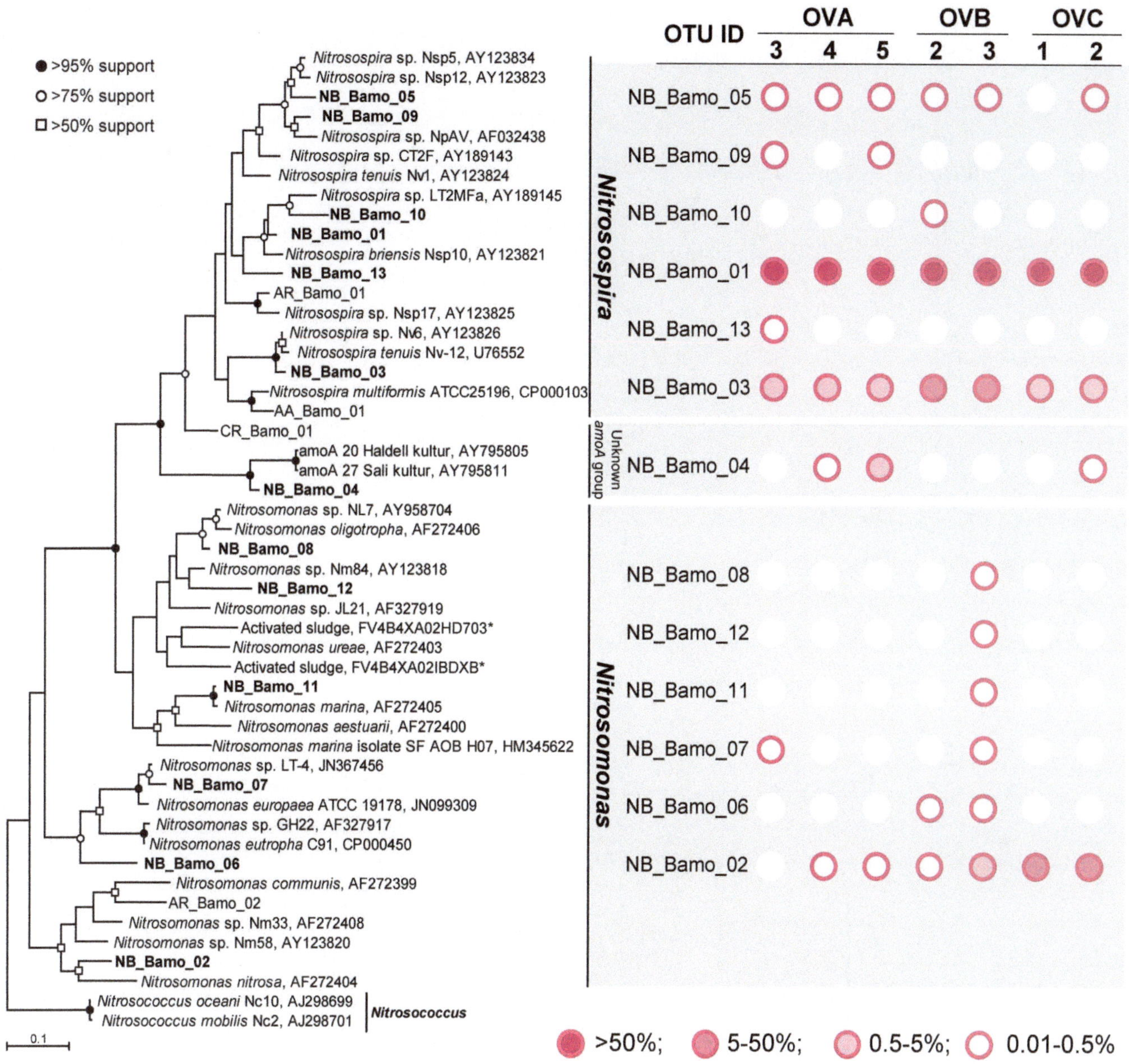

Figure 2. Distance matrix tree of bacterial *amoA* gene sequences retrieved from niter-bed soils based on the maximum likelihood method. Boldface indicates the sequences obtained in this study. The *Nitrosococcus amoA* gene sequences (AJ298699 and AJ298701) were used as outgroups. The bar indicates 10% base substitution. Branching points supported probabilities >95%, >75%, and >50% by bootstrap analyses (based on 1,000 replicates, estimated using the maximum likelihood method) are indicated by solid circle, open circles, and open square, respectively.

MEGA program [29] based on the neighbor-joining algorithm [34]. PCoA and jackknife clustering analyses were performed on weighted and normalized data with QIIME and the Fast UniFrac program [33], respectively. To identify OTUs that exhibited significant differences in abundance between different soil samples, Welch's *t*-test (confidence interval method: Welch's inverted, [$p < 0.05$] for two groups [definition of each group was described in the Results and Discussion]) having possibly unequal variances was performed with STAMP software [35,36].

The distance matrix tree of the bacterial and archaeal *amoA* genes, which is based on the maximum likelihood method, was constructed with the MEGA program [29]. The topology of the trees was estimated by 1,000 bootstrap replicates [37]. Previously reported *amoA* pyrotag sequences retrieved from soil environments in Austria, Costa Rica, Greenland, and Namibia were downloaded from the Sequence Read Archive (SRA) at NCBI (accession no. SRA047303) and used for the statistical and phylogenetic comparisons [14]. Subclusters within the "*Ca.* Nitrososphaera" and "*Ca.* Nitrosopumilus" were defined according to Pester *et al.* [14].

Nucleotide sequence accession numbers

The pyrosequence data obtained in this study have been deposited under DDBJ/EMBL/GenBank accession no. DRA001202.

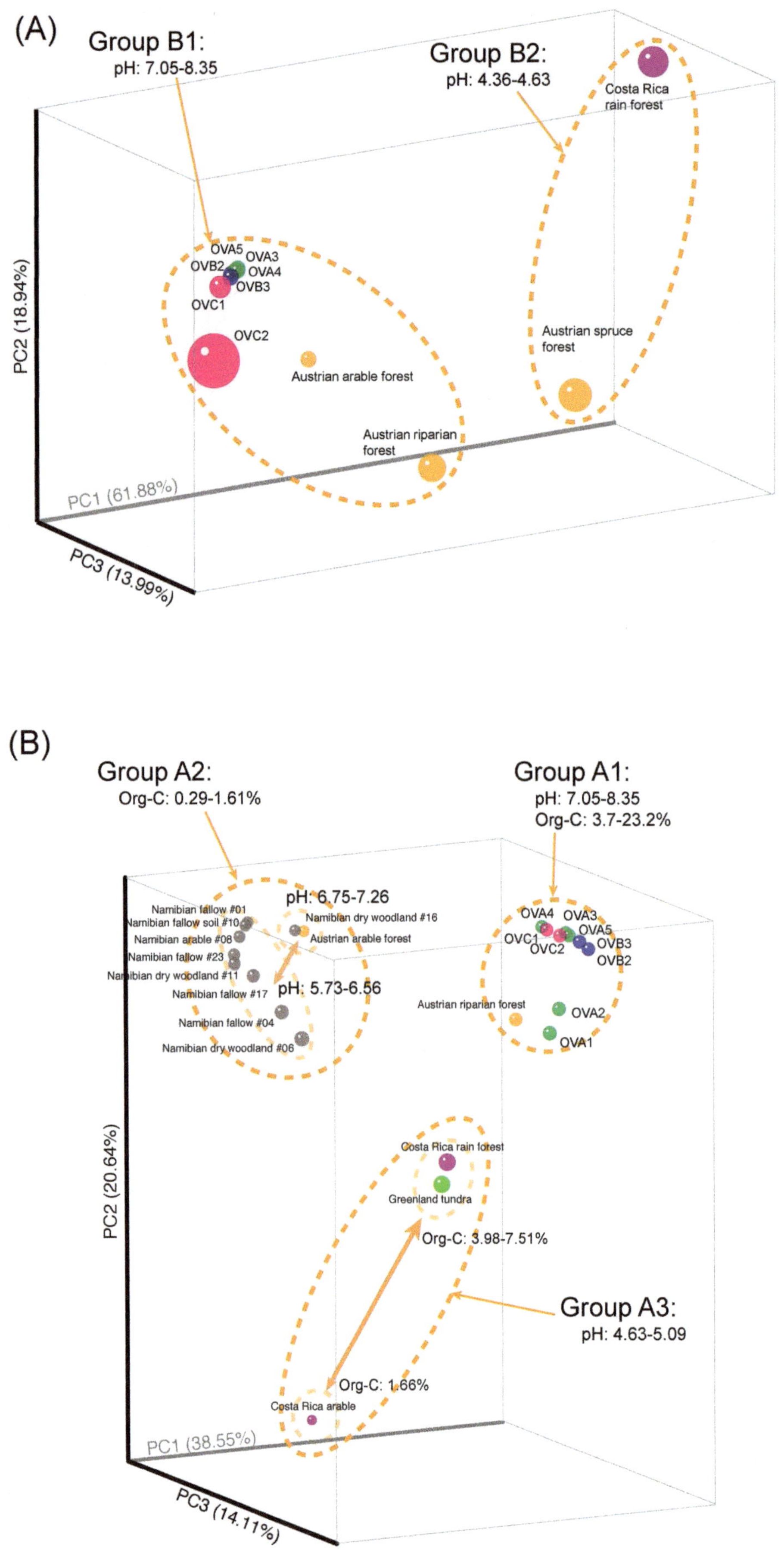
(A)
Group B1:
pH: 7.05-8.35
Group B2:
pH: 4.36-4.63
Costa Rica rain forest
OVA5
OVA3
OVB2
OVA4
OVB3
OVC1
OVC2
Austrian arable forest
Austrian spruce forest
Austrian riparian forest
PC2 (18.94%)
PC1 (61.88%)
PC3 (13.99%)
(B)
Group A2:
Org-C: 0.29-1.61%
Group A1:
pH: 7.05-8.35
Org-C: 3.7-23.2%
pH: 6.75-7.26
Namibian fallow #01
Namibian fallow soil #10
Namibian arable #08
Namibian fallow #23
Namibian dry woodland #11
Namibian fallow #17
Namibian fallow #04
Namibian dry woodland #06
Namibian dry woodland #16
Austrian arable forest
pH: 5.73-6.56
OVA4
OVA3
OVC1
OVA5
OVC2
OVB3
OVB2
Austrian riparian forest
OVA2
OVA1
Costa Rica rain forest
Greenland tundra
Org-C: 3.98-7.51%
Group A3:
pH: 4.63-5.09
Org-C: 1.66%
Costa Rica arable
PC2 (20.64%)
PC1 (38.55%)
PC3 (14.11%)

Figure 3. Principal coordinate analysis based on the abundances of (A) bacterial and (B) archaeal *amoA* gene OTUs (weighted UniFrac) and separating soils according to soil pH and organic carbon contents. For this analysis, *amoA* pyrotag libraries reported by Pester et al. [14] were used for the comparison, and bacterial and archaeal *amoA* OTUs observed here were normalized to 2,320 and 2,130 reads per soil, respectively.

Results and Discussion

16S rRNA gene pyrotag analysis

In this study, nine soil samples from three gassho-style houses, referred to as OVA (samples OVA1–5), OVB (samples OVB2 and OVB3), and OVC (samples OVC1 and OVC2), were used for microbial community analysis using 454 pyrosequencing. We retrieved a total of 173,016 16S rRNA gene pyrotags and found 3,146 OTUs with the criterion of >97% sequence identity (Table 1). Although the rarefaction curves showed that the accumulation of the 16S rRNA gene OTUs was insufficient to achieve the goal of saturation (Fig. S1A), the coverage values were calculated to be 96.99–98.59%, suggesting that the OTUs retrieved here were sufficient for estimating the microbial diversity in the relict niter-bed ecosystems. According to the Chao1 nonparametric estimators, the soil samples contained approximately 1.38–1.63-fold more OTUs than what was detected in our analyses.

The OTUs belonging to the phyla *Proteobacteria* (especially the classes *Alphaproteobacteria* and *Gammaproteobacteria*), *Actinobacteria*, *Bacteroidetes*, *Chloroflexi*, *Firmicutes*, *Gemmatimonadetes*, and *Planctomycetes* predominated and were detected in all samples (Table S3 and S4). Such microbial taxa are frequently found in soil environments [38]. In each sample, only a few number of the 16S rRNA gene pyrotags (<0.8% of the total populations) were associated with previously known ammonia-oxidizing microorganisms at the genus level: for AOB, three OTUs (nos. 1267, 2435, and 2701) of the *Nitrosococcus*, one OTU (no. 2080) of the *Nitrosomonas*, and one OTU (no. 673) of the *Nitrosospira*; for AOA, four OTUs (nos. 1207, 1897, 1946, and 3119) of the "*Ca.* Nitrososphaera" (Table S5). Given that soil samples from the relict niter-beds have retained microbial nitrification activity [13], these

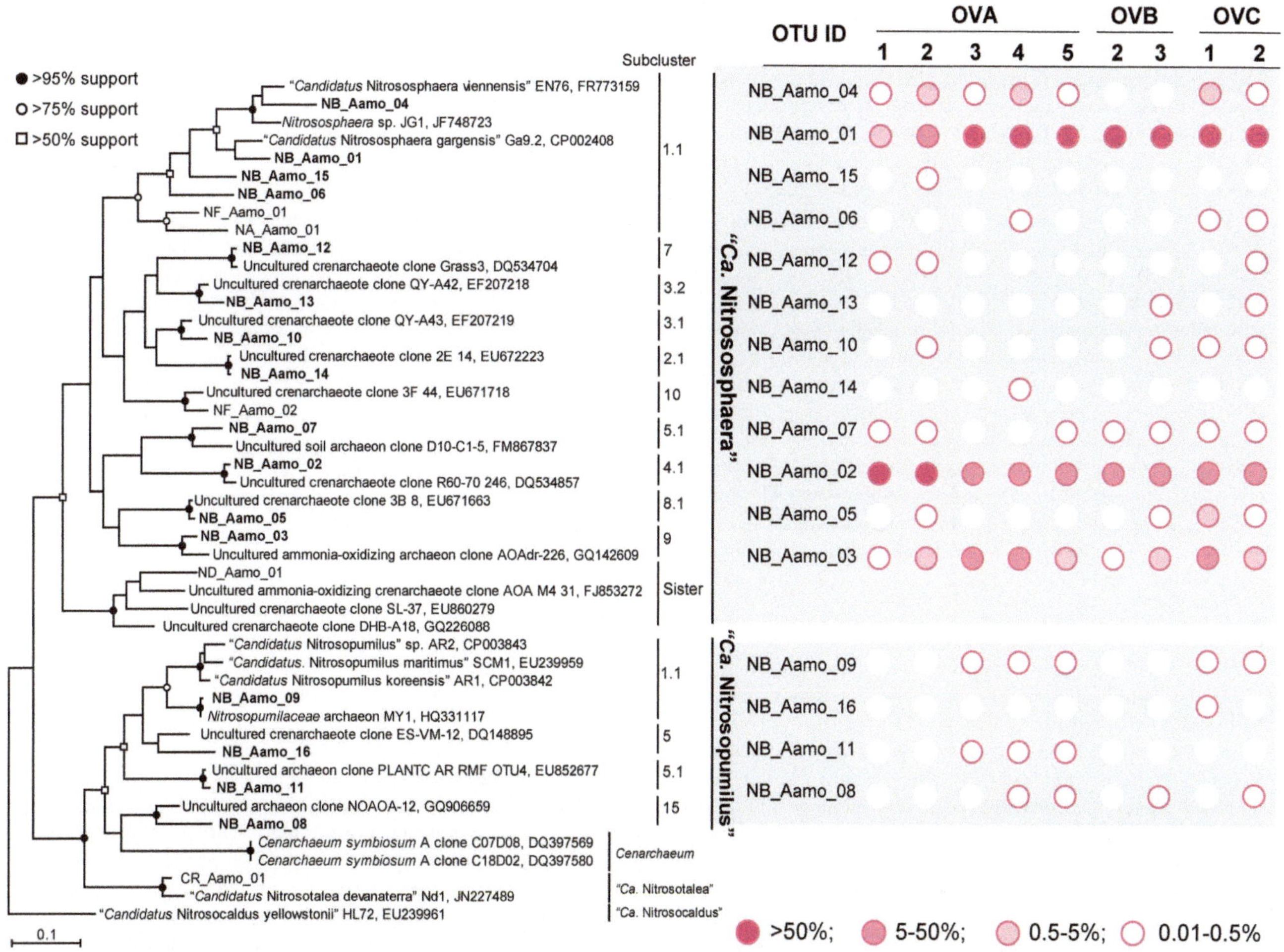

Figure 4. Distance matrix tree of archaeal *amoA* gene sequences retrieved from niter-bed soils based on the maximum likelihood method. Boldface indicates the sequences obtained in this study. The *amoA* gene sequence of the "*Ca.* Nitrosocaldus yellowstonii" strain HL72 (EU239961) was used as outgroup. The bar indicates 10% base substitution. Branching points supported probabilities >95%, >75%, and >50% by bootstrap analyses (based on 1,000 replicates, estimated using the maximum likelihood method) are indicated by solid circle, open circles, and open square, respectively.

nitrifying microbes may contribute to the nitrification at low population densities; alternatively, there may have been unidentified nitrifying microbes in the relict niter-bed soils. Although the relative abundances of nitrifying microbes were in good agreement with the microbial compositions of several types of soil ecosystems [38], we could not fully elucidate the biodiversity and ecological niche of ammonia-oxidizing microbes in the relict niter-bed ecosystems based on the results of 16S rRNA gene pyrotag analysis (e.g., there are no significant relationship between the ratio of the abundance ratio of AOB to AOA and clustering of the communities in PCoA and jackknife analysis described below) (Table S5). These observations emphasize the need to implement further analysis targeting *amoA* gene that enable specific detection of ammonia-oxidizing microbes.

The similarity of microbial compositions among the nine soil samples was compared on the basis of 16S rRNA gene pyrotag sequences. According to the PCoA and jackknife analyses, the microbial community structures of the relict niter-bed soil were classified into three groups (Fig. 1 and S2A). Group S1, consisting of OVA1 and OVA2, was distinct from the other two groups. Based on previously reported physicochemical characterization [15–17], relatively high contents of NO_3^-, Cl^-, K^+, and Na^+ were found in OVA1 and OVA2 compared with the samples clustered into groups S2 and S3 (Table S1). Because the soils in group S1 were not subjected to the extraction step of saltpeter production, the ion contents were higher than those in the other relict niter-bed soil samples [17]. In addition, the NO_3^- and Cl^- contents were relatively low in the soils clustered in group S3 compared with those in group S2. Based on Welch's *t*-test, we found a total of 21 OTUs that were significantly different in relative abundances among the groups and were associated with dominant taxa (e.g., *Alphaproteobacteria*, *Gammaproteobacteria*, *Actinobacteria*, *Bacteroidetes*, *Chloroflexi*, *Firmicutes*, and *Gemmatimonadetes*) (Fig. S3). Among them, *Gammaproteobacteria*-type OTU (NB_16S_507) and *Chloroflexi*-type OTU (NB_16S_1415) were involved in the differentiation of groups S1, S2, and S3. The samples in group S2 were further separated according to their moisture contents (groups S2-1 and S2-2), and the relative abundances of six OTUs (NB_16S_1093, 1144, 1409, 1415, 2046, and 2058) were significantly affected by the moisture contents (Fig. S3). Several types of ion species and moisture contents have been recognized as potential factors in the determination of soil microbial constituents [39–42]. Thus, the NO_3^-, Cl^-, K^+, Na^+, and moisture contents may affect the microbial community structures of relict niter-bed ecosystems.

Bacterial *amoA* gene pyrotag analysis

As described above, the proportions of 16S rRNA gene pyrotags associated with AOB and AOA were very low in the relict niter-bed soils. To further elucidate the microbial constituents responsible for ammonia oxidation, we performed pyrotag analysis targeting bacterial and archaeal *amoA* genes. As a result, we obtained a total of 59,404 bacterial *amoA* pyrotags and found 13 OTUs in relict niter-bed soils (Table 1). However, the bacterial *amoA* gene fragment could not be amplified from OVA1 and OVA2 using the primer set described in this study, suggesting that there were either low levels of AOB in the samples or there were some AOB with significantly different types of *amoA* genes. Instead, the archaeal *amoA* gene fragments were successfully amplified from OVA1 and OVA2 samples, suggesting that AOA may be a major player in ammonia oxidation in these soils (details described below). Although the rarefaction curves for the bacterial *amoA* libraries shorted to reach a plateau (except for OVC1 and OVC2), the coverage values were calculated to be 99.96–100%, indicating that the OTUs retrieved here were sufficient for estimating the biodiversity of AOB in seven relict niter-bed soils (Table 1 and Fig. S1B).

The most abundant AOB belonged to the genus *Nitrosospira* (6 OTUs, 68.13–99.96% of the total reads in each sample), and *Nitrosospira briensis*-type OTU (NB_Bamo_01) predominated in all samples (except for OVA1 and OVA2, in which there were no bacterial *amoA* gene amplification) (Table S6). This result is in good accordance with 16S rRNA gene pyrotag analysis that found a *Nitrosospira briensis*-type OTU (no. 673) across the samples. An OTU (NB_Bamo_01) related to the *Nitrosospira briensis* strain Nsp10 (95.5% *amoA* gene sequence identity) was detected as a major AOB population in all niter-bed soils (Fig. 2 and Table S6). *Nitrosospira briensis* are enriched in and have been isolated from soil samples taken from sandy and agriculturally uncultivated areas of Europe [43]. To the best of our knowledge, only two reports have described the clear appearance of *Nitrosospira briensis* populations: water columns of the Chesapeake Bay [44] and soil samples of the Cascade Mountains in Oregon [45] in the U.S. Interestingly, Bollmann *et al.* [46] reported that under ammonia-starved conditions, the cells of *Nitrosospira briensis* had the ability to immediately recover ammonia oxidation activity upon exposure to ammonia-rich conditions. Although the mechanism behind this phenomenon is unclear, *Nitrosospira briensis*-type AOB may reserve their activity in ammonia-limited environments, such as the relict niter-bed soils. Within the genus *Nitrosospira*, an OTU (NB_Bamo_03) that is closely related to the *Nitrosospira tenuis* strain Nv-12 (96.2% sequence identity) was also detected in all bacterial *amoA* gene pytotag libraries; however, its ecophysiological traits (other than ammonia oxidation) remain unclear.

In addition to the *Nitrosospira* populations, six *Nitrosomonas*-type OTUs were found in seven niter-bed soils (Fig. 2 and Table S6). In contrast to the *Nitrosospira*-type OTUs, 16S pyrotags related to *Nitrosomonas* were found in OVA3 and OVA4; however, the proportions were quite low (0.005–0.006% of total 16S rRNA gene pyrotags). This finding may also support the utility of *amoA*-based pyrotag analysis for relict niter-bed soils. An OTU (NB_Bamo_02) associated with *Nitrosomonas nitrosa* (87.2% sequence identity) was observed with significant abundance in OVC1 and OVC2 (5.6% and 31.8%, respectively) (Table S6). According to previous reports, *Nitrosomonas nitrosa*-type AOB have frequently been found in estuary soils [47], nitrification bioreactors [48,49], and freshwater ecosystems [50]. In addition to these taxa, an OTU (NB_Bamo_04) related to the unidentified *amoA* gene cluster consisting of clones retrieved from potash-polluted marsh soil [51] was found in three relict niter-bed soils (Fig. 2 and Table S6). This cluster has a significantly low identity with previously reported bacterial *amoA* gene sequences (e.g., NB_BamoA_04 had 77.4% identity with *Nitrosospira multiformis* strain C-71, accession no. X90822), suggesting that phylogenetically novel AOB may play a role in ammonia oxidation in some niter-bed ecosystems.

The similarity of AOB compositions in relict niter-bed soils was compared with previously reported *amoA* gene-based microbial community analysis [14]. PCoA and jackknife analyses showed that soil bacterial *amoA* libraries were separated into two groups according to their pH: seven niter-bed soils were clustered into group B1 with two Austrian arable and riparian forest soils (pH 7.05–8.35), and the remaining two soils, from an Austrian spruce forest and a Costa-Rican rain forest, were clustered into group B2 (pH 4.36–4.63) (Fig. 3A). pH is generally to be a key parameter in defining the abundance of AOB populations in soil ecosystems [52,53] and bioreactors [54,55]. The niter-bed soils in group B1 were further divided into two subgroups (Fig. S2B and S3A);

however, there was no clear explanation for the differentiation between OVC2 and the other soils based on the physicochemical parameters analyzed herein. In short, soil pH is one of the important factors affecting the AOB populations in relict niter-bed ecosystems.

Archaeal *amoA* gene pyrotag analysis

A total of 33,087 archaeal *amoA* gene sequences were retrieved from nine relict niter-bed soils, and we found 16 OTUs with the criterion of >85% sequence identity. The archaeal *amoA* diversity was almost fully covered according to the rarefaction curves, coverage values, and Chao1 diversity index (Table 1 and Fig. S1C).

Members of the genus "*Ca.* Nitrososphaera" were detected as major AOA populations (12 OTUs, 99.23–100% of the total reads in each sample), whereas those of "*Ca.* Nitrosopumilus" were absent or few (<0.28% of the total reads in each sample) in all samples tested (Fig. 4 and Table S6). Based on the 16S rRNA gene pyrotag analysis, only a few OTUs (accounting for 0.16–1.1% of the total prokaryotic population) associated with "*Ca.* Nitrososphaera" were identified as AOA populations in relict niter-bed soils (Table S5). Thus, the results from *amoA*-based pyrotag analysis expanded the biodiversity of AOA in relict niter-bed soils. The most abundant OTU in OVA1 and OVA2 soils was NB_Aamo_02, which was categorized into subcluster 4.1 of "*Ca.* Nitrososphaera" and was distantly related to previously characterized AOA (77.8% sequence identity with "*Ca.* Nitrososphaera viennensis" strain EN76, accession no. FR773159), whereas the most abundant OTU in the other seven samples (NB_Aamo_01) was closely related to "*Ca.* Nitrososphaera gargensis" strain Ga9.2 (90.6% sequence identity, accession no. CP002408) and categorized into subcluster 1.1. This marked difference in a major AOA population within OVA soils was clearly associated with the maximum nitrification rate of bulk soil (Table S1) [13]. This finding also implies that the nitrification activity and/or affinity for ammonia may differ between subclusters 1.1 and 1.4. The *amoA* fragments associated with "*Ca.* Nitrososphaera" subcluster 1.4 have been observed in diverse environments, including soil [20,56,57], salt marsh sediment [58], and marine sediment [20,59]. In addition to the major AOA, several subclusters of "*Ca.* Nitrososphaera" were detected as minor AOA populations. For example, an OTU (NB_Aamo_03) assigned to subcluster 9 was commonly observed in the relict niter-bed soils (Fig. 4). Although information regarding the ecological traits of "*Ca.* Nitrososphaera" subcluster 9 organisms is limited, these organisms were widely distributed in natural environments, such as marine sediment [60] and soil [20]. Further studies using metagenomic and single-cell genomic approaches are necessary to elucidate the ecological functions of unidentified AOA in relict niter-bed ecosystems.

PCoA and jackknife analyses clearly showed that archaeal *amoA* pyrotag libraries were separated into three clusters according to soil pH and organic carbon concentration (Fig. 3B and S2C). The niter-bed soils formed group A1 with an Austrian riparian forest soil. Group A1 soils had a weak alkaline pH (7.05–8.35) and relatively high organic carbon content (3.7–23.2%). In addition, relict niter-bed soils were further separated into two subgroups according to their NO_3^-, Cl^-, K^+, and Na^+ contents in addition to the 16S rRNA gene-based comparison (Fig. S4B). The samples clustered into group A2 were mainly composed of Namibian soils, which had relatively low organic carbon contents (0.29–1.61%) and a weakly acidic pH (5.7–6.8). The samples categorized into group A3 (soil samples from a Costa-Rican rain forest, a Costa-Rican arable forest, and Greenland tundra) were defined as having a relatively high organic carbon content (1.66–7.51%) and an acidic soil pH (4.63–5.09). Interestingly, there was a significant correlation between the Chao1 diversity index and organic carbon content, with higher richness at lower levels of organic carbon (Fig. S5A). Moreover, a relatively weak correlation between the Chao1 index and soil pH was also observed, with the highest diversity at a pH of approximately 6 (Fig. S5B). These results were in good agreement with the previously reported *amoA* gene-based pyrotag analysis [14], and they provided further evidence of the importance of soil pH and organic carbon for defining the ecological niche of the AOA community in soil ecosystems, including relict niter-beds.

Conclusions

16S rRNA and *amoA* gene-based pyrotag analyses revealed that microbes responsible for ammonia oxidation (both AOB and AOA) still persist in relict niter-bed soils, although saltpeter production ceased more than 100 years ago. Our study provided the first insights into the microbial diversity in relict niter-bed ecosystems. Our main conclusions are as follows. (i) Based on 16S rRNA gene pyrotag analysis, members of the phyla *Proteobacteria*, *Actinobacteria*, *Bacteroidetes*, *Chloroflexi*, *Firmicutes*, *Gemmatimonadetes*, and *Planctomycetes* were major microbial constituents in the niter-bed soil samples. The NO_3^-, Cl^-, K^+, Na^+, and moisture contents were potential factors for the microbial compositions of the relict niter-bed soils. (ii) Members of *Nitrosospira* and "*Ca.* Nitrososphaera" were the major AOB and AOA populations, respectively, in the niter-bed soils. Soil pH, organic carbon content, and NO_3^-, Cl^-, K^+, and Na^+ contents were important factors for the ecological niche of AOB and AOA in the relict niter-bed soils.

Supporting Information

Figure S1 Rarefaction curves of (A) 16S rRNA, (B) bacterial *amoA*, and (C) archaeal *amoA* gene sequences of nine niter-bed soils.

Figure S2 Jackknife clustering of (A) 16S rRNA, (B) bacterial *amoA*, and (C) archaeal *amoA* gene pyrotag libraries from the relict niter-bed soils (weighted UniFrac, normalized to 7,770, 2,320, and 2,130 reads per soil, respectively).

Figure S3 Extended error bar plot showing the OTUs that have significantly (p<0.05) different abundances in the group. Groups were defined on the basis of principle coordinate analysis (Fig. 1) and Jackknife clustering of 16S rRNA gene pyrotag analysis (Fig. S2A). The OTUs overrepresented in upper group have a positive difference between relative abundances, and those overrepresented in bottom group have a negative difference between relative abundances. Only OTUs with mean proportions more than 0.01% in the total populations were shown in the plot.

Figure S4 Principal coordinate analysis based on the abundances of (A) bacterial and (B) archaeal *amoA* gene OTUs (weighted UniFrac) only for niter-bed soils. For this analysis, bacterial and archaeal *amoA* OTUs observed here were normalized to 2,320 and 2,130 reads per soil, respectively.

Figure S5 Correlation analysis of organic carbon contents and soil pH to Chao1 diversity indexes. Previously reported archaeal *amoA* pyrotag data from various soil environments [14] were combined with those from niter-bed soils.

Table S1 Physicochemical parameters of niter-bed soil samples.

Table S2 Primer sequences for pyrotag analysis.

Table S3 The 16S rRNA gene pyrotag libraries of relict niter-bed soil.

Table S4 The OTU table of 16S rRNA gene pyrotag libraries.

Table S5 The relative abundances of ammonia-oxidizing bacteria and archaea in 16S rRNA gene pyrotag libraries.

Table S6 Phylogenetic identification of OTUs derived from bacterial and archaeal *amoA* gene sequences.

Acknowledgments

The relict niter-bed soils were kindly provided by the education board of Shirakawa-Village, Gifu, Japan. We are grateful to Ms. Kiriko Ikeba (National Institute of Advanced Industrial Science and Technology, Japan) for technical assistance.

Author Contributions

Conceived and designed the experiments: TN HT YK SH TM. Performed the experiments: TN AA. Analyzed the data: TN KT KN. Contributed reagents/materials/analysis tools: TN TM. Contributed to the writing of the manuscript: TN HT YK SH.

References

1. Williams AR (1975) The production of saltpetre in the middle ages. Ambix 22: 125–133.
2. Lockyer WJS (1908) Modern nitre beds. Nature 77: 513–515.
3. Kyeser K (1967) Umschrift und Ubersetzung nebst Erlauterungen (German translation). Düsseldorf, Germany: Verlag des Vereins Deutscher Ingenieure.
4. Kyeser K (1967) Facsimile-Ausgabe der Pergamenthandschrift, Cod. Ms. philos. 63 der Universititsbibliothek, Gottingen (German translation). Düsseldorf, Germany: Verlag des Vereins Deutscher Ingenieure.
5. Multhauf RP (1971) The French Crash Program for Saltpeter Production, 1776–94. Technol Cult 12: 163–181.
6. Kunnas J (2007) Potash, saltpeter and tar - Production, exports and use of wood in Finland in the 19th century. Scand J Hist 32: 281–311.
7. Anonymous (1912) The Indian saltpetre industry. Nature 88: 330–331.
8. Thorpe TE (1917) Indian saltpetre. Nature 99: 447–448.
9. Maji T (2009) Transaction on saltpeter of Shirakawa Village in the second half of the 18th century described by old letters found from Shinshoji-temple. Annu Rep Fac Edu, Gifu Univ, Humanit Soc Sci (in Japanese with English abstract) 57: 121–126.
10. Maji T, Maji A (2007) The Unknown History of Shirakawago: The Circulation of Nitrogen in Relation with Urine, Saltpeter and Rice, and The Daily Life in the Edo Era. Nagoya, Japan: Fubaisha Press.
11. Anonymous (1912) The condition of the Chilean saltpeter industry. J Ind Eng Chem-Us 4: 687–687.
12. Jimura T (2014) The impact of world heritage site designation on local communities - A case study of Ogimachi, Shirakawa-mura, Japan. Tourism Manage 32: 288–296.
13. Maji T (2009) Efficiency of nitrification and activity of nitrite production by soil in niter-bed. Sci Rep Fac Ed Gifu Univ Natur Sci (in Japanese with English abstract) 33: 103–108.
14. Pester M, Rattei T, Flechl S, Grongroft A, Richter A, et al. (2012) *amoA*-based consensus phylogeny of ammonia-oxidizing archaea and deep sequencing of *amoA* genes from soils of four different geographic regions. Environ Microbiol 14: 525–539.
15. Maji T (2006) A study of saltpeter production at Shirakawa Village in Edo era. Part 3: an estimation of contribution of urinay nitrogen to nitrate-N in niter-bed. Kagakushi (in Japanese with English abstract) 33: 1–14.
16. Maji T (2005) A study of saltpeter production at Shirakawa Village in Edo era. Part 2: productivity of saltpeter from artificial niter-bed soil: an estimation from extraction yield of nitrite. Kagakushi (in Japanese with English abstract) 32: 137–143.
17. Maji T (2005) A study of saltpeter production at Shirakawa Village in Edo era. Part 1: an analysis of constituents of soil gathered from niter-bed and the starting materials. Kagakushi (in Japanese with English abstract) 32: 75–84.
18. Baker GC, Smith JJ, Cowan DA (2003) Review and re-analysis of domain-specific 16S primers. J Microbiol Methods 55: 541–555.
19. Rotthauwe JH, Witzel KP, Liesack W (1997) The ammonia monooxygenase structural gene amoA as a functional marker: Molecular fine-scale analysis of natural ammonia-oxidizing populations. Appl Environ Microbiol 63: 4704–4712.
20. Francis CA, Roberts KJ, Beman JM, Santoro AE, Oakley BB (2005) Ubiquity and diversity of ammonia-oxidizing archaea in water columns and sediments of the ocean. Proc Natl Acad Sci USA 102: 14683–14688.
21. Caporaso JG, Kuczynski J, Stombaugh J, Bittinger K, Bushman FD, et al. (2010) QIIME allows analysis of high-throughput community sequencing data. Nat Methods 7: 335–336.
22. Reeder J, Knight R (2010) Rapidly denoising pyrosequencing amplicon reads by exploiting rank-abundance distributions. Nat Methods 7: 668–669.
23. Edgar RC (2010) Search and clustering orders of magnitude faster than BLAST. Bioinformatics 26: 2460–2461.
24. Caporaso JG, Bittinger K, Bushman FD, DeSantis TZ, Andersen GL, et al. (2010) PyNAST: a flexible tool for aligning sequences to a template alignment. Bioinformatics 26: 266–267.
25. Haas BJ, Gevers D, Earl AM, Feldgarden M, Ward DV, et al. (2011) Chimeric 16S rRNA sequence formation and detection in Sanger and 454-pyrosequenced PCR amplicons. Genome Res 21: 494–504.
26. Wang Y, Qian PY (2009) Conservative fragments in bacterial 16S rRNA genes and primer design for 16S ribosomal DNA amplicons in metagenomic studies. PLoS One 4: e7401.
27. Cole JR, Wang Q, Cardenas E, Fish J, Chai B, et al. (2009) The Ribosomal Database Project: improved alignments and new tools for rRNA analysis. Nucleic Acids Res 37: D141–D145.
28. Camacho C, Coulouris G, Avagyan V, Ma N, Papadopoulos J, et al. (2009) BLAST+: architecture and applications. BMC Bioinformatics 10: 421.
29. Tamura K, Peterson D, Peterson N, Stecher G, Nei M, et al. (2011) MEGA5: Molecular Evolutionary Genetics Analysis Using Maximum Likelihood, Evolutionary Distance, and Maximum Parsimony Methods. Mol Biol Evol 28: 2731–2739.
30. Edgar RC, Haas BJ, Clemente JC, Quince C, Knight R (2011) UCHIME improves sensitivity and speed of chimera detection. Bioinformatics 27: 2194–2200.
31. Colwell RK (2013) EstimateS: Statistical estimation of species richness and shared species from samples. Version 9 and earlier. User's Guide and application.
32. Good IJ (1953) The population frequencies of species and the estimation of population parameters. Biometrika 40: 237–264.
33. Hamady M, Lozupone C, Knight R (2010) Fast UniFrac: facilitating high-throughput phylogenetic analyses of microbial communities including analysis of pyrosequencing and PhyloChip data. ISME J 4: 17–27.
34. Saitou N, Nei M (1987) The neighbor-joining method - a new method for reconstructing phylogenetic trees. Mol Biol Evol 4: 406–425.
35. Welch BL (1947) The generalization of students problem when several different population variances are involved. Biometrika 34: 28–35.
36. Parks DH, Beiko RG (2010) Identifying biologically relevant differences between metagenomic communities. Bioinformatics 26: 715–721.
37. Felsenstein J (1985) Confidence-limits on phylogenies - an approach using the bootstrap. Evolution 39: 783–791.
38. Mao YJ, Yannarell AC, Mackie RI (2011) Changes in N-transforming archaea and bacteria in soil during the establishment of bioenergy crops. PLoS One 6: e24750.
39. Xue K, Wu LY, Deng Y, He ZL, Van Nostrand J, et al. (2013) Functional gene differences in soil microbial communities from conventional, low-input, and organic farmlands. Appl Environ Microbiol 79: 1284–1292.
40. Grossman JM, O'Neill BE, Tsai SM, Liang BQ, Neves E, et al. (2010) Amazonian anthrosols support similar microbial communities that differ distinctly from those extant in adjacent, unmodified soils of the same mineralogy. Microb Ecol 60: 192–205.
41. Corneo PE, Pellegrini A, Cappellin L, Roncador M, Chierici M, et al. (2013) Microbial community structure in vineyard soils across altitudinal gradients and in different seasons. FEMS Microbiol Ecol 84: 588–602.

42. Van Horn DJ, Van Horn ML, Barrett JE, Gooseff MN, Altrichter AE, et al. (2013) Factors controlling soil microbial biomass and bacterial diversity and community composition in a cold desert ecosystem: role of geographic scale. PLoS One 8: e66103.
43. Watson SW (1971) Reisolation of *Nitrosospira briensis* Winogradsky, S and Winogradsky, H 1933. Arch Mikrobiol 75: 179–188.
44. Ward BB, Eveillard D, Kirshtein JD, Nelson JD, Voytek MA, et al. (2007) Ammonia-oxidizing bacterial community composition in estuarine and oceanic environments assessed using a functional gene microarray. Environ Microbiol 9: 2522–2538.
45. Mintie AT, Heichen RS, Cromack K, Myrold DD, Bottomley PJ (2003) Ammonia-oxidizing bacteria along meadow-to-forest transects in the oregon cascade mountains. Appl Environ Microbiol 69: 3129–3136.
46. Bollmann A, Schmidt I, Saunders AM, Nicolaisen MH (2005) Influence of starvation on potential ammonia-oxidizing activity and *amoA* mRNA levels of *Nitrosospira briensis*. Appl Environ Microbiol 71: 1276–1282.
47. Li XR, Xiao YP, Ren WW, Liu ZF, Shi JH, et al. (2012) Abundance and composition of ammonia-oxidizing bacteria and archaea in different types of soil in the Yangtze River estuary. J Zhejiang Univ-Sc B 13: 769–782.
48. Fukushima T, Whang LM, Chiang TY, Lin YH, Chevalier LR, et al. (2013) Nitrifying bacterial community structures and their nitrification performance under sufficient and limited inorganic carbon conditions. Appl Microbiol Biotechnol 97: 6513–6523.
49. Shore JL, M'Coy WS, Gunsch CK, Deshusses MA (2012) Application of a moving bed biofilm reactor for tertiary ammonia treatment in high temperature industrial wastewater. Bioresour Technol 112: 51–60.
50. Avrahami S, Jia ZJ, Neufeld JD, Murrell JC, Conrad R, et al. (2011) Active autotrophic ammonia-oxidizing bacteria in biofilm enrichments from simulated creek ecosystems at two ammonium concentrations respond to temperature manipulation. Appl Environ Microbiol 77: 7329–7338.
51. Eilmus S, Rosch C, Bothe H (2007) Prokaryotic life in a potash-polluted marsh with emphasis on N-metabolizing microorganisms. Environ Pollut 146: 478–491.
52. Gubry-Rangin C, Nicol GW, Prosser JI (2010) Archaea rather than bacteria control nitrification in two agricultural acidic soils. FEMS Microbiol Ecol 74: 566–574.
53. Shen JP, Zhang LM, Zhu YG, Zhang JB, He JZ (2008) Abundance and composition of ammonia-oxidizing bacteria and ammonia-oxidizing archaea communities of an alkaline sandy loam. Environ Microbiol 10: 1601–1611.
54. Tarre S, Shlafman E, Beliavski M, Green M (2007) Changes in ammonia oxidiser population during transition to low pH in a biofilm reactor starting with *Nitrosomonas europaea*. Water Sci Technol 55: 363–368.
55. Yasuda T, Waki M, Kuroda K, Hanajima D, Fukumoto Y, et al. (2013) Responses of community structure of amoA-encoding archaea and ammonia-oxidizing bacteria in ammonia biofilter with rockwool mixtures to the gradual increases in ammonium and nitrate. J Appl Microbiol 114: 746–761.
56. Leininger S, Urich T, Schloter M, Schwark L, Qi J, et al. (2006) Archaea predominate among ammonia-oxidizing prokaryotes in soils. Nature 442: 806–809.
57. Jia ZJ, Conrad R (2009) *Bacteria* rather than *Archaea* dominate microbial ammonia oxidation in an agricultural soil. Environ Microbiol 11: 1658–1671.
58. Moin NS, Nelson KA, Bush A, Bernhard AE (2009) Distribution and diversity of archaeal and bacterial ammonia oxidizers in salt marsh sediments. Appl Environ Microbiol 75: 7461–7468.
59. Dang HY, Zhang XX, Sun J, Li TG, Zhang ZN, et al. (2008) Diversity and spatial distribution of sediment ammonia-oxidizing crenarchaeota in response to estuarine and environmental gradients in the Changjiang Estuary and East China Sea. Microbiology-SGM 154: 2084–2095.
60. Park HD, Wells GF, Bae H, Criddle CS, Francis CA (2006) Occurrence of ammonia-oxidizing archaea in wastewater treatment plant bioreactors. Appl Environ Microbiol 72: 5643–5647.

21

The Kill Date as a Management Tool for Cover Cropping Success

María Alonso-Ayuso, José Luis Gabriel, Miguel Quemada*

School of Agriculture Engineering, Technical University of Madrid, Madrid, Spain

Abstract

Integrating cover crops (CC) in rotations provides multiple ecological services, but it must be ensured that management does not increase pre-emptive competition with the subsequent crop. This experiment was conducted to study the effect of kill date on: (i) CC growth and N content; (ii) the chemical composition of residues; (iii) soil inorganic N and potentially mineralizable N; and (iv) soil water content. Treatments were fallow and a CC mixture of barley (*Hordeum vulgare* L.) and vetch (*Vicia sativa* L.) sown in October and killed on two different dates in spring. Above-ground biomass and chemical composition of CC were determined at harvest, and ground cover was monitored based on digital image analysis. Soil mineral N was determined before sowing and after killing the CC, and potentially mineralizable N was measured by aerobic incubation at the end of the experiment. Soil water content was monitored daily to a depth of 1.1 m using capacitance sensors. Under the present conditions of high N availability, delaying kill date increased barley above-ground biomass and N uptake from deep soil layers; little differences were observed in vetch. Postponing kill date increased the C/N ratio and the fiber content of plant residues. Ground cover reached >80% by the first kill date (~1250°C days). Kill date was a means to control soil inorganic N by balancing the N retained in the residue and soil, and showed promise for mitigating N losses. The early kill date decreased the risk of water and N pre-emptive competition by reducing soil depletion, preserving rain harvested between kill dates and allowing more time for N release in spring. The soil potentially mineralizable N was enhanced by the CC and kill date delay. Therefore kill date is a crucial management variable for maximizing the CC benefits in agricultural systems.

Editor: Anil Shrestha, California State University, Fresno, United States of America

Funding: This work was funded by the Spanish Comisión Interministerial de Ciencia y Tecnología (project AGL 2011-24732) and the Regional Government of Madrid (Project AGRISOST, S2009/AGR-1630). The funders had no role in study design, data collection and analysis, decision to publish, or preparation of the manuscript.

Competing Interests: The authors have declared that no competing interests exist.

* Email: miguel.quemada@upm.es

Introduction

The potential for cover crops (CC) to provide ecological services has been documented in diverse cover cropping systems and environments [1]. Replacing fallow periods with CC may enhance soil aggregate stability [2], water retention capacity [3], nutrient supply [4] and disease suppression [5]. Moreover, plant residues covering the soil following crop kill can improve soil protection over time and help control weeds, preserve soil moisture, ameliorate compacted soils, and reduce soil erosion [6]. However, improperly managed CC may have a detrimental effect on the cash crop either by competing for water and nutrients, building up diseases or retarding seed germination [6,7]. Overall, there are two crucial management factors that determine CC success: species selection and CC termination [8]. In this manuscript, we focus on the termination of CC incorporated into field crop rotations, particularly on kill date. Previous experiments have shown that changing kill date by a few weeks in spring may have a large effect on the water and N pre-emptive competition with the subsequent cash crop, nitrate leaching control, and soil moisture conservation [7,9–12], but there is no agreement on the advantages and limitations of postponing CC termination. Therefore, a better understanding of the effect of the living CC and its dead mulch on the dynamics of N and water might improve our ability to choose a kill date that maximizes the benefits of CC in the system.

The effect of kill date on soil water availability is a balance between the water extracted by the living CC and the evaporation prevented by the residue mulch [9]. Integrating CC into the cropping system increases above-ground biomass and resistance to decomposition, which enhances the development of a dead mulch covering the soil surface [13]. Some authors have found that, despite the soil water depletion, moisture conservation was improved when the CC was killed later [9,11]. However, other experiments have shown that an earlier CC kill reduced water pre-emptive competition, preserved top soil moisture, and increased water availability for the subsequent crop [7,12]. Similarly, postponing kill date enhanced N pre-emptive competition as soil available N is depleted by the CC uptake and only slowly replenished by the mineralization of a high C/N residue [6]. Legume species may compensate for this effect by fixing more N_2 when killed late [10]. The water and N balance is driven by various factors that are all affected by kill date, such as above-ground biomass accumulation, ground cover (GC), and N uptake

[6]. The chemical composition of crop residues, such as lignin content and the C/N ratio, are important characteristics governing the decomposition process and determining the persistence of the residue layer covering the soil [14]. However, there are few experiments that combine multiple approaches to study these variables, and to our knowledge, none that involves continuous monitoring of soil water dynamics.

We hypothesize that a combined study of CC growth, GC and water and N dynamics would contribute to understanding the effect of kill date and improve management practices. This experiment was conducted over two years to study the effect of kill date on the following factors: (i) the growth and N content of the CC; (ii) the chemical composition and residue quality of the CC; (iii) soil inorganic N dynamics and its potentially mineralizable N; and (iv) the soil water content. A CC mixture of barley and vetch was selected because mixing legumes and non-legumes is known to be an efficient technique to merge N and water management benefits of the individual species [9,15].

Materials and Methods

1. Field experiment

The experiment was conducted from October 2011 to November 2013 at La Chimenea field station (40°03′N, 03°31′W, 550 m a.s.l.), which is located in the central Taxus river basin in Aranjuez (Madrid, Spain). The soil is a silty clay loam (Typic Calcixerept) [16] with a high content of organic matter and carbonates, pH ~8 and low stone content throughout the profile. The most relevant soil characteristics are presented in Table 1. Total organic C was measured by the Walkley-Black method [17]. Calcium carbonate was determined by a volumetric calcimeter method [18]. The climate is semiarid Mediterranean [19] with high interannual variability (Figure 1). Annual rainfall averages ~350 mm with less rain in the summer and more in the autumn and a mean annual temperature of 14.2°C. Weather data were recorded throughout the experimental period by a data logger (CR10X. Campbell Scientific Ltd, Shepshed, UK) located ~100 m from the field site.

The experiment consisted of twelve plots (15 m×12 m) randomly assigned to three treatments (fallow and two different CC kill dates) with four replications. Cover crop treatments were sown with a barley (*Hordeum vulgare* L., cv. Vanessa) + vetch (*Vicia sativa* L., cv. Aitana) mixture in early October (October 6, 2011; October 8, 2012). The mixture, 30% (54 kg ha^{-1}) barley seeds and 70% (45 kg ha^{-1}) vetch, was broadcast by hand and buried (~5 cm) with a shallow cultivator. All treatments received ~18 mm irrigation on the sowing date with a sprinkle irrigation system (9.5 mm h^{-1}) to ensure uniform CC establishment. Plots from the first kill treatment (FK) were killed in mid-March (March 13, 2012; March 14, 2013) when barley was at the end of the booting growth stage and vetch at the stem elongation stage. Plots from the second kill treatment (SK) were killed in mid-April (April 9, 2012; April 13, 2013) when barley was at the emerging inflorescence growth stage and vetch was at the stem elongation stage. Cover crops were killed by application of glyphosate (N-phosphonomethyl glycine, 0.7 kg a.e. ha^{-1}) followed by a shredder. The last day of April was chosen as a hypothetical planting date (HPD) for both years, as it is a typical date for planting summer cash crops in the area. The HPD represented the beginning of the mulch period during which the CC residue remained on the ground. To compare the residue mulch effect on water conservation, SWC was homogenized by adding water to the drier plots with a drip irrigation system (30.5 mm to FK treatment and 35.5 mm to SK) in April 2012. As rainfall was abundant in March 2013, no supplemental irrigation was applied. Fallow plots were kept free of weeds by hand weeding during the trial period. The experiment was conducted in a field that was cultivated with triticale during the two previous years and had not received organic amendments or N fertilizer for four years prior to the beginning of the trial.

2. Cover crop analysis

To measure the above-ground biomass and GC in each plot, four 0.5 m×0.5 m squares were marked after sowing the CC. Ground cover was monitored in each square based on digital photos taken from a nadir perspective at 1.2 m height every other week [20]. The images were taken with a Nikon S210 Coolpix camera that had a lens resolution of 5 Mpixels. The images were saved at a resolution of 1200×800 pixels and the GC percentage was determined with the software SigmaScan Pro (Systat software, Chicago, IL, USA) by means of the "Turf analysis" macro developed by Karcher and Richardson [21]. The process consisted of creating a pixel layer by selecting the hue and saturation ranges that identified the surface covered by the crop. The GC was determined to be the number of pixels of the layer divided by the total number of pixels constituting the image of the marked squares [20]. The GC evolution was adjusted to the Gompertz function [22], and the thermal times (°C days) until the GC reached 30% (t_{30}) and 80% (t_{80}) were calculated.

Just before killing the CC, the above-ground biomass of the marked squares was hand harvested by cordless grass shear at the ground level, separated by species, oven dried for 48 h at 65°C,

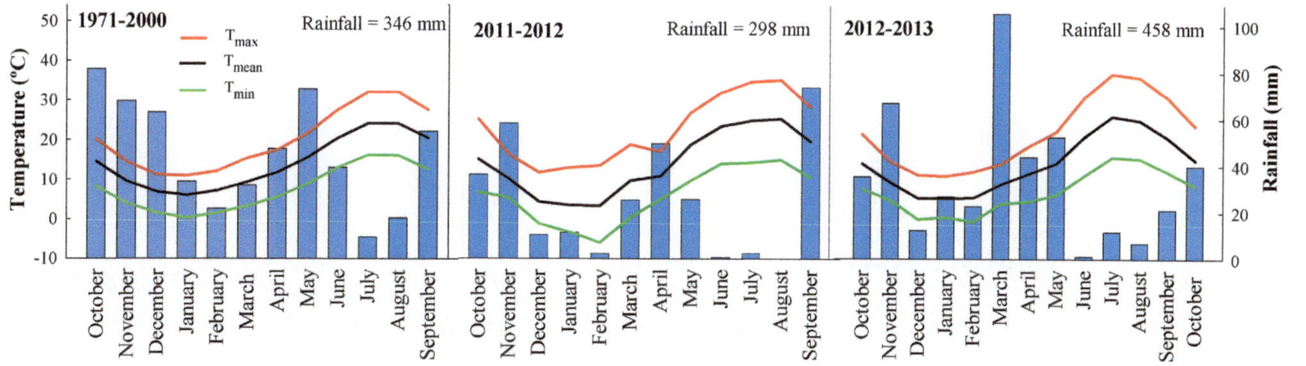

Figure 1. Monthly rainfall (bars) and average maximum, mean and minimum temperatures from 1971 to 2000 and during the two experimental seasons at Aranjuez (Madrid, Spain).

Table 1. Soil properties at the beginning of the experiment.

Depth (cm)	0–23	23–40	40–70	70–120
pH (1:2.5)	8.16	8.06	8.02	7.84
Organic Matter (g kg^{-1})	31.8	29.2	21.9	22.3
CO_3 (g CO_3^{2-} kg^{-1})	198.0	201.3	159.0	181.0
Sand (g kg^{-1})	260	250	250	250
Silt (g kg^{-1})	490	510	520	460
Clay (g kg^{-1})	250	240	230	290

and weighed to determine biomass of barley, vetch, and mixture in kg ha^{-1}. The C and N concentrations in the above-ground biomass were determined for a subsample of each species from each plot using the Dumas combustion method (LECO CHNS-932 Analyzer, St. Joseph, MI, USA). Total N content was calculated for each specie as the product of above-ground biomass times N concentration. The atmospheric N_2 fixation by the legume was estimated by the natural abundance method [23] based on the $\delta^{15}N$ (‰) determination (Europe Scientific 20–20 IRMS Analyzer, Crewe, UK) for subsamples from the vetch and a barley reference cultivated as a sole crop in an adjacent field. Cover crop residue quality was assessed by measuring neutral detergent fiber (NDF), acid detergent fiber (ADF), and lignin (L) with the Goering and Van Soest method [24] in subsamples of each crop species from each plot.

3. Soil inorganic nitrogen content (N_{min})

Four soil cores were taken from each plot to a depth of 1.2 m in 0.2 m intervals with an Eijkelkamp helicoidal auger (Eijkelkamp Agrisearch Equipment, Geisbeek, Netherlands) just before sowing the CC, after the second kill date each year, and at the end of the experiment. Soil cores were combined by depth to provide a composite profile of six samples. For each plot, soil N_{min} was calculated for each layer. Soil samples were placed in a plastic box and immediately firmly closed then transported, and refrigerated (4–6°C). Within five consecutive days, a soil subsample from each box was extracted with 1 M KCl (~30 g of soil: 150 ml of KCl), centrifuged, decanted, and a subsample of the supernatant volume was stored in a freezer until later analysis. Nitrate concentration in the extracts was determined by the Griess-Ilosvay method [25] after reduction of NO_3^- to NO_2^- with a Cd column. Ammonium in the soil extracts was determined by the salicylate-hypochlorite method [26].

4. Soil nitrogen mineralization potential (N_0)

Soil N mineralization potential was estimated by adapting the procedure proposed by Stanford and Smith [27]. A subsample of the top soil layer (0–0.2 m) free from residue contamination was collected from each plot after the second kill date of the second year, air-dried, and sieved (<6 mm). A homogeneous mixture of 30 g of that soil and an equal weight of sand was packed to a depth of 5 cm in plastic syringes (3.5 cm diameter, 10 cm long) to achieve a bulk density of 1.25 g cm^{-3}. Thin glass wool layers were placed between the soil and the bottom of the syringe and over the soil sample to avoid soil loss during leaching and to minimize moisture loss. Initial soil inorganic N was removed by leaching

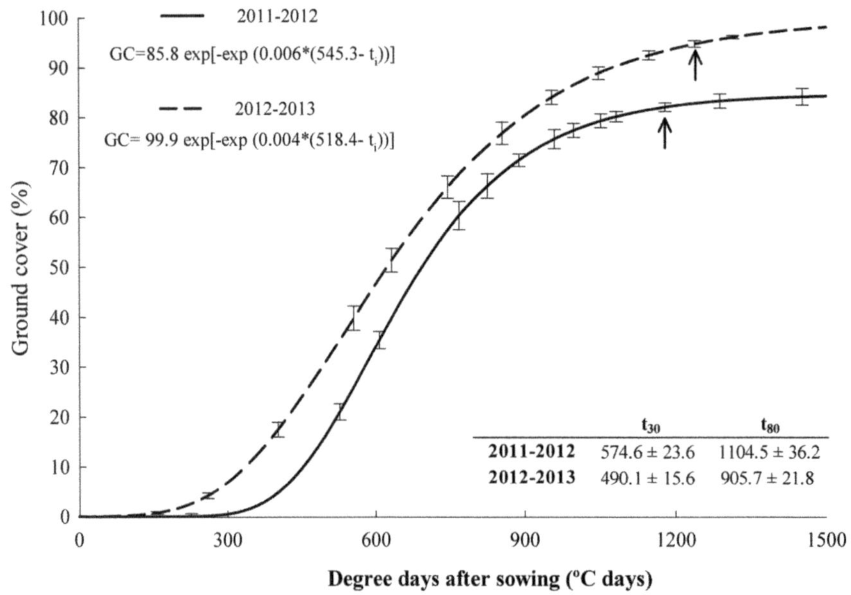

Figure 2. Ground cover (GC) development during the cover crop growth period in both experimental seasons. Arrows show the first kill date. Lines represent the Gompertz model adjusted to the observed values. Fitted models and the thermal time until the ground cover reaches 30 and 80% (t_{30} and t_{80}) are shown.

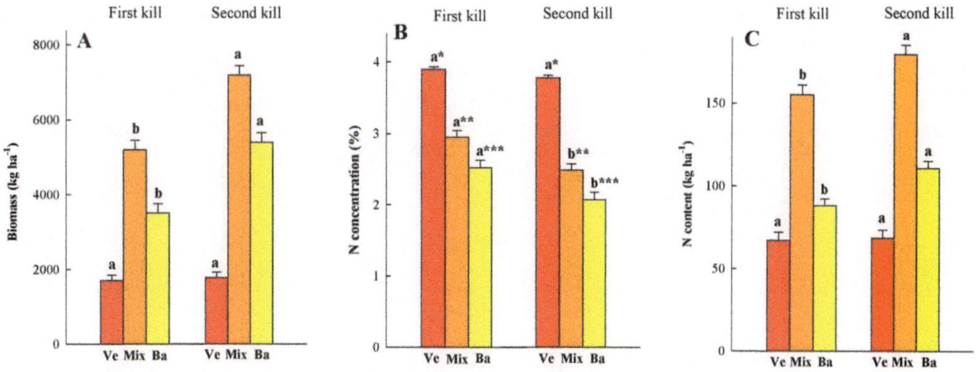

Figure 3. Biomass (A), N concentration (B) and N content (C) for vetch (Ve), barley (Ba) and the mixture (Mix) measured for the first and second kill dates. Values are the mean of two cropping seasons. Letters above bars show significant differences between kill dates for each species, and asterisks between species for each kill date. Small bars represent the standard error.

with 100 ml of 0.01 M $CaCl_2$ followed by 20 ml of a N-free nutrient solution (0.0095 M $CaSO_4$, 0.000047 M KH_2PO_4, 0.00138 M K_2SO_4, 0.0003 M $MgSO_4$). The excess water was removed using a vacuum pump until a weight within 2 g of that measured before leaching was reached. The syringes were covered with a porous parafilm and incubated aerobically at 35°C. Syringes were removed from the incubator at 14, 28, 42, 56, and 60 d after preparation and were leached with 100 ml of 0.01 M $CaCl_2$ solution followed by 20 ml of N-free solution. Leachates were made up to 100 ml with $CaCl_2$, and subsamples were stored in a freezer at −25°C until later analysis. After the leaching procedure, the cores were allowed to drain under vacuum until a weight within 2 g of that measured before leaching was reached. Nitrate concentrations in the leachates were determined by the Griess-Ilosvay method, ammonium by the salicylate-hypochlorite method and total N by the Dumas combustion method, as described above. The N_0 and the mineralization rate constant (k) were estimated after fitting a non-linear regression model ($Nt = N_0 \exp(-k\ t)$) for describing the cumulative N mineralized (Nt) with time (t) in each soil sample.

5. Soil volumetric water content (SWC)

The SWC was monitored daily during the field trial using the EnviroScan capacitance probes (Sentek Pty Ltd, Stepney, Australia) that has been described in detail elsewhere [28]. Seventy-two capacitance sensors were mounted on twelve plastic extrusions (four repetitions per treatment), introduced into access tubes located in the middle of each plot and connected to three data loggers. Sensor readings were automated and stored in the data loggers and downloaded weekly. To ensure the measurement reliability, a normalization procedure was conducted that obtained reference readings by exposing each sensor to air and water (~20°C). The sensors were centered at 10, 30, 50, 70, 90, and 110 cm below the soil surface in each plastic extrusion, and normalized readings were registered every 6 h. A daily average of the four readings from the 0–20 (10), 20–40 (30), 40–60 (50), 60–80 (70), 80–100 (90), 100–120 (110) cm-deep soil layers was transformed into SWC using a calibration equation that was obtained at the experimental site [29]. The SWC data set was comprised of two CC seasons, which started before sowing in October and lasted until August each year, and used to study the effect of the living CC and the dead mulch on soil moisture.

6. Statistical analysis

Analyses of variance (ANOVA) and t-test were performed in order to determine differences between kill dates for each variable. Year was considered a random effect and, interaction between kill date and year was also tested. Means were separated by Tukey's test at the 0.05 probability level ($P \leq 0.05$). Least significant differences (LSD) were calculated for SWC ($P \leq 0.05$). Prior to conducting the ANOVA, tests were conducted to verify if the assumptions of ANOVA were met. The Gompertz model was

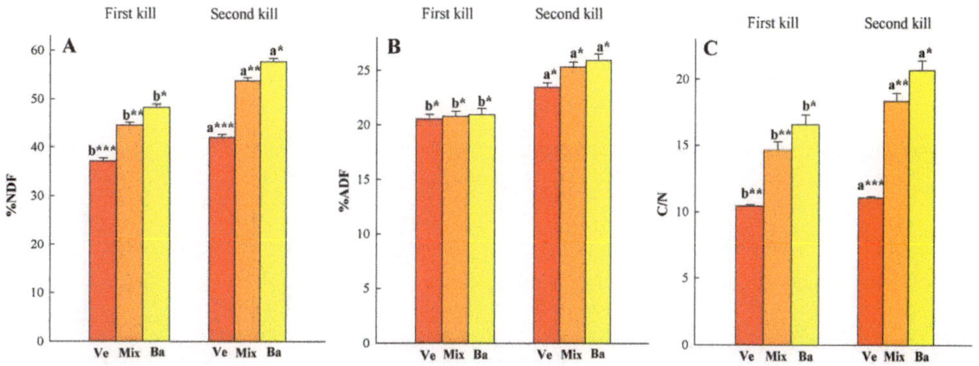

Figure 4. Neutral detergent fiber-NDF- (A) and acid detergent fiber -ADF- (B) fractions, and C/N ratio (C) in vetch (Ve), barley (Ba) and the mixture (Mix), measured for the first and second kill dates. Values are the mean of two cropping seasons. Letters above bars show significant differences between kill dates for each species, and asterisks between species for each kill date. Small bars represent the standard error.

Table 2. Soil inorganic N content (kg N ha^{-1}) in the different soil layers and the entire profile at different sampling dates for the fallow (Fa), first kill (FK) and second kill (SK) treatments.

	Oct 2011	Apr 2012	Oct 2012	Apr 2013	Oct 2013
0–40 cm					
Fa	94 (13.7)	126.8 (7.9) a	135.3 (15.4)	29.7 (3.8) b	71.54 (5.38) c
FK	94 (13.7)	46.9 (7.1) b	121.5 (16.6)	56.9 (3) a	119.37 (28.48) b
SK	94 (13.7)	37 (8.2) b	154 (8)	24.8 (4.5) b	171.72 (2.36) a
40–80 cm					
Fa	24.7 (9.5)	71.6 (10.6) a	86.1 (4.6) a	40.1 (1.8) a	72.22 (7.43)
FK	24.7 (9.5)	30.4 (5.3) b	48.3 (4.4) b	42.6 (3.5) a	65.38 (19.89)
SK	24.7 (9.5)	25.4 (5.1) b	48.1 (7.1) b	27.1 (4.1) b	39.39 (2.68)
80–120 cm					
Fa	26.1 (11.1)	60.6 (3.5)	66.7 (11)	68.8 (3.3) a	100.37 (11.74) a
FK	26.1 (11.1)	56.1 (24.5)	61.5 (29.3)	47 (7.4) b	61.59 (17.91) ab
SK	26.1 (11.1)	34.5 (3.7)	33.5 (6.3)	32.3 (8.3) b	39.24 (5.83) b
Soil profile (0–120 cm)					
Fa	144.8 (22.8)	259 (8.0) a	288.1 (25.4)	138.6 (5.1) a	244.13 (19.05)
FK	144.8 (22.8)	133.32 (27.5) b	231.3 (31.2)	146.6 (10.7) a	246.34 (52.87)
SK	144.8 (22.8)	96.9 (15.4) b	235.6 (14.2)	84.2 (13.3) b	247.34 (7.9)

Means with standard error in parentheses.
Within a date and depth, means with the same letter are not significantly different between kill dates at $P<0.05$.

fitted to the GC and the N mineralization potential model was fitted to the cumulative N mineralized using a non-linear regression procedure. The models were evaluated for their ability to simulate the observed data by comparing the mean of the lack of fit to the mean square due to pure error by using the variance ratio, or F-test. When the lack of fit was significantly smaller than the pure error, the model fitted the data. Further discussion regarding this evaluation procedure can be found elsewhere [30,31]. Statistical analyses were accomplished using the PASW Statistics Software (SPSS, Chicago, IL, USA).

Results

1. Weather conditions

The first experimental year was drier than the second (Figure 1). In the first year, rainfall was substantially lower (298 mm) than the 30-year average although in the second, it was greater (408 mm). Differences mainly occurred in spring; from March to May, the rainfall was 202 mm in 2013 compared with 101 mm in 2012. The rest of the season was rather similar for both years except for intense precipitation in late September 2012. Air temperature followed a typical Mediterranean seasonal distribution over the two years. Winter was cooler in the first year with an average minimum temperature of −3 and −5°C in January and February, respectively.

2. Cover crop: GC, above-ground biomass and N content

The GC followed a classical Gompertz model (Figure 2). The F-test comparing the mean squares due to pure error and lack of fit was not significant at 0.01 level, therefore, the model was adjusted to the GC observations. The first kill occurred between 157 and 159 d after sowing (~1200°C days) whereas the second year occurred between 184 and 186 d after sowing (~1500°C days). The GC>30%, a crucial threshold for erosion control [32], was attained between 37 and 41 d after sowing (>490°C days). The GC>80%, considered to be full cover for erosion control and direct soil evaporation [32], was reached between 95 and 130 d after sowing (>900°C days). Differences were observed in the maximum GC attained, 85% in 2012 and 100% in 2013. In both

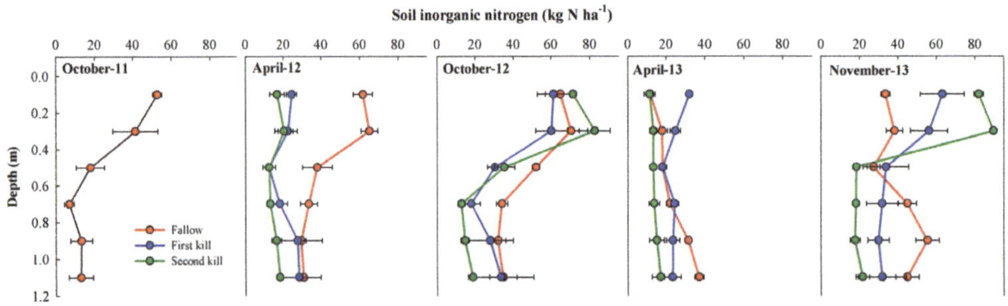

Figure 5. Soil inorganic nitrogen (kg N ha^{-1}) in the upper 1.2 m of the soil profile for the fallow (Fa), first kill (FK) and second kill (SK) treatments at different sampling times.

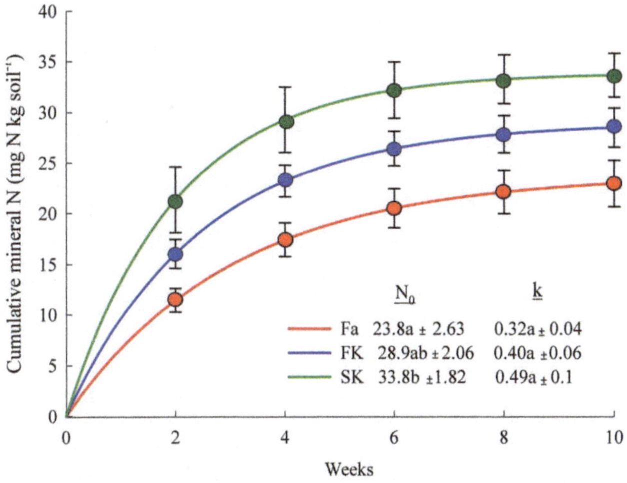

Figure 6. Cumulative N mineralization in soils from the fallow (Fa), first kill (FK) and second kill (SK) treatments during a 10-week aerobic incubation. Soil N mineralization potential (N_0) and N mineralization rate (k) were calculated by fitting a non-linear regression model ($Nt = N_0 \exp(-k\ t)$). Values followed by different letters are significantly different between treatments at ≤0.05 by Tukey's test. Bars represent the standard error.

years, maximum GC was already attained by the first kill date. The ground continued to be covered by the CC residue mulch more than six months after CC killing. Above-ground biomass increased from the FK to the SK by ~2000 kg ha^{-1}, and most of the increase was due to barley (Figure 3). Although barley biomass was greater at the SK than at the FK, no differences were found in vetch biomass between kill dates. Vetch above-ground biomass was double in 2013 than in 2012 whereas barley above-ground biomass did not vary by year. At the end of the mulch period, residues left on the soil surface were greater in the SK (2214 kg ha^{-1} in 2012, and 4014 in 2013) than in the FK treatment (966 kg ha^{-1} in 2012, 2151 in 2013. The ground was fully covered by the residues in both treatments.

Nitrogen concentration in the above-ground biomass was higher in the FK than in the SK. Nitrogen concentration in the mixture decreased from 2.7 to 2.35% in 2012 and from 3.2 to 2.6% in 2013. As expected, vetch had a higher N concentration than barley whereas the N concentration of the mixture was intermediate (Figure 3). Differences in N content between the FK and SK, calculated as the product of above-ground biomass and N concentration, were not significant for the mixture or vetch (Figure 3). However, barley N content in the SK was higher than in the FK; N uptake increased from 79 to 107 kg N ha^{-1} in 2012 and from 97 to 115 kg N ha^{-1} in 2013. The vetch reached high N_2 atmospheric fixation rates during the experiment. In the first season, N_2 atmospheric fixation was >80% of N content and ~100% in the second. No differences in N_2 atmospheric fixation were found between kill dates.

Residue quality fractions varied between kill dates (Figure 4). The NDF and ADF fractions increased when the CC was killed later. An interaction effect between treatment and year occurred for lignin content as differences in the lignin fraction between kill dates were only significant in barley and in the mixture in 2012. Lignin fraction in the mixture increased from 2.2 to 3.6% and in the barley from 1.2 to 2.9% in 2012. The C/N of barley, vetch and mixture residues was always lower for the FK than for the SK.

3. Soil inorganic nitrogen content

At the beginning of the experiment, average N_{min} of the upper layers of the soil profile was 145 kg N ha^{-1} (Figure 5, Table 2). During the first CC growing period, soil N_{min} increased in the fallow treatment, particularly in the upper 0.8 m, while it decreased in the remaining depths. During the mulch period, soil N_{min} in the fallow treatment increased slightly whereas, in the CC treatments, N_{min} increased largely in the upper soil layers, but there was no difference between the treatments by October in 2012.

During the second CC growing season, soil N_{min} was depleted in all treatments. After the CC kill in April 2013, the N_{min} was 84 kg N ha^{-1} in the SK treatment and ~140 kg N ha^{-1} in the others. The FK treatment accumulated more N_{min} in the upper layers, most likely due to the early mineralization that occurred between the CC kill and soil sampling (~4 weeks). At the end of the mulch period in November 2013, a large increase in soil N_{min} occurred in all treatments. Although no difference was observed for average N_{min} (~245 kg N ha^{-1}) over the entire profile, differences between treatments were observed at some soil depths. In the SK, most of the N_{min} (~70%) was in the top layer (0–40 cm) whereas it was ~29% in the fallow treatment. Similarly, while >40% of soil N_{min} was in the bottom layer in the fallow, < 16% was in the bottom layer in the SK. The N_{min} distribution in the FK treatment was intermediate between the others.

4. Soil nitrogen mineralization potential

The one-pool exponential model fit the cumulative mineral N from the aerobic incubation, as the F-test comparing the mean squares due to pure error and lack of fit was not significant at 0.01 level. The N_0 was higher for soils from the SK treatment (34 mg N kg $soil^{-1}$) than those from the fallow (24 mg N kg $soil^{-1}$). The N_0 from the FK treatment was intermediate. No differences between treatments were observed in the N mineralization (Figure 6).

5. Soil water content

Soil water content was affected by the presence of CC and by the kill date with differences observed in the upper layers and the whole soil profile (Figure 7, Table 3). During both CC growth periods, SWC followed a similar pattern. At sowing, the three treatments started with low SWC in the entire profile (~220 mm) in both seasons, and only small differences appeared in the top layers. The precipitation during the three months following CC sowing recharged the soil profile, reaching SWC>300 mm in 2013 due to abundant rainfall (Figure 7a). No differences were observed between treatments during this period in any season. However, during the next three months, CC extracted water from the upper layers, and by the first kill date, SWC was higher in the fallow than in the CC treatments. After the first kill date, the SWC in the FK treatment varied compared to the SK depending on the annual weather conditions. During the first season, no differences were observed between the FK and SK treatments, most likely due to low precipitation, and by the second kill date, both had similar SWC and were lower than the fallow (Table 3). During the second season, the high precipitation between both kill dates recharged the soil profile, and the FK treatment SWC was similar to the fallow except in the deeper soil layer (100–120 cm, Figure 7e) where the fallow treatment remained wetter. For the second kill date of 2013, the SK treatment SWC was lower, and differences with the FK were obvious down to 80 cm and down to 120 cm with the fallow (Table 3).

Although initial differences in SWC were observed between the two years of the study during the mulch period, such differences were not evident toward the end of the season. After water

Table 3. Soil water content (mm) in each of the soil layers and the entire profile at different sampling dates for the fallow (Fa), first kill (FK) and second kill (SK) treatments.

		0–20		20–40		40–60		60–80		80–100		100–120		0–120	
		mm													
2011–2012															
07/10/2011	Fa	23.9	a	40.3	a	38.1	a	40.8	a	37.4	a	40.4	a	221	a
CC sowing	FK	18.7	a	30.8	b	39.4	a	41.7	a	44.2	a	43.4	a	218	a
	SK	18.5	a	35.1	ab	42.6	a	41.0	a	37.0	a	41.0	a	215	a
13/03/2012	Fa	25.2	a	42.7	a	40.5	a	37.5	a	34.1	a	34.9	a	215	a
First kill date	FK	12.4	b	28.7	b	33.3	a	35.5	a	35.5	a	36.5	a	182	b
	SK	11.4	b	31.7	b	35.5	a	32.2	a	33.1	a	35.5	a	179	b
09/04/2012	Fa	33.9	a	42.2	a	40.1	a	38.3	a	35.4	ab	36.3	a	226	a
Second kill date	FK	18.3	b	28.1	b	33.2	a	36.0	a	36.3	a	37.5	a	189	b
	SK	12.6	c	28.6	b	33.8	a	29.8	a	30.8	b	36.4	a	172	c
30/04/2012	Fa	9.4	c	27.6	c	32.6	a	36.2	a	34.1	ab	36.4	a	251	a
HPD	FK	10.6	b	27.4	b	36.9	a	42.7	a	41.5	a	44.0	a	266	a
	SK	12.8	a	32.0	a	38.7	a	37.3	a	37.4	b	43.6	a	269	a
12/09/2012	Fa	9.4	a	27.6	a	32.6	a	36.2	a	34.1	b	36.4	a	176	a
Mulch period end	FK	10.6	a	27.4	a	36.9	a	42.7	a	41.5	a	44.0	a	203	a
	SK	12.8	a	32.0	a	38.7	a	37.3	a	37.4	ab	43.6	a	202	a
2012–2013															
08/10/2012	Fa	43.9	b	55.3	a	31.3	a	33.6	a	32.0	b	34.4	a	231	a
CC sowing	FK	48.5	ab	55.7	a	36.3	a	39.7	a	38.5	a	40.5	a	259	a
	SK	52.7	a	55.2	a	39.0	a	34.7	a	35.3	ab	41.0	a	258	a
15/03/2013	Fa	50.3	a	56.0	a	58.8	a	56.4	a	44.7	a	39.1	a	305	a
First kill date	FK	43.4	b	47.3	b	41.1	b	43.8	b	40.8	a	38.3	a	255	b
	SK	41.2	b	42.2	c	39.3	b	39.3	b	37.4	a	39.2	a	239	b
11/04/2013	Fa	49.4	b	56.1	a	57.7	a	60.0	a	55.9	a	47.7	a	327	a
Second kill date	FK	53.9	a	59.1	a	61.2	a	58.2	a	43.8	b	39.8	b	316	a
	SK	39.1	c	50.4	b	41.0	b	40.1	b	37.8	b	39.3	b	248	b
30/04/2013	Fa	44.2	b	52.6	a	54.6	a	57.6	a	55.6	a	51.8	a	311	a
HPD	FK	52.7	a	55.4	a	57.6	a	59.5	a	50.0	a	42.4	b	307	a
	SK	40.3	c	41.7	b	42.2	b	40.8	b	38.8	b	40.5	b	244	b
01/09/2013	Fa	7.8	a	37.7	a	47.0	a	56.2	a	56.5	a	54.9	a	315	a
Mulch period end	FK	8.5	a	32.5	ab	45.3	a	50.9	a	50.1	ab	49.2	a	318	a
	SK	8.8	a	27.5	b	41.4	a	44.0	a	43.3	b	47.2	a	244	b

Within a date and depth, means with the same letter are not significantly different at P<0.05.

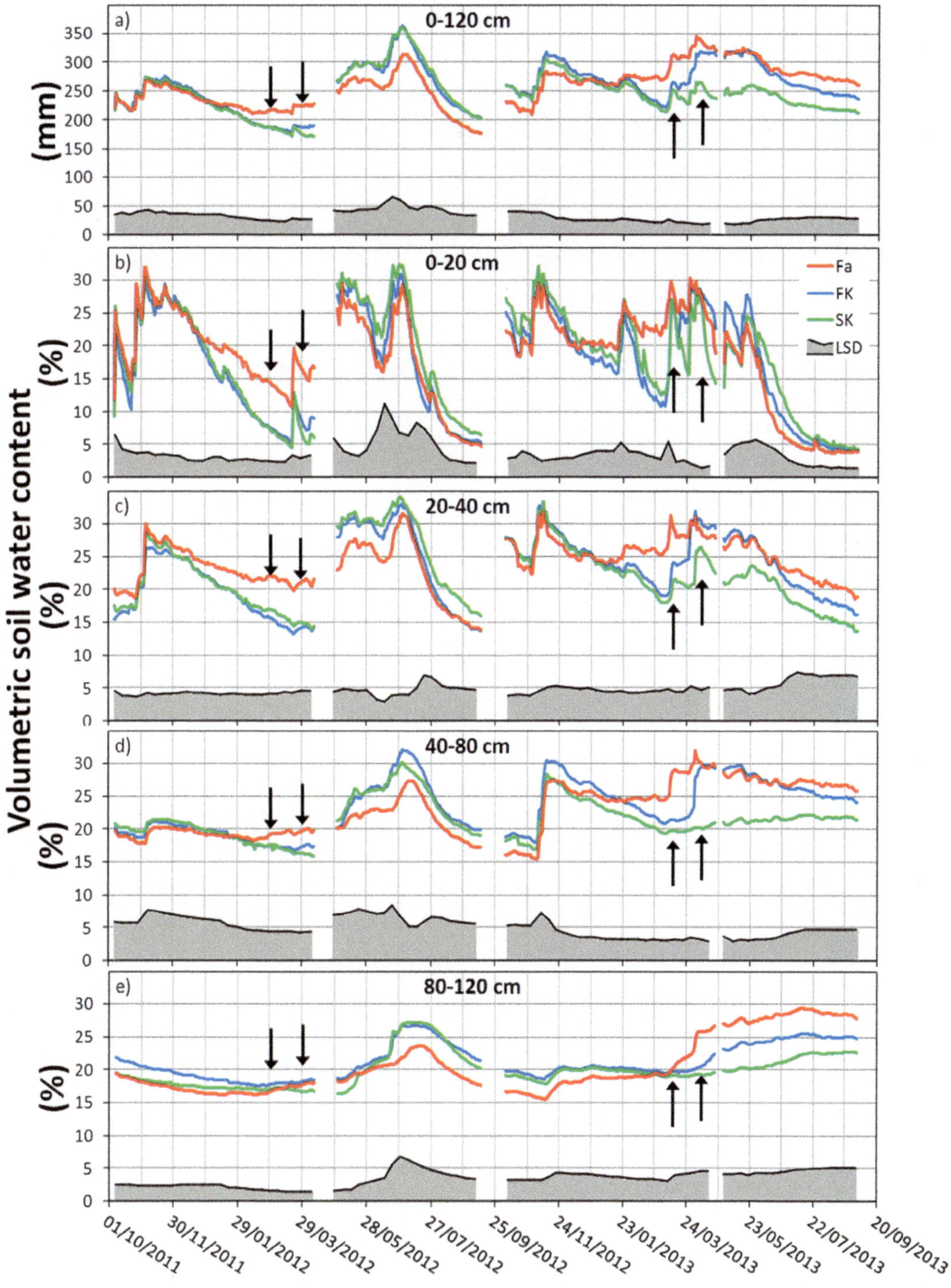

Figure 7. Soil water content monitored with the capacitance sensors at different depths for the fallow (Fa), first kill (FK) and second kill (SK) treatments during the experimental period. Arrows represent the first and second kill dates. The shadowed area in the bottom represents the LSD at 0.05 probability level.

application in April 2012, the SWC in the entire profile at the HPD was similar in all treatments (~262 mm; Table 3). During the wet/dry cycles that occurred in spring and summer, the CC treatments always retained more moisture than the fallow which showed the ability of the residue mulch to reduce soil water evaporation losses (Figure 7). After mid-June, when the soil was left to dry under the summer heat, differences in SWC between treatments disappeared, first in the upper layers and later in the entire soil profile. Even though slower drying was observed in the CC treatments, no differences were observed between treatments by the end of September (Table 3). In 2013, no water was applied, and the SWC in the entire soil profile was lower for the SK than for the FK and fallow treatments at the HPD (Table 3). At this time, no differences in SWC were observed between the FK and fallow treatments in the entire soil profile. However, the FK upper layer was wetter than the fallow because of the reduction in soil

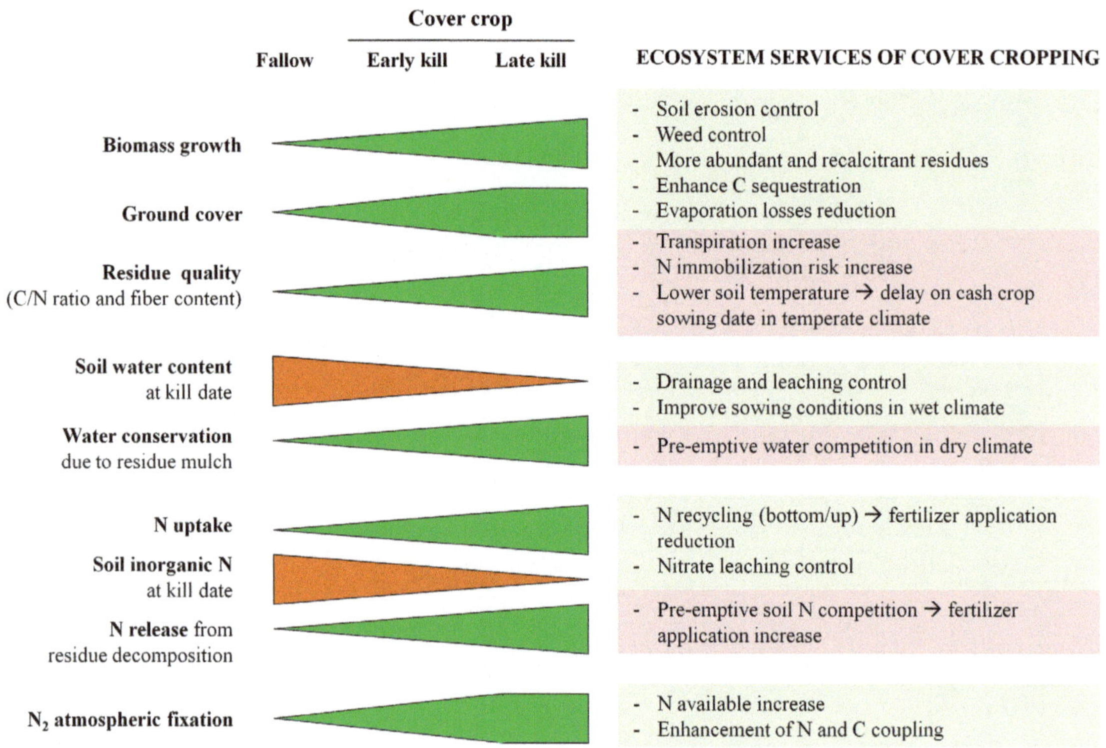

Figure 8. Schematic representation of the effect of kill date on environmental variables and ecosystem services with fallow as a reference condition.

water evaporation caused by the residue mulch (Table 3, Figure 7b). During summer, water losses in the top layer were greater in the fallow than the CC treatments, showing the effect of the mulch. By the end of the experiment, no differences in the SWC of the entire profile were observed between treatments, but some SK subsoil layers remained drier as the result of water depletion caused by the living CC.

Discussion

This study confirmed that CC kill date is an important management strategy as it affected key variables of the soil-plant system, influencing the success of the subsequent cash crop. Previous studies have focused mainly on the effect of N on the succeeding crop [6] and on the CC residue quality [14], but there is a lack of information on the combined effects of CC growth, N dynamics, and water dynamics. This information is necessary to optimize CC benefits and for development of more accurate simulation models.

Delaying kill date did not increase GC whereas the year had an effect on the maximum cover attained. We could not find data to compare the effect of kill date on GC. Nevertheless, our results agreed with those of previous researchers that showed the capacity of the asymptotic Gompertz function to capture the characteristics of the GC evolution of several CC [22,33]. Ground cover is an important variable used in studies related to soil erosion [32], evaporation [34], weed control [35,36] or radiation interception [22]. In addition, the evolution of digital technologies allows for non-destructive monitoring using reliable and efficient techniques [20]. For all of these reasons, we propose that GC should become a common variable for the characterization of CC that facilitates comparisons between species, varieties and management strategies in different regions. In the present study, GC differed with seasonal weather conditions; the dry 2011–2012 season was less favorable for CC growth than the wet and mild 2012–2013 season. In both seasons, GC>80% was already attained by the first kill date, so no additional benefits in terms of soil erosion or weed control were expected by the second kill date [32,35].

As expected, delaying kill date increased the above-ground biomass of the barley/vetch mixture. This increase was mainly due to barley as there was no difference in vetch biomass between kill dates. The barley/vetch ratio, expressed as seed weight at sowing, was 3:7, but barley was a stronger competitor due to faster establishment and growth rate [20]. The observed increase in biomass is common in many other studies [8,13], but the relationship between the legume and the non-legume differed. In Central Italy, a barley/vetch mixture with a 1:1 sowing rate yielded between 48 and 66% vetch biomass 160 d after planting which increased vetch in the sowing mixture up to 75% and produced between 53 and 76% vetch biomass at harvest [37]. A possible explanation for the large barley dominance in our study is that soil available N was not a limiting factor. Other authors reported similar superior dry matter values for the non-legume compared with the legume in mixtures, and this dominance became stronger with higher levels of N [38]. This hypothesis is corroborated by the plant and soil available N results in our study. Most of the N content in the vetch/barley mixture accumulated before the FK date. Nitrogen content of barley was greater at SK than FK (Figure 3), and in the SK treatment, the inorganic N in the soil was depleted in April, particularly from the deeper layers (Table 2, Figure 5). The high competition with barley for soil N forced the vetch to rely mainly on N_2 fixation for its N requirement, which explains the large degree of N_2 fixation observed in the legume. We were expecting the vetch to fix a significant amount of N_2 between both kill days, as observed in other experiments [10,13], but no increase in vetch N content was observed as in the study of Benincasa et al. [8]. The strong competition with barley for light explains the low vetch growth between kill dates and reinforces the need for sowing mixtures with a large proportion of vetch seeds (>50%) to favor legume activity. Most authors have found that the N content in CC and mixtures increased as kill date was delayed [10,13], but our results show that the main effect of delaying kill date increased above-ground biomass rather than N content.

The effect of postponing kill date on chemical residue composition was in agreement with previous findings in the literature [8,12,39]. Nitrogen concentration decreased from the first to the second kill date, being greater for vetch than for barley (Figure 3). These differences were reinforced in the C/N ratio as a slight increase in C content was observed in the SK samples compared with the FK (Figure 4). Vetch N concentration was similar to that found by other authors, but barley values were greater than those reported in other CC experiments [10,40,41]. In the FK samples, barley N concentration was ~2.9% and ~2.2% in the SK which agrees with data from barley cultivated as a cash crop using synthetic N fertilizer [42]. These data confirm that barley did not suffer from lack of N. The increase in fiber fractions, which is related to a decrease in the labile fraction, from delaying kill date was consistent with other studies and was expected to lead to slower C and N mineralization [43]. Increases in lignin fraction were only significant in barley in the first season, but this is not surprising as there were only three weeks between kill dates, and lignin content is less sensitive to change [39]. In the drier season (2011–2012), barley residue and the mixture had significantly lower fiber and lignin content than in the wetter season for both kill dates. A reduction in NDF, ADF and lignin cereal crop straw in dry years was previously noted in a study conducted over several years and locations in Washington State [44]. Our results confirmed that delaying kill date increased the amount of CC residue and the residues were more recalcitrant to decomposition. This is particularly interesting when the goal is preserving soil moisture [11], controlling weeds [35], or increasing the slow release of the N pool [45]. In the present experiment, a detailed study of CC residue decomposition was not conducted primarily because the lowest part of the residue layer stuck to the soil when sampling, and we decided that it was reliable for determining the amount of the remaining residue but not its quality.

The topsoil N_0 was also affected by the kill date two years after starting the experiment. Soil N mineralization potential assess the capacity of a soil to mineralize N and is a good indicator of land use as it is very sensitive to small changes in labile C and N fractions [46]. The increase of N_0 in the CC treatments proves that CC contribute to the enhancement of labile soil organic pools. The low contribution of CC to total organic C is a known fact as residues contain low quantities of recalcitrant organic matter [2]. However, CC add substrate for soil microorganisms during their growth period (root exudation, root turnover), and residue decomposition increases the organic matter labile fractions [6]. Many studies of N mineralization from CC residues have been published, but there is a lack of studies that focus on the effects on the soil in which they were cropped. In our study, no differences were observed in total soil C, but delaying kill date increased the C supplied by CC and, as a consequence, topsoil N_0. Nitrogen mineralization rate, a parameter that mainly depends on environmental conditions, was not different between treatments and was within the range found in the literature [47].

Consistent with other studies, CC absorbed most of the available soil inorganic N during their growth period, modifying the degree of pre-emptive competition and the risk of N loss with respect to the fallow (Table 2, Figure 5). This study also showed that kill date affected these parameters and could be a means to adapt soil N availability to the requirements of a particular agroecosystem. Delaying kill date increased the risk of pre-emptive competition, a negative effect that was reflected not only in the deeper layers as is common [4,48], but in the upper layers during the second season as well. The early kill date allowed more time for N release from CC residues, and by the HPD, more N was available in the topsoil. This positive effect of an early kill date on reducing pre-emptive competition has not been shown before and was evident when comparing FK with the SK and the fallow in the wet year, in which N_{min} was washed out of the soil during the fall and winter. The well-known CC effect of recycling N in the soil-plant system by uptaking inorganic N from deeper layers and subsequently releasing it to the top layers was clear in this study [6]. In the first season, the soil N_{min} depleted in April by the CC uptake was replenished by October by the N released from the residue mineralization. In the second season, CC prevented most of the available N from being lost from the soil, as happened in the fallow, and released the N by November. The difference in November was not in the amount of N_{min} but its distribution in the soil profile. In the fallow, N_{min} was mostly in the deeper layers and had greater potential for leaching during fall and winter. However, in the SK treatment, N_{min} was in the top layers making it readily available for uptake by the subsequent crop. The N_{min} profile of the FK soil was intermediate. Organic N remained in the residue layer at the end of the experiment, but as mentioned before, we could not obtain a reliable measure of N concentration.

Even if N uptake by CC depleted the soil by April in the first season and seemed to increase the risk of pre-emptive competition with respect to the fallow, we propose that this may be a positive effect in many agricultural systems as it may help control N losses. In a tomato drip irrigation study, it was observed that high soil N_{min} levels at spring planting were linked to large leaching losses as crops are often overwatered to ensure establishment [49]. Excessive irrigation during the crop establishment period may represent up to 80% of total nitrate leaching losses [34,50]. Therefore, keeping soil N_{min} at low levels and correcting fertilizer needs based on crop nutritional status during the cropping season holds promise for controlling N losses and increasing the efficiency of N use [51].

The larger above-ground biomass observed in the SK affected SWC. The CC extracted soil water by transpiration, increasing evapotranspiration losses when compared with the fallow [34]. Spring growth of the vetch/barley mixture was vigorous and soil water depletion was enhanced between the first and the second kill date. The differences in the whole profile SWC at the first kill date between CC treatments and the fallow were >35 mm in the first season and >55 in the second. At the second kill date, the SWC in the FK treatment was similar to the fallow. By the HPD and during the rest of the mulch period, the treatment effect on water availability was variable. While in the first season there was no difference, most likely because the soil was replenished by irrigation, the SWC was 60 mm lower in the SK than in the other treatments during the second season. Similar results were found in California [52,53], where reductions in spring SWC were up to 80 mm due to CC extraction. In Minnesota, the difference in SWC in the upper 0.6 m caused by delaying the rye kill date by three weeks (from April to May) was 27 mm [12]. Our results confirm that late killing of CC increases pre-emptive water competition compared with early killing, a risk that might be mitigated in rainy years or by irrigation if water is available.

A relevant aspect of the increased water uptake by CC is the effect on the upper layers' SWC. Sufficient top layer moisture at planting is crucial to ensure crop establishment and plant survival in many crops [49]. In the present study, the difference in SWC between the fallow and the CC treatments at the first kill date due to water depletion ranged between 13–19 mm in the upper 0.2 m of the soil and was enhanced between the fallow and the SK by the second kill date. However, the effect of CC residue mulching led to a higher SWC in the upper layers of the FK treatment than in the fallow at the HPD. When comparing CC treatments, the results are less clear; more water in the upper layers was retained in the SK than in the FK in the first season, but the SK was drier in the second. The FK treatment always increased water retention in the upper layers with respect to the fallow, but the result varied by year in the SK treatment. The top layer SWC was influenced by CC water uptake during winter, the residue mulch effect, and the amount of rain between the CC kill date and cash crop planting [9]. To clarify the discrepancy between the soil water depletion by CC uptake and conservation by the mulched CC residue, an experiment comparing sheltered and un-sheltered plots in Maryland showed that delaying kill date of a vetch CC produced more biomass and resulted in better soil moisture conservation in the upper 0.3 m [11]. However, recent experiments in Minnesota reported that delaying the rye kill date from the tillering to the booting stage came at the cost of SWC depletion in the top 0.3 m even if the residues continued to cover the soil after termination [12]. In Indiana, early kill of the CC resulted in better water conservation in the topsoil (0.1 m) than late kill. In drought years, early kill date produced an increase in maize yield but had a negative impact in normal precipitation years due to a lowering of soil temperature and workability at the maize planting date [7]. Our results indicate that early kill of the CC enhanced spring topsoil water moisture with respect to late kill, particularly by preserving soil water replenished by precipitation between kill dates.

A limitation found when comparing our study with others that focus on CC killing was that kill dates were always expressed as days after sowing. Given the variety of climatic regions in which cover cropping is practiced, the number of days after sowing is not a valid variable. We propose using degree-days accumulated and growth stage as more suitable variables for comparison between different regions. Soil type also has a major influence on the water balance and N cycling. Simulation models that can take multiple environmental factors into account might allow to overcome these limitations and increase the applicability of this and others studies to different locations and years.

One criticism of this study might be that a crop was not planted after the CC to obtain real data on crop water uptake. We are aware of this situation, but we think that this type of dataset is necessary to advance our understanding of the complex effect of CC kill date on crop resources. Separating the mulching and the effect of transpiration might allow for the calibration of key parameters, proper simulation of both processes, and finally, of the whole system. Replacing fallows by CC is known to provide many benefits, but the agronomic application needs to account for the requirements of specific agroecosystems. Kill date is a management tool to regulate the effects of CC on many environmental variables as presented in Figure 8. In dry environments, residue moisture conservation is the main goal of CC [9] whereas the risk of delaying sowing because of soils that are too wet is a major concern in cooler climates [7]. Depleting soil N enhances pre-emptive competition with the subsequent crop [6] but also controls N losses and creates opportunities for rational fertilizer application [34]. Late termination of CC maximizes their biomass production and the formation of a thick mulch, enhancing C sequestration, water conservation and weed control [36]. Our work does not close the debate on kill date but contributes to more rational decision-making for the successful use of CC.

Conclusions

Kill date is a major management variable with CC that may lead to increasing water and N use efficiency and controlling N losses and pre-emptive competition with the subsequent crop. The ground cover of CC was >80% by the first kill date so no more benefits in terms of soil erosion or weed control derived from the living CC were expected by the second kill date. However, delaying kill date increased above-ground biomass, providing more CC residues with a higher fiber content and C/N ratio that were more recalcitrant to decomposition and, therefore, more suitable to protect the soil and enhance the slow release of the N pool.

The early kill date decreased the risk of pre-emptive competition by diminishing the N uptake of CC, and allowing more time for N release from CC residues. Cover crops generally uptake inorganic N from deeper layers to be released on the topsoil through residue mineralization. Delaying kill date enhanced this recycling effect that prevented losses of available N by keeping soil inorganic N at low levels. Kill date was a means of controlling this process and showed promise in mitigating N losses and increasing N use efficiency. The N mineralization potential in topsoil was enhanced by the presence of CC and by delaying kill date, proving that CC contribute to the enhancement of labile organic pools.

Delaying kill date increased pre-emptive competition of water, as the water extracted by CC from the first to the second kill date was greater than the water conservation due to the extra residue generated by the SK treatment. The soil water content in the upper layers at the time of planting the subsequent crop increased by the first kill date with respect to the fallow by preserving rain harvested between kill dates, but the soil water content in the late kill could be even lower than in the fallow depending on the climatic conditions of the year.

Supporting Information

Table S1 Biomass, N content and residue left. Above-ground biomass and N content of cover crops at the first kill (FK) and second kill (SK) dates and residue covering the soil at the end of the mulch period (~7 months after cover crop killing). Means with standard error in parentheses. Within a row, means with the same letter are not significantly different between kill dates at $P<0.05$.

Table S2 Chemical composition of cover crops. Cover crop residue neutral detergent fiber (NDF), acid detergent fiber (ADF) and lignin (L) fractions and C/N ratio for the first kill (FK) and second kill (SK) dates. Means with standard error in parentheses. Within a row, means with the same letter are not significantly different between kill dates at $P<0.05$.

Image S1 Aerial view of the experimental site. Plots are marked as fallow (FA), first kill (FK) nd second kill (SK) plus the replication number. Image taken between kill dates (March 29, 2012).

Image S2 Ground level view of the experimental site. Image taken between kill dates (April 4, 2012).

Acknowledgments

We would like to thank the staff of La Chimenea field station (IMIDRA) for their helpful assistance.

Author Contributions

Conceived and designed the experiments: MQ. Performed the experiments: MA JLG MQ. Analyzed the data: MA JLG MQ. Contributed to the writing of the manuscript: MA JLG MQ.

References

1. Tonitto C, David MB, Drinkwater LE (2006) Replacing bare fallows with cover crops in fertilizer-intensive cropping systems: A meta-analysis of crop yield and N dynamics. Agriculture, Ecosystems & Environment 112: 58–72.
2. Kuo S, Sainju UM, Jellum EJ (1997) Winter cover crop effects on soil organic carbon and carbohydrate in soil. Soil Science Society of America Journal 61: 145–152.
3. Quemada M, Cabrera ML (2002) Characteristic moisture curves and maximum water content of two crop residues. Plant and Soil 238: 295–299.
4. Gabriel JL, Quemada M (2011) Replacing bare fallow with cover crops in a maize cropping system: Yield, N uptake and fertiliser fate. European Journal of Agronomy 34: 133–143.
5. Abawi GS, Widmer TL (2000) Impact of soil health management practices on soilborne pathogens, nematodes and root diseases of vegetable crops. Applied Soil Ecology 15: 37–47.
6. Thorup-Kristensen K, Magid J, Jensen LS (2003) Catch crops and green manures as biological tools in nitrogen management in temperate zones. Advances in Agronomy 79: 227–302.
7. Stipesevic B, Kladivko EJ (2005) Effects of winter wheat cover crop desiccation times on soil moisture, temperature and early maize growth. Plant Soil and Environment 51: 255–261.
8. Benincasa P, Tosti G, Tei F, Guiducci M (2010) Actual N Availability from Winter Catch Crops Used for Green Manuring in Maize Cultivation. Journal of Sustainable Agriculture 34: 705–723.
9. Clark AJ, Meisinger JJ, Decker AM, Mulford FR (2007) Effects of a grass-selective herbicide in a vetch-rye cover crop system on corn grain yield and soil moisture. Agronomy Journal 99: 43–48.
10. Tosti G, Benincasa P, Farneselli M, Pace R, Tei F, et al. (2012) Green manuring effect of pure and mixed barley-hairy vetch winter cover crops on maize and processing tomato N nutrition. European Journal of Agronomy 43: 136–146.
11. Clark AJ, Decker AM, Meisinger JJ, McIntosh MS (1997) Kill date of vetch, rye, and a vetch-rye mixture: 2. Soil moisture and corn yield. Agronomy Journal 89: 434–441.
12. Krueger ES, Ochsner TE, Porter PM, Baker JM (2011) Winter Rye Cover Crop Management Influences on Soil Water, Soil Nitrate, and Corn Development. Agronomy Journal 103: 316–323.
13. Clark AJ, Decker AM, Meisinger JJ, McIntosh MS (1997) Kill date of vetch, rye, and a vetch-rye mixture: 1. Cover crop and corn nitrogen. Agronomy Journal 89: 427–434.
14. Wagger MG, Cabrera ML, Ranells NN (1998) Nitrogen and carbon cycling in relation to cover crop residue quality. Journal of Soil and Water Conservation 53: 214–218.
15. Tosti G, Benincasa P, Farneselli M, Tei F, Guiducci M (2014) Barley-hairy vetch mixture as cover crop for green manuring and the mitigation of N leaching risk. European Journal of Agronomy 54: 34–39.
16. Soil Survey Staff (2014) Keys to Soil Taxonomy. Washington, DC, USA: USDA, Natural Resources Conservation Service.
17. Nelson DW, Sommers LE (1996) Total carbon, organic carbon and organic matter. In: Sparks D, editor. Methods of soil analysis, part 3: chemical methods. Madison, WI, USA: ASSA and SSSA. 961–1010.
18. Loeppert RD, Suarez DL (1996) Carbonate and Gypsum. In: Sparks D, editor. Methods of soil analysis, part 3: chemical methods. Madison, WI, USA: ASA and SSSA. 437–474.
19. Papadakis J (1966) Climates of the world and their agricultural potentialities. Rome, Italy: DAPCO.
20. Ramirez-Garcia J, Almendros P, Quemada M (2012) Ground cover and leaf area index relationship in a grass, legume and crucifer crop. Plant Soil and Environment 58: 385–390.
21. Richardson MD, Karcher DE, Purcell LC (2001) Quantifying turfgrass cover using digital image analysis. Crop Science 41: 1884–1888.
22. Bodner G, Himmelbauer M, Loiskandl W, Kaul HP (2010) Improved evaluation of cover crop species by growth and root factors. Agronomy for Sustainable Development 30: 455–464.
23. Unkovich M, Herridge D, Peoples M, Cadisch G, Boddey B, et al. (2008) Measuring plant-associated nitrogen fixation in agricultural systems. Camberra, Australia: Australian Centre for International Agricultural Research. 258 p.
24. Goering HK, Van Soest PJ (1970) Forage fiber analysis: apparatus, reagents, procedures, and some applications. Washington, DC: Agricultural Research Service. Handbook no. 379. US Government Printing Office.
25. Keeney DR, Nelson DW (1982) Nitrogen–inorganic forms. In: Page AL, editor. Methods of soil analysis Part 2: Chemical and microbiological properties. Madison, WI, USA: ASA and SSSA. 643–698.
26. Crooke WM, Simpson WE (1971) Determination of ammonium in Kjeldahl digests of crops by an automated procedure. Journal of the Science of Food and Agriculture 22: 9–10.
27. Stanford G, Smith SJ (1972) Nitrogen mineralization potentials of soils. Soil Science Society of America Proceedings 36: 465–472.
28. Paltineanu IC, Starr JL (1997) Real-time soil water dynamics using multisensor capacitance probes: Laboratory calibration. Soil Science Society of America Journal 61: 1576–1585.
29. Gabriel JL, Lizaso JI, Quemada M (2010) Laboratory versus field calibration of capacitance probes. Soil Science Society of America Journal 74: 593–601.
30. Quemada M, Cabrera ML (1995) CERES-N model predictions of nitrogen mineralized from cover crop residues. Soil Science Society of American Journal 59: 1059–1065.
31. Whitmore AP (1991) Method for assessing the goodness of computer simulation of soil processes. Journal of Soil Science 42: 289–299.
32. Quinton JN, Edwards GM, Morgan RPC (1997) The influence of vegetation species and plant properties on runoff and soil erosion: results from a rainfall simulation study in south east Spain. Soil Use and Management 13: 143–148.
33. Ramirez-Garcia J, Gabriel JL, Alonso-Ayuso M, Quemada M (2014) Quantitative characterization of five cover crop species. Journal of Agricultural Science. doi:10.1017/S0021859614000811.
34. Gabriel JL, Muñoz-Carpena R, Quemada M (2012) The role of cover crops in irrigated systems: Water balance, nitrate leaching and soil mineral nitrogen accumulation. Agriculture Ecosystems & Environment 155: 50–61.
35. Teasdale JR, Mohler CL (1993) Light transmittance, soil-temperature, and soil-moisture under residue of hairy vetch and rye. Agronomy Journal 85: 673–680.
36. Saini M, Price AJ, Van Santen E, Arriaga FJ, Balkcom KS, Raper RL (2008) Planting and termination dates affect winter cover crop biomass in a conservation-tillage corn-cotton rotation: Implications for weed control and yield. In: Endale DM editor. Tifton, GA, USA. Southern Conservation Agricultural Systems. 137–141.
37. Tosti G, Benincasa P, Guiducci M (2010) Competition and Facilitation in Hairy Vetch-Barley Intercrops. Italian Journal of Agronomy 5: 239–248.
38. Ofori F, Stern WR (1987) The combined effects of nitrogen-fertilizer and density of the legume component on production efficiency in a maize cowpea intercrop system. Field Crops Research 16: 43–52.
39. Wagger MG (1989) Time of desiccation effects on plant composition and subsequent Nitrogen release from several winter annual cover crops. Agronomy Journal 81: 236–241.
40. Clark AJ, Decker AM, Meisinger JJ (1994) Seeding rate and kill date effects on hairy vetch cereal rye cover crop mixtures for corn production. Agronomy Journal 86: 1065–1070.
41. Benincasa P, Guiducci M, Tei F (2011) The nitrogen use efficiency: Meaning and sources of variation-case studies on three vegetable crops in Central Italy. HortTechnology 21: 266–273.
42. Arregui LM, Quemada M (2008) Strategies to improve nitrogen use efficiency in winter cereal crops under rainfed conditions. Agronomy Journal 100: 277–284.
43. Quemada M, Cabrera ML (1995) Carbon and nitrogen mineralized from leaves and stems of four cover crops. Soil Science Society of America Journal 59: 471–477.
44. Stubbs TL, Kennedy AC, Reisenauer PE, Burns JW (2009) Chemical composition of residue from cereal crops and cultivars in dryland ecosystems. Agronomy Journal 101: 538–545.
45. Quemada M, Cabrera ML, McCracken DV (1997) Nitrogen release from surface-applied cover crop residues: Evaluating the CERES-N submodel. Agronomy Journal 89: 723–729.
46. Cookson WR, Abaye DA, Marschner P, Murphy DV, Stockdale EA, et al. (2005) The contribution of soil organic matter fractions to carbon and nitrogen mineralization and microbial community size and structure. Soil Biology and Biochemistry 37: 1726–1737.
47. Quemada M, Diez JA (2007) Available nitrogen for corn and winter cereal in Spanish soils measured by electro-ultrafiltration, calcium chloride, and incubation methods. Communications in Soil Science and Plant Analysis 38: 2061–2075.
48. Thorup-Kristensen K (1994) The effect of nitrogen catch crop species on the nitrogen nutrition of succeeding crops. Fertilizer Research 37: 227–234.
49. Vázquez N, Pardo A, Suso ML, Quemada M (2005) A methodology for measuring drainage and nitrate leaching in unevenly irrigated vegetable crops. Plant and Soil 269: 297–308.
50. Salmerón M, Isla R, Cavero J (2011) Effect of winter cover crop species and planting methods on maize yield and N availability under irrigated Mediterranean conditions. Field Crops Research 123: 89–99.
51. Vázquez N, Pardo A, Suso ML, Quemada M (2006) Drainage and nitrate leaching under processing tomato growth with drip irrigation and plastic mulching. Agriculture Ecosystems & Environment 112: 313–323.
52. McGuire AM, Bryant DC, Denison RF (1998) Wheat yields, nitrogen uptake, and soil moisture following winter legume cover crop vs. fallow. Agronomy Journal 90: 404–410.
53. Mitchell JP, Peters DW, Shennan C (1999) Changes in soil water storage in winter fallowed and cover cropped soils. Journal of Sustainable Agriculture 15: 19–31.

22

Why Some Plant Species Are Rare

G. W. Weiger Wamelink[1]*, Paul W. Goedhart[2], Josep Y. Frissel[1]

1 Alterra, Wageningen UR, Wageningen, the Netherlands, 2 Biometris PRI, Wageningen UR, Wageningen, the Netherlands

Abstract

Biodiversity, including plant species diversity, is threatened worldwide as a result of anthropogenic pressures such as an increase of pollutants and climate change. Rare species in particular are on the verge of becoming extinct. It is still unclear as to why some plant species are rare and others are not. Are they rare due to: intrinsic reasons, dispersal capacity, the effects of management or abiotic circumstances? Habitat preference of rare plant species may play an important role in determining why some species are rare. Based on an extensive data set of soil parameters we investigated if rarity is due to a narrow habitat preference for abiotic soil parameters. For 23 different abiotic soil parameters, of which the most influential were groundwater-table, soil-pH and nutrient-contents, we estimated species responses for common and rare species. Based on the responses per species we calculated the range of occurrence, the range between the 5 and 95 percentile of the response curve giving the habitat preference. Subsequently, we calculated the average response range for common and rare species. In addition, we designed a new graphic in order to provide a better means for presentation of the results. The habitat preferences of rare species for abiotic soil conditions are significantly narrower than for common species. Twenty of the twenty-three abiotic parameters showed on average significantly narrower habitat preferences for rare species than for common species; none of the abiotic parameters showed on average a narrower habitat preference for common species. The results have major implications for the conservation of rare plant species; accordingly management and nature development should be focussed on the maintenance and creation of a broad range of environmental conditions, so that the requirements of rare species are met. The conservation of (abiotic) gradients within ecosystems is particularly important for preserving rare species.

Editor: Hans Henrik Bruun, University Copenhagen, Denmark

Funding: This work was financed by the Dutch Ministry of Economical Affairs. The funders had no role in study design, data collection and analysis, decision to publish, or preparation of the manuscript.

Competing Interests: The authors have declared that no competing interests exist.

* Email: wieger.wamelink@wur.nl

Introduction

Biodiversity, including plant species richness, is threatened worldwide [1], [2], [3], as a result of anthropogenic pressures such as an increase of pollutants [4] and climate change [5]. Rare species in particular are on the verge of becoming extinct. It is still unclear as to why some plant species are rare and others are not [6], [7], [8], [9]. Are they rare due to intrinsic reasons, dispersal capacity, the effects of management or abiotic circumstances? Habitat preference of rare plant species may play an important role in determining why some species are rare and others are not.

Species occurrence can also be limited by dispersal capacity, for instance when all habitat requirements are met but the species is not yet able to reach suitable habitat [10], [11], [12]. Human influence is another major factor that impacts on the occurrence of rare species; for example, through changes in habitats caused by construction of infrastructure and built development, water-related management or intensified agricultural land use [10], [13], [14], [15], [16]. Nature management can be applied to conserve the habitats necessary for the rare species and thus prevent them from becoming extinct [16], [17]. Understanding the differences in habitat requirements between rare species and common species is also likely to be an important factor in protecting rare species [6].

All plant species establish their own niches in their preferential habitat and this is a major determinant of their spatial distribution. The occurrence of species is being determined at different scales, ranging from the biogeographical till the habitat scale. On the biogeographical scale species occurrence is probably mostly limited by climatic parameters such as temperature and precipitation. The realized distribution of the biogeographical niche is often referred to as the climate envelope of a species and used to predict the effects of climate change on species occurrence [12], [18]. Within the climate envelope some habitats are suitable and some are not. Species occurrence on the habitat scale can be limited by factors such as the vegetation types that are present as a result of management, the absence of suitable abiotic conditions including soil pH and nutrient availability, the status of the groundwater table (both water quality and availability) or fine scale heterogeneity in for example vegetation structure, soil gradients or management related gradients [17], [19], [20].

The specific topic of this paper is to consider the influence of abiotic (soil) conditions in relation to rarity of species. We hypothesised that rare species have narrower habitat preferences for abiotic soil conditions than common species. Based on an extensive data set of measured soil parameters we therefore investigated if rarity of plant species is constrained by their restricted habitat preference for abiotic soil parameter.

Material and Methods

Database and abiotic soil parameter selection

Species responses for abiotic soil parameters were estimated based on field measurements in just over 8000 plots in the Netherlands, mostly collected from literature. For each plot species composition was recorded and a mixed soil sample was taken from the upper soil layer (mostly upper 10 cm) and analysed in a laboratory. Plot sizes ranged from 1 m^2 for grasslands till 100 m^2 for forest. Plots were recorded following the Brown-Blanquet method [21]. The abiotic values of the plot were linked to all the species in the plot, including mosses, lichens shrubs and trees. The data cover the period from 1936 till present day and the database is still expanding. Each plot is accompanied by at least one abiotic soil measurement (e.g. pH or nutrient availability). The most frequently measured soil parameter is pH, with well over 5,000 entries. The database is part of the European metadata database for vegetation plots (the Ecological Conditions Database; GIVD ID EU-00-006) [22].

Abiotic ranges for an individual species were only estimated when at least 25 positive findings in relation to the abiotic parameter combination were present in the database. The procedure was carried out for the 23 different soil parameters (Table 1).

Species response

Responses per species were modelled by means of logistic regression employing a penalized B-spline [23] to estimate the curve (details in Wamelink *et al.*) [24]. Applying the spline function has the advantage that the responses of the species are independent of the number of findings. By dividing the abiotic (x) axes in 25 parts and subsequently calculating the chance of occurrence per part, a bias due to the number of findings is prevented. For each species an indicator value was derived as the mean of the response curve. We could have used the spline to estimate the range of the species. However, this would give ranges for only a limited number of species and based on a relative small number of relevés (just over 7000 for soil pH and over 1000 for the other abiotic parameters). The species indicator values were therefore used to calculate the mean abiotic values for a calibration dataset with 160,000 relevés, representative for the Dutch vegetation [25]. However, this was never tested and relevés were made for all kind of purposes and are therefore not random divided over Dutch nature. From the resulting responses for the abiotic parameters per species the 5 and 95 percentile were retrieved from a re-estimated species response curve, employing the full calibration set of 160,000 relevés, giving the range at which each species occurs (Figure 1). By applying this calibration dataset we are able to increase the number of species for which we can estimate a response. The length of the range of the species was defined as the difference between the 5 and 95 percentile of the response curve. The full method and the database used are described in Wamelink *et al.* [19], [22], [24].

To counteract the effect of non-random selection of the plots we used the 5 and 95 percentiles of the ranges of the fitted species response curve (Figure 1), instead of simply the abiotic value of the relevés at which the species occurs. To understand why this works consider the case in which relatively more narrow abiotic values are present in the database, or for a part of the abiotic range the plots are overrepresented. Using only the abiotic values for which a species is present will inevitably result in percentiles which are too narrow; the overrepresented plots will dominate the range. Since the effect of this might be different for the 5 and 95 percentile, this will also affect the range. However, the fitted species response curve, employing presence and absence data, will in principle not be affected by the non-random selection of the abiotic values as long as there are data for the full range of the abiotic values. Consequently, the range as estimated from the fitted curve will not be heavily affected.

Rare species selection

The species list with responses was split in two: one part containing rare species and one containing the other species. The latter group will subsequently be called common species. Rare species were defined as species fulfilling the red list criteria [26] and as such indicated in the Dutch flora [27]. The rarity of the species is based on their frequency and trend as measured on the Dutch national 5 by 5 km grid, based on the period 1980–1990. There are five levels for the Dutch red list:

1. Red list category 0: species extinct in the Netherlands
2. Red list category 1; Species occurs in 1–12 grids, with a decrease in grid frequency of at least 50%, or species occurs in 13–40 grids, with a decrease in grid frequency of at least 75%.
3. Red list category 2: Species occurs in 1–12 grids, with a decrease in grid frequency of 25–50%, or species occurs in 13–40 grids, with a decrease in grid frequency of 50–75%, or the species occurs in 41–225 grids, with a decrease in frequency of at least 75%.
4. Red list category 3: the species occurs in 13–40 grids, with a decrease of 25–50%, or the species occurs in 41–225 grids, with a decrease of 25–75%.
5. Red list category 4: the species occurs in 1–60 grids, its occurrence is more or less stable and the species is not under immediate threat of becoming extinct.

All the species that fulfil one of the five above given criteria were merged into the set of rare species, in total 190 species out of 973 species.

Statistical analysis

Means and standard errors for the ranges were calculated for the rare species group and the common species groups. A two sided student t-test was used to test for a difference between the averages of both groups for each abiotic soil parameter. All calculation, including the estimation of the spline functions were done in GenStat version 15 [28].

To test the results we looked within the red list species in order to establish whether ranges of species depend on the rarity of the species. To this end the red list species were divided into four pools: species that are not rare but made the list because of their decline (a), species that are rare (r), species that are very rare (rr) and species that are almost extinct (rrr, see Appendix S1, rarity is also based on Meijden [27]). Note: these categories were used in preference to the list above because the first list is a combination of rarity and the trend of a species, while the list presented here only is based on the rarity of the species and the effect of the latter is investigated here. To test for differences between the pools we simply ranked the average ranges of the four 'rarity groups' from one till four per abiotic parameter and then calculated the average ranking over the 23 abiotic parameters.

Results

Species responses

Species were only included in further analyses if ranges could be estimated for a minimum of 20 abiotic parameters. Responses were thus estimated for 973 species (783 common and 190 rare species, for an example see Figure 2). The newly designed figures give clearly the difference in ranges between the common species *Agrostis canina* (brown bent) and the rare species *Allium oleraceum* (field garlic) (Figure 2a versus 2b). The bars of the spokes of the wheels of the species show that the ranges (bars) for most of the abiotic parameters for *A. oleraceum* are narrower than

Table 1. Species ranges for 23 abiotic soil parameters for rare and common species, including standard error, the difference between both groups and the p-values for the difference (two tail student t-test, n.s.; not significant, * $p<0.05$, ** $p<0.01$, *** $p<0.001$).

Parameter	unit	Common species		rare species		Δ	against rare	p
		range	s.e.	range	s.e.		%	
C/N	-	4.95	1.97	4.47	2.19	0.48	10.63	0.004^{**}
C_{total}	mg/kg	10.92	3.80	9.80	3.59	1.12	11.45	0.000^{***}
Ca_{H2O}^{1}	mg/kg	5092.36	2028.77	4707.72	1865.53	384.64	8.17	0.017^{*}
$CaCO_3$	mg/kg	2.77	1.11	2.48	1.29	0.29	11.86	0.002^{**}
Cl	mg/kg	209.86	199.28	168.42	199.15	41.44	24.60	0.010^{*}
Electronic conductivity	mS	518.06	205.58	481.61	219.12	36.45	7.57	0.031^{*}
K_{H2O}^{1}	mg/kg	102.05	32.03	87.47	27.61	14.58	16.67	0.000^{***}
Mg_{H2O}^{1}	mg/kg	151.42	62.91	142.84	66.85	8.58	6.01	$0.096^{n.s.}$
MHG^{2}	cm -bs	47.74	16.19	47.50	19.57	0.24	0.51	$0.860^{n.s.}$
MLG^{3}	cm -bs	57.54	21.59	56.01	27.20	1.52	2.72	$0.409^{n.s.}$
MSG^{4}	cm -bs	40.38	12.16	34.44	10.86	5.94	17.24	0.000^{***}
Moisture	%	19.36	7.08	18.00	7.14	1.36	7.57	0.018^{*}
Na_{CaCl2}^{5}	mg/kg	56.88	30.22	47.67	28.05	9.20	19.31	0.000^{***}
Na_{H2O}^{1}	mg/kg	263.51	142.98	209.74	131.69	53.78	25.64	0.000^{***}
$NH4_{CaCl2}^{5}$	mg/kg	6.64	2.61	5.67	2.61	0.96	16.96	0.000^{***}
$NO3_{CaCl2}^{5}$	mg/kg	19.72	6.96	13.69	7.27	6.03	44.03	0.000^{***}
N_{total}	mg/kg	3896.27	1845.54	3467.11	1526.65	429.16	12.38	0.003^{**}
Organic matter	mg/kg	9.04	4.33	8.43	4.41	0.60	7.16	$0.086^{n.s.}$
P_{CaCl2}^{5}	mg/kg	1.62	0.66	1.25	0.73	0.37	29.20	0.000^{***}
$P_{citric\ acid}$	mg/kg	290.97	99.87	218.46	82.64	72.52	33.19	0.000^{***}
P_{total}	mg/kg	434.98	167.16	359.89	162.22	75.09	20.86	0.000^{***}
pH_{H2O}	-	1.16	0.41	0.96	0.42	0.20	20.88	0.000^{***}
pH_{KCl}	-	1.18	0.49	0.99	0.49	0.18	18.65	0.000^{***}

Results per species are given in Appendix S1.
[1]In water extract,
[2]mean highest groundwater level,
[3]mean lowest groundwater level,
[4]mean spring groundwater level,
[5]in $CaCl_2$ extract.

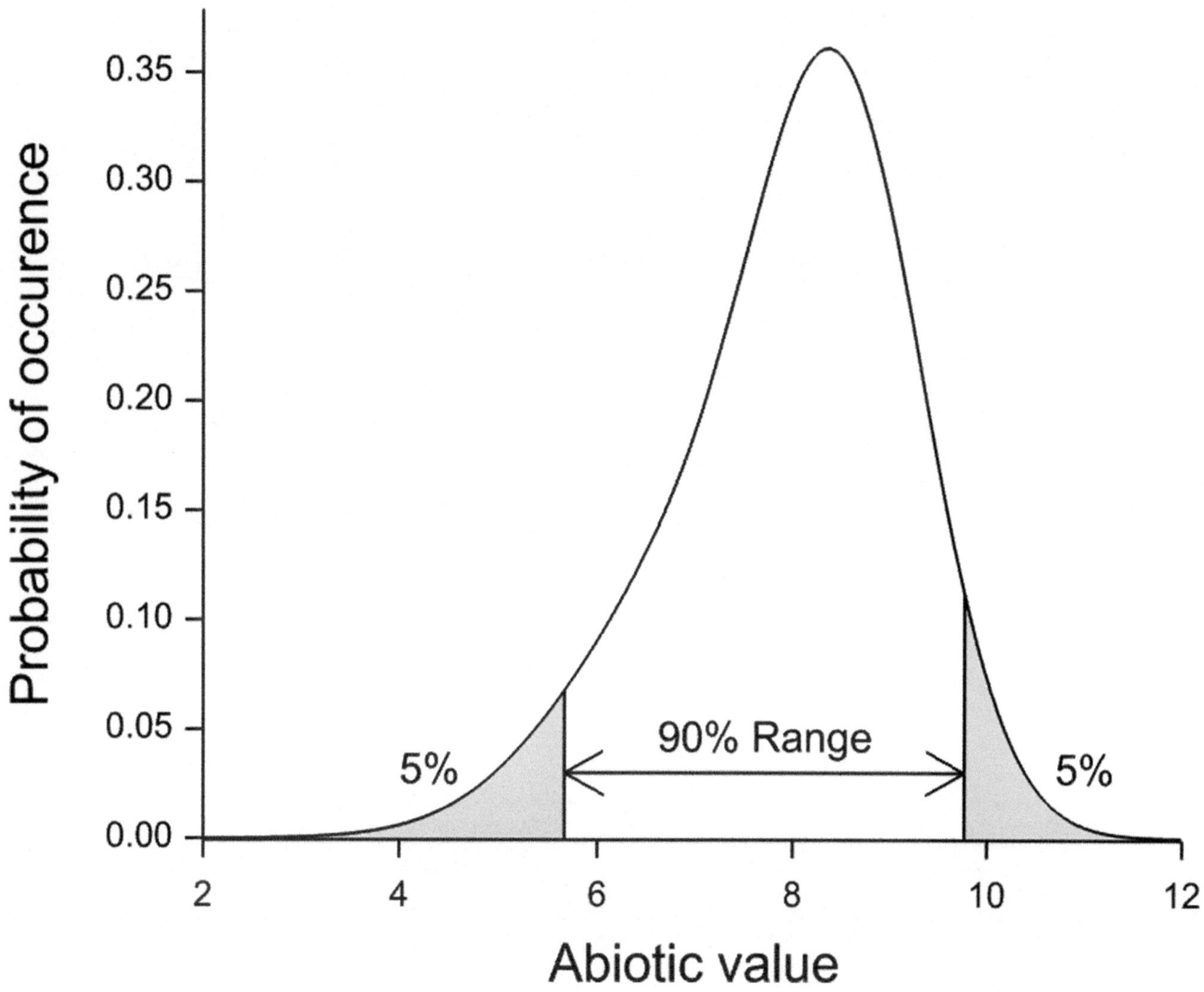

Figure 1. Hypothetical response curve (p-spline) for a hypothetical abiotic parameter, defining the range used in this research.

for *A. canina*. 190 species fulfilled the red list criteria and were identified as rare species, leaving 783 common species for comparison.

Differences in ranges

For nineteen out of the twenty-three abiotic soil parameters that were tested, the ranges for rare species were significantly narrower than for common species. No significant difference was found for the average highest and lowest groundwater table and for magnesium and organic matter content of the soil. Most of the differences in average range lay between 10 and 25% of the range (calculated against the average range of the rare species), including the important soil parameters pH, chloride ammonium concentration, potassium concentration and mean spring groundwater table. The biggest difference in range length is for nitrate concentration; up to 44%. The length of the range of a species may thus be an indicator for rare species.

Within the red list species, in general the rarer a species is, the narrower its range is likely to be (Appendix S2). The difference between the declining species 'a' and the rare species 'r' is narrow and not significant, but ranges for the common declining species 'a' is slightly narrower. All other possible combinations show statistically significant differences (Appendix S2). The difference between rare species in the category 'rr' and 'rrr' is bigger than the difference between 'a' and 'r'. The species in the categories 'rr' and 'rrr' have, on average, clearly narrower ranges than both other categories. The rarest species ('rrr') have on average the narrowest ranges of all. In general, the rarer the species the narrower its range.

Discussion

We clearly show that, on average, rare species have narrower ranges than common species for the majority of the examined abiotic soil parameters. None of the rare species showed significantly larger ranges. Obviously, this difference has a major impact on the species fundamental and realized niche and thus it occurrence. A wider habitat preference gives a species an in principle higher tolerance and therefore as well a higher resilience compared to species with narrower habitat preferences. This will be reflected in its spatial distribution, but also have an effect on the occurrence of the species as a result of natural changes of abiotic circumstances in time. The latter both within a season but also between years, since for both abiotic circumstances may change.

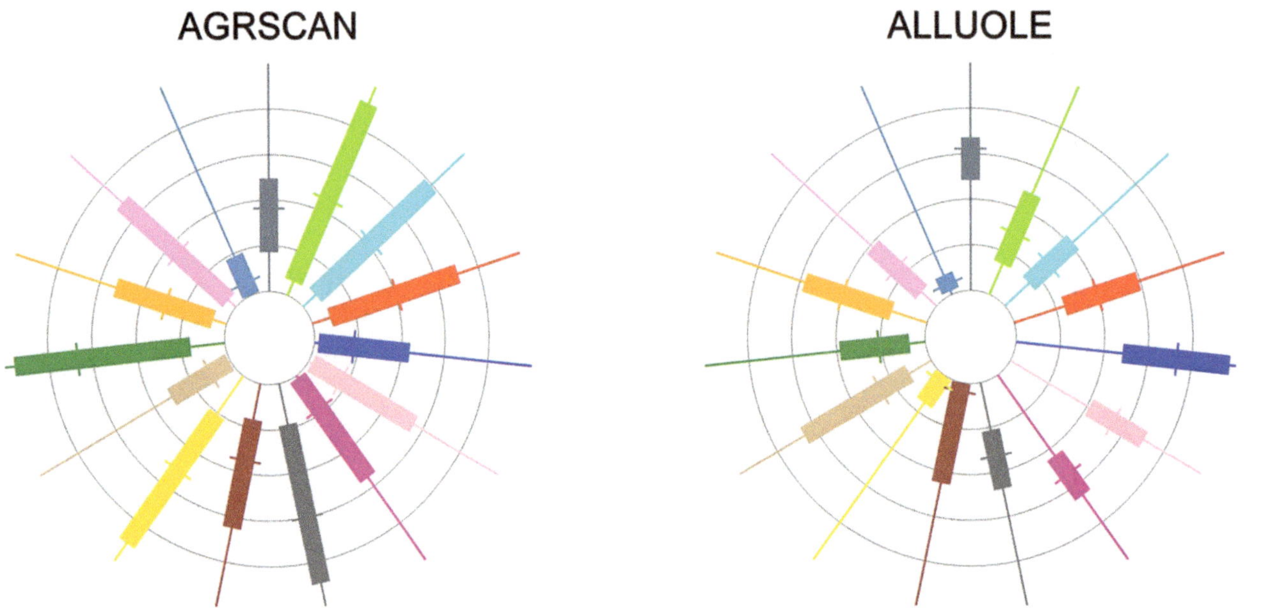

pH	Ca	Mg	K	gvg	ghg	glg	vocht	C_N	NH4	NO3	Ntot	PO4	Ptot	Cl
4.370	139.4	21.23	21.79	-11.02	-12.87	13.28	2.508	12.44	7.133	7.108	520.0	130.3	71.68	22.16
7.554	11419	643.0	243.8	89.67	123.2	210.8	54.74	28.13	34.2	82.44	11878	808.2	1399	756.9
-	mg/kg	mg/kg	mg/kg	cm-mv	cm-mv	cm-mv	%	-	mg/kg	mg/kg	mg/kg	mg/kg	mg/kg	mg/kg

Figure 2. Abiotic ranges for *Agrostis canina* (brown bent, left) and *Allium oleraceum* (field garlic, right). The bar gives the range based on the 5 and 95 percentile of the occurrence of the species. Each spoke of the wheel represents a different abiotic parameter. The values per parameter are standardised with the minimum absolute value set to 0% and the maximum value set at 100%.The circles indicate the 0, 20th, 40th, 60th and 80th percentile. The wheels make it possible to compare the ranges between the species. With: pH: pH in water extract, Ca: calcium in water extract, Mg: magnesium in water extract, K: potassium in water extract, gvg: spring groundwater level, ghg: highest groundwater level, glg: lowest groundwater level, vocht: moist content of the soil, C_N: C/N ratio, NH4: ammonium content in CaCl2 extract, NO3: nitrate content in CaCl2 extract, Ntot: total nitrogen content, PO4: phosphate content, Ptot: total phosphor content and Cl: chloride content.

Unfavourable circumstances during a part of the season or in some years may cause that species with narrow habitat requirement breadth cannot occur, whereas species with a wider habitat requirement breath can. For the most part, these ranges do not depend on the number of findings per species. The results therefore indicate that rare species have narrower habitat preferences for abiotic soil parameters than common species. Spitale [29] also found that rare spring bryophytes species had narrow habitat requirement breadth. Macandza *et al.* [30] found that for three wild grazer species of the African savannah, the rare sable antelope and the common buffalo and zebra, habitat requirement breadth and resource availability played a role in the difference in occurrence and consequently rarity of species. They concluded that rare species may persist by specializing and precisely selecting favourite patches to graze. Though these are animals, the results are similar to our results for plant species.

This research has been carried out for single responses; the interaction between abiotic parameters is not included, also due to lack of data. It is often speculated that interaction exists on a major scale [31]. Including interactions in relation to species responses may give different results; e.g. for a single response a species may have a wide range for soil pH but, for example, when phosphorus content is low it may only be able to grow at intermediate pH levels. How this will influence the results presented here is so far unknown to us. The biggest difference between rare and common species is present for nitrate concentration of the soil (44%). One of the major pressures on the occurrence of plant species in the Netherlands, but also in many other places in Europe, North America and Asia is an excessive nitrogen deposition [32], [33]. Nitrogen deposition leads to higher nitrogen concentrations in the soil and has all kinds of effects on plants [34]; some species benefit more than others by e.g. outgrowing them. We believe that our results indicate at least one of the reasons, why many rare species are threatened in the Netherlands: they have narrow ranges for nitrogen concentration and cannot cope with the input of anthropogenic nitrogen. Nitrate concentration in the soil is a result of release from biological processes (e.g. mineralisation and nitrification) and uptake by plants and bacteria and denitrification. This makes the nitrate concentration highly variable and thus resulting in a relative high uncertainty in the relation between plant species and their preferred nitrate concentration in the soil. However, this applies for both groups examined here. Furthermore we try to minimize this uncertainty by collecting as much data as possible, covering the whole growth season and under different environmental conditions (e.g. temperature, moist content of the soil, vegetation structure).

The occurrence of plant species depends on many factors, of which dispersal [11], [12], habitat availability [13], [14], [15], [16], grazing and vegetation management [16], [17] and abiotic conditions (including climate)[12], [16], [17], [18], [19], [20] are among the most important. It is known that some rare species besides narrow niches (as is shown here) also have poor dispersal capacity [11], [12]. However, how big the importance of the above mentioned factors is on the species occurrence remains unknown. This can probably only be established after a multivariate analyses applied on a database not only containing abiotic measurements but also information on the dispersal capacity of the species, the reaction of the species towards management and grazing. It may be clear however that the results presented here give a strong signal that abiotic habitat widths of rare species are of importance explaining their occurrence.

The results obtained here are relevant to the recent past and circumstances in relation to present climate. Changes in climate may give changes in species preferences and ranges, also due to interactions in relation to abiotic soil parameters, which are not included in the analyses. In principle it is possible that some rare species may benefit from climate change and thus become less rare; other species with broader ranges may become rare. However, in general we postulate that species with narrow abiotic preferences are likely to suffer more from changes in environmental circumstances than species with broad habitat requirements, because new (favourable) circumstances may be out of the current range of the rare species. In such situations resilience of species will be less and they will therefore have more difficulty in reacting to temporary or permanent changes.

The results have major implications in relation to management and nature development. They indicate that, in order to preserve rare species, management has to be focussed on the maintenance and creation of habitats fulfilling the specific requirements of rare species, both in space and time. An example of this may be the way excessive nitrogen has been mitigated in heathlands in the Netherlands. In the past sod cutting took place on a large scale, removing narrow scale differences in (for example) altitude, vegetation, light and moisture. This led to low (poor) nutrient levels, but also removed narrow scale variation from the fields, where common heather (*Calluna vulgaris*) recovered, but many other species did not. It is particularly important for ecosystem functioning [35] and ecosystem resilience [36] to conserve (abiotic) gradients within ecosystems. It will be easier to reach tipping points for rare species than for common species, making them more vulnerable to changes and as a consequence to local extinction.

Supporting Information

Appendix S1 Ranges per species for 23 abiotic soil parameters including rarity of the species, with c: common species but on the red list due to the trend, r: rare species and on the red list, rr: very rare species and on the red list and rrr: almost extinct and on the red list.

Appendix S2 Average range of the red list species per rarity category and the ranking (1–4) of the category per abiotic factor including the overall averages of the abiotic parameters for the ranking.

Acknowledgments

Earlier versions of the manuscript were corrected by Dr. L. Jones-Walters. Much of the data used were collected in the field by others, to whom we are greatly indebted. Part of the article is based on discussions with Dr. M.M.P.D. Heijmans. The manuscript improved due to the help of two anonymous reviewers.

Author Contributions

Conceived and designed the experiments: GWWW PWG. Performed the experiments: GWWW JYF. Analyzed the data: GWWW PWG JYF. Wrote the paper: GWWW PWG.

References

1. Díaz S, Fargione J, Chapin FS, Tilman D (2006) Biodiversity Loss Threatens Human Well-Being. PLoS Biol 4: 1300–1305.
2. Tilman D, Reich PB, Knops JMH (2006) Biodiversity and ecosystem stability in a decade-long grassland experiment. Nature 441: 629–632.
3. Butchart SHM, Walpole M, Collen B, Strien A van, Scharlemann JPW, et al. (2010) Global Biodiversity: Indicators of recent declines. Science 328: 1164.
4. Fagerli H, Gauss M, Benedictow A, Griesfeller J, Jonson JE, et al. (2010) Transboundary acidification, eutrophication and ground level ozone in Europe in 2009. EMEP Report 1/2011. Oslo: Norwegian Meteorological Institute.
5. Reich PB, Knops J, Tilman D, Craine J, Ellsworth D, et al. (2001) Plant diversity enhances ecosystem responses to elevated CO2 and nitrogen deposition. Nature 410: 809–812.
6. Bevill RL, Louda SM (1999) Comparisons of Related Rare and Common Species in the Study of Plant Rarity. Conservation Biology 13: 493–498.
7. Prendergast JR, Quinn RM, Lawton JH, Eversham BC, Gibbons DW, et al. (1993). Rare species, the coincidence of diversity hotspots and conservation strategies. Nature 365: 335–337.
8. Tittensor DP, Mora C, Jetz W, Lotze HK, Ricard D, et al. (2010) Global patterns and predictors of marine biodiversity across taxa. Nature 466: 1098–1101.
9. Gravel D, Bell T, Barbera C, Bouvier T, Pommier T, et al. (2011) Experimental niche evolution alters the strength of the diversity-productivity relationship. Nature 469: 89–92.
10. Bakker JP, Berendse F (1999) Constraints in the restoration of ecological diversity in grassland and heathland communities. Trends in Ecology and Evolution 14: 63–68.
11. Ozinga WA, Schaminée JHJ, Bekker RM, Bonn S, Poschlod P, et al. (2005). Predictability of plant species composition from environmental conditions is constrained by dispersal limitation. Oikos 108: 555–561.
12. Normand S, Ricklefs RE, Skov F, Bladt J, Tackenberg O, et al. (2011) Postglacial migration supplements climate in determining plant species ranges in Europe. Proceedings of the Royal Society B: Biological Sciences 278: 3644–3653.
13. Foley JA, DeFries R, Asner GP, Barford C, Bonan G, et al. (2005) Global Consequences of Land Use. Science 309: 570–574.
14. Potts SG, Biesmeijer JC, Kremen C, Neumann P, Schweiger O, et al. (2010) Global pollinator declines: Trends, impacts and drivers. Trends in Ecology and Evolution 25: 345–353.
15. Condé S, Jones-Walters L, Torre-Marín A, Romao C (2010) EU 2010 Biodiversity Baseline. Copenhagen: EEA Technical report No 12/2010. EEA.
16. Wesche K, Krause B, Culmsee H, Leuschner C (2012) Fifty years of change in Central European grassland vegetation: Large losses in species richness and animal-pollinated plants. Biological Conservation 150: 76–85.
17. Ruprecht E, Enyedi MZ, Eckstein RL, Donath TW (2010) Restorative removal of plant litter and vegetation 40 years after abandonment enhances re-emergence of steppe grassland vegetation. Biological Conservation 143: 449–456.
18. Austin MP, Niel KP van (2011) Improving species distribution models for climate change studies: Variable selection and scale. Journal of Biogeography 38: 1–8.
19. Wamelink GWW, Goedhart PW, Dobben HF van, Berendse F (2005) Plant species as predictors of soil pH: replacing expert judgement by measurements. Journal of vegetation science 16: 461–470.
20. Richardson PJ, MacDougall AS, Larson DW (2012) Fine-scale spatial heterogeneity and incoming seed diversity additively determine plant establishment. Journal of Ecology 100: 939–949.
21. Braun-Blanquet J (1964) Pflanzensoziologie: Grundzüge der Vegetationskunde. 3. ed. Springer, Wien.
22. Wamelink GWW, Adrichem MHC van, Dobben HF van, Frissel JY, Held M den, et al. (2012) Vegetation relevés and soil measurements in the Netherlands; a database. Biodiversity and Ecology 2012: 125–132.
23. Eilers HC, Marx BD (1996) Flexible smoothing with Bsplines and penalties. Statistical Science 11: 89–121.
24. Wamelink GWW, Goedhart PW, Malinowska AH, Frissel JY, Wegman RJM, et al. (2011) Ecological ranges for the pH and NO3 of syntaxa: A new basis for the estimation of critical loads for acid and nitrogen deposition. Journal of Vegetation Science 22: 741–749.
25. Schaminee JHJ, Hennekens SM, Ozinga W (2012) The Dutch national vegetation database. Biodiversity and Ecology 2012: 201–209.

26. Mace GM, Stuart SN (1994) Draft IUCN Red List Categories, version 2.2. Species 21–22: 13–24.
27. Meijden R Van der (1996) Heukels flora van Nederland. Groningen: Wolters-Noordhoff.
28. GenStat. Available: www.vsni.co.uk.
29. Spitale D (2012) A comparative study of common and rare species in spring habitats. Ecoscience 19: 80–88.
30. Macandza VA, Owen-Smith N, Cain JW (2012) Habitat and resource partitioning between abundant and relatively rare grazing ungulates. Journal of Zoology 287: 175–185.
31. Laughlin DC, Abella SR (2007) Abiotic and biotic factors explain independent gradients of plant community composition in ponderosa pine forests. Ecological Modelling 205: 231–240.
32. Galloway JN, Aber JD, Erisman JW, Seitzinger SP, Howarth RW, et al. (2003) The Nitrogen Cascade. Bioscience 53: 341–356.
33. Wamelink GWW, Knegt B de, Pouwels R, Schuiling C, Wegman RMA, et al. (2013) Considerable environmental bottlenecks for the Habitats and Birds directives Species in the Netherlands. Biological Conservation 165: 43–53.
34. Stevens CJ, Manning P, Berg LJL van den, Graaf MCC de, Wamelink GWW, et al. (2011) Ecosystem responses to reduced and oxidised nitrogen inputs in European terrestrial habitats. Environmental Pollution 159: 665–676.
35. Isbell F, Calcagno V, Hector A, Connolly J, Harpole WS, et al. (2011) High plant diversity is needed to maintain ecosystem services. Nature 477: 199–201.
36. Scheffer M, Bascompte J, Brock WA, Brovkin V, Carpenter SR, et al. (2009). Early-warning signals for critical transitions. Nature 461: 53–59.

23

A Multi-Criteria Index for Ecological Evaluation of Tropical Agriculture in Southeastern Mexico

Esperanza Huerta[1]*, Christian Kampichler[2,3], Susana Ochoa-Gaona[4], Ben De Jong[4], Salvador Hernandez-Daumas[1], Violette Geissen[5,6]

1 El Colegio de la Frontera Sur, Unidad Campeche, Dpto. Agroecología, Campeche, México, 2 Universidad Juárez Autónoma de Tabasco, División de Ciencias Biológicas, Villahermosa, Tabasco, México, 3 Sovon Dutch Centre for Field Ornithology, Natuurplaza (Mercator 3), Nijmegen, The Netherlands, 4 El Colegio de la Frontera Sur, Unidad Campeche, Dpto. Sustainability Sciences, Campeche, México, 5 University of Bonn - INRES, Bonn, Germany, 6 Wageningen University and Research Center – Alterra, Wageningen, Gelderland, Netherlands

Abstract

The aim of this study was to generate an easy to use index to evaluate the ecological state of agricultural land from a sustainability perspective. We selected environmental indicators, such as the use of organic soil amendments (green manure) versus chemical fertilizers, plant biodiversity (including crop associations), variables which characterize soil conservation of conventional agricultural systems, pesticide use, method and frequency of tillage. We monitored the ecological state of 52 agricultural plots to test the performance of the index. The variables were hierarchically aggregated with simple mathematical algorithms, if-then rules, and rule-based fuzzy models, yielding the final multi-criteria index with values from 0 (worst) to 1 (best conditions). We validated the model through independent evaluation by experts, and we obtained a linear regression with an $r^2 = 0.61$ ($p = 2.4e\text{-}06$, $d.f. = 49$) between index output and the experts' evaluation.

Editor: Yong Deng, Southwest University, China

Funding: Financial support was obtained from the Conacyt-SEMARNAT Project "Uso sustentable de los recursos naturales en la frontera sur de México" (Sustainable use of natural resources on the southern border of Mexico) (code SEMARNAT-2002-C01-1109). El Colegio de la Frontera Sur provided infrastructural resources. The funders had no role in study design, data collection and analysis, decision to publish, or preparation of the manuscript.

Competing Interests: The authors have declared that no competing interests exist.

* Email: ehuerta@ecosur.mx

Introduction

In the past 60 years, degradation and deforestation of tropical forests worldwide have occurred much faster and more extensively than in any other period in history [1], [2]. Furthermore, countries like Mexico have been undergoing drastic land use changes. In the tropics of Mexico large parts of its lowland rainforest areas have been converted into pasture and cropland. In the state of Tabasco, for example, only 3.4% of the state is covered with original forest [3], whereas 76.4% of the surface was used for cattle production in 2000 [4] and 15.6% was used for agriculture, principally sugarcane and fruit plantations [3]. The ecological consequences of these land-use changes in Tabasco are well documented. Soil losses in hilly regions are very high up to 200 t ha^{-1} year^{-1} [5]; high pesticide and fertilizer inputs to crops that have replaced forests have caused considerable environmental contamination [6], [7], and soil fertility is decreasing [8], [5], [9].

It is of the utmost importance to identify sustainable land use strategies which are economically attractive for the region's farmers and which may also reconcile the need for food production with that of soil conservation. In order to assist a variety of stakeholders at the local and regional level in making land use decisions, simple evaluation tools are needed. This is even more needed since a high percentage of the population consists of immigrants from other Mexican states, who are unfamiliar with the conditions of the humid tropics and use intensive techniques for farming the land. This is mostly due to large-scale agricultural development projects, such as "Plan Balancán-Tenosique," named after the two municipalities involved, in which, in the 1970 s, over 1100 km^2 of lowland rain forest was destroyed and converted into crop and pasture land. Ecological values of the 1970 s were very different from those of the present, and government representatives were willing to deforest in order to grant land to farmers [10].

The direct impact of farming is difficult to measure due to methodological difficulties (impossibility of measurement, complexity of the system) or practical reasons (time, costs) [11]. Therefore, the use of indicators appears to be an alternative way of guiding land use decisions [12], [13]. However, the "indicator explosion" [14], that is, the use of an exaggerated number of indicators aimed at assessing environmental impacts of agricultural activities, has been of little use to local decision-makers. Particularly in the tropics, land use decisions are still based on the informal opinion of local experts rather than on implementation of Decision Support Systems for environmentally sound resource management [15]. On the one hand, this is due to farmers' restricted access to modern communications and information technologies; on the other, application of indicators is often beyond the capability of local farmers. For example, the agricultural sustainability index proposed by Nambiar et al. [16], which aims to measure agricultural sustainability as a function of biophysical, chemical, economic, and social indicators, would require considerable training of government stakeholders and farmers in order to be applied, and such training is rarely available.

Thus, any tool which local farmers or regional decision makers may use to support their decision making must be as simple as possible. Agroecosystems (like any other ecosystem) are too complex to be precisely measured and evaluated [17]. We agree with the view of Darnhofer et al. [18] in favour of developing less precise rules of thumb which may be used by farmers as well as to guide local land-use decisions toward a more environmentally friendly system of agriculture.

This index may provide farmers and others involved with a tool to evaluate the sustainability of their management of crop land which is oriented toward diminishing soil damage and conserving soil fertility. The index, exclusively based on terms which describe the environmental conditions of the crop system, is accessible to most farmers, and may be calculated using a simple internet application. In this paper we present a simple, easy-to-use index in order to evaluate the ecological state of farms in south-eastern Mexico. We applied the indicator system to 52 crop production systems in south-eastern Mexico and compared the results of the indicator system with expert opinions.

Human knowledge of how to efficiently and sustainably manage complex systems (including agricultural systems) is incomplete and much of what is thought to be known about this topic is actually incorrect. Yet, decisions must be made by policy makers, agricultural extension agents, and farmers despite uncertainty and knowledge gaps [19]. Therefore, tools to support local decision-makers must be flexible, should not enter into too much detail or precision, and should allow for an adaptive strategy which promotes "learning through management" [19]. Consequently, our rationale for developing an index which aids farmers in making environmentally friendly land-use decisions is based on basic, simplified ecological concepts, i.e. the presence of trees, since trees within an agroecosystem enhance soil microclimate in terms of radiation partitioning (shading), evapotranspiration partitioning, and rain interception/redistribution [20]. These factors all help to retain soil moisture. Branches, bark, roots, and living and dead leaf surfaces provide shelter [21] for soil micro-, meso-, and macro-invertebrates. Tree cover for instance, enhances above- and below-ground diversity, serving to support agricultural sustainability [22].

Materials and Methods

1. Rationale of index composition

We define conventional agriculture as a cropping system, typically promoted by government development programs, that is "capital-intensive, large-scale, highly mechanized agriculture with monocropping and extensive use of synthetic fertilizers, and pesticides" [23], [24]. Furthermore, we acknowledge that farming systems are sustainable only if "they minimize the use of external inputs and maximize the use of internal inputs already present on the farm" [25], [26]. The strategy most frequently linked to sustainability is reduction or elimination of agrochemicals, particularly chemical fertilizers and pesticides [25], [27], [28], [29], [30], [31]. Another key to sustained productivity of agricultural systems is the maintenance of soil functions, such as organic matter and nutrient cycling [32], based on organic inputs [33], above-and below-ground biodiversity [22], and diversifying crop systems with nitrogen-fixing legumes [34]. The principal role of the index we propose is to characterize methods of tillage, external inputs, and crop structure.

2. Primary indicators

We chose 12 field variables as primary indicators related to the above mentioned aspects of ecologically sound agricultural land use based on farmer's practices. These are easy to evaluate in the field and characterize plot structure (primary indicators: tree cover, tree density, tree diversity), crop structure and crop conditions (primary indicators: crop type, crop rotation, crop density, crop colour), tillage (primary indicators: type of tillage, timing of tillage), the use of fertilizers (chemical versus organic) and pesticide application.

2.1 Tree cover. Tree cover is defined as the canopy of trees, measured in the field, and recorded as percentage classes of tree cover in three height classes (trees >15 m, 10–15 m, <10 m). Thus, this variable characterized one aspect of agroecosystem management: the farmers' decision to maintain the canopy of the trees in his or her agroecosystem.

2.2 Tree density. Tree density is defined as the number of trees per area. To measure this, we distinguished three categories of tree density: high density (abundant), medium density, and low density (isolated or no trees). A high number of trees per area guarantee carbon sequestration [35] while a stable microclimate is maintained. This variable, measured in the field, is one the variables that characterize the effect of trees in the agroecosystem.

2.3 Tree diversity. This variable was measured in the field by counting the number of trees species within the plot. In agroecosystems, biodiversity may; (i) contribute to constant biomass production and reduce the risk of crop failure in unpredictable environments, (ii) restore disturbed ecosystem services such as water and nutrient cycling, and (iii) reduce risks of pests and diseases through enhanced biological control or direct pest control [36], [20], [22].

2.4 Crop type. This variable indicates whether the crop is annual, seasonal, or perennial. We obtained this information by observing the type of crop. Annual crops in general have higher environmental impacts, ie: greenhouse emissions, and nutrient leaching, than perennial crops [37].

2.5 Crop rotation. In sustainable farm systems leguminous crops are increasingly used in crop rotations as a source of nutrients, particularly nitrogen for crop growth [38], [39], nitrogen-fixing legumes, contribute to maintaining biodiversity above and in the soil, contribute nitrogen to the soil/plant system, and help avoid the build-up of pest populations [34]. In this study, we asked farmers whether they planted another crop before planting the main crop and whether they practice crop rotation, as crop rotation may assist with weed and pest control [40], [34]. According to Bellon [41], an activity which leads into the maintenance or increase of renewable resources in agroecosystems, is considered as an ecological technology. This variable helped us to characterize the technology used in the agroecosystem.

2.6 Crop density. Crop density is defined as the number of plants (individuals) per area. Three categories were recorded in the field: abundant (high density: 3,000 plants/ha), medium density (1000–1600 plants/ha), and sparse (<1000 plants/ha). This variable also indicated the level of technology applied to the crop, as less intensive techniques typically yield lower densities [42].

2.7 Crop colour. The colour of a crop indicates the nutritional status of the plants; green plants generally have sufficient nutrients, while yellow plants lack nitrogen [43].

2.8 Type of tillage. Type of tillage was categorized into no tillage, manual tillage, and mechanical tillage using machinery, the latter of which generally indicates high disturbance of the soil surface and rapid loss of soil organic carbon and other nutrients

[44]. In the field, we asked the farmers how they prepared the land for crop planting.

2.9 Frequency of tillage. This variable indicates frequency of tillage - every year or every 2 years. With this information, it was possible to estimate the frequency of soil disturbance due to tillage.

2.10 Chemical input. Searching for sustainable productions, it is recommended no or low or use of inorganic fertilizers and pesticides [45], [46]. Long term use of some pesticides, as glifosate can decrease earthworm species number, density and biomass [47], [48]. This information allowed us to estimate the amount of chemical fertilizers and pesticides applied within a given area. These variables (chemical fertilizers and pesticides) are included in the indicator for chemical disturbance.

2.11 Green manure. Is defined as the presence or absence of leguminous crops mixed with the principal crop, generally used to increase total soil nitrogen content. Green manures should always be intercropped, as it has been proven that growing legumes with cereal crops decreases N_20 emissions [49], and therefore is a sustainable, environmentally friendly practice. Examples of green manure use have been observed in traditional Mesoamerican cultures, for example, intercropping beans, as well as other edible plants, within the *milpa* or traditional maize cropping system [50].

3. Index development

Primary indicators were hierarchically aggregated into higher levels, forming intermediate variables, which in turn are structured into a single index that evaluates the ecological condition of a given plot on a scale from 0 (worst) to 1 (best). An index close to zero either would mean that a more environmentally friendly farming techniques need to be implemented or that the plot should be subjected to a fundamental change in land-use, e.g., reforestation, in order to return to a more ecologically sound state. An index close to one, on the other hand, would indicate an ecologically sound land-use. Methods applied in indicator aggregation were (i) simple mathematical operations, ii) sets of if-then rules, and (iii) sets of if -then rules combined with fuzzy logic.

3.1. Aggregation through mathematical operations. Mathematical operations include calculating averages, weighed averages, minimum values, and maximum values. For example, if primary indicator A has the value a and primary indicator B has the value b, then the value x of the intermediate variable X is determined as $x = (a+b)/2$, $x = (w_1*a+w_2*b)$, where $w1$ and $w2$ are weights, or $x = \min(a, b)$ or $\max(a, b)$.

3.2. Aggregation by IF-THEN rules. If primary indicator A can have the discrete values a_1, a_2, and a_3, and primary indicator B can have the discrete values b_1, b_2, and b_3, then the value x of the intermediate variable X is determined by a set of nine (number of levels of A x number of levels of B) rules. For example, IF $A = a_1$ and $B = b_1$ THEN $X = x_1$.

3.3. Aggregation by IF-THEN rules and fuzzy logic. We used small fuzzy rule-based models for aggregation in the case of non-linear interactions among indicators using a continuous numerical scale or an ordinal scale with a large number of possible values. In classic set theory, an object can either be a member (membership = 1) or not (membership = 0) of a given set. The central idea of fuzzy set theory is that an object may have a partial membership of a set, which consequently may possess all possible values between 0 and 1. The closer an element is to 1, the more it belongs to the set; the closer the element is to 0, the less it belongs to the set. To apply the fuzzy set theory, three steps are involved in calculating the model's output. First, for any observed value of the primary indicators, its corresponding membership value in the fuzzy set domain is calculated (*fuzzification*); second, the memberships of the intermediate variable X are calculated, applying the rules in the fuzzy set theory (*fuzzy inference*); third, the fuzzy results are converted into a discrete numerical output (*defuzzification*; see Wieland 2008 for an introduction to fuzzy models). Fuzzy rule-based models have become popular in ecological modelling [51], [52], and several examples exist of its usefulness in the context of ecosystem evaluation, bioindication, and sustainable management [15], [53], [54]. Here, if both primary indicators A and B can have numerical values from 0 to 1, then the value x of the intermediate variable X is determined by a series of fuzzy set rules representing the linguistic variables "low A", "medium A", "high A", and "low B", "medium B", and "high B", as well as the output "low X", "medium X", and "high X". The value of the intermediate variable X is determined by nine levels of A x the number of levels of B; for example: If A = low and B = low Then X = [low, medium, or high].

To maintain the number of rules as well as their complexity as low as possible, we aggregated only two variables at a time. A simple example shows the reasoning behind this decision; if there are three primary variables, A, B, and C with three categories for each variable, a single rule node requires 3*3*3 = 27 If-Then rules of the type "If $A = a_1$, a_2, or a_3 and If $B = b_1$, b_2, or b_3 and if $C = c_1$, c_2, or c_3 Then...", whereas if an additional intermediate variable X is introduced to the model, only 18 rules are needed: 3*3 = 9 rules to aggregate A and B to X (If $A = a_1$, a_2, or a_3 and If $B = b_1$, b_2, or b_3), and 3*3 = 9 rules to aggregate X and C (If $X = x_1$, x_2, or x_3 and If $C = c_1$, c_2, or c_3).

4. Study area and application of the index

The state of Tabasco in south-eastern Mexico is characterized by a humid tropical climate with a mean annual rainfall between 1200 and 4000 mm and a mean annual temperature of 27°C [55]. Predominant soils are Gleysols and Fluvisols over alluvial sediments in the plains, Vertisols, Cambisols, Luvisols, and Acrisols over Miocene or Oligocene sediments, and Leptosols and Regosols over limestone mountains [9], [56]. We chose the municipalities Balancán and Tenosique in western Tabasco (17°81′50″–18°81′00″ N, 91°80′10″–91°84′60″ W) as a study area (Figure 1), we worked with private, *ejidal* (multipurpose land, where owners can or cannot sell the property according to their legal status in the National Agrarian File), and communal lands (coordinates of each plot are shown in Table 1). The region is mainly a plain towards the North (67% of area has an elevation <20 m. a. s. l.) with hills (29%, 20–200 m. a. s. l.), and mountains (4%, max. 640 m. a. s. l.) in the South, comprising a total area of 5474 km^2. These municipalities have undergone a high degree of land use change over the past 40 years. Until the early 1970 s, this region was still covered by lowland rain forest. The principal form of land use is pastureland, and covers 60% of the land [57]. An additional 30% is cropland, mainly cultivated under small-medium holder systems with seasonal conventional agriculture using high levels of agrochemical inputs (Table 2 & 3). Common crops are maize (*Zea mays*), a variety of hot peppers (*Capsicum* sp), cucumbers (*Cucurbita argyrosperma*), watermelon (*Citrullus lanatus*), perennial fruit crops such as papaya (*Carica papaya*), and biannual crops such as sugar cane (*Saccharum officinarum*) [58].

We chose 52 farms in the study area (Table 1), and selected those farms whose main economic activity (100%) is agriculture and chose one agricultural plot from each farm for evaluation. There were annual and biannual crops with or without trees (Table 2). Average plot size was 32.4±55.1 ha, and the average time that the plot had been used for a given crop system was 2.5±3.0 years.

Table 1. Characterization of plots.

Plot #	Municip	Lat	Long	Altitud (mls)	Plot size (ha)	Type of property	Original-vegetation
1	Bal	2007880	652445	43	18	Ejidal	A
2	Bal	1964368	642505	10	0.5	Private	TRF
3	Bal	1970936	647701	11	160	Private	TRF
4	Bal	1957459	667496	14	2	Ejidal	TRF
5	Bal	1964107	660656	14	7	Private	PT
6	Bal	1960297	658989	22	15	Private	TRF
7	Bal	1976300	676424	40	30	Private	TRF
8	Bal	1990326	665947	40	28	Private	TRF
9	Bal	1964590	667264	11	3	Private	RV
10	Bal	1972260	672382	28	2.5	Ejidal	TRF
11	Bal	1950339	666409	49	40	Private	TRF
12	Bal	1966112	682994	40	80	Private	PT
13	Ten	1931754	704474	65	14	Private	TRF
14	Ten	1912951	690381	130	2	Communal	TRF
15	Ten	1932644	665521	15	3	Ejidal	TRF
16	Ten	1926163	678608	56	68	Private	TRF
17	Ten	1933663	674227	59	35	Private	A
18	Ten	1926570	669486	35	8.5	Ejidal	TRF
19	Ten	1942431	663885	21	1.5	Ejidal	TRF
20	Ten	1942068	671156	31	7.25	Ejidal	TRF
21	Ten	1943167	650736	82	14	Ejidal	A
22	Ten	1923838	669841	118	0. 25	Private	TRF
23	Bal	1973301	695548	105	40	Ejidal	TRF
24	Bal	1982392	701383	74	20	Ejidal	A
25	Bal	1965354	709053	44	11	Ejidal	A
26	Bal	1960496	706464	51	44	Ejidal	TRF
27	Bal	1942484	709731	66	10	Ejidal	A
28	Ten	1944148	673673	16	320	Ejidal	TRF
29	Ten	1930350	659650	29	20	Private	TRF
30	Ten	1924549	656910	41	53	Ejidal	TRF
31	Ten	1926226	652978	65	3	Ejidal	TRF
32	Bal	1955194	688843	45	44	Private	TRF
33	Bal	1960980	679265	49	29	Ejidal	TRF
34	Bal	1942552	696222	48	10	Private	TRF
35	Ten	1938120	686470	49	5	Private	TRF

Table 1. Cont.

Plot #	Municip	Lat	Long	Altitud (mls)	Plot size (ha)	Type of property	Original-vegetation
36	Bal	1968750	695595	60	2	Private	A
37	Bal	1977546	699417	53	200	Ejidal	TRF
38	Bal	1949770	708854	53	23	Ejidal	TRF
39	Bal	1951216	697890	46	2	Ejidal	TRF
40	Ten	1931041	667544	31	17	Private	TRF
41	Ten	1927447	664223	3	3	Ejidal	TRF
42	Ten	1928226	664551	31	3	Ejidal	TRF
43	Ten	1930579	691785	54	15	Communal	TRF
44	Ten	1931219	677895	59	63	Private	TRF
45	Ten	1931525	676500	60	6.75	Ejidal	TRF
46	Ten	1908445	675040	140	40	Ejidal	TRF
47	Ten	1910875	669750	222	45	Ejidal	TRF
48	Ten	1924222	704702	42	30	Ejidal	A
49	Ten	1935076	712227	88	20	Ejidal	TRF
50	Ten	1914876	686285	113	4	Ejidal	TRF
51	Ten	1911365	684878	389	22	Ejidal	TRF
52	Ten	1911503	698606	139	10	Ejidal	TRF

Municip: municipality, Bal: Balancan, Ten: Tenosique. TRF: tropical rain forest, PA: pasture, A: acahual (secondary vegetation). Ejidal: multipurpose land, where owners can or cannot sell the property according to their legal status in the National Agrarian File.

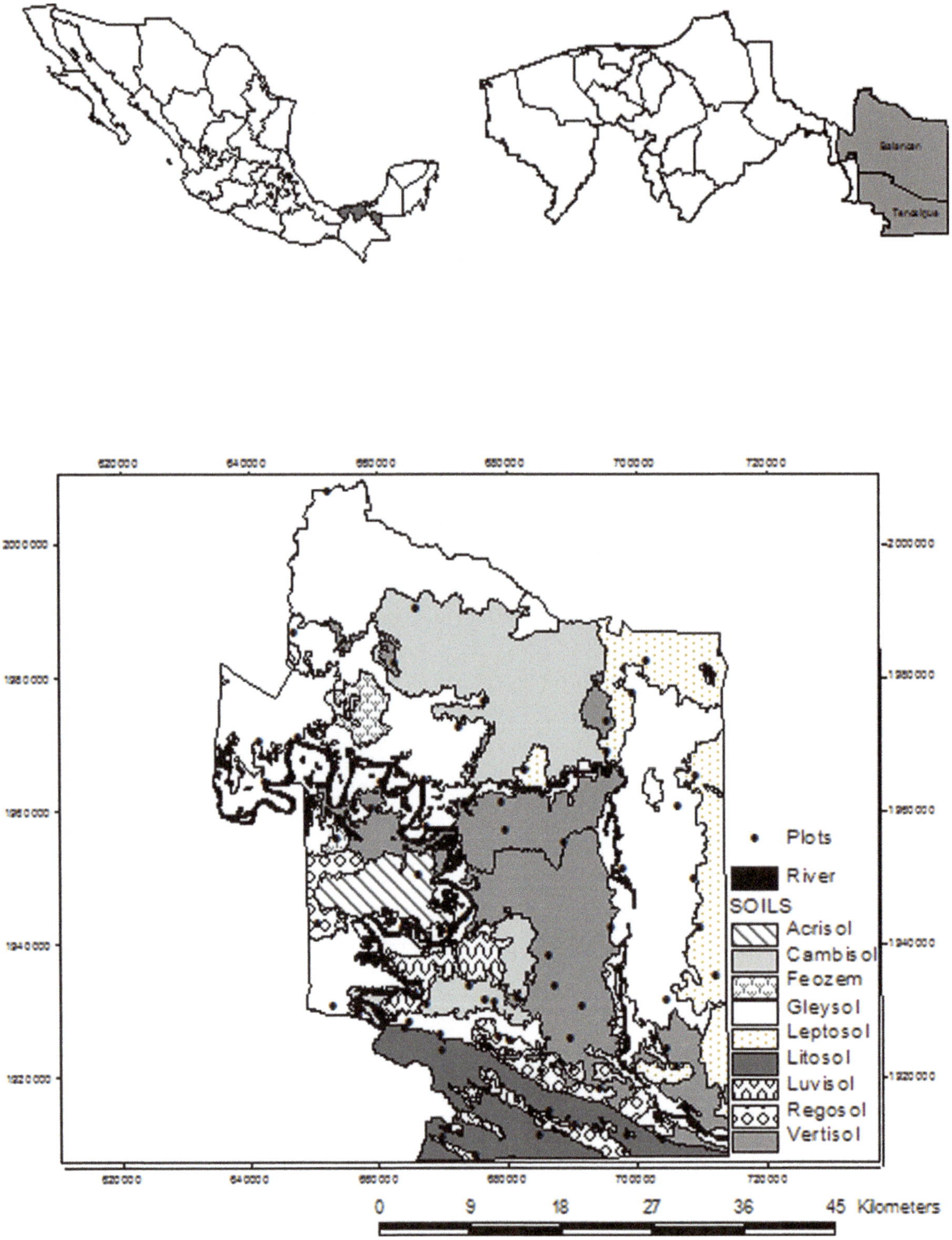

Figure 1. Distribution of 52 evaluated agricultural plots in tropical South-East Mexico.

We questioned farmers as to frequency, amount and type of chemical fertilizers and pesticides applied per area of cropland. After their verbal consent, farmers gave us their complete name and signed next to their name on the record sheet, validating all the information. Due to the fact that we only asked about the use of fertilizers and management of the land, the procedure approval from the Ethics Committee was not required.

Between March and October 2004, for each plot, the values of the primary indicators were determined and the index of ecological condition was calculated. Prior to index calculation, the plots were also evaluated by experts (2 scientists, each with

Table 2. Crop characterization.

Plot	Cycle	Main seasonal or perennial crop	Trees
1	s-s	*Zm, Pv, Cs, Cl*	*Tr, Pa, Cpa*
2	s-s	*Zm, Pv, Cs, Cl*	*Sh, Dg*
3	w-s	*Sv*	*Sa*
4	a-w	*Zm, Pv, Cpa*	*0*
5	Y	*Zm, Pv, Cl, Cm*	*0*
6	Y	*Zm*	*Cpa, Eg, Tr*
7	s-a	Ca	Tr
8	Y	*Cp*	*Cp*
9	Y	*Zm*	*0*
10	Y	*Zm*	*0*
11	w-s	*Zm, Pv, Cl*	*Sm, Cpa*
12	Y	*Co*	*Co*
13	s-s	*Zm*	*0*
14	s-s	*Le, Se*	*0*
		Cpa, Zm, Pv	
15	s-s	*Cl, Ta, Mp,*	*Fruit trees*
16	Y	*Jj*	*Gs, Hc, Cc, Tr*
17	Y	*So*	*0*
18	Y	*So*	*0*
19	Y	*So*	*0*
20	Y	*Zm*	*Mi*
21	s-s	*Zm*	*0*
22	Y	*Csi, Cli, Js*	*Co, Pa, Bc, Cn, Mi*
23	s-s	*Zm, Cpa*	*Cpe*
24	s-s	*Zm*	*Cpa*
25	s-s	*Zm*	*0*
26	s-s	*Zm, Cpa*	*0*
27	s-s	*Zm, Ca*	*Sm*
28	Y	*So*	*0*
29	s-s	*Zm, Pv, Cs, Cl*	*0*
30	Y	*Zm, Pv, Cs, Cl, Ta*	*0*
31	Y	*Zm, Pv, Cs*	*0*
		Cl, Os, Le	
32	s-s	*Zm, Pv, Cpa, Cl, Os*	*Tr*
33	Y	*Zm, Pv*	*Sm, Bg*
34	s-s	*Zm, Pv, Os*	*Cpa*
			Tr
35	Y	*Zm, Cl*	*Tr*
36	Y	*Zm, Pv, Cpa, Cl, Ca*	*0*
37	s-s	*Zm, Pv, Ca*	*Sm, Pa, Sh*
38	Y	*Zm, Pv, Cpa, Cl, Ta*	*Co, Sh*
39	Y	*Zm, Pv, Cpa, Cl, Ca*	
		Cs, Cm	
40	Y	*So*	*Tr, Cpe*
			Co, D a
41	Y	*So*	*Tr, Cpe*
			Co, D a
42	Y	*So*	*Tr, Cpe*
			Co, D a

Table 2. Cont.

Plot	Cycle	Main seasonal or perennial crop	Trees
43	Y	*So*	*Tr, Cpe*
			Co, Da
44	Y	*So*	*Tr, Cpe*
			Co, Da
45	Y	*So*	*Tr, Cpe*
			Co, D a
46	Y	*Zm, Pv, Cpa, Me*	*Sm*
47	Y	*Zm, Pv, Me, Ib*	*Bc, Mi, Dg, Cs, Ll*
48	Y	*Zm, Pv, Cpa, Cl*	*0*
49	Y	*Zm, Pv, Ca*	*Tr, Gs*
50	Y	*Zm, Pv, Cpa, Cl*	*Sm, Co, Ma*
		Cs, Ib, Me	0
51	Y	*Zm, Pv, Cpa, Cs*	*Fruit trees*
52	Y	*Zm, Pv, Cpa, Cl, Cs*	*Co, Sh, Bc*
		Mp, Ca, Me	0

Cycle: s–s: spring-summer, w-s: winter-summer, s-a: summer-autumn, Y: all the year. **Main seasonal or perennial crop**: Ca: *Cucurbita argyrosperma* (cushaw pumpkin); Cp: *Carica papaya* (papaya); Cl: *Citrullus lanatus* (watermelon); Cli: *Citrus limon* (lemon); Cm: *Cucumis melo* (muskmelon); Cpa: *Cucurbita pepo* (squash); Cs: *Capsicum* sp. (pepper); Csi: *Citrus sinensis* (orange tree); Ib: *Ipomoea batatas* (sweet potato); Jj: *Jarthropha jurcas* (oil palm); Js: *Jobo spondia* (plum); Le: *Lycopersicon esculentum* (tomato); Me: *Manihot esculenta* (cassava); Mp: *Musa paradisiaca* (banana); Os: *Oryza sativa* (rice); Pv: *Phaseolus vulgaris* (bean), So: *Saccharum officinarum* (sugarcane); Se: *Sechium edule* (pear squash); Sv: *Sorghum vulgare* (milo); Ta: *Triticum aestivum* (wheat); Zm: *Zea mays* (maize). **Trees**: *Bc: Byrsonima crassifolia* (nanche); *Cc: Crescentia cujete* (calabash tree); *Co: Cedrela odorata* (Mexican cedar); *Cpa: Carludovica palmate* (toquilla palm); *Cpe: Ceiba pentandra* (kapok); Cn: *Cocos nucifera* (Coconut); *Dg: Dialium guianense* (wild tamarind); Eg: *Eucalyptus grandis* (Eucalyptus); *Gs: Gliricidia sepium* (Cocoite, gliricidia, cacao de nance); *Ll: Leucaena leucocephala* (white lead tree, jumbay); *Ma: Mammea americana* (mamey apple); *Mi: Mangifera indica* (mango); *Pa: Persea americana* (avocado); *Sa: Sterculia apetala* (camoruco, manduvi tree); *Sh: Swietenia humilis* (small mahagoni); *Sm: Spondias mombin* (Yellow plum, bai makok); *Tr: Tabebuia rosea* (savannah oak, macuilis).

over 15 years of experience in agroecology) on a scale from 0 (poor condition) to 1 (good condition) in order to test the correlation between the experts' opinions and the index.

Data was normalized for carrying out multiple regression with plot size, altitude of plot site, or previous vegetation cover and the ecological index.

Finally we located the index in this public web address: http://201.116.78.102/~modelo/Index.html, where farmers in the near future can introduce independent data sets and evaluate or monitor their own agro ecosystems.

Results

1. Index structure

Ten nodes were used to aggregate primary indicators to the final index of ecological conditions of agricultural systems (Figure 2). Six of these used simple mathematical operations; two were based on rule sets, and two on fuzzy rule-based models (Table 4).

2. Index application

2.1. General characterization of sampled plots. A total of 67% of the plots were cultivated after cutting of primary lowland forest, 24% after cutting secondary forests of various ages, and 1.5% after cutting riparian vegetation, the rest conversion from pastureland (livestock production) to crop system. On 63% of the farms, maize, beans and pumpkins were cultivated, 13% sugar cane, 12% watermelon, 9% rice and 3% pepper. On 68.5% of the plots, trees were scattered among the crops. Conventional tillage was used on 55.5% of the plots, and pesticides were used on 79.4%, green manure was used on 21% of the plots and chemical fertilizers on 67%. All farmers understood the meaning of the variables evaluated for their plots (for a complete list of plots, see Table 5).

2.2. Plot evaluation with the index. Ecological condition of the plots ranges from 0.0 to 0.8125 (Figure 3; Table 5). One plot which was intercropped with timber and fruit trees presented the highest value. This site was characterized by an absence of agrochemical use, use of green manure, presence of annual crop rotation, and high tree diversity. We found 18 plots with index values of 0–0.25, 7 plots with values of 0.3–0.5, and 26 plots with values of 0.5–0.7. Thus, the majority of the plots evaluated in this study had an intermediate index value. 50% of the plots with this intermediate index are *ejidal* property, 38% private land, and 12% is under another type of ownership (smallholder or communal). 80% of these intermediate plots were lowland rain forest before being converted into agricultural land, 12% were secondary vegetation, and 8% were pastureland (Table 1). One might believe that a prior forest condition implies a more environmentally friendly agroecosystem that allows for preserving a more diverse system. However, plots with low index values were also lowland rain forest before being turned into agricultural land (Table 5). In this case, the land managers or owners decided to deforest the land to subsequently plant annual crops. 13 plots were larger than 40 ha (Table 1). All of these were previously covered with lowland rain forest; the ecological index ranges from 0.5–0.56 for 46% of these plots, and another 23% have an ecological index of 0.37–0.43. 10 plots fell under the smallest size category (0.25–3 ha) and had an ecological index of 0.5–0.68 (4 plots), 0 to 0.18 (5 plots), and 0.39 (1 plot).

Carrying out multiple regression with normalized data, we did not observe significant correlations between plot size, altitude of

Table 3. Crop land preparation and inputs characterization.

Plot	Tillage	Chemical Fertilization	Pesticides Use	Gm	Oa
1	Ma & Me	NPK 17:17:17 & U	Endosulfan**	0	0
2	Ma	0	Chlorpyrifos*	0	0
3	Me	Pholiar	0	0	c s m
4	Ma	0	0	0	0
5	Ma & Me	NPK 17:17:17	Chlorpyrifos**	Y	Y
6	Ma & Me	NPK 17:17:17	Chlorpyrifos*	0	0
7	Me	0	ID	0	0
8	Me	ID	ID	Y	0
9	Ma & Me	0	Carbofuran*	0	0
10	Me	NPK 18-46-0	ID	0	0
11	Me	NPK 18- 46- 0, U & P	ID	Y	0
12	Me	NPK 18-46-0	ID	0	0
13	Me	NPK 17:17:17 & U	Chlorothalonil**	0	0
			Chlorpyrifos*		
14	Ma	0	Endosulfan*	0	0
15	Me	0	0	Y	0
16	Ma	Urea	0	0	0
17	Me	NPK 19-19-19	ID	0	0
18	Me	NPK 19-19-19	ID	0	0
19	Me	Urea	ID	0	bg
20	Me	0	Methilic*	0	0
21	Me	NPK 17:17:17 & U	Zeta**	0	0
22	Ma	0	0	0	0
23	Me	NPK 17:17:17 & U	Chlorpyrifos**	0	lcf
24	Me	Urea & P	Zeta*	Y	lcf
25	Me	Urea	(2,4-D AMINA)*	Y	0
26	Me	NPK 17:17:17 & U	(Z)-(1R,3R)**	0	0
27	Me	0	Zeta*	N	0
28	Me	NPK 17:17:17	(RS)*	0	0
29	Me	NPK 17:17:17 & U	Chlorpyrifos**	0	0
30	Ma	0	Chlorpyrifos**	Y	0
31	Me	NPK 17:17:17 & U	Chlorpyrifos*	0	0
32	Me	0	0	Y	0
33	Me	0	0	0	0
34	Me	0	0	0	0
35	Me	NPK 17:17:17 & U	0	0	0
36	Me	NPK 17:17:17 & U	Chlorpyrifos*	Y	0
37	Me	0	Urea	0	0
38	Me	NPK 17:17:17 & U	Carbofuran*	Y	0
39	Me	NPK 17:17:17 & U	Chlorpyrifos**	0	0
40	Me	NPK 17:17:17	(Z)-(1R,3R)*	0	0
41	Me	NPK 17:17:17 & U	(Z)-(1R,3R)*	0	0
42	Me	NPK 17:17:17 & U	(Z)-(1R,3R)*	0	0
43	Me	NPK 17:17:17 & U	(Z)-(1R,3R)*	0	0
44	Me	NPK 17:17:17 & U	(Z)-(1R,3R)*	0	0
45	Me	NPK 17:17:17 & U	(Z)-(1R,3R)*	0	0
46	Ma	0	0	Y	0
47	Me	U	0	0	0
48	Ma &Sb	Urea	Chlorpyrifos**	0	0

Table 3. Cont.

Plot	Tillage	Chemical Fertilization	Pesticides Use	Gm	Oa
49	Me	Urea & P	Zeta*	0	0
50	Ma	Urea	Endosulfan*	0	0
51	Ma	0	ID	0	0
52	Me	NPK 17:17:17 & U	Chlorpyrifos*	0	0

Tillage: Ma: Manual; Me: Mechanized. **Pesticides**: Endosulfan: 6,7,8,9,10,10-Hexachlor- 1,5,5a,6,9,9a-Hexahidro-6,9-metane-2,4,3-benzodioxatiepin-3-oxide; Chlorpyrifos: 0,0-dimetil 0-(3,5,6-trichlore-2-piridinil) fosforotioate (33.8%), Permetrine: 3-fenoxibenzil (1RS)-cis, trans-3-(2,2 diclorovinil)-2,2 dimetil ciclopropane-carboxilate (4.8%); Carbofuran: 2,3 Dihidro-2,2-dimetil-7-benzofuranil metil carbamate; Chlorothalonil: Tetrachloroisoftalonitrile; Zeta: Zeta-cipermetrine a-ciano-3-(fenoxifenil) metil (±) cis-trans; (Z)-(1R,3R): (Z)-(1R,3R)-3-(2-chloro-3,3,3-trifluoroprop-1-enil)-2,2-dimetilcichlopropanecarboxilate de (S)-α-ciano-3-fenoxibencile & (Z)-(1S,3S)-3-(2-chloro-3,3,3-trifluoroprop-1-enil)-2,2-dimetilcichlopropanecarboxilate de (R)-α-ciano-3-fenoxibencile; (RS): (RS)- alfa- ciano-3-fenoxybencil(1RS)-cis-trans-3-(2,2-dichlorvinil)- 1,1-dimetilcichlopropanecasrboxilate. * Once per year, ** 2 per year. **Chemical Fertilization**: U: Urea, P: Phosphorus. **Gm**: Green manure, Y: Yes, 0: null application. **Oa**: Organic amendments: csm: cow, sheep manure, bg: burned grass, lcf: last crop fallow, 0: null application.

plot site, or previous vegetation cover and the ecological index (Kendall's Tau, $T = 0.016$ $p = 0.98$, $T = 0.11$ $p = 0.90$, $T = -0.31$ $p = 0.75$, respectively). Therefore, in this study, it seems that neither size nor location of the plot determines the type of plot management; rather, plot management is likely determined by government development programs and traditional farming techniques.

3. Correlation between index and expert opinion

We obtained a Pearson correlation coefficient of $r^2 = 0.61$ ($p = 2.4e\text{-}06$, $d.f. = 49$) between the index and the values determined by independent experts, indicating a satisfactory correspondence (Figure 4). However, the experts systematically awarded higher scores to the plots than did the index. Moreover, they suggested to include additional variables to the index which would yield better information regarding (i) type of organic inputs to the crop system, (ii) types of pest and disease control used, (iii) number of native plant species among the crop, (iv) origin of crop seeds, (v) vegetation surrounding the crop, (vi) presence of vertebrate fauna, and (vii) diversity of soil macroinvertebrates.

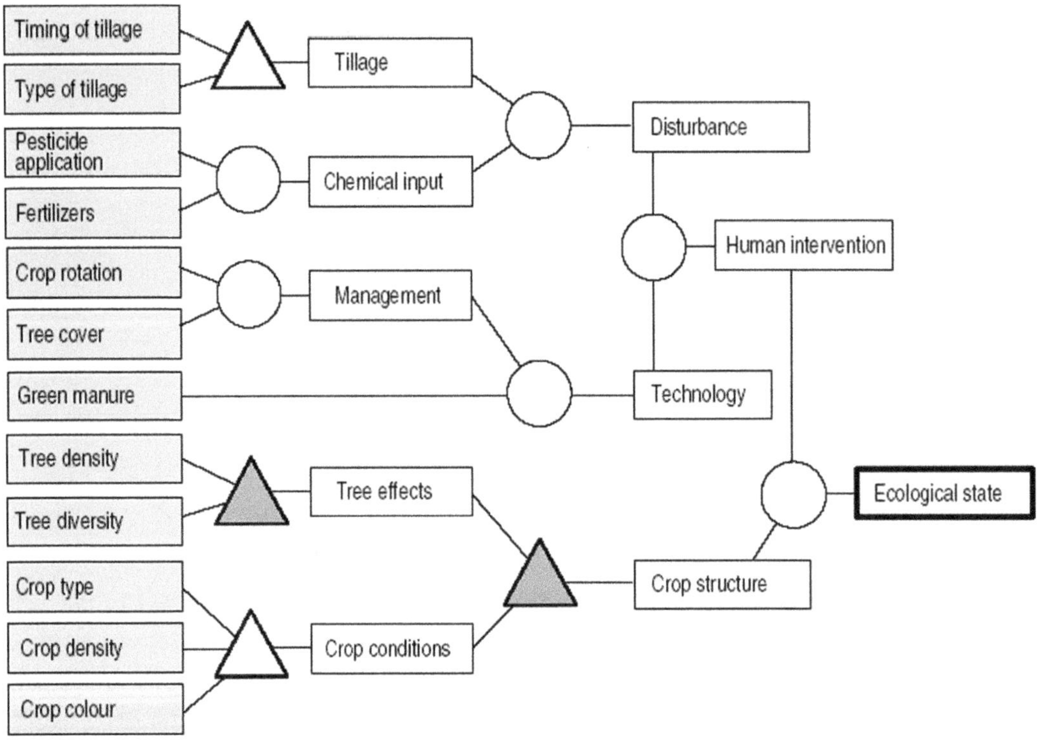

Figure 2. Structure of the Index of ecological condition of tropical agroecosystems. Primary indicators are shaded in grey. Circles represent simple mathematic algorithms, white triangles represent rule sets, and grey triangles represent rule sets, and grey triangles represent rule sets based on fuzzy logic. This index was presented together with other indexes within a frame of Indicators of environmentally sound land use in the humid tropics [15].

Table 4. Description of the measurement levels of the variables.

			Best	Intermediate			Worst
	Groupal variables	**Field data variables**		**5 levels**			
			1	0.75	0.5	0.25	0
	Management	Tree cover	all year	presence polyculture	present	rare	null
	Tree effect	Tree density	high	medium	low	isolated	null
				4 levels			
			1	0.66	0.33		0
		Tree diversity	>4 species	4 species	2 species		1 species
Crop structure	Crop conditions	Crop type	perennial polyculture	perennial monoculture	biannual		annual
				3 levels			
			1	0.33			0
	Plough (tillage)	Tillage time	null	frequent			constant
		Tillage form	null	manual			technical
Disturbance	Chemical inputs	Pesticides	null	frequent			constant
		Fertilizers	null	frequent			constant
	Management	Crop rotation	constant	frequent			null
Technology		Green manure	constant	frequent			null
				2 levels			
			1				0
	Crop conditions	Crop density	abundant				disperse
		Crop coleur	green				yellow

Table 5. Normalized Ecological Index for plots evaluated (52).

Plot	TT	TF	P	F	GM	RC	CA	DA	DivA	CT	CD	Ccol	Ev	Index
1	n	m	c	c	n	n	>1	l	>4	pm	a	g	0.24	0.56
2	n	n	c	n	n	n	>1	i	>4	a	a	g	0.35	0.5
3	c	m	n	c	c	n	>1	i	>4	a	a	g	0.26	0.56
4	n	n	n	n	n	n	n	n	1	a	a	g	0.01	0.37
5	n	n	c	c	c	c	>1	i	>4	a	a	g	0.09	0.62
6	n	m	c	c	n	n	>1	l	>4	a	a	g	0.22	0.5
7	c	t	c	n	n	n	>1	i	>4	a	d	b	0	0.15
8	c	t	c	c	c	c	n	n	1	pp	d	g	0.1	0.12
9	n	t	c	n	n	n	n	n	1	a	a	g	0.09	0.25
10	c	t	c	c	n	n	n	n	1	a	d	b	0	0
11	c	t	c	c	n	c	>1	i	>4	pm	ID	ID	0.03	0.21
12	c	t	c	c	n	n	>1	h	>4	ba	a	g	0.2	0.5
13	n	t	c	c	n	n	>1	i	>4	a	a	g	0.03	0.5
14	n	n	c	n	n	n	n	n	1	pp	a	g	0.29	0.37
15	n	t	n	n	n	c	>1	h	>4	pp	a	g	0.2	0.81
16	n	n	c	c	n	n	>1	l	>4	ba	a	g	0.25	0.5
17	c	t	c	c	n	n	n	n	1	a	a	g	0	0.25
18	c	t	c	c	n	n	n	n	1	a	d	g	0	0
19	c	t	c	c	c	n	n	n	1	a	d	g	0	0.06
20	n	t	c	n	n	n	>1	i	>4	a	a	g	0.1	0.5
21	c	t	c	c	n	n	n	n	1	a	a	g	0	0.25
22	n	n	n	n	n	n	>1	i	>4	pp	d	g	0.33	0.62
23	n	t	c	c	c	n	>1	i	>4	a	a	g	0.15	0.56
24	n	t	c	c	c	c	>1	l	>4	a	a	g	0.2	0.62
25	n	n	c	c	n	c	n	n	1	a	d	g	0.01	0.06
26	n	m	c	c	n	n	n	n	1	a	a	g	0.25	0.25
27	n	t	c	n	n	n	>1	i	>4	a	a	g	0.01	0.5
28	n	t	c	c	n	n	n	n	1	a	a	g	0	0.25
29	c	t	c	c	n	n	n	n	1	pm	a	g	0.01	0.37
30	n	m	c	n	n	c	n	n	1	pm	a	g	0.1	0.43
31	c	t	c	c	n	n	n	n	1	a	a	g	0.01	0.25
32	c	t	n	n	n	c	>1	i	>4	a	d	y	0.09	0.34
33	c	t	n	n	n	n	>1	h	>4	a	a	g	0.15	0.62
34	c	t	n	n	n	n	>1	i	>4	ba	a	g	0.13	0.62
35	c	t	n	c	n	n	>1	l	>4	a	a	g	0.01	0.5

Table 5. Cont.

Plot	TT	TF	P	F	GM	RC	CA	DA	DivA	CT	CD	Ccol	Ev	Index
36	c	t	c	c	n	c	n	n	1	a	a	g	0	0.31
37	c	t	c	n	n	n	>1	i	>4	pm	a	y	0.01	0.15
38	c	t	c	c	n	c	>1	h	>4	pm	a	g	0.2	0.68
39	c	t	c	c	n	n	>1	l	>4	a	a	b	0.06	0.18
40	c	t	c	c	n	n	>1	l	>4	a	a	g	0.19	0.5
41	c	t	c	c	n	n	>1	l	>4	a	a	g	0.19	0.5
42	c	t	c	c	n	n	>1	l	>4	a	a	g	0.17	0.5
43	c	t	c	c	n	n	>1	l	>4	a	a	g	0.2	0.5
44	c	t	c	c	n	n	>1	l	>4	a	a	g	0.2	0.5
45	c	t	c	c	n	n	>1	l	>4	a	a	g	0.2	0.5
46	n	n	n	n	n	c	>1	l	>4	pp	a	b	0.6	0.37
47	n	t	n	c	n	n	>1	l	>4	pp	d	g	0.18	0.5
48	c	n	c	c	n	n	n	n	1	a	a	b	0	0
49	c	t	c	c	n	n	>1	l	>4	a	a	b	0.07	0.18
50	c	n	c	c	n	n	>1	l	>4	pm	a	g	0.2	0.56
51	c	n	c	n	n	n	>1	l	>4	pm	a	g	0.15	0.56
52	c	t	c	c	n	n	>1	l	>4	pm	a	ID	0.24	0.18

TT: Tillage Time: c: >once per year; n: once per year; TF: Tillage Form: n: null, m: manual, t: technical; P. Pesticide use: c: constant, n: null, F: Use of chemical Fertilizers: c: constant, n: null; GM: Green Manure, c: constant, n: null; RC: Crop Rotation: c: constant, n: null; CA: tree cover: >1 year, n: null; DA: tree density: h: high, l: low, i: isolated, n:null; DivA: tree diversity: >4 species, 1 species, CT: crop type: a: annual, ba: biannual, pp: perennial polyculture, pm: perennial monoculture; CD: crop density: a: abundant, d: disperse; Ccol: crop colour: g: green, b: brown, y: yellow; EV: evaluation.

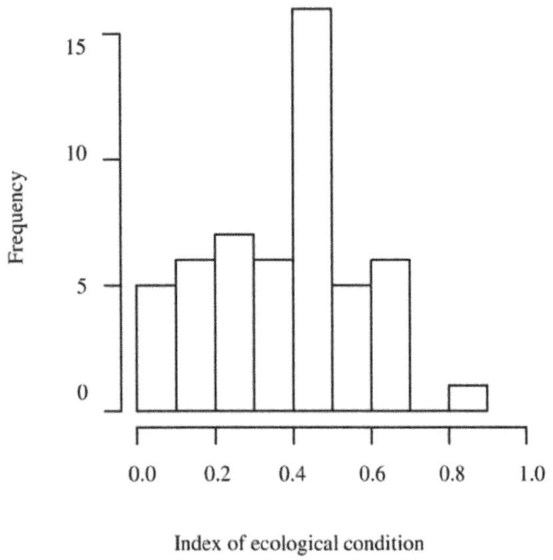

Figure 3. Frequency histogram of values of the index of ecological condition applied to 52 plots in South-East Mexico.

Discussion

Since the Rio Earth Summit, there has been a concerted effort to construct indicators to monitor progress toward sustainable development [59]. Most of these indicators have been developed in Europe (10) and Asia (2) [60]. In Latin America, sustainable indicators have been developed by Astier et al. [61]; this index mainly focuses on subsistence level agriculture, and evaluates the sustainability of a system.

Our index is geared toward small and medium scale producers who principally grow for market and have a fairly large crop area (32.4 ha on average). According to Bockstaller and Girardin [62], indicators must be elaborated according to a scientific approach, and one of the important steps in this elaboration is validation.

Our index was developed according to the consensus of a group of scientists, with knowledge in agroecology, and was validated independently by 2 scientists, each with over15 years of experience in agroecology. The evaluation included 3 important steps: design validation, output validation, and end use validation [62].

In previous studies, only seven indicators have been used to evaluate farm systems: crop diversity, crop succession, pesticide use, nitrogen level, phosphorus level, soil organic matter, and irrigation methods [63]. All of these practices depend on the farmer's decision, and to a large extent they impact the environment.

Our index is based on qualitative and quantitative concrete data and includes most of these 7 indicators, except nitrogen and phosphorus, both of which are observed indirectly via plant health, through the crop colour indicator (according to whether plants have a greenish or yellowish colour); organic matter, which is indirectly characterized by the technology applied in the system (green manure indicator, Figure 2); and irrigation, which in this study was not evaluated, given that all plots evaluated were only used for seasonal rainfed agriculture, according to local rainfall patterns.

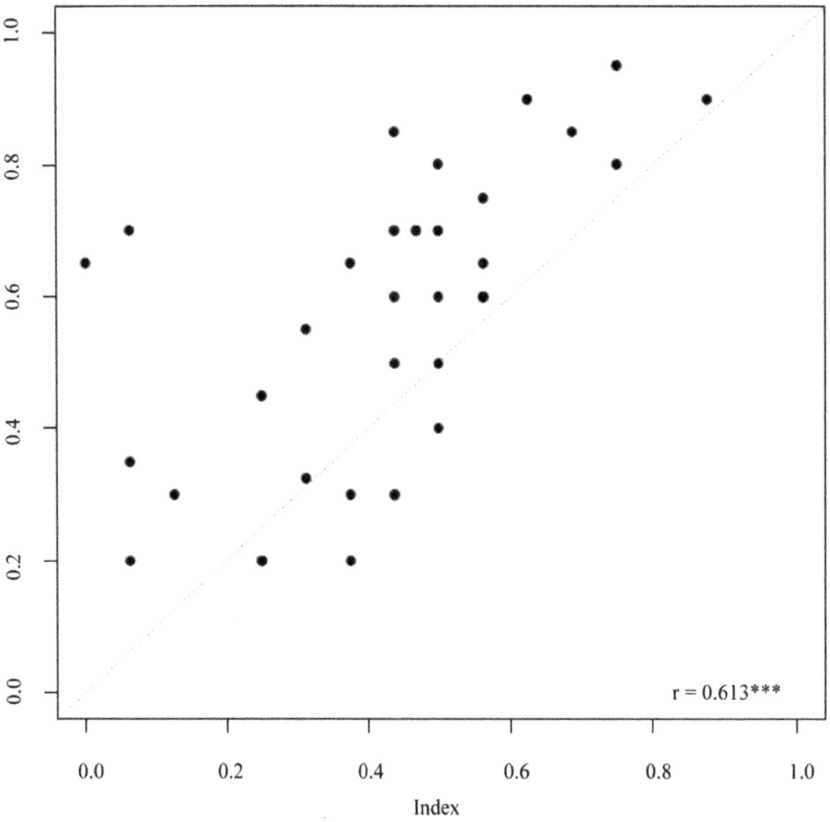

Figure 4. Scatterplot of the values of the index of ecological condition applied to 52 plots in South-East Mexico (x-axis) versus quality values between 0 (worst) and 1 (best) assigned by experts to the same plots (y-axis).

At the farm level there are indicators that evaluate the environmental impact of the agricultural practices and indicators that evaluate the effect of those practices at the local and global level [60].

Our index includes both types of indicators; evaluating agricultural practices: those variables taken in the field: type and frequency of tillage, pesticide application, fertilizer use, crop rotation, tree cover and density, green manure, crop density (see Table 2 & 3). Effect indicators used were disturbance (mainly soil disturbance) measured by tillage (frequency and type) and chemical inputs, technology used, and crop structure (see index, Figure 2). Some existing indexes focus on evaluation or evolution of environmental performance, thus encouraging environmentally sound practices [60], such as crop rotation, organic fertilizers, and no-tillage. Meanwhile, our index identifies the indicator that has the highest environmental impacts in each plot evaluated, with the idea that the farmer could potentially improve these with a given practice, ie. to monitor the soil ecological condition via the use of a soil macroinvertebrates index [64], where the lack of macroinvertebrates informs of a severe negative activity as pollution or conventional tillage.

In developing the variables to be included in our index, we reviewed bibliographic studies and carried out field work obtaining data which we hoped would reflect the negative and environmentally friendly practices commonly used in agro ecosystems of southeastern Mexico. The index can be used by farmers, using the following web address: http://201.116.78.102/~modelo/Index.html.

However, we did not evaluate certain indicators such as water use, and water quality, as did - for example - the index (monitoring tool) of ecological indicators used on a Flemish dairy farm [65]. Nor did we evaluate environmental impacts due to energy consumption [66]. Within a tropical framework, in south-eastern Mexico, the priorities were to identify those practices that were soil perturbing and environment polluting, practices that can be modified by the farmers, by an attitude changing. Nambiar et al. [16], proposed an agricultural sustainability index (ASI) to measure sustainability as a function of biophysical, chemical, economic, and social indicators, our index only measures the ecological state of the agroecosystem, and provides easy to use tools for improving those practices which negatively impact the environment. Van der Werf and Petit [60], state that indicators based on farmer practices cost less in data collection but do not allow for an actual evaluation of environmental impact. In the case of our study, the experts' evaluation correlated satisfactorily with the index, although the index rendered more penalizing scores than did the experts. Some improvements must be made to our index, relating quality and quantity of the applied inputs for instance; the index should specify the kind of manures used and then to evaluate their effect when added to the systems. The consulted experts found important to integrate this information into the index, they also found that the possible relations among different crops and environmental effects of using cow manure, vermicompost, or traditional compost have to be considered. Another variable which the experts suggested should be added to the index is the presence of natural vegetation surrounding the crop. Farmers see advantages of having crops surrounded by secondary forest, diversity in the agricultural area can be increased, ie when different pollinators arrive.

The advantages of using this index is that the common agricultural practices (mechanized land preparation and use of common pesticides ie. Carbofuran, Chlorpyrifos), evaluated in this study as indicators are used throughout the world; over time, through practice, the index may be improved. Farmers and other land owners may realize which of the practices they use are disturbing the environment, due to the fact that these practices generate a value in each of the evaluated indicators. The variables obtained in the field contribute to the information of each of the indicators, and the index is the compendium of all the indicators. Our index doesn't give a sustainability measure, because it does not include socioeconomic indicators of the farms. Further studies are required in order to observe the acceptance of this index by farmers in a regional scale.

Supporting Information

Figure S1 Indicators of environmentally sound land use in the humid tropics [15].

Acknowledgments

We thank Lorena Soto and Guillermo Jimenez for participating in the evaluation of the plots. We thank Simon Hernandez de la Cruz, Alejandra Sepulveda Lozada, Lauritania Ibarra Hernandez, and Marcelo Rodriguez Ricardes, who helped to collect data in the field. We are also thankful to Lorena Reyes for bibliographic support. El Colegio de la Frontera Sur provided infrastructural resources. Language revision was done by Ann Greenberg.

Author Contributions

Conceived and designed the experiments: EH CK SOG BDJ SHD VG. Performed the experiments: EH CK. Analyzed the data: EH CK. Contributed reagents/materials/analysis tools: EH CK SOG BDJ SHD VG. Contributed to the writing of the manuscript: EH CK SOG BDJ VG. Figures elaboration: CK SOG.

References

1. Houghton RA (1994) The world wide extent of land-use change. BioScience 44: 305–313.
2. Tilman D, Lehman C (2001) Human-cause environmental change: impacts on plant diversity and evolution. Proceedings of the National Academy of Sciences 98: 5433–5440.
3. INEGI (2008) Cuaderno Estadístico Municipal de Tenosique. Villahermosa, Tabasco: Gobierno del Estado de Tabasco.
4. Grande D, de Leon F, Nahed J, Perez-Gil F (2010) Importance and Function of Scattered Trees in Pastures in the Sierra Region of Tabasco, Mexico. Research Journal of Biological Sciences 5: 75–87.
5. Geissen V, Morales-Guzman G (2006) Fertility of tropical soils under different land use systems-a case study of soils in Tabasco, Mexico. Appl Soil Ecol 841: 1–10.
6. Melgar C, Geissen V, Cram S, Sokolov M, Bastidas P, et al. (2008) Pollutants in drainage channels following long-term application of Mancozeb to banana plantations in southeastern Mexico. Journal of Plant Nutrition and Soil Science 171: 597–604.
7. Aryal DR, Geissen V, Ponce-Mendoza A, Ramos-Reyes RR, Becker M (2012) Water quality under intensive banana production and extensive pastureland in tropical Mexico. Journal of Plant Nutrition and Soil Science 175(4): 553–559.
8. Ortiz SM, Anaya G, Estrada BW (1994) Evaluación, Cartografía y Políticas Preventivas de la Degradación de la Tierra. Chapingo, Mexico.
9. Geissen V, Sanchez-Hernandez R, Kampichler C, Ramos-Reyes R, Sepulveda-Lozada A, et al. (2009) Effects of land-use change on some properties of tropical soils – An example from Southeast Mexico. Geoderma 151: 87–97.
10. Moreno-Unda AA (2011) Environmental effects of the National Tree Clearing Program, Mexico, 1972–1982 Cologne: Cologne University of Applied Sciences 119 p.
11. Bockstaller CGL, Keichinger O, Girardin P, Galan MB, Gaillard G (2009) Comparison of methods to assess the sustainability of agricultural systems. A review. Agron Sustain Dev 29: 223–235.
12. Mitchell GMA, McDonald A (1995) PICABUE: a methodological framework for the development of indicators of sustainable development. Int J Sust Dev World 104–123.

13. Bockstaller C, Guichard L, Makowski D, Aveline A, Girardin P, et al. (2008) Agri-Environmental Indicators to Assess Cropping and Farming Systems: A Review Sustainable Agriculture. In: Lichtfouse E, Navarrete M, Debaeke P, Véronique S, Alberola C, editors. Springer Netherlands. 725–738.
14. Riley J (2001) The indicator explosion: local needs and International challenges. Agr Ecosyst Environ 87: 119–120.
15. Kampichler C, Hernández-Daumás S, Ochoa-Gaona S, Geissen V, Huerta-Lwanga E, et al. (2010) Indicators of environmentally sound land use in the humid tropics: The potential roles of expert opinion, knowledge engineering and knowledge discovery. Ecological Indicators 10: 320–329.
16. Nambiar KK, Gupta AP, Fu Qinglin, Li S (2001) Biophysical, chemical and socio-economic indicators for assessing agricultural sustainability in the Chinese coastal zone Agriculture Ecosystems and Environment 87: 209–214.
17. Cabell JF, Oelofse M (2012) An indicator framework for assessing agroecosystem resilience. Ecology and Society 17: 18 http://dx.doi.org/10.5751/ES-04666-170118.
18. Darnhofer I, Bellon S, Dedieu B, Milestad R (2010) Adaptiveness to enhance the sustainability of farming systems. A review. Agronomy for Sustainable Development 30: 545–555.
19. Allen CR, Fontaine JJ, Pope KL, Garmestani AS (2011) Adaptive management for a turbulent future. Journal of Environmental Management 92: 1339–1345.
20. Malezieux E, Crozat Y, Dupraz C, Laurans M, Makowski D, et al. (2009) Mixing plant species in cropping systems: concepts, tools and models. A review. Agron Sustain Dev 29: 43–62.
21. Jones CG, Lawton J, Shachak M (1997) Positive and negative effects of organisms as physical ecosystem engineers. Ecology and Society 78: 1946–1957.
22. Brussaard L, Caron P, Campbell B, Lipper L, Mainka S, et al. (2010) Reconciling biodiversity conservation and food security: scientific challenges for a new agriculture. Current opinion in Environmental sustainability 2: 34–42.
23. Knorr D, Watkins TR (1984) Alterations in Food Production Knorr DW, Watkins, T.R., editor. New York: Van Nostrand Reinhold.
24. Seufert V, Ramankutty N, Foley JA (2012) Comparing the yields of organic and conventional agriculture. Nature 485: 229–232.
25. Carter H (1989) Agricultural sustainability: an overview and research assessment. Calif Agric 43: 1618–1637.
26. Tellarini V, Caporali F (2000) An input/output methodology to evaluate farms as sustainable agroecosystems: an application of indicators to farms in central Italy. Agriculture, Ecosystems and Environment 77: 111–123.
27. Stinner BR, House GJ (1987) Role of ecology in lower-input, sustainable agriculture: an introduction. Am J Alternative Agric 2: 146–147.
28. Lockeretz W (1988) Open questions in sustainable agriculture. Am J Alternative Agric 3: 174–181.
29. Hauptli H, Katz D, Thomas BR, Goodman RM (1990) Biotechnology and crop breeding for sustainable agriculture. In: Edwards CA, Lal R, Madden P, Miller RH, House G, editors. Sustainable Agricultural Systems Soil and Water Conservation Society: Ankeny, Iowa. 141–156.
30. Madden P (1990) The economics of sustainable low-input farming systems. In: Francis CA, Flora CB, King LD, editors. Sustainable Agriculture in Temperate Zones. New York: John Wiley & Sons. 315–341.
31. Dobbs TL, Becker DL, Taylor DC (1991) Sustainable agriculture policy analyses: South Dakota on-farm case studies. J Farming Systems ResExt 2: 109–124.
32. Blair GJ, Lefroy RD, Lisle L (1995) Soil carbon fractions based on their degree of oxidation, and the development of a carbon management index for agricultural systems. Aust J Soil Res 46: 1459–1466.
33. Ouédraogo E, Mando A, Zombré NP (2001) Use of compost to improve soil properties and crop productivity under low input agricultural system in West Africa. Agriculture, Ecosystems and Environment 84: 259–266.
34. Pretty J, Toulmin C, Williams S (2011) Sustainable intensification in African agriculture. International Journal of Agricultural Sustainability 9: 5–24.
35. Lal R (2004) Soil carbon sequestration impacts on global climate change and food security. Science 304: 1623–1627.
36. Gurr GM, Wratten SD, Luna JM (2003) Multi-function agricultural biodiversity: pest management and other benefits. Basic Appl Ecol 4: 107–116.
37. Börjesson P (1999) Environmental effects of energy crop cultivation in Sweden I: Identification and quantification. Biomass and Bioenergy 16: 137–154.
38. Rommelse R (2001) Economic assessment of biomass transfer and improved fallow trials in western Kenya In: ICRAF Natural Resource Problems PaPP, editor. Natural Resource Problems. Nairobi (Kenya): International Centre for Research in Agroforestry (ICRAF).
39. Nyende P, Delve R (2004) Farmer participatory evaluation of legume cover crop and biomass transfer technologies for soil fertility improvement using farmer criteria, preference ranking and logit regression analysis. Exp Agric 40: 77–88.
40. Koocheki A, Nassiri M, Alimoradi L, Ghorbani R (2009) Effect of cropping systems and crop rotations on weeds. Agronomy for Sustainable Development 29: 401–408.
41. Bellon M (1995) Farmers 'Knowledge and Sustainable Agroecosystem Management: An Operational Definition and an Example from Chiapas, Mexico. Human Organization 54: 263–272.
42. Belalcázar S, Espinosa J (2000) Effect of Plant Density and Nutrient Management on Plantain Yield. Better Crops International 14: 12–15.
43. Shaahan MM, El-Sayed AA, Abou El-Nour EAA (1999) Predicting nitrogen, magnesium and iron nutritional status in some perennial crops using a portable chlorophyll meter. Scientia Horticulturae 82: 339–348.
44. Agbede TM (2008) Nutrient availability and cocoyam yield under different tillage practices. Soil Tillage Research 99: 49–57.
45. Edwards CA (1989) The Importance of Integration in Sustainable Agricultural Systems. Agriculture, Ecosystems and Environment 27: 25–35.
46. Edwards CA, Grove TL, Harwood RR, Pierce Colfer CJ (1993) The role of agroecology and integrated farming systems in agricultural sustainability. Agriculture, Ecosystems and Environment 46: 99–121.
47. García-Pérez JA, Alarcón-Gutiérrez E, Perroni Y, Barois I (2014) Earthworm communities and soil properties in shaded coffee plantations with and without application of glyphosate. Applied Soil Ecology 83: 230–237.
48. Correia FV, Moreira JC (2010) Effects of Glyphosate and 2,4-D on Earthworms (*Eisenia foetida*) in Laboratory Tests. Bull Environ Contam Toxicol 85: 264–268.
49. Dick J, Kaya B, Soutoura M, Skiba U, Smith R, et al. (2008) The contribution of agricultural practices to nitrous oxide emissions in semi-arid Mali Soil. Use and Management 24: 292–301.
50. Morales H, Perfecto I (2000) Traditional knowledge and pest management in the Guatemalan highlands. Agriculture and Human Values 17: 49–63.
51. Li BL, Rykiel EJ (1996) Introduction. Ecological Modelling 90: 109–110.
52. Salski A (1996) Introduction Fuzzy Logic in Ecological Modelling. Ecological Modelling 85: 1–2.
53. Mendoza GA, Prabhu R (2003) Fuzzy methods for assessing criteria and indicators of sustainable forest management. Ecological Indicators 3: 227–236.
54. Kampichler C, Platen R (2004) Ground beetle occurrence and moor degradation: modelling a bioindication system by automated decision-tree induction and fuzzy logic. Ecological Indicators 4: 99–109.
55. INEGI (2000) Cuaderno Estadístico Municipal de Tenosique. Gobierno del Estado de Tabasco; INEGI, editor. Villahermosa, Tabasco.
56. INEGI (1985) Carta Edafológica, Villahermosa In: E15-8, editor. 1: 250,000. Aguascalientes, México: Instituto Nacional de Estadística, Geografía e Informática.
57. Manjarrez-Muñoz B (2008) Ordenamiento territorial de la ganadería bovina en Balancán y Tenosique, Tabasco. Villahermosa Tabasco: El Colegio de la Frontera Sur. 105 p.
58. Isaac-Márquez R (2008) Análisis del Cambio de Uso y Cobertura del Suelo en los Municipios de Balancán y Tenosique, Tabasco, México. Villahermosa, Tabasco, México: El Colegio de la Frontera Sur.
59. Rigby D, Woodhouse P, Young T, Burton M (2001) Constructing a farm level indicator of sustainable agricultural practice. Ecological Economics 39: 463–478.
60. van der Werf HMG, Petit J (2002) Evaluation of the environmental impact of agriculture at the farm level: a comparison and analysis of 12 indicator-based methods. Agriculture, Ecosystems and Environment 93: 131–145.
61. Astier M, Speelman S, López-Ridaura S, Masera O, Gonzalez-Esquivel CE (2011) Sustainability indicators, alternative strategies and trade-offs in peasant agroecosystems: analysing 15 case studies from Latin America. International Journal of Agricultural Sustainability 9: 409–422.
62. Bockstaller C, Girardin P (2003) How to validate environmental indicators. Agr Syst 76: 639–653.
63. Bockstaller C, Girardin P, Van der Werf HGM (1997) Use of agroecological indicators for the evaluation of farming systems. Eur J Agron 7: 261–270.
64. Huerta E, Kampichler C, Geissen V, Ochoa-Gaona S, de Jong B, et al. (2009) Towards an ecological index for tropical soil quality based on soil macrofauna. Pesq agropec bras 44 (8): 1056–1062.
65. Meul M, Nevens F, Reheul D (2009) Validating sustainability indicators: Focus on ecological aspects of Flemish dairy farms. Ecological Indicators 9: 284–295.
66. Pervanchon F, Bockstaller C, Girardin P (2002) Assessment of energy use in arable farming systems by means of an agro-ecological indicator: the energy indicator. Agricultural Systems 72: 149–172.

24

Vertical Profiles of Soil Water Content as Influenced by Environmental Factors in a Small Catchment on the Hilly-Gully Loess Plateau

Bing Wang[1], Fenxiang Wen[1], Jiangtao Wu[1], Xiaojun Wang[1]*, Yani Hu[2]

1 College of Environmental Science and Resources, Shanxi University, Taiyuan, China, **2** Library, Hebei University of Science and Technology, Shijiazhuang, China

Abstract

Characterization of soil water content (SWC) profiles at catchment scale has profound implications for understanding hydrological processes of the terrestrial water cycle, thereby contributing to sustainable water management and ecological restoration in arid and semi-arid regions. This study described the vertical profiles of SWC at the small catchment scale on the hilly and gully Loess Plateau in Northeast China, and evaluated the influences of selected environmental factors (land-use type, topography and landform) on average SWC within 300 cm depth. Soils were sampled from 101 points across a small catchment before and after the rainy season. Cluster analysis showed that soil profiles with high-level SWC in a stable trend (from top to bottom) were most commonly present in the catchment, especially in the gully related to terrace. Woodland soil profiles had low-level SWC with vertical variations in a descending or stable trend. Most abandoned farmland and grassland soil profiles had medium-level SWC with vertical variations in varying trends. No soil profiles had low-level SWC with vertical variations in an ascending trend. Multi-regression analysis showed that average SWC was significantly affected by land-use type in different soil layers (0–20, 20–160, and 160–300 cm), generally in descending order of terrace, abandoned farmland, grassland, and woodland. There was a significant negative correlation between average SWC and gradient along the whole profile ($P<0.05$). Landform significantly affected SWC in the surface soil layer (0–20 cm) before the rainy season but throughout the whole profile after the rainy season, with lower levels on the ridge than in the gully. Altitude only strongly affected SWC after the rainy season. The results indicated that land-use type, gradient, landform, and altitude should be considered in spatial SWC estimation and sustainable water management in these small catchments on the Loess Plateau as well as in other complex terrains with similar settings.

Editor: Andrew C. Singer, NERC Centre for Ecology & Hydrology, United Kingdom

Funding: This work was supported by the Natural Science Foundation of China (No. 41201277 and 41101025) and the Natural Science Foundation of Shanxi Province of China (No. 2014011034-2). The funders had no role in study design, data collection and analysis, decision to publish, or preparation of the manuscript.

Competing Interests: The authors have declared that no competing interests exist.

* Email: xjwang@sxu.edu.cn

Introduction

Soil water content (SWC) is a critical factor for plant growth and a determinant of plant distribution in arid and semiarid areas such as China's Loess Plateau [1, 2]. Vertical distribution of SWC can greatly affect soil water movement [3], thereby greatly affecting the biomass production and water use efficiency of plants (e.g., switchgrass) under water stress [4]. Plant-available water stored in the soil profile has a buffering capacity, which, in deep layers, prolongs or alleviates the effects of seasonal or inter-annual drought on plant growth and soil water flux to the atmosphere [5–7]. Research has provided strong evidence that deep soil water depletion plays a key role in sustainable agriculture, ecological restoration, and terrestrial water cycling on the Loess Plateau [8–10]. However, measurement of SWC profiles has been frequently conducted at different spatial scales. The results thus need to be converted before comparison analysis or practical uses. The SWC profile in small catchment is considered to be at a moderate scale for data exchanging. In particular, small catchment is thought to be the basic unit for integrated soil and water loss management in complicated terrain of the Loess Plateau [11, 12]. Characterization of SWC profiles and evaluation of relevant influencing factors at the small catchment scale have implications for hydrological modeling of soil water dynamics and sustainable management of soil water resources in similar areas.

Classical statistics is frequently used to analyze the variability of SWC profiles at the small catchment scale, which involves the estimation of descriptive parameters such as average (mean), variance, standard deviation (STD), and coefficient of variation (CV). Average SWC at individual soil depth intervals or across the whole soil profile is extensively determined. The CV of SWC is also routinely calculated as the temporal variable in a certain period of time or the spatial variable across a specific area. The SWC profile can be divided into distinct intervals by considering its average and CV which exhibit complex spatial-temporal relationships in several plots or watersheds [13–16]. Additionally, ranking method, clustering method, and semivariogram model have been applied for the division of SWC profile [3, 17–20]. However, the above-mentioned methods cannot clearly reflect the variation trend in SWC profiles. Thus, great effort has been made

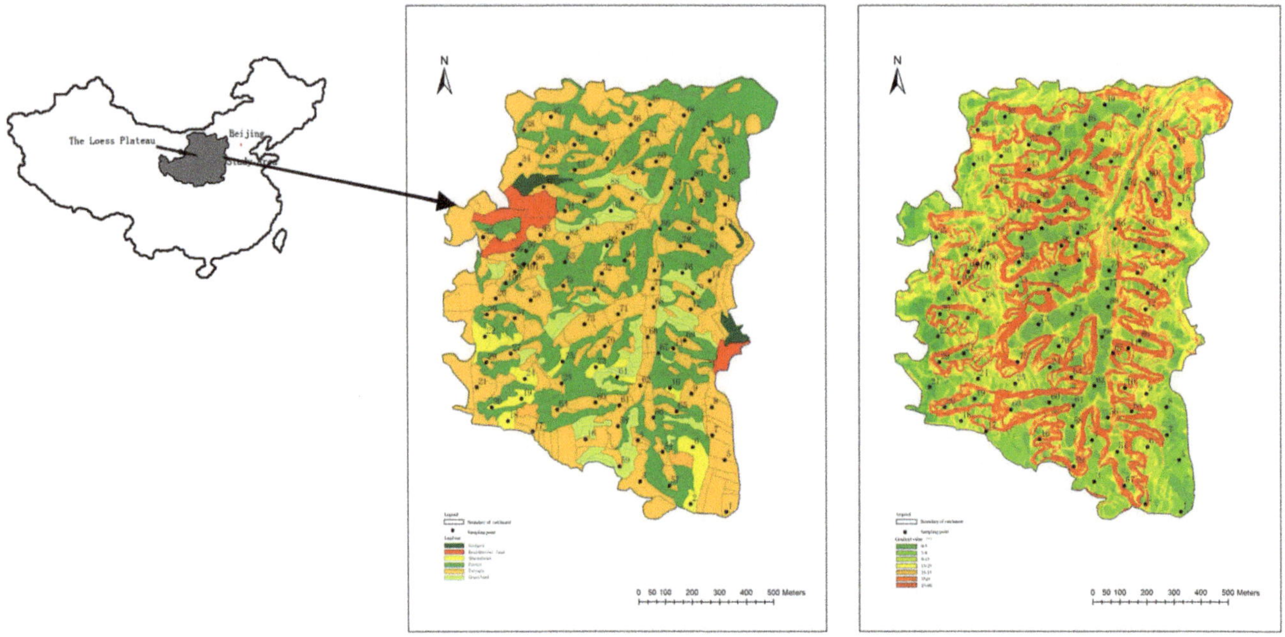

Figure 1. Location of Sanyanjing catchment and distribution of 101 sampling points in the Sanyanjing catchment.

to describe the vertical profiles of SWC through comparing variation curves or variation ranges, in small watersheds related to different land-use types, vegetation species, and/or terrain factors [3, 17, 18, 20–22]. If a massive sample size is involved, however, it becomes difficult to distinguish the vertical profiles and major influencing factors of SWC by direct comparisons.

In recent decades, a great number of studies have been conducted on the spatiotemporal variability of SWC and related influencing factors worldwide. Canton [23] pointed out that wasteland-scale spatial variability of SWC is mainly controlled by surface cover and soil properties in a semi-arid region of Spain, where surface cover counteracts the influence of terrain factors (including gradient, aspect, topographic wetness index, and distance from the river) on SWC distribution. Burnt [24] reported a topographic index which can simulate changes in high-level SWC in a humid climate zone of Devin County, UK. O'loughlin [25] estimated the spatial pattern of SWC distribution in a small catchment using humidity index model based on digital terrain dataset. Hawley [15] discovered that topography is the major factor responsible for the spatial distribution of SWC in an agricultural region, where resultant SWC variation is diminished by vegetation in a moist climate zone in Chickasha, USA. In arid and semi-arid areas, catchment-scale distribution of SWC is strongly affected by land-use/vegetation and topographic indices, e.g., land-use type, soil organic matter content, tillage, soil physical properties, gradient, and aspects [6, 17–19, 21, 26].

Many researchers have focused on the quantification of environmental parameters such as topographic factor, vegetation type, soil texture, and land-use type, in attempt to evaluate their impacts on the variability of SWC. At the small catchment scale, little information is available on the major factors affecting vertical profiles of SWC in cinnamon soil (Haplic Lixisols, FAO) zone on the hilly and gully Loess Plateau [3, 19]. As a regional water reservoir experiencing depletion, the plateau region requires measurements and characterization of deep SWC profiles for the thick soil layer. However, soil sampling for SWC profile analysis at the catchment scale has been commonly conducted at <200 cm depth [18–20, 23]. Wang et al. [21] exceptionally examined SWC along the 0–21 m soil profile on the Loess Plateau, but the reliability of their tests might be affected by a small sample size (11 sites). Deep soil sampling at a larger number of sites and statistical analysis of parameters involving soil depth information will contribute to better understanding of the vertical profiles and influencing factors of SWC.

In the present study, we characterized the vertical profiles of SWC in a small catchment on the Loess Plateau by cluster analysis of two descriptive parameters (mean and regression gradient). Sampling was carried out in the 0–300 cm profile at 101 points throughout the catchment before and after the rainy season, to meet the demand for deep depth, spatial representativeness and temporal comparability. The influences of selected environmental factors (land-use type, topographic factors, and landform) on average SWC were examined by multi-linear regression [6, 26–32]. The results were discussed in order to provide new insights to the vertical profiles and influencing factors of SWC on the Loess Plateau, further providing reference data for sustainable management of water resource in small catchment areas in the semi-arid region with complex terrain.

Material and Methods

1 Site description

This study was conducted in Sanyanjing catchment (112°2′13″, 37°46′23″), which is located on the east margin of the Loess Plateau in Shouyang county, mid-east Shanxi province, China (Figure 1). The catchment has a total area of 1.32 km^2 and the elevation ranges from 1001 to 1160 m. It is a hilly and gully area with mostly deep gully erosion slopes. The landform consists of ridge and gully.

The catchment area has a semi-arid continental climate (Cwa by Koppen Climate Classification) with an average annual precipitation of 474.2 mm (1967–1999). Snow in the winter accounts for ~8% and rainfall in July to September for ~73% of annual precipitation. Monthly average precipitation, potential evapotranspiration and precipitation in 2013 are shown in Figure 2. Annual mean temperature in this area is 8.1°C, with a maximum of 34.7°C and a minimum −20.6°C. The soil type is cinnamon soil (Haplic lixisols, FAO), which consists of 54–62% silt and 10.95–30.15% sand with the bulk density of 1.3–1.4 g/cm^3. Soil texture was measured using a particle size analyzer (SEDIMAT 4–12, UGT, Germany). Soil bulk density was determined through sampling with cutting rings (inner diameter 5.0 cm, volume 100 cm^3) and drying in an oven (105°C, 24 h). The profile of soil texture related to different landforms is listed in Table 1. Maximum soil depth is mostly down to 300 cm on the ridge and bare rock could be rarely seen only on the northern margin of the gully area.

The distribution of land-use types across the catchment is shown in Figure 1. Terrace is the dominant land-use type, accounting for about 60% of the study area. Few terraces had been abandoned for natural restoration of vegetation because of the Grain for Green project since 2000. Grassland is mainly covered with herbs and semi-shrubs, which had never been reclaimed for several decades. About 80% of the woodland is covered with semi-shrubs at steep slopes unsuitable for sampling.

2 Soil sampling

Ethics statement. Sampling activities at the farmland were allowed by the owners. No specific permissions were required at other locations because they were not privately-owned or protected in any way and the field activities did not involve any endangered or protected species.

A total of 101 sampling points were designed in a 150 m×150 m grid throughout the catchment area by considering major land-use types, including terrace (83), abandoned farmland (9), grassland (3), and woodland (6). Soil sampling was carried out during two periods in 2013, from April 29 to May 4 (before the rainy season) and from October 28 to November 1 (after the rainy season). No precipitation occurred during the two sampling periods or a week before sampling. Each sample was taken at 20 cm intervals along the 0–300 cm soil profile using an auger (inner diameter 5.0 cm). The samples were kept in capped aluminum boxes for transportation. Measurement of SWC was conducted using an oven-drying method (105°C, 24 h). At the majority of the sampling points, soils were collected along a vertical profile over 300 cm. A few exceptions were in the north of the gully at the lowest altitude where weathered rock was occasionally encountered. Background information of the 101 sampling points is summarized in Table 2.

3 Data analysis

To identify the variability of SWC profiles at the catchment scale, descriptive parameters were calculated for each profile. Further, we calculated the linear regression coefficient (K value) between SWC and soil depth to represent the variation trend of SWC vertical profiles and the mean value to describe the average level of SWC along the 0–300 cm profile.

The SWC profiles were classified using a combined cluster analysis of the K and mean values. Cluster analysis is the process of grouping a set to data objects into multiple groups (or clusters), so that objects within a cluster share high similarity but are dissimilar to those in other clusters. In this approach, dissimilarities and similarities are assessed based on the attribute values describing the objects and often involve distance measures. Cluster analysis is a statistical classification method for discovering whether the individuals of a population gall into different groups by making quantitative comparisons of multiple characteristics [34]. Here a combined cluster analysis was conducted in three steps: 1) cluster of the mean to three groups, which present the average level of SWC along the vertical soil profile, 2) cluster of K to three groups, which reflect the variation trend of SWC profiles (top to bottom), and 3) combination of the two sets of groups into nine new groups using the between-groups linkage method with squared Euclidean distance criteria [34].

For group comparisons, SWC profiles (0–300 cm) of the same group were averaged and re-plotted. The average curves of SWC were compared between groups to identify the major factors influencing SWC in individual soil layers (0–20, 20–160, and 160–300 cm). On the basis of cluster analysis, the influences of land-use type, topography and landform on average SWC in individual soil layers were examined by multi-regression analysis. The independent variables were land-use type, landform type, Sin(gradient), Sin(aspect), flow accumulation (calculated cell numbers to a grid cell from surrounding cells with the ArcGIS hydrology analysis

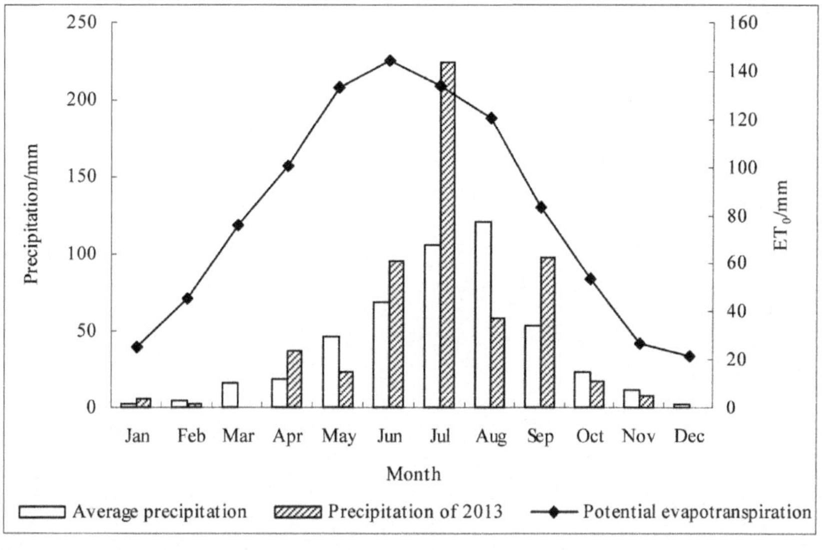

Figure 2. Average annual precipitation and potential evapotranspiration of 1967–1999 and precipitation in 2013 in the Sanyanjing catchment, Shanxi province, China.

Table 1. Soil texture in vertical profiles related to different landforms in the Sanyanjing catchment in Shanxi province, China.

Point description	Soil depth (cm)	Sand (%)		Silt (%)		Clay (%)
		>0.05 mm (%)	0.05–0.02 mm (%)	0.02–0.0063 mm (%)	0.0063–0.002 mm (%)	<0.002 mm (%)
	0–20	26.05	30.40	20.60	7.00	15.95
	20–40	21.35	34.80	20.80	6.90	16.15
	40–60	30.25	25.90	21.60	4.50	17.75
	60–80	28.25	28.80	21.70	2.80	18.45
	80–100	18.45	38.00	20.80	6.70	16.05
	100–120	26.45	30.10	20.40	5.20	17.85
Terrace at	120–140	26.95	27.90	22.00	4.40	18.75
ridge	140–160	16.35	32.60	24.20	8.10	18.75
(Point 33)	160–180	27.75	22.40	22.70	7.90	19.25
	180–200	36.55	17.80	20.60	8.00	17.05
	200–220	18.45	29.00	24.30	9.00	19.25
	220–240	20.45	28.50	23.20	8.60	19.25
	240–260	22.05	27.30	23.90	8.80	17.95
	260–280	31.35	20.30	22.20	9.40	16.75
	280–300	24.65	29.80	20.70	5.90	18.95
	0–20	30.15	20.00	23.20	6.40	20.25
	20–40	27.75	21.80	21.90	6.10	22.45
	40–60	25.95	26.80	20.50	4.50	22.25
	60–80	23.85	27.09	21.20	1.20	26.65
	80–100	0.95	31.00	20.10	20.10	27.85
	100–120	10.95	23.40	18.40	18.40	28.85
Terrace at	120–140	7.45	27.30	19.20	19.20	26.85
ridge	140–160	17.45	19.50	18.90	18.90	25.25
(Point 39)	160–180	10.45	25.80	18.20	18.20	27.35
	180–200	15.75	24.70	17.00	17.00	25.55
	200–220	13.35	21.40	18.60	18.60	28.05
	220–240	20.85	8.80	22.00	22.00	26.35
	240–260	20.25	12.40	20.00	20.00	27.35
	260–280	12.15	23.70	19.10	19.10	25.95
	280–300	24.95	29.00	20.50	0.50	25.05

module), and elevation. The former two factors were categorical variables converted into dummy variables before introduced into the regression analysis; and the latter four factors were continuous variables produced using digital elevation model at 1-m resolution.

SWC data were statistically analyzed in SPSS13.0 (SPSS Inc., Chicago, IL, USA), and topographic features were analyzed in ArcGIS 10.0 (ESRI, Redlands, CA, USA). SWC profiles were drawn in Microsoft Excel 2010 (Microsoft Corp., Redmond, WA, USA) and then clustered in SPSS 13.0 by considering descriptive parameters (maximum, minimum, mean, CV, STD, and K). Multi-regression analysis was performed in SPSS 13.0, with a *P*-value less than 0.05 considered statistically significant.

Results

1 Vertical profiles and descriptive parameters of SWC

The vertical profiles (0–300 cm) of SWC at 101 sampling points before the rainy season were drawn (Figure 3). These SWC profiles showed dynamic variations across the catchment study area, with substantial differences in the soil layers below 100 cm. At a few sampling points, there were obvious soil water depletion (e.g., 67, 75, 85, and 94) and an increasing trend (top to bottom) of SWC (e.g., 12 and 36). High degrees of soil desiccation were rarely detected in the lower soil layers, and low SWC was mainly found in the lower soil layers of woodland.

Descriptive parameters such as maximum, minimum, mean, and CV, STD are commonly used to reveal the spatial-temporal variability of SWC. However, these parameters cannot reflect the variation trend of SWC vertical profiles. To this end, the K value of SWC to profile depth was introduced for quantification of variation trend of SWC vertical profiles (Figure 4). Results showed that before the rainy season, SWC substantially varied between 5.87% and 34.72%, whereas the mean, STD, CV, and K values respectively ranged from 10.57% to 21.76%, 0.47 to 4.53, 3% to 24%, and −0.0405 to 0.0274 along the vertical soil profile (0–

Table 2. Background information of 101 soil sampling points in the Sanyanjing catchment study area in Shanxi province, China.

Land-use type	Vegetation	Landform type	Soil profile/cm	Sampling points
Terrace (n = 83)	Maize	Ridge	300	1, 3-5, 7-9, 12-18, 21-22, 25-26, 28, 32-35, 37, 51, 53, 91, 95, 96, 98
			260	99
		Gully	300	29, 30, 38-46, 48-50, 54-60, 62, 63, 65, 66, 68-74, 77-84, 87-90
			280	47
			260	31, 61
			220	52
			160	86
	Millet	Ridge	300	11
	Maize +five-year-walnut	Gully	300	27
Abandoned farmland	Subshrubs + herbs	Ridge	300	6, 10, 23
(n = 9)			240	2
	Subshrubs + herbs + few ulmus pumila	Gully	220	19
	Robinia peseudoacacia + subshrubs + herbs	Ridge	300	24
	Robinia peseudoacacia+ subshrubs + herbs	Gully	300	64
	Herbs + few almond-apricot	Gully	300	20
	Poplar + subshrubs +herbs	Gully	300	75
Grassland	Subshrubs +herbs	Ridge	300	92, 93
(n = 3)		Gully	300	76
Woodland	Poplar	Gully	300	97
(n = 6)			280	85
	Poplar + herbs	Gully	140	101
	Poplar + subshrubs	Gully	280	100
	Poplar + subshrubs + herbs	Gully	300	67, 94

300 cm). The ranges of the parameters after the rainy season were generally similar with those before the rainy season.

According to the division criteria of Nielson [24], CV in the range of 10–100% indicates moderate variability. Thus, the vertical variability of SWC at all sampling points in Sanyanjing catchment (Figure 4) can be classified to the medium degree. K is the linear regression coefficient between SWC and soil depth. A positive value of K indicates that SWC increases with increasing soil depth. Inversely, a negative value of K indicates that SWC decreases with increasing soil depth. The positive and negative K values of SWC data (Figure 4) are indicative of different variation trends of SWC vertical profiles in the catchment.

2 Clustering of SWC profiles

The vertical profile of SWC across the catchment can be described more clearly using cluster analysis. The 101 SWC profiles before the rainy season were classified into the first three groups by considering the mean value of SWC (Figure 5a), and the second three groups by considering the K value of SWC to soil depth (Figure 5b). The mean value of SWC ranged from 10.57% to 13.13% (low level), 14.15% to 16.86% (medium level), and 17.13% to 21.76% (high level) in the first three groups, whereas the K value of SWC to soil depth ranged from −0.0405 to 0.0106 (decreasing trend), 0.0144 to 0.0163 (stable trend), and 0.0194 to 0.0274 (increasing trend) in the second three groups. By combining the two cluster series, we obtained nine groups of SWC profiles (Table 3).

Before the rainy season, vertical profiles of SWC in groups 1–3 featured low-level SWC (Table 3). In group 1, SWC decreased along the vertical profile (0–300 cm) in woodland (2) and abandoned farmland (1) located in the gully area. In group 2, SWC remained stable along the vertical profile in terrace (2) and grassland (1) located on the ridge as well as woodland (1) located in the gully (Table 4). No sampling points were classified into group 3 with increasing SWC along the vertical profile.

Vertical profiles of SWC in groups 4–6 featured medium-level SWC (Table 3). In group 4, SWC decreased along the vertical profile on the ridge related to terrace (4) as well as in the gully related to abandoned farmland (1) and woodland (1) located in the gully. In group 5, SWC remained stable along the vertical profile in terrace (7) mostly located in the gully, few terrace (3) and abandoned farmland (1) located on the ridge, and grassland (1) and woodland (1) located in the gully. In group 6, SWC increased along the vertical profile in terrace (2) located on the ridge (Table 4).

Vertical profiles of SWC in groups 7–9 featured high-level SWC (Table 3). In group 7, SWC decreased along the vertical profile in terrace (5), grassland (1), and abandoned farmland (1) located on the ridge, and terrace (2), abandoned farmland (1) and woodland (1) located in the gully. In group 8, SWC remained stable along

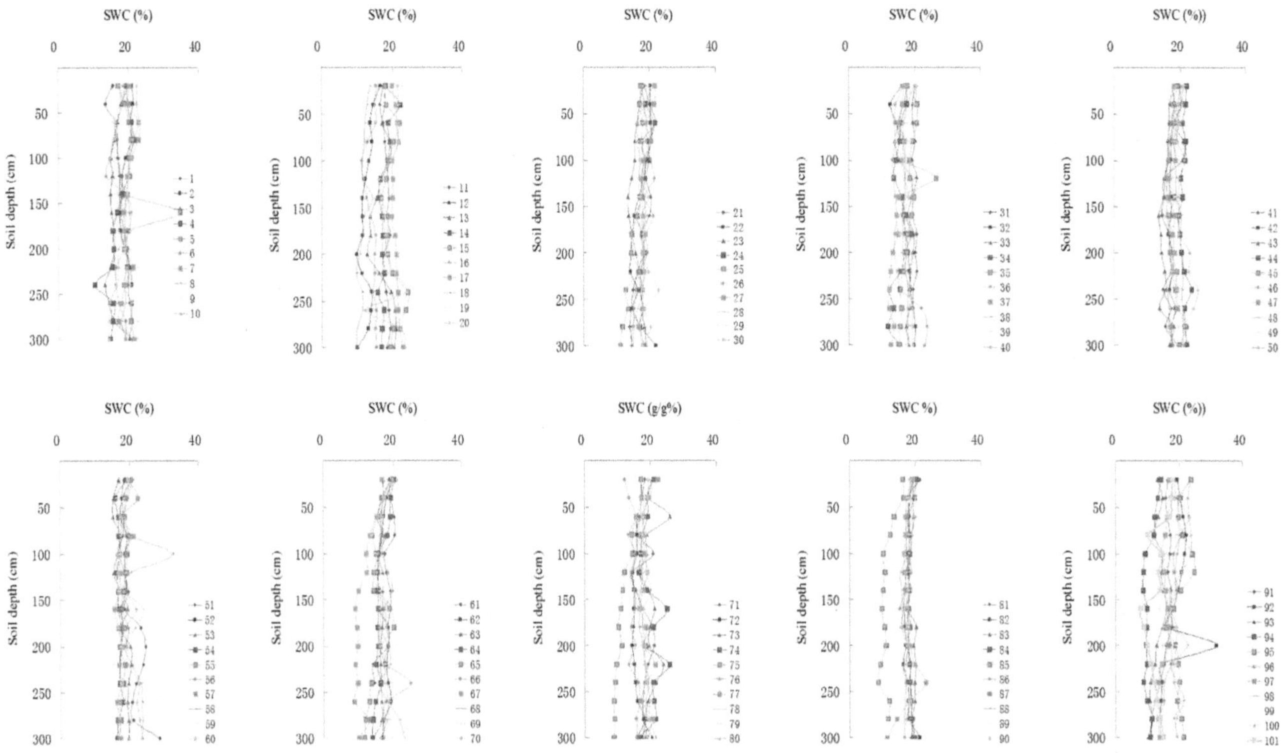

Figure 3. Vertical profiles of soil water at 101 sampling points in the Sanyanjing catchment.

the vertical profile at up to 58 of 101 sampling points, far more than other groups. Most sampling points of group 2 were located in the gully related to terrace (41), and few were on the ridge related to terrace (14) and abandoned farmland (3). In group 9, SWC increased along the vertical profile in terrace located on the ridge (3) and in the gully (1) (Table 4).

Similar grouping of SWC profiles was obtained with data collected after the rainy season. Overall, soil profiles of group 8 with high-level SWC in a stable trend were most commonly present in the catchment, more after the rainy season than before the rainy season. Group 3 of SWC profiles with low level and increasing trend was absent in the study area.

3 The relationships between average SWC and selected environmental factors

According to cluster analysis, there were nine combinations of SWC profiles in terms of average level and variation trend. We averaged SWC profiles of the same group and plotted the average curves (Figure 6), to examine differences of SWC profiles among various types. From Figure 4, we divided the whole soil profile (0–30 cm) into three layers (0–20, 20–160, and 160–300 cm) for multiple linear regression analysis. The results showed that selected environmental factors had significant linear correlations with average SWC at individual layers of 0–20 cm ($P<0.001$, $R^2=0.30$; $P<0.001$, $R^2=0.37$), 20–160 cm ($P=0.01$, $R^2=0.19$; $P<0.001$, $R^2=0.39$), and 160–300 cm ($P<0.001$, $R^2=0.32$; $P<0.001$, $R^2=0.43$; Table 5).

Before the rainy season, average SWC in the lower soil layer (10–20 cm) was significantly lower in grassland and woodland than in terrace, with no significant difference between abandoned farmland and terrace ($P_{D51}=0.109$, $P_{D52}=0.003$, $P_{D53}=0.047$, $P_{X1}=0.013$, and $P_{D61}=0.005$; Table 5). Additionally, average SWC decreased with increasing gradient, with higher levels on the ridge than in the gully.

In the lower soil layer (20–160 cm), average SWC decreased significantly with increasing gradient ($P=0.026$; Table 5), and was significantly lower in woodland than in the other three types of land-use types, with no significant differences among the latter three types. Other environmental factors had no significant linear correlation with average SWC ($P>0.05$).

In the deeper soil layer (160–300 cm), there also existed a significantly negative correlation between Sin(gradient) and average SWC ($P=0.001$; Table 5). Average SWC obviously increased with increasing gradient and was significantly higher in terrace than in abandoned farmland, grassland, and woodland (in descending order).

Similar results can be seen in the data collected after the rainy season. That is, land-use type was the major environmental factor affecting average SWC, whereas landform and altitude strongly affected average SWC only in specific periods and soil layers (Table 5). In the whole vertical profile (0–300 cm), SWC occurred at high levels from upper to deeper layers in terrace, with the lowest level in woodland. Compared with data of terrace, average SWC was relatively low in grassland and woodland in the top (0–20 cm) and deeper soil layers (160–300 cm), with significantly low levels in abandoned farmland soils in the deeper layer only. In the deeper soil layer (160–300 cm), average SWC varied with different land-use types in descending order of terrace > abandoned farmland > grassland > woodland.

Discussion

1 Vertical profiles of SWC at the catchment scale

According to Wang [21], the variability of SWC (as indicated by the CV) varies notably across the whole Loess Plateau, i.e., 15% in

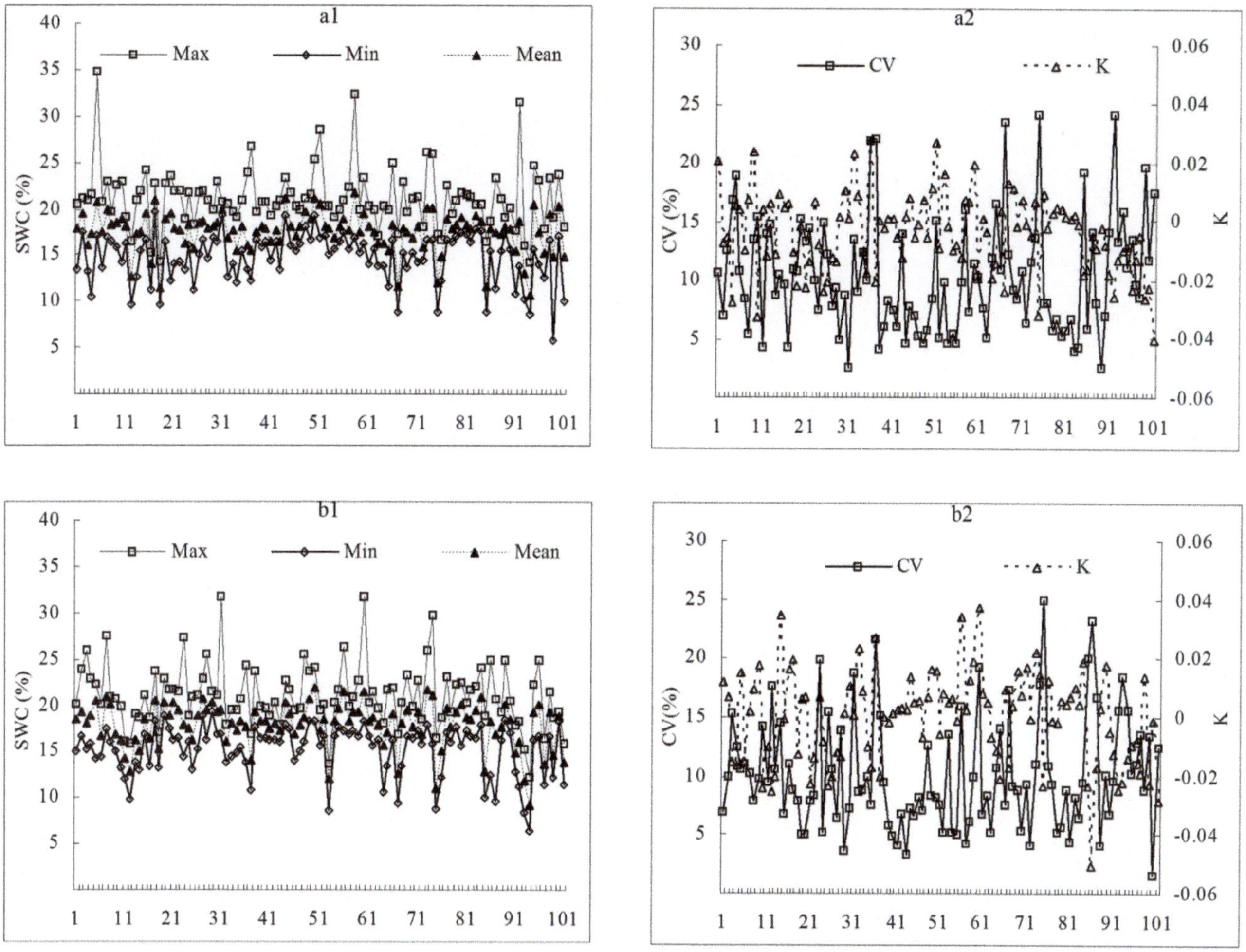

Figure 4. Statistical parameters of soil water content at 101 sampling points across the Sanyanjing catchment. (a. before the rainy season; and b. after the rainy season.)

Changwu and 55% in Shenmu. In the small catchment of Sanyanjing, SWC profiles exhibited weak and medium degrees of variability at 0–300 cm depth [33], with CV in the range of 3–24% (Figure 2). The lower variability of SWC profiles in our study area may be related to the higher SWC levels across the catchment (Pearson correlation coefficient between average SWC and CV,

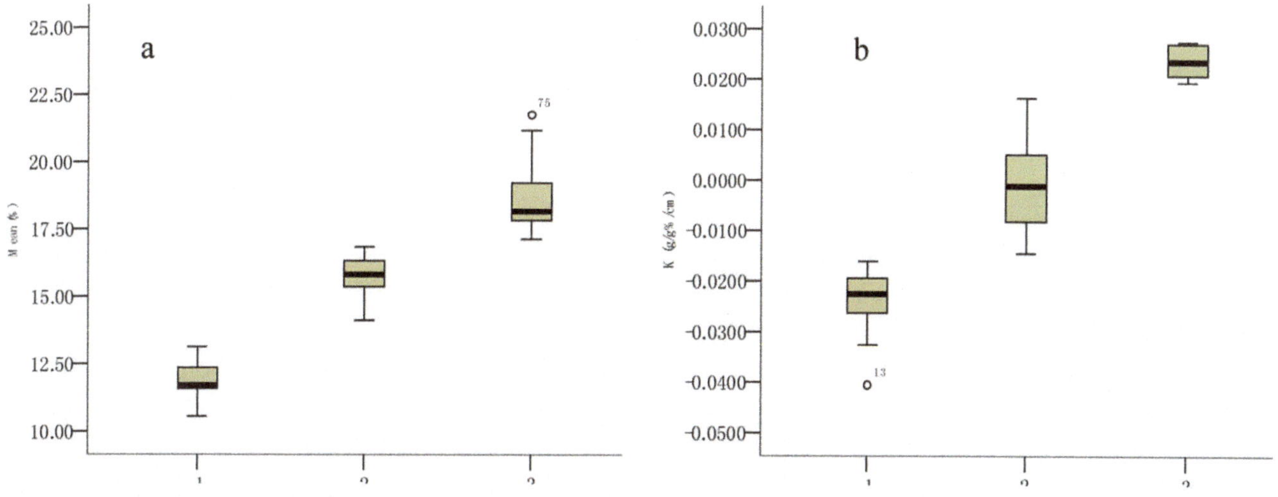

Figure 5. Grouping of 101 vertical soil water profiles in the Sanyanjing catchment before the rainy season by cluster analysis of the mean value (a) and regression gradient (K, b).

Table 3. Combined grouping of 101 vertical profiles of soil water content (0–300 cm) in Sanyanjing catchment by cluster analysis of the mean value and regression gradient.

Cluster by mean	Cluster by K	Combined grouping	Quantity of points		Point Nos.	
			Before the rainy season	After the rainy season	Before the rainy season	After the rainy season
1	1	1	3	0	65,67, 85	-
	2	2	4	1	12,18,93,94	94
	3	3	0	0	-	-
2	1	4	6	15	25,37,64,91,99,101	11,12,18,37,53,64,67,75,76,85,91,93,97,99,101
	2	5	13	0	3,16,23,34,41,43,56,63,65,70,72,76,97	3,16,23,34,41, 43,56,63,65,70,72,76,97
	3	6	2	1	32,36	14
3	1	7	11	1	4,10,19,21,26,35,61,86,92,96,100	86
	2	8	58	80	2,5–8,11,13–15,17, 20,22,24,27–31,33, 38–40,42,44–50,52, 53,55,56–59,62,66, 68,69,71,73,74, 77–84,87–90,95,98	2–10,13,15,16, 17,19–36,38–52,54,55,57–59, 61–63,65,66,68,69–74, 77–84, 87–90,92,95,96,98,100
	3	9	4	3	1,9,51,60	1, 51,60

Table 4. Grouping richness of 101 vertical profiles of SWC (0–300 cm) in relation to different land-use types in Sanyanjing catchment before the rainy season.

Land-use type	Grouping								
	1	2	3	4	5	6	7	8	9
Terrace	0	2	0	4	10	2	7	54	4
Abandoned	1	0	0	1	1	0	2	0	0
Grassland	0	1	0	0	1	0	1	0	0
Woodland	2	1	0	1	1	0	1	0	0

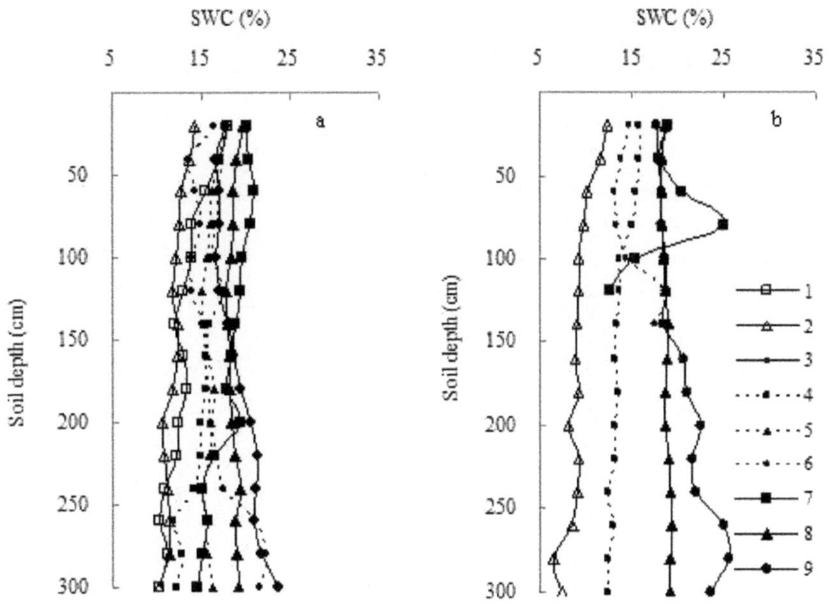

Figure 6. Vertical soil water profiles in relation to different groups in the Sanyanjing catchment study area before and after the rainy season (a. before the rainy season; and b. after the rainy season).

−0.40; $P<0.01$). Qiu [17] found that wetter soil with greater vertical variations in an increasing trend along the SWC profile (mean 13.03%; and STD, 2.3%) is representative in a dry year in Danangou catchment on the Loess Plateau, where the land-use pattern (including slope farmland, terrace, and orchard) differs from that in our study area.

Cluster analysis of the mean and K values provides a clear description for the overall variability of SWC in the vertical profile. Based on combined grouping, the 101 vertical SWC profiles were classified into nine groups with high, medium, and low levels associated with increasing, stable, and decreasing trends (Table 3). More than half of the SWC profiles were obtained from terrace soils in the gully and classified into group 8 (58/101 before the rainy season and 80/101 after the rainy season) with high-level SWC in a stable trend (Tables 3, 4). Despite that all sampling points of woodland were also located in the gully, their average SWC remained the lowest among different land-use types and mostly descended along the vertical profile (Table 4). The above differences can be attributed to the lower soil water consumption by maize crop in the terrace, which generally has shallower root distribution and less above-ground biomass than trees in the woodland. Our observations coincide with previous findings on the Loess Plateau that soil water conditions of terrace, gully farmland, and dam land are better than that of artificial woodland. The latter land-use type is associated with soil desiccation, especially in deep soil layers [2, 22, 35, 36].

Although the cluster analysis divided vertical SWC profiles into nine groups, only eight types were present in the Sanyanjing catchment and no sampling points were classified into group 3 (i.e., low-level SWC with an increasing trend from top to bottom). According to previous research in semi-arid regions, if SWC occurs at low level in the upper soil layer, deep-root crops, shrubs, and trees will consume more soil water in the deeper soil layers through root extraction [37–40]. Additionally, it is hard to achieve soil water recharge in the deeper soil layers by precipitation infiltration because the depth of soil water infiltration is shallow. Therefore, soil desiccation exists in the lower soil layer in case of no groundwater recharge [8–10]. These mechanisms explain the absence of high-level SWC with an increasing trend along the vertical profile in the small catchment of Sanyanjing (Table 3).

2 Effects of environmental factors on average SWC at the catchment scale

Consistent with cluster analysis (Table 3), multiple regression analysis showed that land-use type had a significant effect on soil water status in the small catchment of Sanyanjing (Table 4, 5). This result coincides with the data previously reported in small catchments on the Loess Plateau [17, 18, 22, 23, 29]. For example, Zhang [23] concluded that average SWC (20–200 cm) descends with different land-use types (farmland > grassland > shrub land > and woodland, n = 80) in the small catchment of Zhifanggou. Bai [42] found that average SWC (0–500 cm) ranges from of 9% to 16% in orchard, gradient farmland, terrace, and grassland, but remains less than 10% in shrub land and most woodland (n = 91) in Nangou catchment in the central area of Loess Plateau, Ansai, Shaanxi. The consistency of the data demonstrates that cluster analysis is a reliable method for characterization of SWC profiles.

The effect of land-use type on SWC can be related to the differences existing in anthropogenic activity and vegetation type [22]. Average SWC was found significantly higher in terrace and abandoned farmland than in grassland and woodland along the 0–300 cm profile (Table 5). Abandoned farmland and terrace are associated with artificial tillage in the surface soil layer, which improves soil porosity and loosens soil structure, further enhancing soil water infiltration [22, 43]. Additionally, soil water consumption by crops is less than that in grassland and woodland due to lower leaf area index [21], contributing to the accumulation of SWC. The above mechanisms account for the greater average of SWC profiles with a stable trend to soil depth in terrace and abandoned farmland.

Difference in root distribution is another factor contributing the effect of land-use type on SWC [40]. In the Sanyanjing catchment, average SWC of woodland was higher in the 0–20 cm soil layer but lower in the 20–160 and 160–300 cm soil

Table 5. Multi-linear regression analysis of soil water content and selected environmental factors in three layers (0–20, 20–160, and 160–300 cm) of the vertical soil profile in the Sanyanjing catchment study area.

				Before the	**rainy season**				
Model		**Y_1**			**Y_2**			**Y_3**	
	Unstandardized coefficients B	Standardized coefficients (Beta)	Sig.	Unstandardized coefficients B	Standardized coefficients (Beta)	Sig.	Unstandardized coefficients B	Standardized coefficients (Beta)	Sig.
Constant	14.714		0.066	6.031		0.498	18.514		0.069
X_1	−2.743	−0.23	0.013	−2.751	−0.22	0.026	−4.471	−0.296	0.001
X_2	−0.112	−0.037	0.68	0.042	0.013	0.889	−0.261	−0.068	0.449
X_3	4.86E−06	0.025	0.783	−7.80E−06	−0.038	0.694	1.20E−05	−0.05	0.578
X_4	0.005	0.081	0.507	0.011	0.177	0.175	0	0.005	0.97
D_{51}	1.09	0.146	0.109	0.019	0.002	0.98	−0.788	−0.084	0.354
D_{52}	−3.472	−0.278	0.003	−0.845	−0.064	0.503	−2.553	−0.162	0.073
D_{53}	−1.715	−0.191	0.047	−3.216	−0.342	0.001	−4.745	−0.384	0
D_{61}	−1.512	−0.348	0.005	−0.607	−0.134	0.307	−0.833	−0.151	0.212
	$R^2 = 0.30$	($P < 0.001$)		$R^2 = 0.19$	($P = 0.001$)		$R^2 = 0.32$	($P < 0.001$)	
				After the	**rainy season**				
Model		Y_1			Y_2			Y_3	
	Unstandardized coefficients B	Standardized coefficients (Beta)	Sig.	Unstandardized coefficients B	Standardized coefficients (Beta)	Sig.	Unstandardized coefficients B	Standardized coefficients (Beta)	Sig.
Constant	11.426		0.111	−5.811		0.452	−9.045		0.41
X_1	−2.01	−0.178	0.041	−2.763	−0.221	0.01	−3.688	−0.206	0.015
X_2	0.313	0.11	0.201	−0.132	−0.042	0.618	0.294	0.063	0.44
X_3	6.30E−06	−0.034	0.692	1.80E−05	−0.089	0.285	1.60E−05	−0.054	0.508
X_4	0.007	0.125	0.278	0.023	0.359	0.002	0.027	0.291	0.009
D_{51}	−0.622	−0.088	0.307	−1.161	−0.149	0.08	−1.81	0.162	0.053
D_{52}	−2.374	−0.2	0.021	−2.372	−0.181	0.032	−4.166	−0.222	0.008
D_{53}	−2.599	−0.305	0.001	−4.807	−0.51	0	−6.862	−0.467	0
D_{61}	−2.082	−0.506	0	−1.512	−0.332	0.004	−3.14	−0.476	0
	$R^2 = 0.37$	($P < 0.001$)		$R^2 = 0.39$	($P = 0.001$)		$R^2 = 0.43$	($P < 0.001$)	

Dependent Variable: Y_1 (soil water content of 0–20 cm layer).
Y_2 (average soil water content of 20–160 cm layer).
Y_3 (average soil water content of 160–300 cm layer).
Independent Variables: X_1 = Sin(gradient), X_2 = Sin(aspect), X_3 = flowaccu, X_4 = elevation.
Dummy Variables: X_5 = terrace, (D_{51}, D_{52}, D_{53}) = (0,0,0); X_5 = abandoned farmland, (D_{51}, D_{52}, D_{53}) = (1,0,0).
X_5 = grassland, (D_{51}, D_{52}, D_{53}) = (0,1,0); X_5 = woodland, (D_{51}, D_{52}, D_{53}) = (0,0,1).
X_6 = ridge, (D_{61}) = 1; X_6 = gully, (D_{61}) = 0.
D represents sub-variable; binary variables 0 and 1 for the absence and presence of some land-use type or landform, respectively.

layers than data of grassland (Table 5). The varying trends of SWC profiles between grassland and woodland can be related to different distribution of root system in individual soil layers and stratified root extraction of soil water. The rooting depth of maize crop is reported to be approximately 100 cm and most maize roots are distributed in the soil layer of 0–20 cm, shallower than average rooting depths in grassland (20–60 cm) and woodland (20–100 cm) [40]. Diverse root distribution patterns can lead to different levels of soil water consumption by plants, contributing to great variability of SWC level.

In addition to land-use type, topographic factors strongly affected SWC in the study area (Table 4). This is because the distribution of wind and solar radiation varies with different topographic conditions, leading to different levels of soil evaporation, runoff on gradient, and soil water infiltration [41]. Gradient negatively affected SWC in the soil layers of 0–20, 20–160, and 160–300 cm (Table 5), possibly due to the increased runoff with increasing gradient and resultant reduction of precipitation infiltration [17, 27, 29, 41]. Other topographic factors including aspect and flow accumulation had no significant effects on average SWC in the three soil layers (Table 5). Similarly, Gómez [29] referred that aspect has no obvious influence on SWC in burned and unburned areas. Shi [19] suggested that aspect and catchment area significantly affect SWC during the wet period only, whereas elevation has a significant effect on SWC in arid and humid periods but not in semi-arid and semi-humid periods. In the present study, we found the effect of elevation on SWC of the three layers varying with the period of time and being significant after the rainy season only.

As for the landform type, location of sampling points significantly affected SWC only in the surface layer (0–20 cm) before the rainy season and throughout all the three layers (0–20, 20–160, and 160–300 cm) after the rainy season, with greater values in the gully than on the ridge (Table 5). The effect of landform type on SWC can be related to different levels of soil evaporation as affected by wind strength and solar radiation and soil physical properties. Similarly, Zhang [22] suggested that average SWC descends with different landforms as gully > terrace > slop land > hill top.

Overall, land-use type is the most significant factor affecting SWC while topographic factors and landform type are interacting jointly at the catchment-scale. Because the impact of environmental factors on SWC varies in different periods, it is necessary to increase the observation frequency, in order to better understand the spatiotemporal distribution and influencing factors of SWC in the small catchment. Such work will provide reference data for selecting reasonable environmental parameters in catchment scale SWC simulation over different periods of time.

Conclusions

In this study, cluster analysis enables catchment-scale characterization of soil water profiles in terms of average level and variation trend along the vertical profile, allowing for simple and clear interpretation of the results. A total of nine groups of soil water profiles are recognized but those with low-level soil water content and a decreasing trend are not present in the Sanyanjing catchment. Land-use type, gradient, landform type, and altitude are the major environmental factors significantly influencing average soil water content in the hilly and gully catchment with complex terrain. The former two factors strongly affect soil water content along the 0–300 cm soil profile, whereas effects exerted by the latter two factors vary by soil layer and season.

Understanding the vertical profile of soil water content and evaluation of related major influencing factors in individual soil layers can help with sustainable land use and water management in catchment areas on the hilly and gully Loess Plateau as well as in arid and semi-arid areas with complex terrain. For better estimation of soil water profiles in small catchments, other factors such as fertilization, coverage, and soil physical properties may be considered with respect to specific soil layers.

Acknowledgments

We give our thanks to Prof./Dr. Wenzhao Liu (Institute of Soil and Water Conservation, Northwest A & F University, Yangling) for suggestions on experimental arrangement.

Author Contributions

Conceived and designed the experiments: BW FW JW XW YH. Performed the experiments: BW FW JW XW YH. Analyzed the data: BW FW JW XW YH. Contributed reagents/materials/analysis tools: BW FW JW XW YH. Wrote the paper: BW FW JW XW YH.

References

1. Engelbrecht BMJ, Comita LS, Condit R, Kursar TA, Tyree MT, et al. (2007) Drought sensitivity shapes species distribution patterns in tropical forests. Nature 447: 80–82.
2. Yang L, Wei W, Mo BR, Chen LD (2011) Soil water under different artificial vegetation restoration in the semi-hilly region of the Loess Plateau. Acta Ecologica Sinica 31: 3060–3068 (in Chinese with English abstract).
3. Xing G, Zhang XM, Fei XL, Wu YX (2012) Study on soil moisture content under different land use types in Sunjiacha basin. Agricultural Research in the Arid Areas 30: 225–229 (in Chinese with English abstract).
4. Li JW, Zuo HT, Li QF, Fan XF, Hou XC (2011) Effect of soil water spatial distribution pattern on switchgrass during first growing season. Acta Agrista Sinica 19: 43–50 (in Chinese with English abstract).
5. Jipp PH, Nepstad DC, Cassel DK, Carvalho C (1998) Deep soil moisture storage and transpiration in forests and pastures of seasonally-dry Amazonia. Climatic Change 39: 395–412.
6. Grassini P, You JS, Hubbard KG, Cassman KG (2010) Soil water recharge in a semi-arid temperate climate in the central US Great Plains. Agricultural Water Management 97: 1063–1069.
7. Markewitz D, Devine S, Davidson EA, Brando P, Nepstad DC (2010) Soil moisture depletion under simulated drought in the Amazon: impacts on deep root uptake. New Phytologist 187: 592–607.
8. Li YS (2001) Fluctuation of yield on high-yield field and desiccation of the soil on dryland. Acta Pedologica Sinica 38: 353–356 (in Chinese with English abstract).
9. Huang MB, Dang TH, Gallichand J, Goulet M (2003) Effect of increased fertilizer applications to wheat crop on soil-water depletion in the Loess Plateau, China. Agricultural Water Management 58: 267–278.
10. Liu WZ, Zhang XC, Dang TH, Zhu OY, Li Z, et al. (2010) Soil water dynamics and deep soil recharge in a record wet year in the southern Loess Plateau of China. Agricultural Water Management 97: 1133–1138.
11. Beldring S, Gottschalk L, Seibert J, Tallaksen LM (1999) Distribution of soil moisture and groundwater levels at patch and catchment scales. Agricultural and Forest Meteorology 98–99: 305–324.
12. Li B, Rodell M (2013) Spatial variability and its scale dependency of observed and modeled soil moisture over different climate regions. Hydrology and Earth System Sciences 17: 1177–1188.
13. Henninger DL, Petersen GW, Engman ET (1976) Surface soil moisture within a watershed: Variations, factors influencing, and relationship to surface runoff. Soil Science Society of American Journal 40: 773–776.
14. Jones EB, Owe M, Schmugge TJ (1982) Soil moisture variation patterns observed in Hand county, South Dakota. Water Recources Bulletin 18: 949–954.
15. Hawley ME, Jackson TJ, Mccuen RH (1983) Surface soil moisture variation on small agricultural watersheds. Journal of Hydrology 62: 179–200.
16. Robinson M, Dean TJ (1993) Measurement of near surface soil water content using a capacitance probe. Hydrological Processes 7: 77–86.
17. Qiu Y, Fu BJ, Wang J, Chen LD (2000) Quantitative analysis of relationships between spatial and temporal variation of soil moisture content and

environmental factors at a gully catchment. Acta Ecologica Sinica 20: 741–747 (in Chinese with English abstract).
18. Zeng C, Shao MA, Wang QJ, Zhang J (2011) Effects of land use on temporal-spatial variability of soil water and soil-water conservation. Acta Agriculturae Scandinavica Section B-Soil and Plant Science 61: 1–13.
19. Shi ZH, Zhu HD, Chen J, Fang NF, Ai L (2012) Spatial heterogeneity of soil moisture and its relationships with environmental factors at small catchment level. Chinese Journal of Applied Ecology 23: 889–895 (in Chinese with English abstract).
20. Chen LD, Huang ZL, Gong J, Fu BJ, Huang YL (2007) The effect of land cover/vegetation on soil water dynamic in the hilly area of the loess plateau, China. Catena 70: 200–208.
21. Wang YQ, Shao MA, Liu ZP, Orton R (2013) Regional-scale variation and distribution patterns of soil saturated hydraulic conductivities in surface and subsurface layers in the loessial soils of China. Journal of Hydrology 487: 13–23.
22. Zhang R, Cao H, Wang YQ, Huang CQ, Tan WF (2012) spatial variability of soil moisture and its influence factors in watershed of gully region on the loess plateau. Research of Soil and Water Conservation 19: 52–58 (in Chinese with English abstract).
23. Canton Y, Sole-benet A, Domingo F (2004) Temporal and spatial patterns of soil moisture in semiarid badlands of SE Spain. Journal of Hydrology 285: 199–214.
24. Burnt TP, Butcher DP (1985) Topographic controls of soil moisture distributions. Journal of Soil Science 36: 469–486.
25. O'loughlin EM (1981) Saturation regions in catchments and their relations to soil and topographic properties. Journal of Hydrology 53: 229–246.
26. Huang J, WU P, Zhao XN (2012) Effects of rainfall intensity, underlying surface and slope gradient on soil infiltration under simulated rainfall experiments. Catena 104: 93–102
27. Qiu Y, FU BJ, Wang J, Chen LD (2003) Spatiotemporal prediction of soil moisture content using multiple-linear regression in a small catchment of the Loess Plateau, China. Catena 54: 173–195.
28. Qiu Y, Fu B, Wang J, Chen L, Meng Q, Zhang Y (2010) Spatial prediction of soil moisture content using multiple-linear regressions in a gully catchment of the Loess Plateau, China. Journal of Arid Environments 74: 208–220.
29. Gómez-Plaza A, Martínez-Mena M, Albaladejo J, Castillo VM (2001) Factors regulating spatial distribution of soil water content in small semiarid catchments. Journal of Hydrology 253: 211–226.
30. Dripps WR, Bradbury KR (2007) A simple daily soil-water balance model for estimating the spatial and temporal distribution of groundwater recharge in temperate humid areas. Hydrogeology Journal 15: 433–444.
31. Yao XL, Fu BJ, Lu YH, Sun FX, Wang S, et al. (2013) Comparison of four spatial interpolation methods for estimating soil moisture in a complex terrain catchment. PLoS One 8(1): e54660.
32. Wang MB, Li HJ (1995) Quantitative study on the soil water dynamics of various forest plantations in the loess plateau region in northwestern Shanxi. Acta Ecologica Sinica 15: 172–184 (in Chinese with English abstract).
33. Wang YQ, Zhang XC, Han FP (2008) Profile variability of soil properties in check dam on the Loess Plateau and its functions. Environmental Science 29: 1020–1026 (in Chinese with English abstract).
34. Jain AK (2010) Data clustering: 50 years beyond K-means. Pattern Recognition Letters 31: 651–666.
35. Huang YL, Chen LD, Fu BJ, Wang YL (2005) Spatial pattern of soil water and its influencing factors in gully catchment of the Loess Plateau. Journal of Natural Resources 20: 483–492 (in Chinese with English abstract).
36. Zou JL, Shao MA, Gong SH (2011) Effects of different vegetation and soil types on profile variability of soil moisture. Research of Soil and Water Conservation 18: 12–17 (in Chinese with English abstract).
37. Kizito F, Dragila M,Se'ne M, Lufafa A, Diedhiou I, et al. (2006) Seasonal soil water variation and root patterns between two semi-arid shrubs co-existing with Pearl millet in Senegal, West Africa. Journal of Arid Environments 67: 436–455.
38. Li J, Chen B, Li XF, Zhao YJ, Ciren YJ, et al. (2008) Effects of deep soil desiccation on artificial forestlands in different vegetation zones on the Loess Plateau, China. Acta Ecologica Sinica 28: 1429–1445 (in Chinese with English abstract).
39. Cheng LP, Liu WZ (2013) Long term effects of farming system on soil water content and dry soil layer in deep loess profile of Loess Tableland in China. Journal of Integrative Agriculture. 13(6): 1382–1392.
40. Wang XZ, Jiao F (2011) Partition of soil moisture profiles based on sequential clustering method. Journal of Northwest A &F University 39: 191–201,196 (in Chinese with English abstract).
41. Fu XL, Shao MA, Wei XR, Wang HM, Zeng C (2013) Effects of monovegetation restoration types on soil water distribution and balance on a hillslope in northern Loess Plateau of China. Journal of Hydrologic Engineering 18: 413–421.
42. Bai TL, Yang QK, Shen J (2009) Soil variability of soil moisture vertical distribution and related affecting factors in hilly and gully watershed region of Loess Plateau. Chinese Journal of Ecology 28: 2508–2514 (in Chinese with English abstract).
43. Lian G, Guo XD, Fu BJ, Hu CX (2006) Spatial variability of bulk density and soil water in a small catchment of the Loess Plateau. Acta Ecologica Sinica 26: 647–654 (in Chinese with English abstract).

Permissions

All chapters in this book were first published in PLOS ONE, by The Public Library of Science; hereby published with permission under the Creative Commons Attribution License or equivalent. Every chapter published in this book has been scrutinized by our experts. Their significance has been extensively debated. The topics covered herein carry significant findings which will fuel the growth of the discipline. They may even be implemented as practical applications or may be referred to as a beginning point for another development.

The contributors of this book come from diverse backgrounds, making this book a truly international effort. This book will bring forth new frontiers with its revolutionizing research information and detailed analysis of the nascent developments around the world.

We would like to thank all the contributing authors for lending their expertise to make the book truly unique. They have played a crucial role in the development of this book. Without their invaluable contributions this book wouldn't have been possible. They have made vital efforts to compile up to date information on the varied aspects of this subject to make this book a valuable addition to the collection of many professionals and students.

This book was conceptualized with the vision of imparting up-to-date information and advanced data in this field. To ensure the same, a matchless editorial board was set up. Every individual on the board went through rigorous rounds of assessment to prove their worth. After which they invested a large part of their time researching and compiling the most relevant data for our readers.

The editorial board has been involved in producing this book since its inception. They have spent rigorous hours researching and exploring the diverse topics which have resulted in the successful publishing of this book. They have passed on their knowledge of decades through this book. To expedite this challenging task, the publisher supported the team at every step. A small team of assistant editors was also appointed to further simplify the editing procedure and attain best results for the readers.

Apart from the editorial board, the designing team has also invested a significant amount of their time in understanding the subject and creating the most relevant covers. They scrutinized every image to scout for the most suitable representation of the subject and create an appropriate cover for the book.

The publishing team has been an ardent support to the editorial, designing and production team. Their endless efforts to recruit the best for this project, has resulted in the accomplishment of this book. They are a veteran in the field of academics and their pool of knowledge is as vast as their experience in printing. Their expertise and guidance has proved useful at every step. Their uncompromising quality standards have made this book an exceptional effort. Their encouragement from time to time has been an inspiration for everyone.

The publisher and the editorial board hope that this book will prove to be a valuable piece of knowledge for researchers, students, practitioners and scholars across the globe.

List of Contributors

Lihua Chen, Pinghua Wang, Yuanjun Yang and Jia He
Key Laboratory of Soil and Water Conservation & Desertification Combat of Ministry of Education, School of Water and Soil Conservation, Beijing Forest University, Beijing, China

Rong Xiao, Qingqing Zhao, Qiongqiong Lu and Junjing Wang
State Key Laboratory of Water Environment Stimulation, School of Environment, Beijing Normal University, Beijing, P. R. China

Junhong Bai
State Key Laboratory of Water Environment Stimulation, School of Environment, Beijing Normal University, Beijing, P. R. China
Wetland Biogeochemistry Laboratory, Soil and Water Science Department, University of Florida, Gainesville, Florida, United States of America

K. Ramesh Reddy
Wetland Biogeochemistry Laboratory, Soil and Water Science Department, University of Florida, Gainesville, Florida, United States of America

Yong Liu, Chao Su, Hong Zhang and Xiaoting Li
Institute of Loess Plateau, Shanxi University, Taiyuan, Shanxi Province, China

Jingfei Pei
School of Mathematical Sciences, Shanxi University, Taiyuan, Shanxi Province, China

Yuguang Zhang, Hui Lu and Diqiang Li
Institute of Forestry Ecology, Environment and Protection, and the Key Laboratory of Forest Ecology and Environment of State Forestry Administration, Chinese

Jing Cong
Institute of Forestry Ecology, Environment and Protection, and the Key Laboratory of Forest Ecology and Environment of State Forestry Administration, Chinese Academy of Forestry, Beijing, China
School of Mineral Processing and Bioengineering, Central South University, Changsha, China

Caiyun Yang
Institute for Environmental Genomics and Department of Microbiology and Plant Biology, University of Oklahoma, Norman, Oklahoma, United States of America

Yunfeng Yang
State Key Joint Laboratory of Environment Simulation and Pollution Control, School of Environment, Tsinghua University, Beijing, China

Jizhong Zhou
Institute for Environmental Genomics and Department of Microbiology and Plant Biology, University of Oklahoma, Norman, Oklahoma, United States of America
State Key Joint Laboratory of Environment Simulation and Pollution Control, School of Environment, Tsinghua University, Beijing, China
Earth Sciences Division, Lawrence Berkeley National Laboratory, Berkeley, California, United States of America

Murielle Ghestem
AgroParis Tech, UMR AMAP, Montpellier, France

Kunfang Cao
Xishuangbanna Tropical Botanical Garden, Chinese Academy of Sciences, Menglun, Mengla, Yunnan, China

Wenzhang Ma
Kunming Institute of Botany, Kunming, Yunnan, China

Nick Rowe
CNRS, UMR AMAP, Montpellier, France

Raphaëlle Leclerc
CNRS, UMR CEFE, Montpellier, France

Clément Gadenne
AgroParisTech, Centre de Nancy, Nancy, France

Alexia Stokes
INRA, UMR AMAP, Montpellier, France

Limei Wang, Longhui Li, Xi Chen and Geping Luo
State Key Laboratory of Desert and Oasis Ecology, Xinjiang Institute of Ecology and Geography, Chinese Academy of Sciences, Urumqi, China

Xin Tian
Institute of Forest Information Resource Techniques, Chinese Academy of Forestry, Beijing, China

Xiaoke Wang
Beijing Urban Ecosystem Research Station, State Key Lab of Urban and Regional Ecology

Research Center for Eco-Environmental Sciences, Chinese Academy of Sciences, Beijing, China

Matthew J. Cleary
Department of Botany, University of Wyoming, Laramie, Wyoming, United States of America

Megan L. Mobley
Department of Botany, University of Wyoming, Laramie, Wyoming, United States of America
Helga Otto Haub School of Environment and Natural Resources, University of Wyoming, Laramie, Wyoming, United States of America

Ingrid C. Burke
Department of Botany, University of Wyoming, Laramie, Wyoming, United States of America
Helga Otto Haub School of Environment and Natural Resources, University of Wyoming, Laramie, Wyoming, United States of America
Program in Ecology, University of Wyoming, Laramie, Wyoming, United States of America

Lan Mu, Yinli Liang and Ruilian Han
Institute of Soil and Water Conservation, Northwest A&F University, Yangling Shaanxi Province, China
Institute of Soil and Water Conservation, Chinese Academy of Sciences and Ministry of Water Resources, Yangling, Shaanxi Province, China

Loredana Canfora, Flavia Pinzari and Anna Benedetti
Consiglio per la Ricerca e la Sperimentazione in Agricoltura, Centro di Ricerca per lo studio delle relazioni tra Pianta e Suolo, Roma, Italy

Giuseppe Lo Papa and Carmelo Dazzi
Dipartimento di Scienze Agrarie e Forestali, Universitá degli Studi di Palermo, Palermo, Italy

Giovanni Bacci
Consiglio per la Ricerca e la Sperimentazione in Agricoltura, Centro di Ricerca per lo studio delle relazioni tra Pianta e Suolo, Roma, Italy
Department of Biology, University of Florence, Florence, Italy

Zhuofei Xu, Martin Asser Hansen, Lars H. Hansen, Samuel Jacquiod and Søren J. Sørensen
Section of Microbiology, Department of Biology, University of Copenhagen, Copenhagen, Denmark

Benjamin Pey
INRA, UR251 « PESSAC », Versailles, France
CESAB/FRB, Aix-en-Provence, France

Gaël Caro, Céline Pelosi and Mickaël Hedde
INRA, UR251 « PESSAC », Versailles, France

Baptiste Laporte
CESAB/FRB, Aix-en-Provence, France

Marie-Angélique Laporte
German Centre for Integrative Biodiversity Research (iDiv), Leipzig, Germany
Bioversity International, Montpellier, France

Johanne Nahmani
Centre d'Ecologie Fonctionnelle et Evolutive (CEFE), CNRS, Université of Montpellier II, Montpellier, France

Apolline Auclerc and Sophie Joimel
Laboratoire Sols et Environnement, UMR 1120 INRA, Vandoeuvre-lés-Nancy, France

Yvan Capowiez
INRA, UR1115 « Plantes et Systémes Horticoles », Avignon, France

Daniel Cluzeau8 Muriel Guernion and Charléne Briard
Université de Rennes 1, UMR CNRS 6553 « EcoBio », Paimpont, France

Jérôme Cortet
Université Paul Valéry Montpellier III, Centre d'Ecologie Fonctionnelle et Evolutive, Laboratoire de Zoogéographie, UMR 5175 CEFE, Montpellier, France

Thibaud Decaëns
UFR Sciences et Techniques, EA 1293 « ECODIV », Université de Rouen, Mont Saint Aignan, France

Florence Dubs
IRD, UMR IEES-P, Bondy, France

Céline Pernin
Université de Lille 1, EA 4515 « Laboratoire Génie Civil & géo Environnement », Lille Nord de France, Ecologie Numérique et Ecotoxicologie, Villeneuve d'Ascq, France

Fabien Grumiaux
Université de Lille 1, EA 4515 « Laboratoire Génie Civil & géo Environnement », Lille Nord de France, Ecologie Numérique et Ecotoxicologie, Villeneuve d'Ascq, France
13 Université Lille Nord de France, Ecole Supérieure du Professorat et de l'Education (ESPE), Arras, France

Alain Pasquet
UR AFPA, Faculté des Sciences et Technologies, Université de Lorraine, Vandoeuvre-lés Nancy, France

Jean-François Ponge and Sandrine Salmon
CNRS, UMR 7179, Muséum National d'Histoire Naturelle, Brunoy, France

Lucia Santorufo
Laboratoire Sols et Environnement, UMR 1120 INRA, Vandoeuvre-lés-Nancy, France
Department of Structural and Functional Biology, University of Naples Federico II, Naples, Italy

Sara Palacio
Instituto Pirenaico de Ecología (IPE-CSIC), Jaca, Huesca, Spain

Matt Aitkenhead and A. H. Jean Robertson
The James Hutton Institute, Aberdeen, Scotland, United Kingdom

Adrián Escudero
Biodiversity and Conservation Group, E.S.C.E.T., Universidad Rey Juan Carlos, Móstoles, Madrid, Spain

Gabriel Montserrat-Martí and Melchor Maestro
Instituto Pirenaico de Ecología (IPE-CSIC), Zaragoza, Spain

Yongsheng Wang, Xusheng Dang and Lei Wang
Key Laboratory of Ecosystem Network Observation and Modeling, Institute of Geographical Sciences and Natural Resources Research, Chinese Academy of Sciences, Beijing, China
University of Chinese Academy of Sciences, Beijing, China

Linsen Li, Huajun Fang and Guirui Yu
Key Laboratory of Ecosystem Network Observation and Modeling, Institute of Geographical Sciences and Natural Resources Research, Chinese Academy of Sciences, Beijing, China

Minjie Xu and Shulan Cheng
University of Chinese Academy of Sciences, Beijing, China

Ana Pintado and Leopoldo G. Sancho
Dept. de Biología Vegetal II, Universidad Complutense de Madrid, Madrid, Spain

María Arró niz-Crespo
Dept. de Biología Vegetal II, Universidad Complutense de Madrid, Madrid, Spain
School of Environment Natural Resources and Geography, Bangor University, Bangor, Wales, United Kingdom

Sergio Pérez-Ortega and Asunción De los Ríos
Dept. de Biología Ambiental, Museo Nacional de Ciencias Naturales, MNCN-CSIC, Madrid, Spain

T. G. Allan Green
Dept. de Biología Vegetal II, Universidad Complutense de Madrid, Madrid, Spain
Biological Sciences, Waikato University, Hamilton, New Zealand

Raúl Ochoa-Hueso
Dept. de Biogeografía y Cambio Global, Museo Nacional de Ciencias Naturales, Madrid, Spain

Miguel Ángel Casermeiro and María Teresa de la Cruz
Departamento Edafología, Universidad Complutense de Madrid, Madrid, Spain

David Palacios
Departamento Geografía Física, Universidad Complutense de Madrid, Madrid, Spain

Ricardo Rozzi
Sub-Antarctic Biocultural Conservation Program, University of North Texas, Denton, Texas, United States of America
Institute of Ecology and Biodiversity, Universidad de Magallanes, Puerto Williams, Chile

Niklas Tysklind
School of Biological Sciences, Bangor University, Bangor, Wales, United Kingdom

Annamaria Bevivino, Patrizia Paganin, Maite Sampedro Pellicer and Claudia Dalmastri
ENEA (Italian National Agency for New Technologies, Energy and Sustainable Economic Development) Casaccia Research Center, Technical Unit for Sustainable Development and Innovation of Agro-Industrial System, Rome, Italy

Alessandro Florio and Anna Benedetti
Consiglio per la Ricerca e la Sperimentazione in Agricoltura - Research Centre for the Soil-Plant System, Rome, Italy

Giovanni Bacci
Consiglio per la Ricerca e la Sperimentazione in Agricoltura - Research Centre for the Soil-Plant System, Rome, Italy
Laboratory of Microbial and Molecular Evolution, Department of Biology, University of Florence, Florence, Italy

Maria Cristiana Papaleo, Alessio Mengoni and Renato Fani
Laboratory of Microbial and Molecular Evolution, Department of Biology, University of Florence, Florence, Italy

Luigi Ledda
Dipartimento di Agraria, University of Sassari, Sassari, Italy

Joseph E. Knelman
Institute of Arctic and Alpine Research, University of Colorado, Boulder, Colorado, United States of America
Department of Ecology and Evolutionary Biology, University of Colorado, Boulder, Colorado, United States of America

Steven K. Schmidt, Ryan C. Lynch and John L. Darcy
Department of Ecology and Evolutionary Biology, University of Colorado, Boulder, Colorado, United States of America

Sarah C. Castle and Cory C. Cleveland
Department of Ecosystem and Conservation Sciences, University of Montana, Missoula, Montana, United States of America

Diana R. Nemergut
Institute of Arctic and Alpine Research, University of Colorado, Boulder, Colorado, United States of America
Environmental Studies Program, University of Colorado, Boulder, Colorado, United States of America

Nicolas Chemidlin Prévost-Bouré and Philippe Lemanceau
Unité Mixte de Recherche 1347 Agroécologie, Institut National de la Recherche Agronomique-AgroSup Dijon-Université de Bourgogne, Dijon, France

Lionel Ranjard
Unité Mixte de Recherche 1347 Agroécologie, Institut National de la Recherche Agronomique-AgroSup Dijon-Université de Bourgogne, Dijon, France
Unité Mixte de Recherche 1347 Agroécologie-Plateforme GenoSol, Institut National de la Recherche Agronomique-AgroSup Dijon-Université de Bourgogne, Dijon, France

Samuel Dequiedt, Mélanie Leliévre and Pierre Plassart
Unité Mixte de Recherche 1347 Agroécologie-Plateforme GenoSol, Institut National de la Recherche Agronomique-AgroSup Dijon-Université de Bourgogne, Dijon, France

Jean Thioulouse
Unité Mixtede Recherche 555 Laboratoire de Biométrie et Biologie Evolutive, Université Lyon 1-Centre National de la Recherche Scientifique, Villeurbanne, France

Nicolas P. A. Saby, Claudy Jolivet and Dominique Arrouays
Unité de Services 1106 InfoSol, Institut National de la Recherche Agronomique, Orléans, France

Jingang Liang, Jun Ji, Fang Meng, Xiaobo Zheng and Zhengguang Zhang
Department of Plant Pathology, College of Plant Protection, Nanjing Agricultural University, and Key Laboratory of Integrated Management of Crop Diseases and Pests, Ministry of Education, Nanjing, China

Shi Sun and Cunxiang Wu
The National Key Facility for Crop Gene Resources and Genetic Improvement (NFCRI), MOA Key Laboratory of Soybean Biology (Beijing), Institute of Crop Science, The Chinese Academy of Agricultural Sciences, Beijing, China

Haiying Wu and Mingrong Zhang
Nanchong Academy of Agricultural Science, Nanchong, China

Takashi Narihiro, Hideyuki Tamaki, Aya Akiba and Satoshi Hanada
Bioproduction Research Institute, National Institute of Advanced Industrial Science and Technology (AIST), Tsukuba, Ibaraki, Japan

Kazuto Takasaki and Koichiro Nakano
FASMAC Co, Ltd., Atsugi, Kanagawa, Japan

Yoichi Kamagata
Bioproduction Research Institute, National Institute of Advanced Industrial Science and Technology (AIST), Tsukuba, Ibaraki, Japan
Bioproduction Research Institute, National Institute of Advanced Industrial Science and Technology (AIST), Sapporo, Hokkaido, Japan

Taizo Maji
Shubun University, Ichinomiya, Aichi, Japan

María Alonso-Ayuso, José Luis Gabriel and Miguel Quemada
School of Agriculture Engineering, Technical University of Madrid, Madrid, Spain

G. W. Weiger Wamelink and Josep Y. Frissel
Alterra, Wageningen UR, Wageningen, the Netherlands

Paul W. Goedhart
Biometris PRI, Wageningen UR, Wageningen, the Netherlands

Esperanza Huerta and Salvador Hernandez-Daumas
El Colegio de la Frontera Sur, Unidad Campeche, Dpto. Agroecología, Campeche, México

Christian Kampichler
Universidad Juárez Autónoma de Tabasco, División de Ciencias Biológicas, Villahermosa, Tabasco, México
Sovon Dutch Centre for Field Ornithology, Natuurplaza (Mercator 3), Nijmegen, The Netherlands

Susana Ochoa-Gaona and Ben De Jong
El Colegio de la Frontera Sur, UnidadCampeche, Dpto. Sustainability Sciences, Campeche, México

Violette Geissen
University of Bonn - INRES, Bonn, Germany
Wageningen University and Research Center - Alterra, Wageningen, Gelderland, Netherlands

Bing Wang, Fenxiang Wen, Jiangtao Wu and Xiaojun Wang
College of Environmental Science and Resources, Shanxi University, Taiyuan, China

Yani Hu
Library, Hebei University of Science and Technology, Shijiazhuang, China

Index

R

S

T

U

W

Z